Eggs, Nests, and Baby Dinosaurs

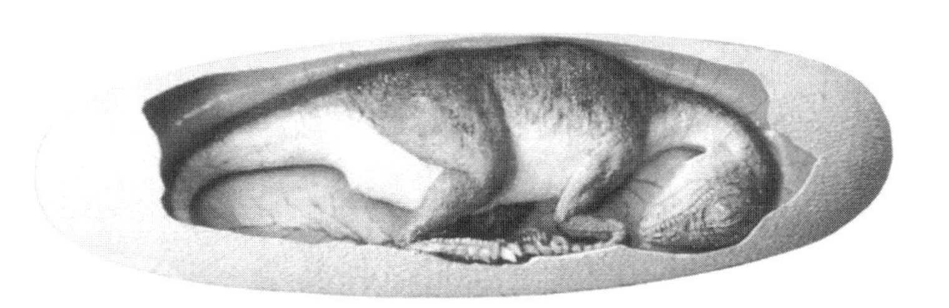

LIFE OF THE PAST
James O. Farlow, Editor

Eggs, Nests, and Baby Dinosaurs

A Look at Dinosaur Reproduction

Kenneth Carpenter

Indiana University Press
Bloomington and Indianapolis

This book is a publication of
Indiana University Press
601 North Morton Street
Bloomington, IN 47404-3797 USA
http://www.indiana.edu/~iupress

Telephone orders 800-842-6796
Fax orders 812-855-7931
Orders by e-mail iuporder@indiana.edu

The paper used in this publication meets the minimum requirements of American National Standard for Information Sciences—Permanence of Paper for Printed Library Materials, ANSI Z39.48-1984.

Manufactured in the United States of America

Library of Congress Cataloging-in-Publication Data

Carpenter, Kenneth, date
Eggs, nests, and baby dinosaurs: a look at dinosaur reproduction / Kenneth Carpenter.
p. cm. — (Life of the past)
Includes bibliographical references and index.
ISBN 0-253-33497-7 (cl: alk. paper)
1. Dinosaurs—Reproduction. 2. Dinosaur—Eggs. 3. Dinosaurs—Infancy. I. Title. II. Series.
QE862.D5C235 1999
567.9—dc21 99-42739

1 2 3 4 5 04 03 02 01 00 99

Dedicated to my two dear friends

Karl F. Hirsch
(1921–1995)

who taught me about fossil eggs and

John R. Horner
(1946–)

who taught me about baby dinosaurs

CONTENTS

Gregory Paul 83/90

Two of the most influential persons in the study of dinosaur eggs: the late Karl Hirsch (left) and Kostantin (Kostya) Mikhailov (right). From a photograph taken during a trip to see various egg sites in North America.

Preface

During the past decade there has been an explosion in our knowledge about dinosaur reproduction, especially about their eggs and babies. My own involvement in the study of dinosaur reproduction is an example of science at work: of ideas, or hypotheses, presented and later modified as new information becomes available. Certainly our ideas about dinosaurs have changed considerably since Richard Owen and Waterhouse Hawkins made a life-sized restoration of a *Megalosaurus* in 1853. Based on a more complete understanding of large theropods, we no longer think that it looked like a cross between a bear and a dog. Today, *Megalosaurus* would be restored as a bipedal animal with short forearms and would look much more like its better-known cousin *Allosaurus*.

A great deal of our new understanding about dinosaur reproduction began with the popularization of the subject by Jack Horner beginning in the mid-1980s. Most of Jack's early work was at a little bump on the prairie in northern Montana that Jack had given the grandiose name of Egg Mountain. It was Jack's work that alerted me to the importance of the bones and teeth of baby dinosaurs I had been collecting in eastern Wyoming. Right from the start we disagreed about what we both had. Jack argued that the sedimentary rocks in which he was finding dinosaur eggs and dinosaur babies, called the Two Medicine Formation, were deposited in upland environments far from the sea. When the Two Medicine Formation was deposited during the Late Cretaceous, North America was split in half by a sea that connected the present Gulf of Mexico with the present Arctic Ocean. Jack felt that the dinosaurs, especially the hadrosaurs, migrated from the low, coastal environments along the western shore of the seaway to the uplands, hundreds of miles inland. I, on the other hand, was collecting baby dinosaurs and eggshell from lowland sediments, called the Lance Formation, deposited only a few tens of miles inland. I argued that the dinosaurs that lived in the lowlands laid their eggs there.

Jack countered that the rarity of eggshell and baby bones in the lowland environments and their abundance in the upland environments showed that the dinosaurs must have migrated. He had me there—for a while. During the years I was collecting vertebrate fossils in the Lance Formation, I became interested in the type of sedimentary rocks the fossils were found in. This interest stemmed from my background training—I studied geology before becoming a paleontologist. Other paleontologists receive primary training in biology, especially in anatomy (they usually end up teaching human anatomy to medical students).

The Lance Formation is mostly drab-colored mudstones and shales in various shades of brown, with some lighter-colored sandstones scattered throughout. The drab color is due to very fine particles of organic carbon, mostly from plant debris. The shades of brown are caused by various amounts of iron in a reduced form (i.e., the iron atoms do not share the electrons in the outer orbit, or shell, of electrons with oxygen atoms; if they did, the iron would be oxidized and a reddish color would result).

I never found eggshell or delicate baby bones in the organic-rich sediments; instead the fossils were usually found in sandstones. This got me wondering if there might be something about the ancient environments represented by the different rock types that might provide a clue to the distribution of the eggshells and bones. The plant debris–rich rocks were an indication of ancient swamps or other waterlogged soils. Studies of such environments today show that stagnant water effectively seals the sediments and plant debris from atmospheric oxygen. This lack of oxygen is why the iron in the Lance Formation is in a reduced state.

In the absence of oxygen, plant debris also decays very slowly and various organic acids are produced. The acidity can be measured with a pH meter and for waterlogged soils, the pH, or acidity, can range from 7 (which is neutral) to 2.8 (which is very acidic; vinegar is around pH 3). It is no wonder that eggshell and baby bones were absent from these organic-rich sediments; they dissolved away. The dissolution of bones has mostly been studied from archeological sites. One of the most famous is the Damendorf Man, found in Denmark. This Iron Age person is known from his skin and leather belt and shoes, the entire skeleton having been dissolved away by the acids of the peat bog in which he was found.

But how weak can acids be and still dissolve eggshell? To test this, I buried some chicken eggshell in leaf litter kept wet. After sixty days the pH was only 6.6, which is a very weak acid, but the eggshell already showed signs of dissolution. This then was another important clue: given enough time eggshell could dissolve away even in very mild acidic environments.

Evidently conditions were acidic enough in the Late Cretaceous lowland environments of the Lance Formation that eggshell and baby bones were dissolved away. Only where there was a constant flow of fresh water to prevent a buildup of acid, as in a river channel, were eggshell and bones preserved. Indeed, all the material I found in the Lance Formation came from sandstone deposits that represent these ancient rivers.

In the Two Medicine Formation where Jack was working, strata are brighter, lacking the drab colors of the Lance Formation. In the vicinity of Egg Mountain, there are even pale reddish rocks caused by an oxidized form of iron. Not surprisingly, plant debris is rare. All of this evidence indicates that during the Late Cretaceous the landscape was well drained. Stagnant pools of water did not accumulate, nor were the soils waterlogged. Instead, atmospheric oxygen circulated into the soil, allowing the rapid growth of microbes which caused decomposition of the vegetation. Because acidic soils or water were rarely produced, the eggshells and baby bones stood a better chance of survival and fossilization.

This example shows science at work: data are collected, a hypothesis is presented, other data challenge the hypothesis, and the hypothesis is modified. Although Jack and I were on opposite sides of the debate, our disagreement was good natured. Ultimately, we were both interested in the same thing: to understand dinosaurs the best we can. We remain friends and even co-edited a technical book on dinosaur eggs and babies with a mutual friend, the late Karl Hirsch.

This book presents what paleontologists know today about dinosaur reproduction. Because there are so many nontechnical books on dinosaurs, I have assumed that anyone reading this book has some knowledge about dinosaurs. If not, then I suggest another book in this series, *The Complete Dinosaur,* edited by James O. Farlow and Mike K. Brett-Surman (Indiana University Press, 1997).

Many of the ideas presented in this book are hypotheses that have been suggested by various paleontologists to explain the evidence they see. Other ideas are my own, although I was probably influenced by others. Undoubtedly new discoveries will alter some of my views, but as with the example I presented earlier, that is science at work.

This book was written with the help of many international colleagues who shared freely in discussions, information, photographs, and access to specimens: Karen Alf (Denver, Colorado), Emily Bray (University of Colorado, Colorado), Brooks Britt (Museum of Western Colorado, Colorado), Daniel Chure (Dinosaur National Monument, Utah), Richard Cifelli (Oklahoma Museum of Natural History, Oklahoma), Philip J. Currie (Royal Tyrrell Museum of Palaeontology, Alberta), Mary Dawson (Carnegie Museum of Natural History, Pennsylvania), Dan Grigorescu (University of Bucharest, Romania), Karl Hirsch (University of Colorado, Colorado), Charlotte Holton (American Museum of Natural History, New York), Jack Horner (Museum of the Rockies, Montana), James Kirkland (Dinamation International Society, Colorado), Kostantin Mikhailov (Paleontology Institute, Russia), John McIntosh (Wesleyan College, Connecticut), Charlie and Florence McGovern (Stone Company, Boulder), Angela Milner (Museum of Natural History, London), Mark Norell (American Museum of Natural History, New York), John Ostrom (Peabody Museum of Natural History, Connecticut), Gregory S. Paul (Baltimore, Maryland), Betty Quinn (Museum of the Rockies, Montana), Peter Robinson (University of Colorado Museum), Karol Sabath (Institute of Paleobiology, Poland), Monique Vianey-Liaud (Laboratoire de Paléontologie, France), Ray Rogers (Cornell College, Iowa), Darla Zelenitsky (University of Calgary, Alberta), and Zhao Zi-Kui (Institute of Vertebrate Paleontology and Paleoanthropology, China).

Translations of Russian papers were provided by Ruth Griffith and of Chinese papers by Yanmin Huang. Earlier drafts of the chapters were reviewed by James Farlow and Darla Zelenitsky, who caught my mistakes and challenged some of my ideas. Nevertheless, any mistakes are my own. Photographs were kindly provided by Karl Hirsch, Jack Horner, Gerald Grellet-Tinner, Terry Manning, Kostya Mikhailov, Altangerel Perle, Karol Sabath, the Stone Company, Monique Vianey-Liaud, and Darla Zelenitsky. Artwork is, in part, by Gregory Paul, Judith Peterson, and Luis Rey.

Eggs, Nests, and Baby Dinosaurs

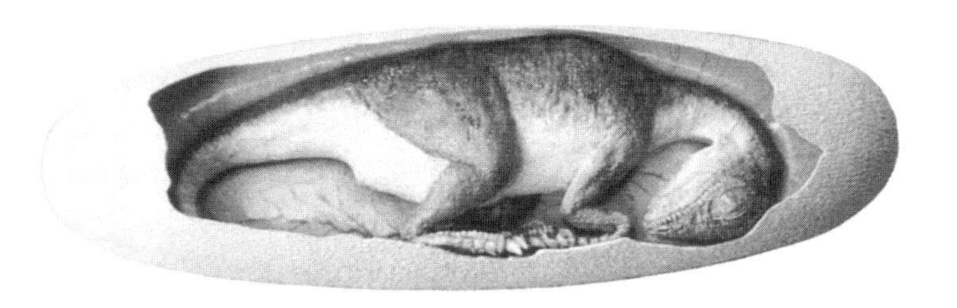

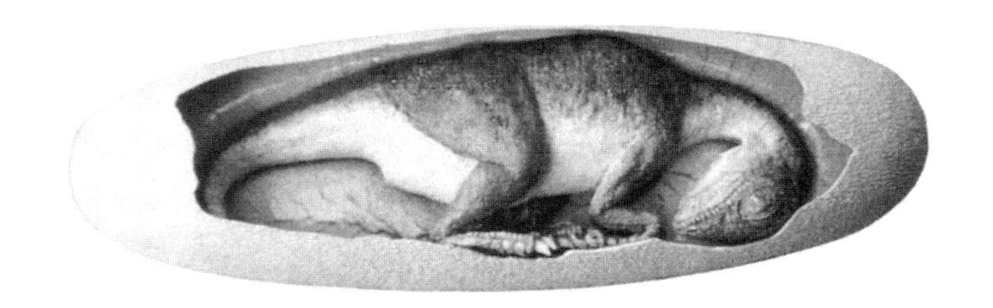

1 • First Discoveries

Birds do it, bees do it. So did dinosaurs. Studying dinosaur reproduction might seem impossible, considering how long ago they became extinct (65 million years ago). True, without a time machine to send Richard Attenborough back in time we will never see a video of how dinosaurs mated. Nevertheless, we have figured out some things about dinosaur mating behavior and reproduction from clues left in the fossil record. For example, the microscopic structure of eggshell tells us something about the egg-forming processes, the pattern of eggs in a clutch tells us about how the eggs were laid, embryos and hatchlings tell us about early development, and ornamental structures of the adults tell us something about mating rituals. We can also use the behavior of living animals as analogs to understand the behavior of extinct animals. Such analogies provide insight into possible ways dinosaurs *might* have behaved, but without seeing the beasts in action we can never be certain that our analogies are correct. Nevertheless, analogies play an important role in paleontology by providing starting points for hypotheses. These hypotheses are later tested and modified or discarded based on additional information. It's fine to be proven wrong, because that is how science advances.

I am not sure who first said that dinosaurs laid eggs, but the inference was there when dinosaur bones were being discovered in England in the early 1800s. When the first dinosaur was named in 1824, William Buckland was clear about the reptilian nature of the bones, calling the dinosaur *Megalosaurus,* or "great lizard." After all, reptiles laid eggs (well, most of them), so it was assumed that dinosaurs laid eggs as well.

The credit for the discovery of the first dinosaur eggs has historically been given to the Central Asiatic Expedition of the American Museum of Natural History. This series of expeditions into Mongolia was led by zoologist Roy Chapman Andrews in the 1920s. Andrews had spent a great deal of time in Asia studying whales for his thesis at Columbia University. Because Columbia had an affiliation with the American Museum, he also spent a great deal of time traipsing around Asia collecting mammal specimens for the museum.

It was after one of these trips in 1919 that Andrews eloquently laid out his idea over lunch to the museum's president, Henry F. Osborn. Osborn was a vertebrate paleontologist by training, and he long held the belief (partially for racist reasons) that Asia, not Africa, was the cradle of human origins. In his account of the meeting, Andrews candidly admits to playing to Osborn's ego (which was quite big):

> When coffee had been served and we were smoking comfortably, he [Osborn] smiled: "Now let's have it Roy. It's another expedition I suppose."
>
> "Yes, that's why I came back. The expedition I've been dreaming about for years. To test your theory of Central Asia. Especially, to try to find evidence of primitive man. Mongolia is the place." (Andrews, 1943, p. 163)

Andrews doesn't write how long the meeting lasted, but in the end he got Osborn's support. Raising the $250,000 for five years of expeditions was the only obstacle, and even that was overcome by Andrews' persuasiveness—and New York's Society vanity:

> "Getting the money will be up to you. What are your ideas on that score?" [asked Osborn.]
>
> "My only chance, I believe, is to make it a 'society expedition' with a big S. You know that New York society follows a leader blindly. If they have the example of someone like Mr. Morgan, for instance, they'll think it is a 'must' for the current season. 'Have you contributed to the Roy Chapman Andrews' expedition? If not, you're not society.'" (Andrews, 1943, p. 165)

In the end, with backers like J. P. Morgan and John D. Rockefeller, Andrews got his money and the expedition sailed for China from New York City in March 1921. But it would be a full year before the expedition left China for Mongolia. During that time, equipment was organized and some reconnaissance was done in China.

Owing to the high cost of sending and maintaining expeditions in Central Asia, each expedition was to squeeze as much as possible out of the trips. An array of scientific specialists were at one time or another part of the teams, including cartographers to map the expedition's routes, geologists to study the rocks, several vertebrate paleontologists, a paleobotanist, a herpetologist who also doubled as an ichthyologist, a zoologist (Andrews) to collect bird and mammal specimens, and an archeologist to study the archeological sites encountered in hopes of proving Osborn's hypothesis.

The expedition finally headed overland from Beijing (then Peking) on April 17, 1922. After winding around through Mongolia and making many stops to collect specimens, the expedition arrived (quite by accident) at Shabarakh Usu (now known as Bayn Dzak), in southern Mongolia. Part of the area was nicknamed the Flaming Cliffs by the Americans because the bright red color of the sandstone bluffs was accentuated in the morning and evening light. Fossils were found in abundance, including a small skull and several fragments of eggshells that were assumed (erroneously as it would turn out) to be bird. Due to time constraints, the expedition left the following morning and so the discovery of whole dinosaur eggs would have to wait another year.

The following summer, the expedition headed back to the Flaming Cliffs because the small skull was identified as a new type of primitive dinosaur related to the horned *Triceratops* of North America. Named *Protoceratops andrewsi,* this small dinosaur had the unusual rostral bone at the front of the upper beak that characterizes the ceratopsians. In addition,

Fig. 1.1. Camel train of supplies for the Central Asiatic Expedition winding its way through southern Mongolia. (Courtesy of A. Perle.)

Fig. 1.2. The Djadokhta Formation at the Flaming Cliffs, Shabarakh Usu (now known as Bayn Dzak), in southern Mongolia. The first eggs were found at the top of the high promontory to the left, called Battlement Bluffs. Eggs and Protoceratops *bones continue to be found all through these rocks. (Courtesy of A. Perle.)*

Fig. 1.3. The first group of eggs found by the Central Asiatic Expedition. (Courtesy of A. Perle.)

Fig. 1.4. Roy Chapman Andrews examining some of the dinosaur eggs found at the Flaming Cliffs, Mongolia. (Courtesy of the Denver Museum of Natural History.)

Protoceratops had the typical frill or collar of bone on the back of its head, but not the horns of its North American relatives.

But before arriving at the Flaming Cliffs, the expedition stopped at Erenhot (known today as Iren Dabasu) near the Chinese–Mongolian border, a site visited the first year. There, numerous dinosaur bones were collected, as well as numerous fragments of eggshell. The identity of these shell pieces was unrecognized at the time, but they were collected nonetheless. Not until whole eggs were collected later that summer at Bayn Dzak was the dinosaurian nature of these fragments realized. Small clusters of eggshell at Erenhot proved to be crushed eggs and several of these have been reconstructed to approximately their original size. About the size of a navel orange, identical eggs were later named *Paraspheroolithus irenensis* (now known as *Spheroolithus irenensis*) by Chinese egg specialist Zhao Zi-Kui in 1979 (1979a).

At Bayn Dzak, the expedition found *Protoceratops* in abundance, as well as several other dinosaurs later to be named *Oviraptor, Velociraptor, Saurornithoides,* and *Pinacosaurus*. But it was the discovery of a nest of eggs that the expedition is best remembered for. As Andrews would later write:

> On July 13, 1923, George Olsen [a technician from the vertebrate fossil laboratory] reported at tiffin [dinner] that he had found some fossil eggs. Inasmuch as the deposit was obviously Cretaceous and too early for large birds, we did not take his story very seriously. We felt quite certain that his so-called eggs would prove to be sandstone concretions or some other geological phenomena. Nevertheless, we were all curious enough to go with him to inspect his find. We saw a small sandstone ledge, beside which were lying three eggs partly broken. The brown striated shell was so egg-like that there could be no mistake. Granger [one of the vertebrate paleontologists] finally said, "No dinosaur eggs have ever been found, but the reptiles probably did lay eggs. These must be dinosaur eggs. They can't be anything else." (Andrews, 1932, p. 208)

But were these in fact the first dinosaur eggs ever found, as Granger said? Actually, no. Archeologists with the expedition found dinosaur eggshells at a late Paleolithic or early Neolithic site near the foot of the Flaming Cliffs. Each piece had been shaped and drilled into jewelry. These early humans collected eggshells several thousand years previous to the 1923 rediscovery, but did not make a media event out of it the way Andrews did. Nor did Father Jean-Jacques Pouech, a French priest and amateur geologist. He had actually found large pieces of eggshell in Upper Cretaceous strata in southern France. This discovery in 1859 preceded Olsen's discovery by over sixty years. Why did Andrews make such a big event out of the American Museum discovery?

The Twenties saw America on top of the world. The wounds of World War I were healing, the economy seemed to be doing really well and everyone was feeling good. American paleontologists in New York City were pleased with themselves for having made major discoveries of dinosaurs in southern Alberta. New dinosaurs with exotic names like *Styracosaurus, Monoclonius,* and *Corythosaurus* were announced fairly regularly.

Faced with such high successes, it was important that the Central Asiatic Expedition be successful as well. It wouldn't do to admit failure when it came to proving that Asia was the center of human origins. Something had to be salvaged and the discovery of these and subsequent eggs gave Andrews his big break. By playing up the discoveries, he made sure that the financial backers felt that they were getting their money's worth. All this attention was rather heady for Andrews and he auctioned an egg as a fundraising stunt. The egg was bought for $5,000 by Austin Colgate, who then donated it to Colgate University. Andrews' stunt backfired when news of the auction reached the Chinese government. They assumed that the eggs had monetary value and that Andrews had not been honest about the purpose of the expeditions. The Chinese had long been suspicious of the Americans' true purpose in China and Mongolia because of the prominent names on the list of financial backers (Andrews, 1943, p. 178). Andrews' stunt only served to deepen the suspicion that the expeditions would enrich the backers. As a result, the Chinese government barred the expedition from returning. A year was lost as Andrews worked very hard at damage control. Eventually, the expedition was able to return to the Flaming Cliffs in 1925 and more eggs were collected.

Surprisingly, except for the media hype, the eggs evoked very little scientific interest. What little study they got was due to the work of Belgian paleontologist Victor van Straelen. In no way, though, did his work reflect the diversity of eggs and shell types collected in Mongolia. Instead, his two short papers were meant to be preliminary accounts based on two eggshell types—one from Erenhot and the other from the Flaming Cliffs. Mostly though, the eggs collected by the Central Asiatic Expedition have been treated as scientific curiosities and they have yet to be formally described. Among them are dinosaur eggs now referred to as *Spheroolithus irenensis* and *Elongatoolithus* sp., and two smaller type eggs, "*Gobiopteryx*" and *Laevisoolithus sochavi,* that might belong to birds. We'll discuss these egg names in another chapter (chapter 9).

In hindsight, it is not surprising that the first study of the Mongolian eggs was conducted by a European. European paleontologists have often been ahead of North American paleontologists in geochemical and microscopic studies of vertebrate fossils. Part of this lead may be due to the relative dearth of vertebrate remains in Europe as compared to North America, with its fossiliferous badlands in the West. As a result, European paleontologists have had to make the most of what they found and this often meant pioneering new methods of study.

The first study of dinosaur eggshells was actually conducted by Father Jean-Jaques Pouech, whom we mentioned earlier. Like many scientists of the eighteenth and nineteenth centuries, Pouech was a self-taught naturalist, with an interest in geology and archeology as well as paleontology. He was head of the Pamiers Seminary in the foothills of the Pyrenees in southern France. During his wanderings in search of fossils, he happened upon what he would eventually identify as eggshells. The objects had a uniform thickness (2 mm), a fibrous structure in cross-section, and gentle curvature. Based on these features he concluded in 1859 that the objects could only be eggshell (Buffetaut and LeLoeuff, 1994). From the curvature, he estimated the egg diameter to be 18 cm (7"). He did not, however, identify the eggshell as dinosaurian, but from a gigantic bird. His identification of the egg-layer is reasonable considering that dinosaurs were poorly known at the time and not yet a household word. In fact the word *dinosaur* had only been coined in 1841 for the incomplete remains of three very different animals, *Megalosaurus, Iguanodon,* and *Hylaeosaurus,* from England.

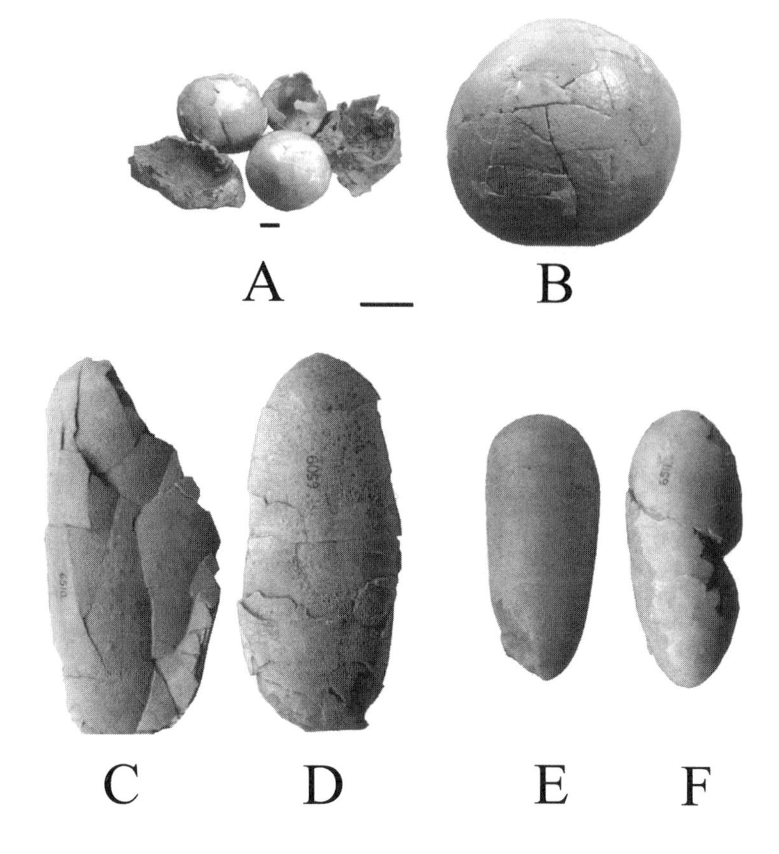

Fig. 1.5. Some of the eggs found by the Central Asiatic Expedition, 1923–1925: A, group of Spheroolithus irenensis *from Erenhot (Iren Dabasu near the Chinese-Mongolian border). B, close-up of a single* Spheroolithus irenensis. *C,* Elongatoolithus frustrabilis *from the Flaming Cliffs. D, possibly a different species of* Elongatoolithus *from the Flaming Cliffs; the top is more rounded. E, F, two examples of* Protoceratopsidovum sincerum *from the Flaming Cliffs. Scales = 2 cm.*

An alternative suggestion as to the identity of the egg-layer was not made until ten years later. In 1869, French geologist Philippe Matheron, working in southern France, discovered some large fossil bones that he named *Hypselosaurus,* which he identified as a giant crocodile (Buffetaut and LeLoeuff, 1994). Near the bones he also found several large shell fragments:

> Along with the remains of the bones I just mentioned [*Hypselosaurus priscus*], are two very enigmatic large segments of spheres or ellipsoids. . . . All things considered, it would seem that they are fragments of eggs. . . . Are the two fragments in question remains of two eggs from a giant bird, or are they perhaps remains of two hypselosaur eggs? This question remains to be answered. (Buffetaut and LeLoueff, 1994, p. 32)

He showed these fragments to his colleague Paul Gervais, who was director of comparative anatomy at the Museum of Natural History in Paris and one of France's leading vertebrate paleontologists. His interest piqued, Gervais undertook solving the identity of these giant shell fragments. As a comparative anatomist, Gervais did what seemed only natural to him. He undertook an exhaustive microscopic comparison of Matheron's eggshells with those of living turtles, geckos, crocodiles, and birds. To do this, he glued pieces of eggshell to microscope slides and ground them so thin that they were translucent. Viewed through a microscope, the structure of the eggshells could be compared. This tedious technique of making thin sections is still important in the study of fossilized eggshell, as we shall see in chapter 8. After several years of study, Gervais concluded that structurally the eggs were most similar to those of turtles:

> I cannot assert that we are dealing here with the eggs of some gigantic species belonging to the order of the tortoises, rather than some reptile of a different group, whose affinities with chelonians remain uncertain. This reservation is necessary because we completely do not know the characteristics of the dinosaurs' egg. (Cousin and others, 1994, p. 56)

Gervais' work was largely forgotten until Andrews made a boastful claim about the Central Asiatic Expedition's egg discovery in the French magazine *L'Illustration.* French geologist and paleontologist Louis Jolead, of the University of Paris, fired off a quick response—after all, French pride was at stake! The only reaction was from Granger, who grudgingly admitted:

> At Rognac in southern France some fragments of what seem to be reptilian egg shells were found in strata bearing dinosaur bones and there is a possibility that these are really bits of dinosaur eggs, but they may also belong to other contemporary reptiles. (Granger, 1936, p. 21)

Andrews chose not to even acknowledge the French eggshells in any of his many books and articles on the Central Asiatic Expedition. Since the discoveries of dinosaur eggs, clutches of eggs and baby dinosaur bones have been found on almost every continent, as we shall see in the next chapter. In fact, dinosaur eggs are so abundant that they now appear for sale as curios and conversation pieces, as well as educational exhibits.

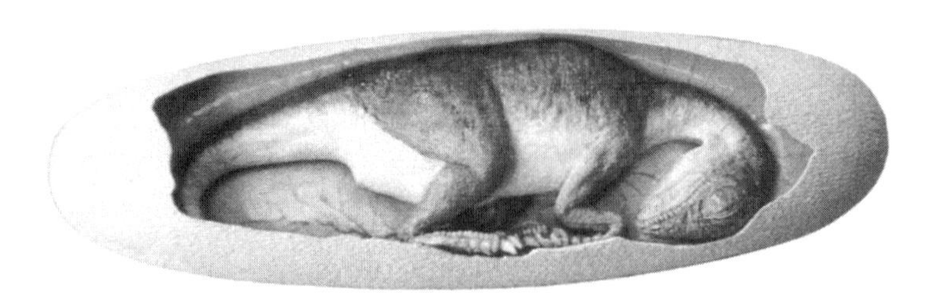

2 • Finding Dinosaur Eggs and Babies

Dinosaur eggshells, whole eggs, clutches of eggs, and even baby dinosaur bones have proven to be more common than was previously realized. A great deal of this is due to paleontologists' deliberately looking for such fossils, as well as recognizing them when they are found accidentally. Let's briefly look at some of the more important discoveries and sites, starting where eggs were first found and moving roughly west. A detailed list of all the sites is presented in Appendix I and plotted in Figures 2.1–2.4.

France

The first eggshells were found in the rugged foothills of the Pyrenees in 1859 (see chapter 1), but the first whole egg was not found until 1930 when it was plowed up in a field in southeastern France. Occasional eggs continued to be found sporadically until 1957, when sites with abundant eggs were found by Albert de Lapparent and his colleagues. Surprisingly though, very little research was conducted on the eggs until recently. During the past decade, French paleontologists from several museums and universities have taken an interest in these long-neglected fossils. Thus far they have discovered over two dozen localities in an arc along the southern and southeastern parts of France in the Languedoc region (Fig. 2.4). All of the sites occur in Upper Cretaceous strata (Campanian and Maastrichtian).

At one time all of these eggs were thought to be of one type because they were big and round, a type referred to as *Megaloolithus* (see chapter 9). Differences in shell thickness led German egg paleontologist Heinrich Erben to suggest that dinosaur extinction was linked to thinning of eggshell (see chapter 13). But recent work by French paleontologist Morris Penner (1983), as well as Monique Vianey-Liaud and her colleagues (Vianey-Liaud and others, 1994; Fig. 2.5), has shown that the differences in shell thickness are also linked to differences in microscopic structure. These differences, then, mean that different egg types or species are present (more on this in chapters 8 and 9, and in Appendix II).

The strata at the French localities are predominantly red, hence they are frequently referred to as the Cretaceous red-beds. Geologists have long used red-colored terrestrial sedimentary rock as an indicator of ancient arid or semiarid environments. The presence of ancient soil horizons, or *paleosols,* in these red-beds also supports such an interpretation for the Cretaceous climate of southern France. Most of the egg sites occur in these ancient soils, as would be expected of land-dwelling creatures. At one site near Rennes-le-Château, Gérard Breton and paleontologists from the Muséum d'Histoire Naturelle mapped the distribution of the eggs in the ground (Breton and others, 1986). Breton and his colleagues laid out a grid system and methodically excavated the site as if it were an archeological dig (Fig. 2.6). The results were used to infer the possible egg-laying behavior of two species of dinosaur (more about this in chapter 10). Despite the hundreds of eggs, not a single embryo has been found, so we really do not know who laid these eggs. Nevertheless, some paleontologists have argued that the large size of the eggs (16 cm or more) means that only a sauropod could have laid these eggs. Maybe, but without proof, that is just supposition.

Yannicke Dauphin (1992; Dauphin and Jaeger, 1990) has conducted more esoteric work by conducting chemical analyses of eggshells. The work provides important clues about the food and water available for the female dinosaur (see chapters 8 and 13).

Spain

Dinosaur eggs were first reported from Spain in 1954. The oldest eggshells are from the Lower Cretaceous of Galve in southeastern Spain (Fig. 2.3). The locality has produced hundreds of shell fragments, although most are turtle and crocodile, and only a few are dinosaurian (Kohring, 1990a). Only a preliminary description of these fragments has appeared and it will be interesting to see how they compare with eggshell of the same age elsewhere in the world.

Large, round *Megaloolithus* eggs identical to those from France are known from the Tremp Formation in the Tremp Basin, northeastern Spain (Fig. 2.4). The Pyrenees Mountains were not present during the Late Cretaceous, allowing free travel of dinosaurs between modern-day Spain and France. Surprisingly, these eggs were not studied until the late 1990s, when French paleontologist Monique Vianey-Liaud and Spanish paleontologist Nieves Lopez-Martinez took them up. Recently, thousands of *Megaloolithus* eggs were found in ancient beach deposits of the Arenisca de Arén Formation in the Tremp Basin. An estimated 300,000 eggs are believed to be preserved in the strata! The eggs occur in clusters of up to seven eggs and are believed to be clutches of eggs laid in a nesting ground (Sanz and others, 1995).

Portugal

Portugal has produced eggs from the western part of the country, but these are Late Jurassic in age (Fig. 2.2). The first egg was found in 1908 associated with the remains of the stegosaur *Dacentrurus armatus.* The specimen, however, was not recognized as an egg until 1957. During all those years it had sat unrecognized in the collections of the Geological

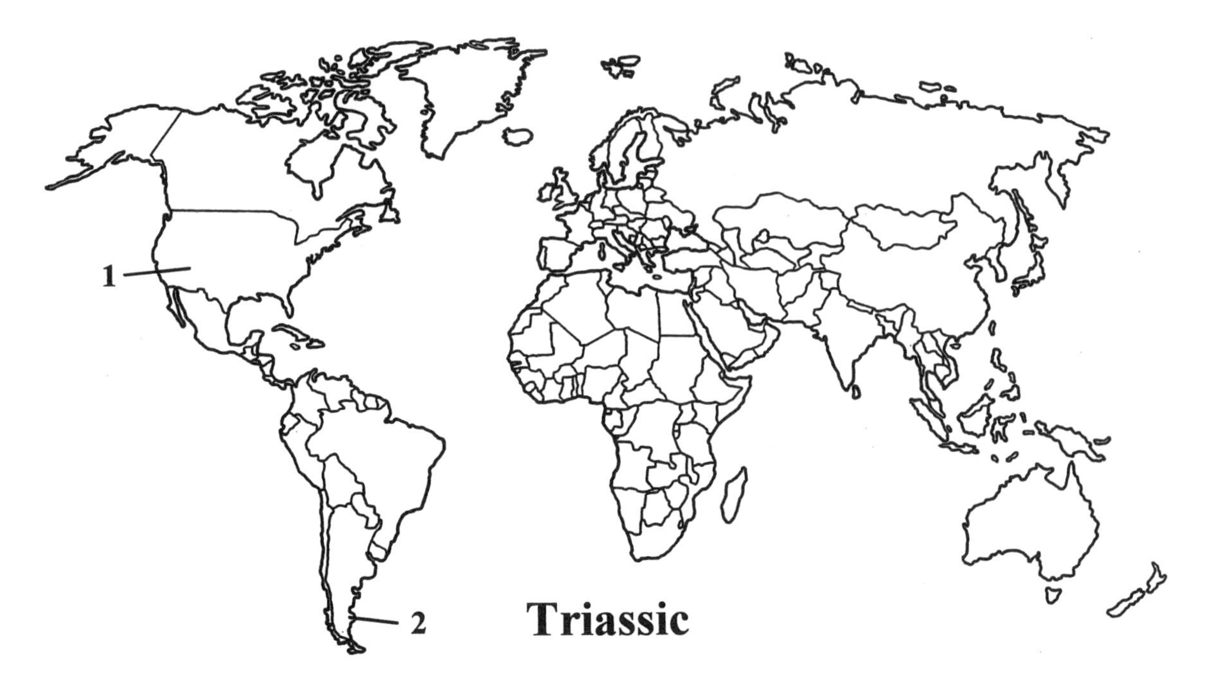

Fig. 2.1. Location of Triassic dinosaur eggs and babies (see Appendix I)

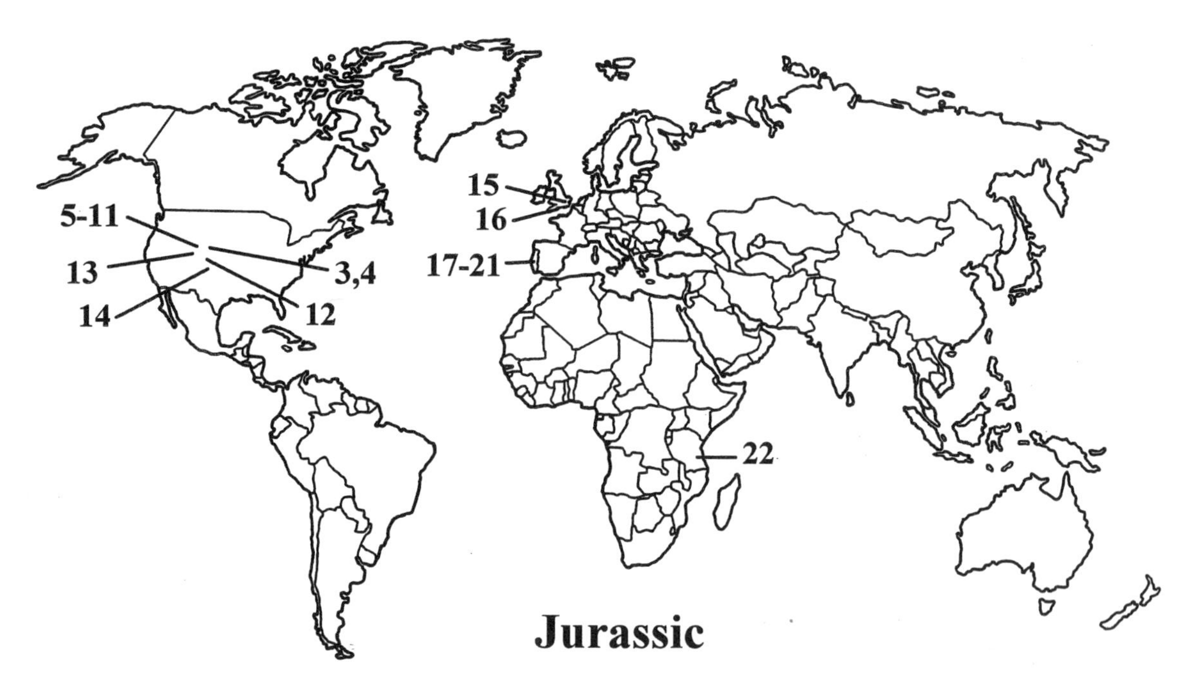

Fig. 2.2. Location of Jurassic dinosaur eggs and babies (see Appendix I)

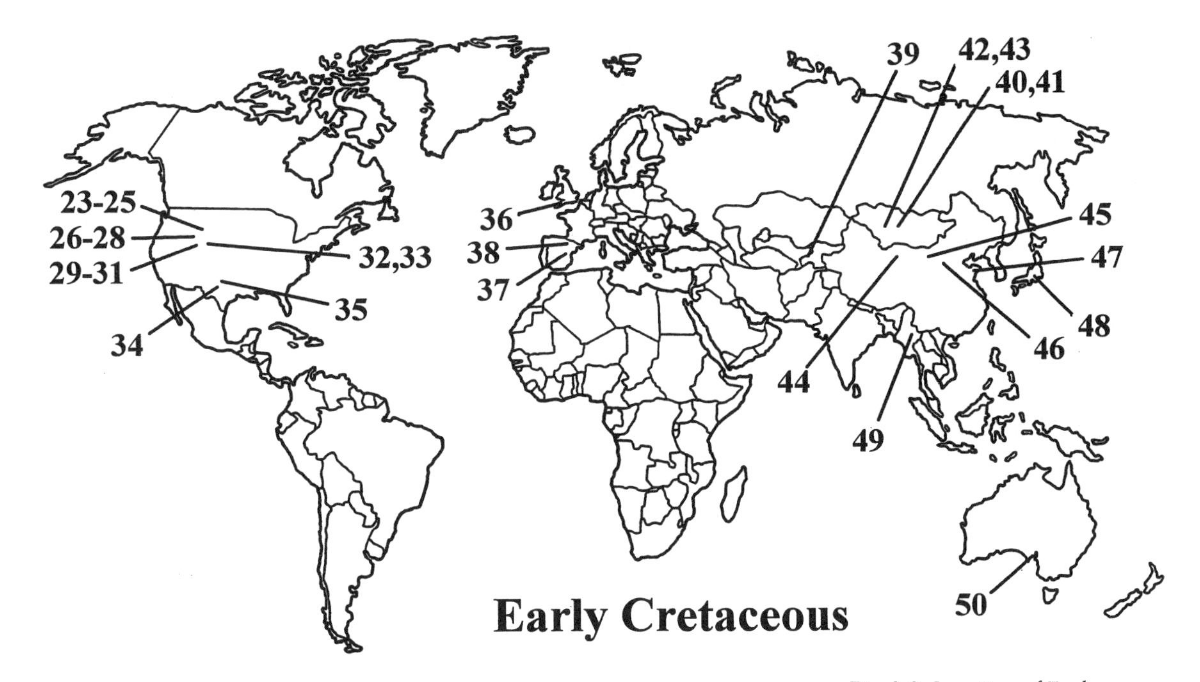

Fig. 2.3. Location of Early Cretaceous dinosaur eggs and babies (see Appendix I)

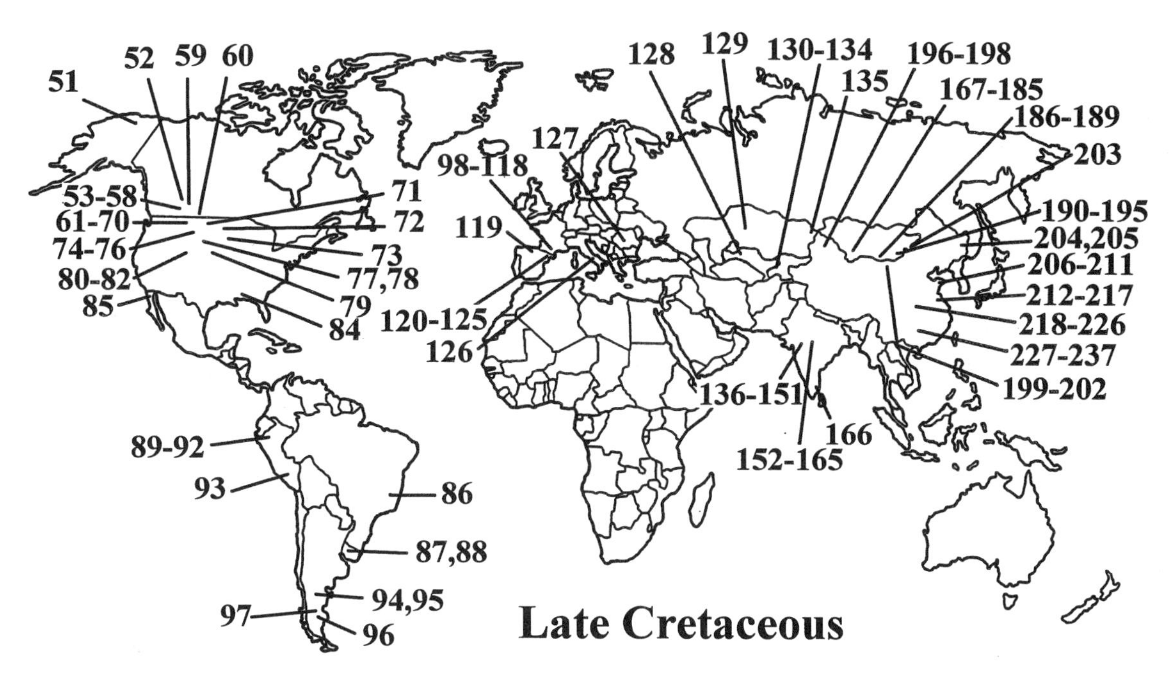

Fig. 2.4. Location of Late Cretaceous dinosaur eggs and babies (see Appendix I)

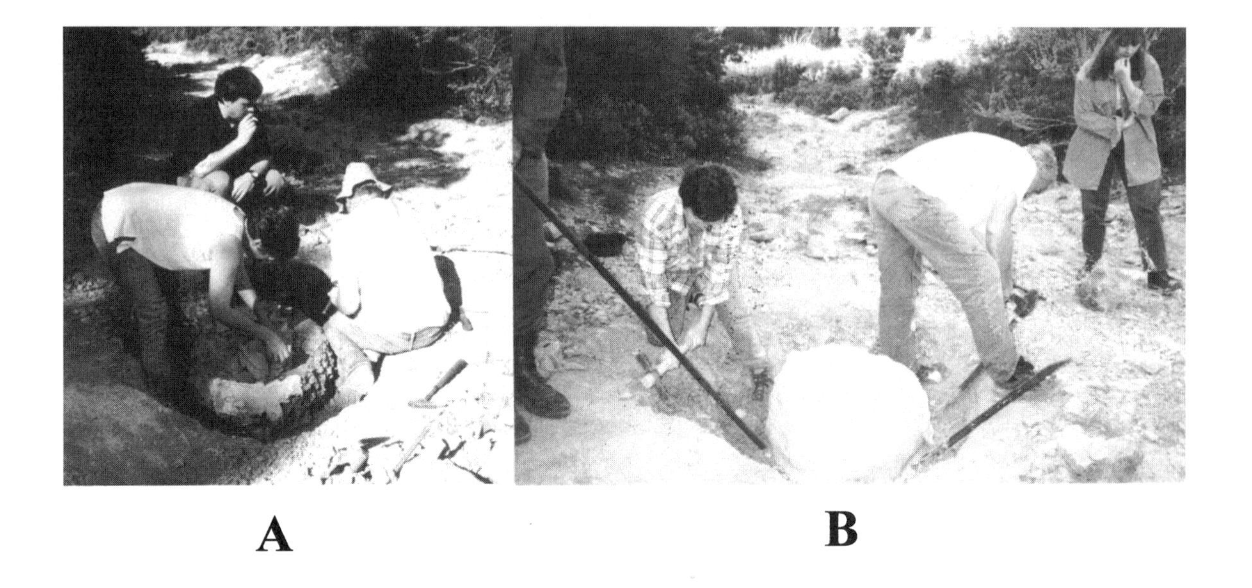

Fig. 2.5. Excavating a clutch of Megaloolithus *eggs in southern France. The surrounding rock was chiseled away (A), and the block encased in plaster of Paris for safe transport. The plastered block is pried free (B). Monique Vianey-Liaud digs with a pick in B. (Courtesy of Vianey-Liaud.)*

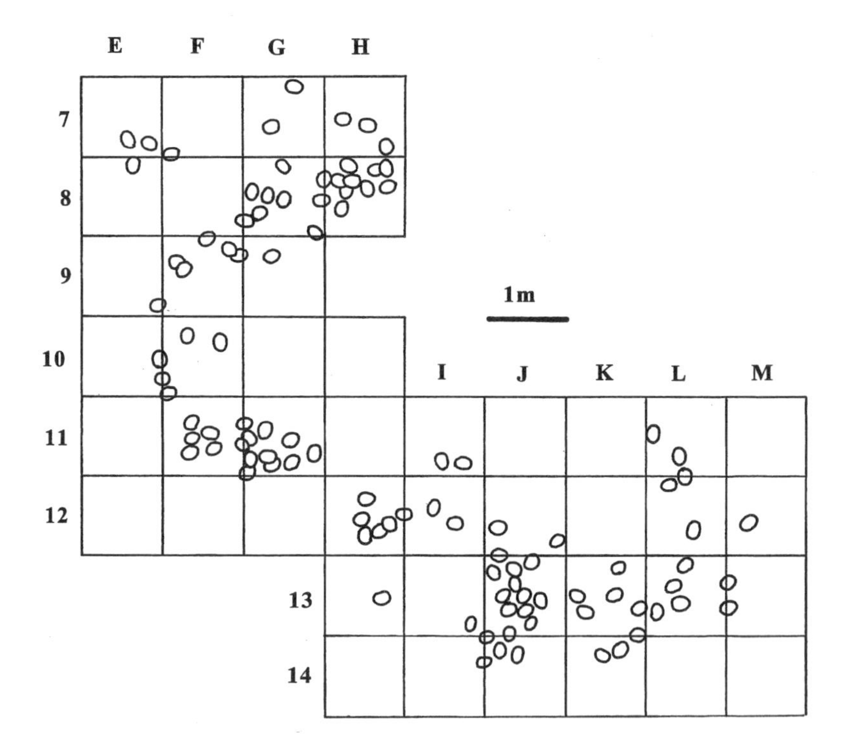

Fig. 2.6. Map showing the distribution of dinosaur eggs (mostly Megaloolithus*) at a locality near Rennes-le-Château at Aude, France. Rémi Cousin and his colleagues conducted an archeological-style excavation using a meter grid system and excavating layer by layer. Note that some eggs occur in clusters. Simplified from Cousin (1995).*

Survey Museum in Lisbon. When finally described, it was thought to be the egg of the *Dacentrurus* by French paleontologist Albert Lapparent and Portuguese paleontologist Georges Zbyszewski (Lapparent and Zbyszewski, 1957).

A more recent study of the egg, as well as others from Portugal, by the late dinosaur egg specialist Karl Hirsch shows that the egg is similar to eggs called *Preprismatoolithus* (Fig. 2.7) from the Upper Jurassic of North America (more in Appendix II). Recently, a clutch of about thirty *Preprismatoolithus*-like eggs were found in Lourinhã with numerous bones of embryonic or hatchling theropods (Fig. 2.8).

Thick eggshells from near the Upper Jurassic–Lower Cretaceous boundary at Porto Pinheiro are similar to the eggshells referred to as *Megaloolithus* from France and India. These eggshells are very important because they are the oldest occurrence of eggshells thought by some to belong to a sauropod.

Fig. 2.7. A Preprismatoolithus *egg from the Upper Jurassic of Portugal.*

England

Eggs were first found in England in 1859, the same year that Pouech found his eggs in France (see chapter 1). The clutch of eggs was found in marine strata (Fig. 2.9) and was briefly described by J. Buckman, who identified them as reptilian. They are small, about 4.5 cm (1¾") long, and oval. They are actually turtle eggs (Hirsch, 1996), but they are important to our story because they were the first eggs to receive a scientific name: *Oolithes bathonicae* meaning "stone egg from Bath." Two more egg species were named in 1871 by William Carruthers from England: *Oolithes sphaericus* ("spherical stone egg") and *Oolithes obtusatus* ("the obtuse stone egg"). Both egg types are small, being 2 cm (¾") and 0.5 cm (3/16")

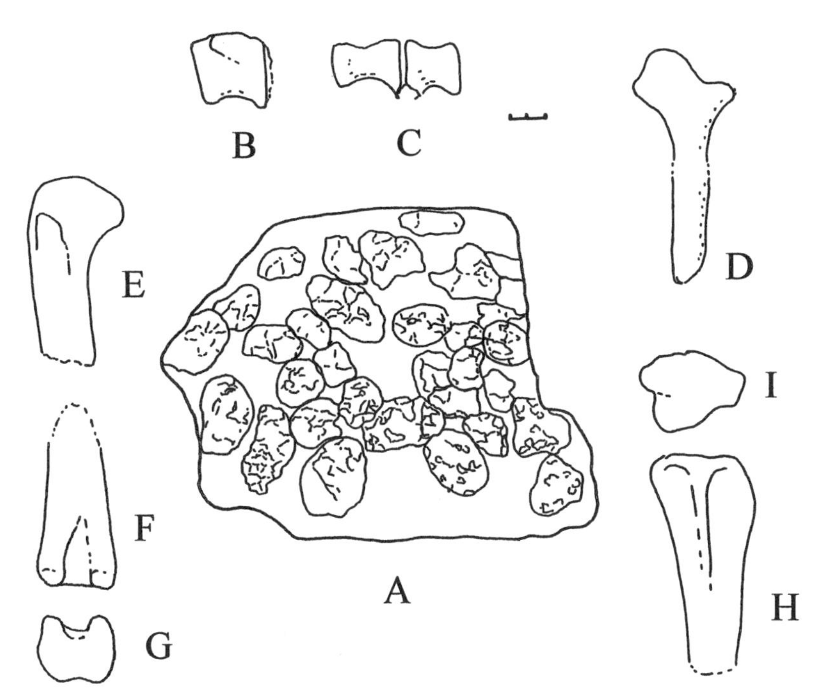

Fig. 2.8. A, sketch of a Preprismatoolithus *clutch, and various embryonic bones collected from the matrix around the clutch; B, dorsal vertebra centrum in side view; C, two caudal centra in side view; D, left humerus in front view; E, proximal half of right femur in front view; F, distal half of femur in rear view; G, end view of femur; H, proximal end of right tibia in side view; I, top view of right tibia. Scale for bones about 2 mm.*

in length respectively. Carruthers remarked that these eggs were rather common in the strata in which they were collected, but no one to my knowledge has ever collected any more. A microscopic study of these eggs by van Straelen (1928) led him to conclude that some of the shells seem to be a carbon film, possibly of a leathery or parchment shelled egg (we'll examine shell types in chapter 9). Karl Hirsch looked at these specimens as well and agreed that they *may* be leathery turtle shelled eggs. Unfortunately, he did not know how one could prove that they are.

Fig. 2.9. Oolithes bathonicae, *the first fossil egg to be named by Buckman in 1859. The rock below the specimen shows impressions of other eggs.*

Recently some dinosaur eggshells were found by Paul Ensom of the Yorkshire Museum. The small pieces were found in the residue left by washing soft, clay-rich rock through window screening stretched across the bottom of a small wooden box. This technique is used extensively to recover the tiny bones of fossil lizards, frogs, and mammals. No scientific study of these pieces has yet appeared. Baby dinosaur bones are rare in England, possibly because no nesting sites have yet been found. One of the few examples of a baby is of the early ankylosaur *Scelidosaurus* (Fig. 2.10) represented by a partial articulated skeleton. Embryonic and other dinosaurling bones may yet turn up in the residue from processing rock through window screening.

Romania

Elsewhere in Europe, a clutch of dinosaur eggs was recently found in Romania by Dan Grigorescu, a rather short, roundish man with a cheerful disposition. These eggs were discovered by accident after a rock fall revealed the eggs in a cliff face (Grigorescu and others, 1994). Grigorescu and his students faced considerable difficulty removing the eggs, but eventually fourteen eggs were recovered. These eggs resemble the *Megaloolithus* of France in size (about 17 cm, 6½" in diameter) and in microscopic structure. Originally, the Romanian eggs were thought to be those of a sauropod related to *Hypselosaurus,* called *Magyarosaurus.* The discovery of some embryonic bones identified as hadrosaur with the eggs, however, has led to the suggestion that they and the eggs might be those of the

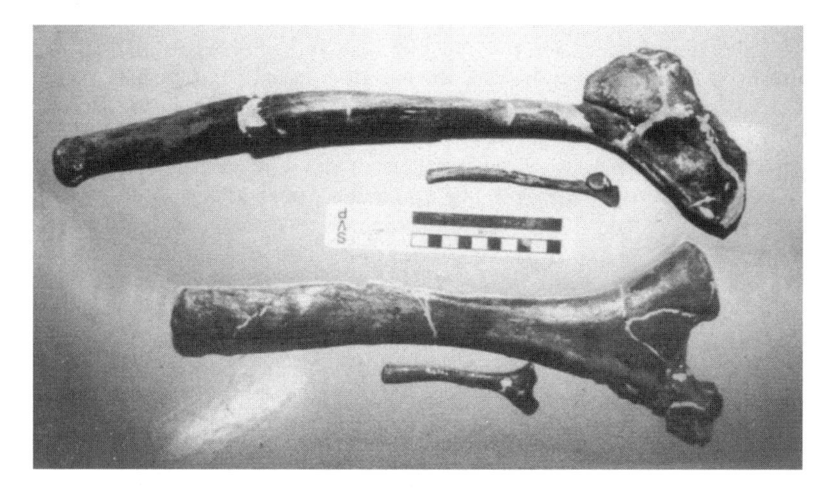

Fig. 2.10. *Pelvic bones of an adult and baby* Scelidosaurus, *an armored dinosaur from the Lower Jurassic of England. Top two bones are the right pubis and the lower two are the ischium.*

hadrosaur *Telmatosaurus* instead (Grigorescu, 1993). I remain unconvinced that the bones have been correctly identified—microscopically, the eggshell does not resemble better-known hadrosaur eggs from Montana or Canada; these other eggs have embryos associated with them (see below). The Romanian eggshell does, however, strongly resemble that of *Megaloolithus* eggs from Argentina, which are known to have sauropod embryos associated with them (see below).

Italy

Italy has never been known for its dinosaurs, so the discovery of a baby theropod (30 cm) was a major event (Fig. 2.11). The skeleton was named *Scipionyx* and was found in marine limestones better known for fossil fishes. The specimen is exceptional among dinosaur fossils because traces of soft tissue are present (Dal Sasso and Signore, 1998).

United States

Across the Atlantic, dinosaur eggshells were first found in the United States in 1913. The site was discovered by Charles Gilmore of the National

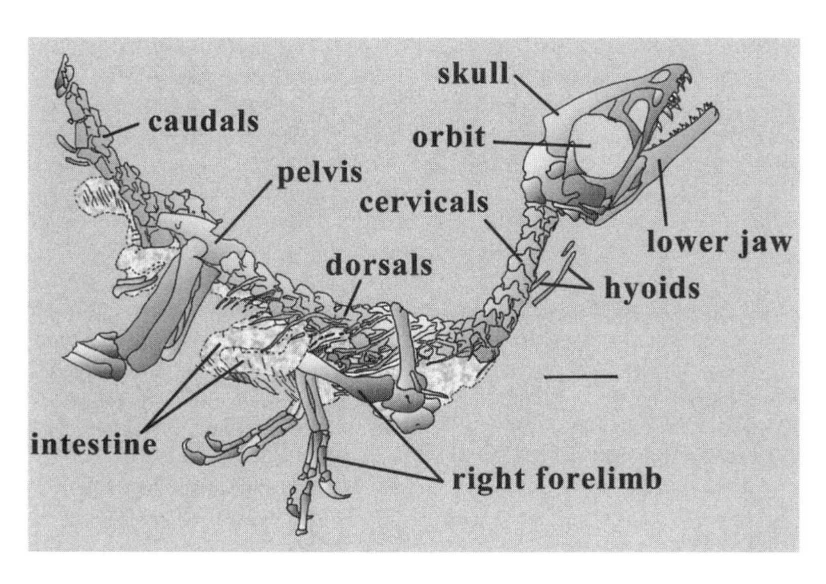

Fig. 2.11. *Drawing of the baby theropod* Scipionyx *from Italy; some traces of soft tissue are preserved. Scale = 2 cm.*

Museum of Natural History, on the Blackfoot Indian Reservation in northern Montana. However, he misidentified the fossils as freshwater clams. Fortunately, he made a small collection, recorded the site in his field notes, and best of all, photographed the site. Many years later, Jack Horner discovered the fossils in a drawer at the National Museum. Spurred on, Jack was able to relocate the site and was astounded! The entire "black clamshell bed" referred to in Gilmore's notes was composed of dinosaur eggshell. The entire bed may have been a nesting site used repeatedly by one species of hadrosaur (all of the eggshell was of one type). Surprisingly, though, these eggshells have not yet been studied.

Since Gilmore's discovery, eggshells and eggs have been found throughout the western United States. As yet, no eggs are known from the extensive Triassic sediments in the Southwest (a possible recent report [Monteith, 1996] may actually be of thin fish bones because chemical analysis indicates the "shells" to be calcium phosphate, same as bone). The oldest definite eggs and shells come from the Upper Jurassic Morrison Formation in Colorado and Utah. These fossils have been studied by Karl Hirsch and Emily Bray (Bray and Hirsch, 1998; Hirsch, 1994), and by Karen Alf (1998), formerly of the Denver Museum of Natural History. Bray and Hirsch (1998) have identified at least four different types, including one that Karl named *Prismatoolithus coloradensis* (the name was recently changed to *Preprismatoolithus* because the name originally proposed by Karl had already been used). The Jurassic egg-layer may have been a small carnivorous dinosaur because the eggshell resembles that of a clutch of eggs from Portugal associated with embryonic theropod bones (see above). At one of the *Preprismatoolithus* sites, Karen Alf also studied how eggshells get distributed by erosion and found that fossil eggshell is rather durable, a conclusion also reached by Tim Tokaryk of the Royal Saskatchewan Museum, Regina, Canada, in experiments with a tumbler mixed with sand, water, and chicken eggshell (Tokaryk and Storer, 1991).

Most of the Morrison sites consist of eggshells, although a circular clutch is known from Cañon City, Colorado (Fig. 2.12). The site is near the famous Marsh-Felch Dinosaur Quarry, where *Allosaurus, Diplodocus,* and *Ceratosaurus* were first found in the late 1800s. Interestingly, the site is just off the road that Marsh traversed in 1886 when he visited Felch. If only he knew. . . . The oddest discovery of a dinosaur egg in the Morrison was at the Cleveland-Lloyd Dinosaur Quarry. This site is a bone bed where the bones of many individuals of *Allosaurus* and other dinosaurs are found jumbled. Among the bones was found a single *Preprismatoolithus* egg (Fig. 2.13), which led to the suggestion by the press that the egg was that of an *Allosaurus*. Maybe, but we really don't know.

Baby dinosaur bones are also known from the Morrison, but are more widely distributed than eggshells, occurring in Wyoming, Utah, Colorado, and Oklahoma. These include sauropods (Fig. 2.14) and the small ornithopods *Dryosaurus* (Fig. 2.15) and *Othnielia* (baby dinosaurs are discussed more in chapter 12). Some baby sauropod bones from Como Bluffs, Wyoming, were briefly described by O. C. Marsh (1883) as a fetal dinosaur (more in chapter 10). As yet none of the large spherical eggs of the kind found in France, and thought to belong to sauropods, has been found in the Morrison Formation. This is all the more remarkable considering that sauropods are the most abundant dinosaurs in the Morrison.

Eggshell and baby bones are known from Lower Cretaceous strata in Utah, Wyoming, Idaho, Montana, Maryland, and Texas. Jim Jensen of Brigham Young University was the first to draw attention to the presence

Fig. 2.12. A, excavating a clutch of Preprismatoolithus coloradensis *eggs from the Morrison Formation of Cañon City, Colorado. B, same eggs after preparation.*

Fig. 2.13. The Preprismatoolithus coloradensis *egg found among* Allosaurus *bones at the Cleveland-Lloyd Quarry, Utah.*

Fig. 2.14. Two baby Apatosaurus humeri *lying atop an adult* Apatosaurus *humerus for comparison. The baby bones are about 24 cm (9") long, whereas the adult is 119 cm (47").*

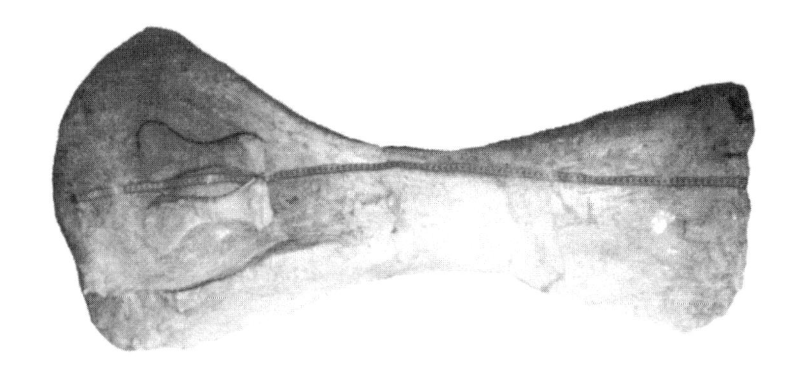

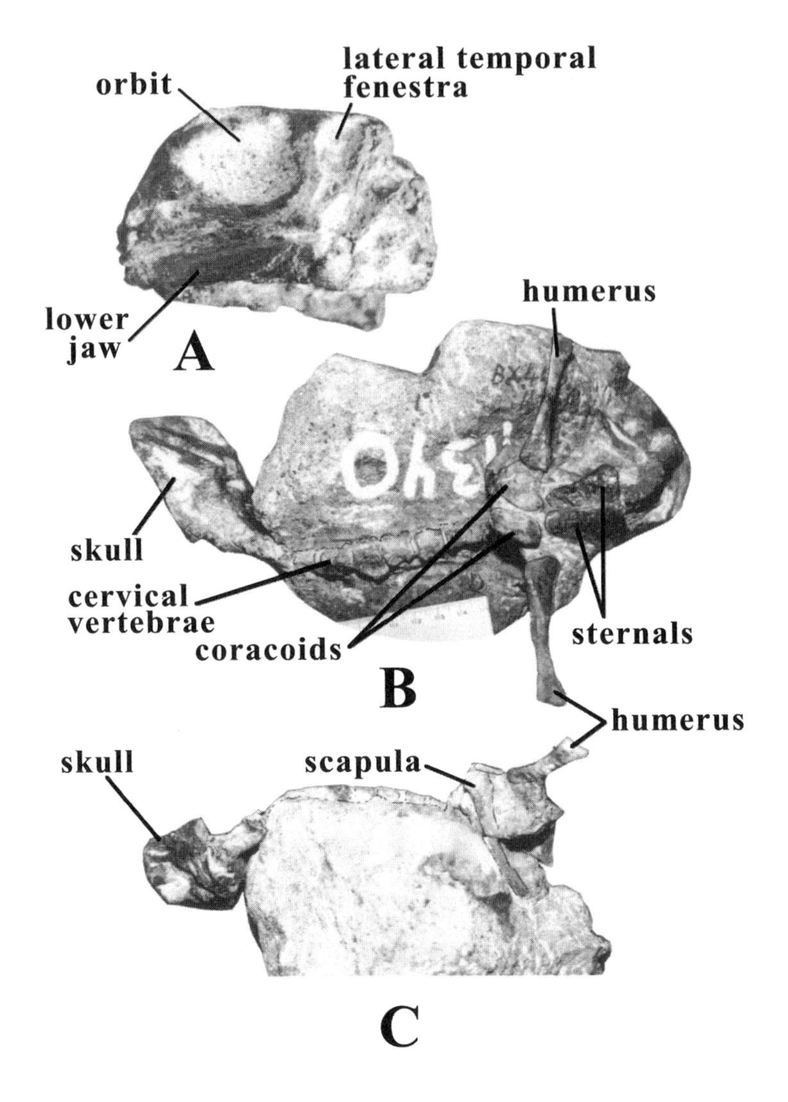

Fig. 2.15. Baby Dryosaurus *from the Morrison Formation, Dinosaur National Monument: A, skull in left lateral view (skull is 5.5 cm long); B, skeleton in ventral view; C, skeleton in lateral view.*

of eggshells in the Cedar Mountain Formation of eastern Utah (Jensen, 1970). More recently other sites have been found, but none has produced whole eggs or a nest yet. Some of the eggshells closely resemble forms known from China and Mongolia (see Appendix II). Eggshell fragments have also been found in the Lower Cretaceous of eastern Idaho (Dorr, 1985), but have only received a preliminary description. Bones of a baby nodosaurid have recently been described from Lower Cretaceous marine rocks of Texas (Jacobs and others, 1994). The specimen had small oyster spats growing on the bones, indicating that the little carcass had drifted out to sea before it sank to the sea floor. Marks on the bones show that sharks and possibly crabs had stripped the carcass of flesh before the oysters grew on the bones. Baby sauropod bones have been found in Maryland not far from the nation's capital and were first described by Othneil Charles Marsh in 1888.

Montana unquestionably has produced the greatest numbers of eggs and babies in the United States, especially from the Upper Cretaceous Two Medicine Formation. Most of the collecting has been done by Jack Horner and his crews from the Museum of the Rockies. The discovery of the eggs and baby bones by Jack in 1978 has been told and retold many times, most notably by Jack and James Gorman in *Digging Dinosaurs* (Horner and Gorman, 1988). What is less well known is that Jack actually found his first egg as a child, but thought it was a crushed fossil turtle. A detailed, preliminary description of these eggs was made by Hirsch and Quinn (1990). They identified at least four different types, including eggs of *Maiasaura* and the small theropod *Troodon* from a site known as Egg Mountain (Fig. 2.16). Baby *Maiasaura* bones were found in a possible nest (Fig. 2.17), and embryonic *Troodon* within eggs (Fig. 2.18). Farther north, baby and embryonic dinosaurs have been found with eggs and eggshells. Most of these bones probably belong to the crest-headed hadrosaur *Hypacrosaurus*.

Other eggshells and baby bones, including the lower jaw of a baby *Troodon*—a small theropod related to *Velociraptor* of *Jurassic Park* fame—have been found in the Lance Formation of Wyoming (Carpenter, 1982). Baby hadrosaur bones have even been found as far north as the North Slope of Alaska!

One of the oddest occurrences of an egg is from the Mooreville Chalk of Alabama. This chalk was deposited in the northern end of the proto–Gulf of Mexico. How this egg came to rest on the bottom of the sea is unknown, especially because the microscopic structure is most similar to an avian or theropod egg.

Canada

Eggshell from neighboring Canada was found in 1964. These specimens were collected along the Milk River in southern Alberta. Since that time, eggshell has been found in various places, including Dinosaur Provincial Park, home of *Gorgosaurus, Chasmosaurus* and other dinosaurs first collected here by Barnum Brown (see chapter 1). The park has also produced a partial baby hadrosaur skull (Sternberg, 1955) that is about the

Fig. 2.16. The bump on the skyline with the grandiose name of Egg Mountain.

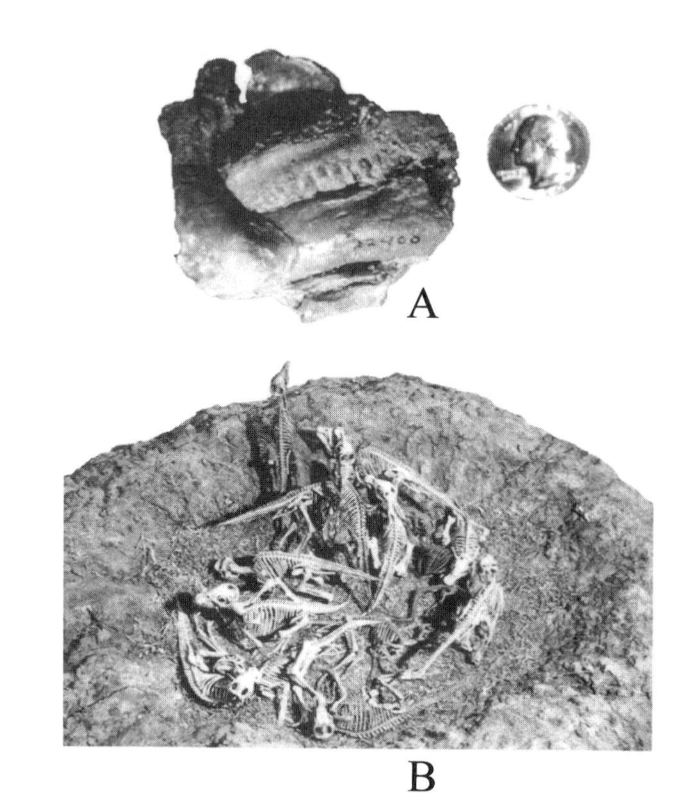

Fig. 2.17. A, one of the original baby Maiasaura *skulls collected by Jack Horner and Bob Makela. B, reconstructed "nest" of* Maiasaura *as inferred by Jack Horner's work.*

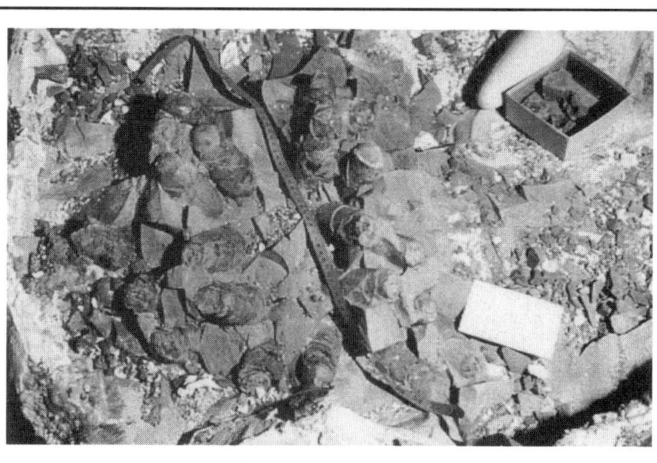

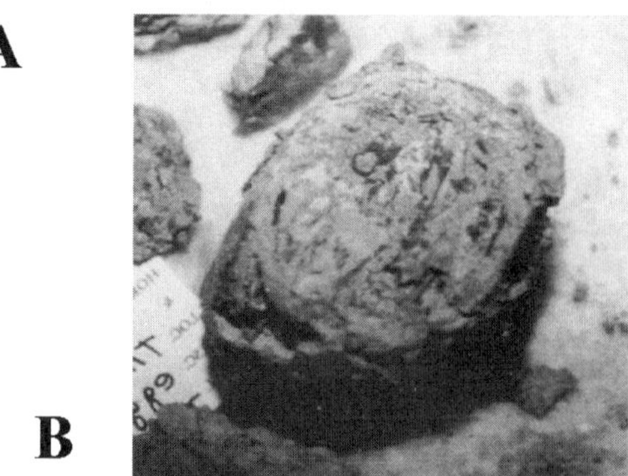

Fig. 2.18. A, one of the Troodon *clutches from Egg Mountain. B, another* Troodon *egg showing embryonic bones.*

same size as that of baby *Maiasaura*. Also found in the park was a partial juvenile ankylosaur *Euoplocephalus* (Fig. 2.19).

More recently, numerous embryonic and baby hadrosaur remains, as well as clutches of hadrosaur eggs, have been found in the Oldman Formation just north of the U.S.–Canadian border at a place called Devil's Coulee (Fig. 2.20A). The site was found in 1986 by high school student Wendy Sloboda while out looking for fossils. The exposures are not very extensive, yet more than a dozen spots have produced clutches of eggs or hundreds of eggshells, many of which were recently described by Darla Zelenitsky and her colleagues (1996, 1997) as those of hadrosaurs, theropods, birds, and turtles. Most important, several groups of hatchling lambeosaurine skeletons have been found (Fig.2.20B) and named by Jack Horner and Phil Currie (1994) as *Hypacrosaurus stebingeri,* one of the crested hadrosaurs. Neither the eggs nor the hatchlings occur in what could be considered bowl-shaped "nests" of the type reported by Jack Horner for *Maiasaura.*

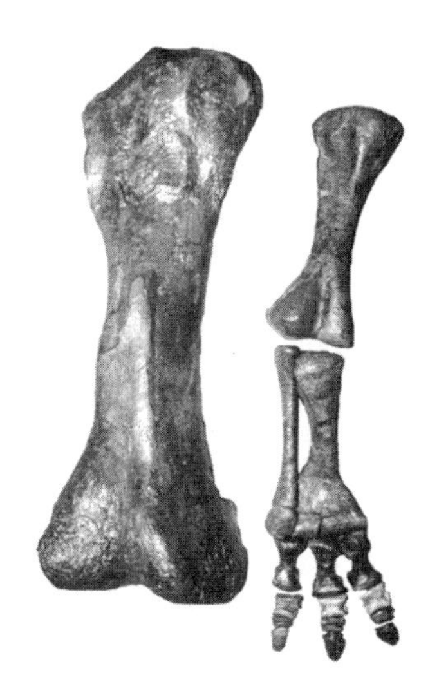

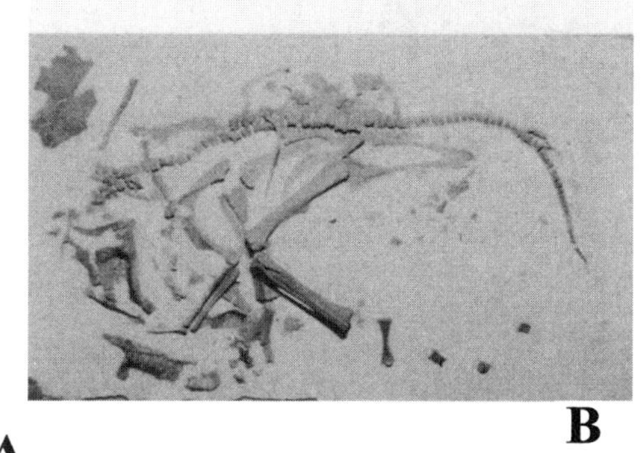

Fig. 2.19. Comparison of a baby Euoplocephalus *leg (right—about 50 cm long) with the femur of an adult (left).*

Fig. 2.20. A hatchling Hypacrosaurus stebingeri *skeleton as reconstructed from bones found near the top of Little Diablo Hill at Devil's Coulee.*

Elsewhere in Alberta, a baby *Pachyrhinosaurus,* a type of ceratopsian, has been found in a bone bed containing numerous individuals of various ages. Why so many individuals of *Pachyrhinosaurus* occur at the site is a mystery, but some catastrophe is believed to have engulfed a herd. Eggshells are known from the Frenchman Formation of neighboring Saskatchewan, although none of these specimens has been described in detail yet.

Mexico

Some eggshells found in 1973 from the Upper Cretaceous of Baja, Mexico, have recently been identified as possibly hadrosaur (Rosa and Armando, 1995). Several hadrosaurs, including parts of a giant crested lambeosaur, were collected in the vicinity by the Los Angeles County Museum (Morris, 1973), so the suggestion is not unreasonable. However, until a detailed description with illustrations has been provided, there is no way of independently verifying the identification. A single whole egg has also been found near Trujeo in northern Mexico.

Brazil

Eggs from South America were first described in 1951. The site, in the Upper Cretaceous Bauru Formation, is located near Brazil's eastern coast, in the rugged Brazilian highlands. The egg was briefly described by Brazilian paleontologist Llewellyn Price (1951). As yet, no detailed description of the egg comparing it with other eggs has been done. The eggs are large and spherical, resembling *Megaloolithus* from France.

Uruguay

Recently, large spherical eggs were described from the Upper Cretaceous by Uruguayan paleontologist Guillermo Faccio (1994). At one site, the eggs occur in a pile suggesting that they had been laid in a hole. The microscopic structure of the shell is poorly preserved, but enough is present to show that the eggs are *Megaloolithus*. The shell is unusually thick, 5 mm on average, thus exceeding *Megaloolithus megadermus* from the Upper Cretaceous of India (see Appendix II). Other *Megaloolithus* eggs have recently been found (Fig. 2.21).

Argentina

Some small spherical eggs from the Triassic have been reported by Argentine paleontologists Bonaparte and Vince (1979). As yet, however, no detailed microscopic study has been conducted to confirm that these are eggs. If they are, there still remains the possibility that they are not dinosaurian, but eggs of some other fossil reptile. The same deposits also produced a small group of baby prosauropods called *Mussaurus* or "mouse reptile" because of their small size (Fig. 2.22; more in chapter 11). Numerous *Megaloolithus* eggs have also been found (Fig. 2.23) at several Upper Cretaceous sites in Argentina, most notably in the recent site discovered near Auca Mahuevo in northwestern Argentina. Thousands of eggs occur here, including many with embryonic bones (Chiappe and others, 1998).

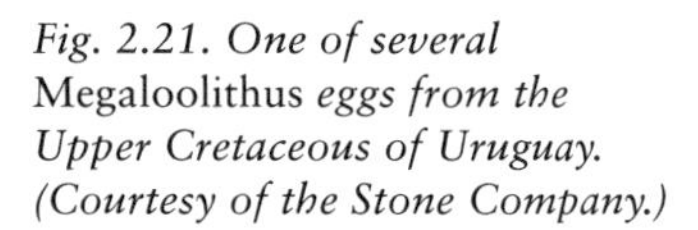
Fig. 2.21. One of several Megaloolithus *eggs from the Upper Cretaceous of Uruguay. (Courtesy of the Stone Company.)*

Fig. 2.22. Baby Mussaurus *skeleton as found.*

Fig. 2.23. Clutch of Megaloolithus *eggs from Argentina. (Courtesy of K. Johnson.)*

Peru

Eggshells have been described from the Upper Cretaceous in the Baga Basin in northern Peru and at Laguna Umayo near the Peruvian–Bolivian border by French paleontologists Philippe Kérourio and Bernard Sigé (1984) and by Karl Hirsch (1988). Some of the shell fragments have an avian structure, meaning that they could belong to either a bird or a small theropod. Other eggshells resemble the *Megaloolithus* eggshell from France and India, while still others resemble modern gecko eggshell. Unfortunately, the identity of the egg-layers is not yet known.

Australia

Recently, bones of a hypsilophodont dinosaur have been found in the southern end of Australia in an area called Dinosaur Cove, Victoria (Rich and Rich, 1989). The bones are all small and immature, and the large eye sockets are characters indicative of babies and young individuals. As yet, however, no eggshells have been reported. Dinosaur Cove is important in that the site was well within the Antarctic Circle during the Early Cretaceous. As we shall discuss further in chapters 11 and 12, this must mean that either the dinosaurlings grew very rapidly and reached adult size by the time "winter" set in, or that the babies were able to endure six months of darkness.

China

China has proven in recent years to be one of the richest places in the world for dinosaur eggs. Almost all of them are Upper Cretaceous in age. The first dinosaur eggs found in China were collected by the Central Asiatic Expedition (see chapter 1) from Iren Dabasu, an outpost near the Chinese–Mongolian border. Several crushed eggs were reconstructed, turning out to be about the size and shape of a navel orange (Fig. 2.24). The eggs were mistaken for fossil ostrich eggs because they were lightweight and whitish, a mistake not corrected until van Straelen (1925) examined them years later. Surprisingly, it would be another thirty years before any other eggs were found. Then came a flurry of discoveries in Laiyang, in northeastern China (Chow 1951; Young, 1954), and in the Nanxiong Basin, in southern China (Young, 1965). This early work has been continued in recent years by Chinese egg specialist Zhao Zi-Kui. Zhao has pioneered the use of what he calls "parataxonomy" for naming eggs (we'll examine this more in chapter 9). Zhao and his colleagues have also examined the chemistry of eggshell. The study was based on the idea that "you are what you eat" and served to determine something about the environment in which the mother lived, as well as to help in the search for possible clues to the extinction of the dinosaurs (more in chapter 13).

Most of the sites are in red mudstones of ancient floodplains (Fig. 2.25). The mud was originally brown as most mud is, and was deposited by seasonal floods in a semiarid environment. When the mud dried, oxygen in

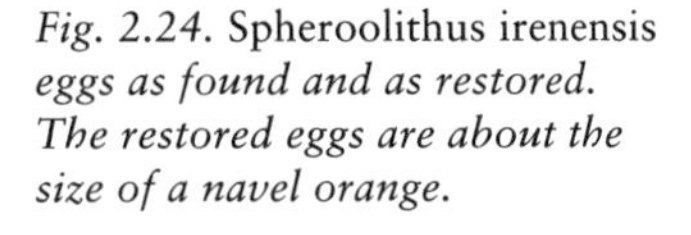

Fig. 2.24. Spheroolithus irenensis *eggs as found and as restored. The restored eggs are about the size of a navel orange.*

the soil destroyed the organic material that gave the mud the brown color and oxidized the naturally occurring iron, producing the reddish tint. All of this process was slow, taking hundreds or thousands of years. Such broad open country was ideal for dinosaurs to lay their eggs, and thousands of eggs have been found preserved where they were laid tens of millions of years ago.

One recent discovery in China originally sounded too good to be true. It made international news as the first occurrence of dinosaur DNA in an egg. The 70–80 m.y. egg contained dark, flaky material identified as organic carbon. The results of the analysis indicated similarities with the DNA of frogs, as well as mammals (Li and others, 1995). Such results seemed odd to Yang Hong (1995), who ran a new analysis. His results indicated that the DNA was, unfortunately, from contamination, probably from plants. Such a conclusion is not surprising—modern plants often seem to seek out fossils because of the wealth of minerals they contain.

Many people in the West first learned about the wealth of dinosaur eggs in China during the early 1990s when thousands of eggs were exported from China for sale. This loss of what the Chinese government considered national treasures was stopped a year or two later by stationing of the People's Liberation Army in the egg-producing regions, especially Henan Province. These troops have since been replaced by peasants hired to guard the sites. The export of the eggs has been a mixed blessing. On one hand, the sale of the eggs was a great financial boost for peasants living at poverty level and it did provide a source of quality eggs for research outside China (Fig. 2.26). On the other hand, locality and stratigraphic data was often minimal, and sites were plundered before the eggs could be studied in their context.

The person who has done the most with these eggs is Terry Manning from England. Terry, a skinny redhead, purchased numerous eggs, selecting eggs that were least crushed and most nearly intact (many of the eggs are sold embedded in a block of matrix, which actually hides the fact that the eggs have hatched or were eaten; see chapter 8). By carefully dissolving away part of the shell and infilled matrix, Terry has revealed embryonic dinosaurs, as well as fly casings from the maggots that had fed on the tiny carcasses. Perhaps most remarkable are the fragments of what appear to be

Fig. 2.25. Charlie McGovern (left) and Philip Currie (right) examine eggshells at Green Mountain, Hebei Province, China. Impression of an egg is seen below Philip's right elbow. (Courtesy the Stone Company.)

Fig. 2.26. Elongatoolithus *eggs from the Upper Cretaceous of China. This clutch is one of hundreds that were brought out of China for sale in the West. (Courtesy the Stone Company.)*

fossilized tissue. As yet no detailed study of Terry's wonderful specimens has been made (more on these specimens in chapters 8 and 11).

Besides the embryonic dinosaurs found by Terry, baby dinosaur skeletons have been reported by Chinese and Canadian paleontologists, mostly from near the Chinese–Mongolian border. The specimens include the ankylosaur *Pinacosaurus* and the ceratopsians *Psittacosaurus* and *Protoceratops* (more in chapter 11). Chinese paleontologist Dong Zhiming and Canadian paleontologist Phil Currie (1996) have recently described a skeleton of an adult *Oviraptor* on a clutch of *Elongatoolithus* eggs found near the Chinese–Mongolian border (Fig. 2.27). The individual may have been brooding the eggs when she was buried by a sandstorm (more in chapter 7).

Mongolia

As mentioned in chapter 1, eggs were found in Mongolia in 1923 by the Central Asiatic Expedition. Over the next few years, the expedition

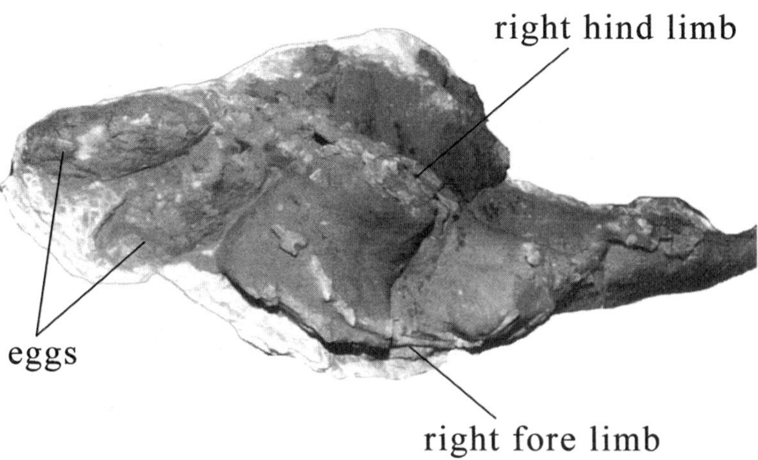

Fig. 2.27. A partial skeleton of an Oviraptor *found associated with eggs, from the Upper Cretaceous of China.*

collected almost two dozen clutches and hundreds of fragments representing at least six different egg types. After the last expedition in 1930, no further paleontological work was done in Mongolia until 1946. That summer, Soviet paleontologists undertook a large expedition retracing, in part, the routes of the Central Asiatic Expedition. This and subsequent expeditions during the late 1940s were remarkable given the state of the Soviet Union at the time. World War II had ended the previous year, most of the major cities—including Moscow and Leningrad—were devastated, and the economic infrastructure was in ruins. To Westerners, conducting an expedition at a time like this might seem frivolous. At least in part, the expedition proceeded because plans had been in the making for a long time before the war (David Unwin, personal communication). But I also suspect that the Soviet Union felt the need to establish a buffer to the east much like the buffer the Warsaw Pact countries provided to the west. What better way to survey the territory than with a scientific expedition?

The Soviet expedition of 1946 and later years discovered a wealth of dinosaur fossils, many of which can be seen today at the museum of the Paleontological Institute in Moscow. Lots of eggs were found in Mongolia (Fig. 2.28), including from new sites not visited by the Americans. Many of the eggs were studied by Ivan Sadov, who worked on modern bird eggshell for his Ph.D. before the "Great Patriotic War," as WW II is known in Russia. Sadov was interested in the microscopic structure of eggshell, and had prepared a manuscript on some of the Mongolian dinosaur eggshells. It was published in 1970 after his death. Other eggs were briefly described by Andrey Sochava. Sochava is actually a limnologist—someone who studies lake deposits. He got interested in the Mongolian eggshells because so many were found during the Mongolian expeditions. Dinosaur eggshells, then, were a hobby for him, just as they were for Karl Hirsch.

A much more thorough job of describing the eggs and eggshells from Mongolia has been done in the past decade by Russian paleontologist Konstantin Mikhailov from material collected by the Joint Soviet-Mongolian Paleontological Expeditions that occurred in the 1970s and 1980s. Mikhailov has had a long interest in eggs, having amassed a collection of

*Fig. 2.28. One of the original egg clutches collected by the Paleontological Institute of Moscow from Mongolia (*Spheroolithus irenensis*).*

bird eggs while a boy. It was only natural, he felt, to expand his interest to dinosaur eggs after graduating from Moscow University. He has described and named many eggshell types from Mongolia, and recently compiled much of the information in a monograph on fossil eggs. Still other Mongolian eggs were studied by Polish paleontologist Karol Sabath from specimens collected during the 1960s and 1970s by the Polish-Mongolian Paleontological Expeditions. Sabath's interests, however, lay not so much in the microscopic distinctions among the various types of eggshell, but rather in how the eggs occurred in the nest and how the eggs behaved as the "house" for the developing embryo.

Embryonic dinosaurs from Mongolia have only recently been discovered in any quantity. A remarkable recent discovery is the partial oviraptorid embryo found within a broken egg collected by a recent American Museum of Natural History Expedition. Associated with the embryo were fragments of an embryonic or hatchling velociraptorine. Was this an example of a nest parasite, much like the cowbird, which lays an egg in the nest of another bird? Or might the velociraptorine have been food for other oviraptorids that had successfully hatched? We may never know. Another important discovery was made by Japanese paleontologists of nine hatchling *Protoceratops* found together. We'll look at other babies in chapter 11.

India

Thousands of eggs have been found in north-central India since their first reported discovery by M. Sahni in 1957. The eggs have only recently been described as various species of *Megaloolithus* (Fig. 2.29) by Indian paleontologists D. M. Mohabey, Ashok Sahni, and Sohn Jain. As yet no embryonic or baby bones have been found; the one report has since proven to be a poorly preserved fossil snake skeleton (Jain, 1989).

The egg-producing strata occur in what are called the Deccan Traps, a rugged landscape containing Late Cretaceous lava flows similar to those along the Columbia River in Oregon. These lava flows record an interval of immense volcanism. Yet during volcanic lulls, sediments were deposited on

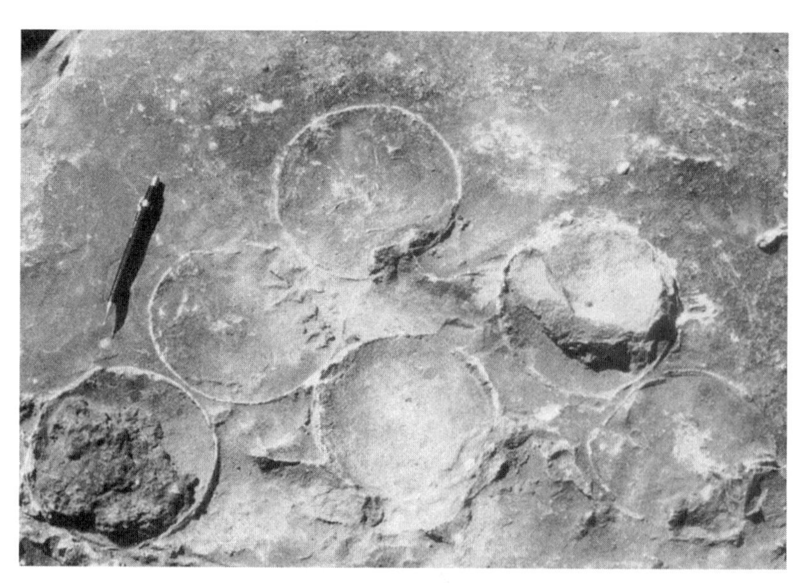

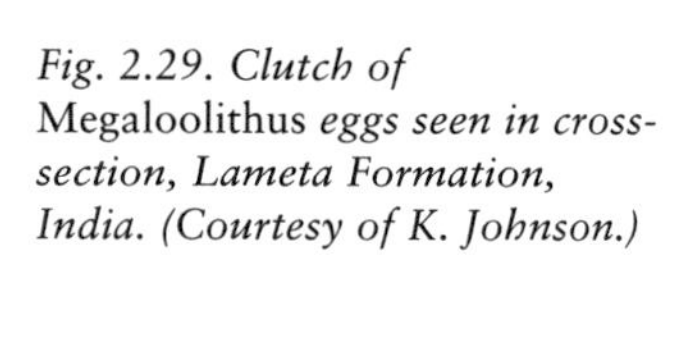

Fig. 2.29. Clutch of Megaloolithus *eggs seen in cross-section, Lameta Formation, India. (Courtesy of K. Johnson.)*

floodplains, plants grew, and dinosaurs returned and nested. Interestingly, during this time India was an island continent much like Australia is today. India had separated from Antarctica and remained isolated for the last 65 million years of the Cretaceous—as much time as separates us today from the dinosaurs at the end of the Cretaceous! How this dinosaur fauna evolved in isolation is still an ongoing area of study by Indian paleontologists.

South-Central Asia

Dinosaur eggshells were reported from Cretaceous rocks of Kazakhstan in 1961. In subsequent years, eggshell was found at several other sites. Most remarkable was the recovery of eggshell from a rock sample recovered while drilling for oil. These specimens, as well as others, have been very briefly described by Soviet paleontologists Lev Nessov and M. Kaznyshkin (1986), and Mikhailov (1994a,b). Nessov and Kaznyshkin also reported on eggshells and eggs from neighboring Uzbekistan and Kyrgyzstan. One site in Kyrgyzstan is reminiscent of Gilmore's "clam" locality in Montana, with eggshell extending almost 800 m (about half a mile) along the outcrop. At another site nearby, the eggshell-rich strata extended for 15 km (9 miles), making it the richest eggshell site in the world. The width of the stratum is not known; nevertheless, there must be millions if not billions of shell fragments representing millions of eggs. This abundance of eggshells suggests that both areas were sites of extensive dinosaur rookeries for perhaps thousands of years. There is still much that needs to be done with the sites, especially excavations in order to determine the extent of the shell beds and to look for whole eggs.

Africa

Dinosaur eggs have also been found in Africa. The oldest eggs were a clutch of six found in the Upper Triassic Elliot Formation of South Africa. The eggs also contain embryonic bones originally believed to be those of a prosauropod, possibly *Massospondylus,* which is common in the beds (Kitching, 1979). Unfortunately, the identification is probably wishful thinking; the eggshell is most certainly crocodilian (Fig. 2.30; Zelenitsky, personal communication). The Triassic crocodiles were different from their modern descendants in that they were long-limbed, terrestrial animals. *Baroqueosuchus* is a crocodile that has been named from the Elliot Formation, but the embryos within the eggs have yet to be compared with it.

A single egg, possibly hatched, was uncovered by the British Museum of Natural History at Tendaguru, Tanzania (Fig. 2.31). These Upper Jurassic beds produced a skeleton of *Brachiosaurus* now mounted at the Natural History Museum in Berlin, Germany. Two other eggs were found in southern Tanzania in 1950. The eggs were found in strata possibly Cretaceous in age by the Tanzanian Geological Survey. Except for a brief popular account by British paleontologist W. Swinton (1950), the eggs have not been studied in detail.

From this brief survey, it is clear that dinosaur eggs and baby bones have been found on almost every continent except Antarctica. I suspect that they will eventually be found there as well. After all, the recently discovered carnivorous dinosaur *Cryolophosaurus,* an Antarctic native, had to have hatched from an egg (live birth in dinosaurs is examined in chapter 11).

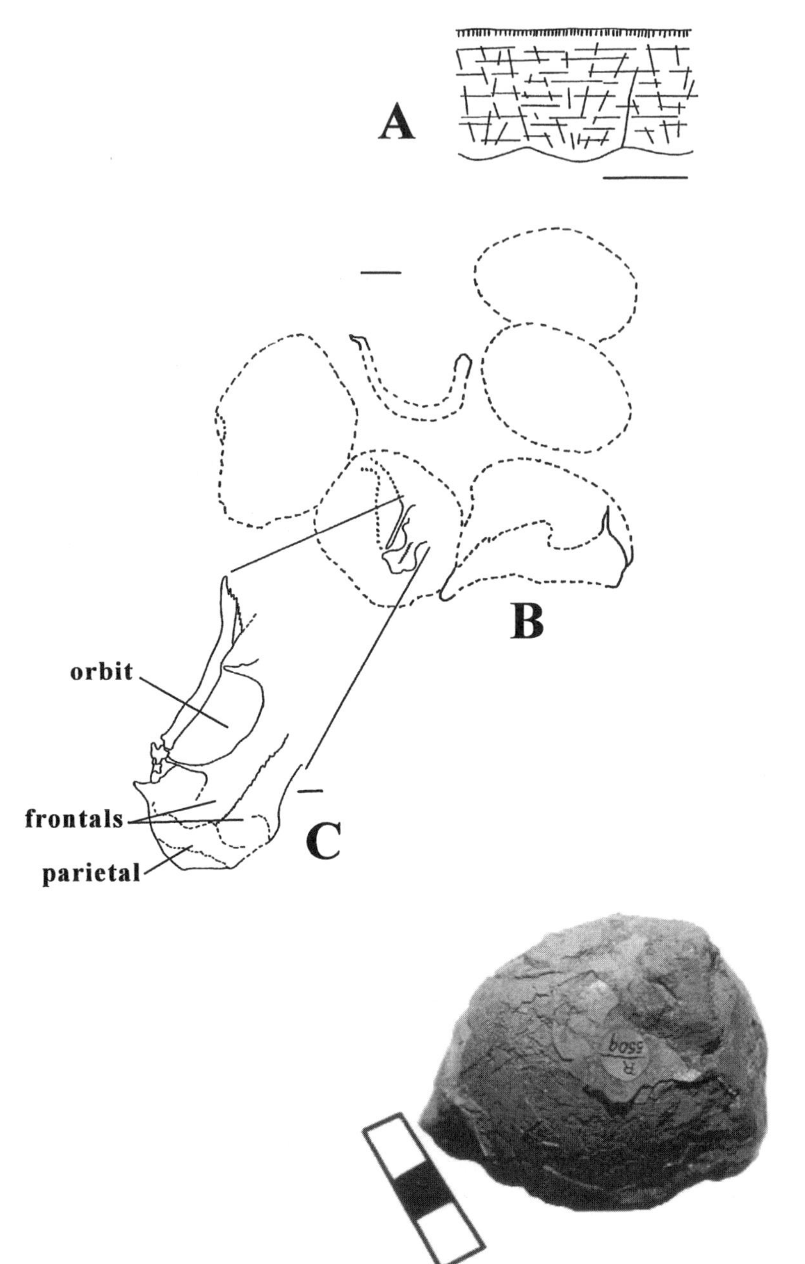

Fig. 2.30. The only Triassic eggs from Africa appear to have the microscopic structure of crocodile eggs (A). The clutch consists of at least six group eggs (B), one of which shows a partial skull of an embryo. C, an enlargement of the skull with the identity of some of the features. Scale in A = 100 microns; in B and C = 1 cm.

Fig. 2.31. An egg (possibly turtle) from the Upper Jurassic of Tendaguru, Tanzania, Africa. The specimen was uncovered by the British Museum (Natural History) expedition in the 1920s.

This dinosaur and fragments of others were collected from the Lower Cretaceous Falla Formation in the Transantarctic Mountains. When these dinosaurs lived, Antarctica had not yet developed its ice covering, so there were plenty of places for dinosaurs to nest. It will be just a matter of time until dinosaur eggs are found on this continent as well.

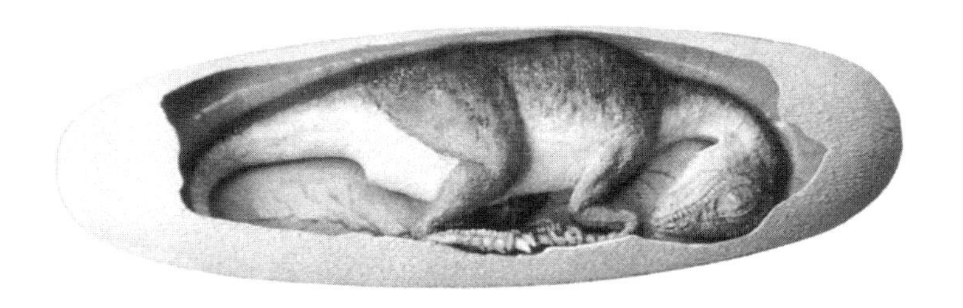

3 • Which Came First, the Lizard or the Egg?

What would Western civilization be like without coffee, bacon, and eggs while reading the newspaper on a lazy Sunday morning? Shelled eggs are Nature's way of packaging food for the embryo or carnivore in a neat, recyclable package. Several ideas have been proposed to explain this packaging, but unfortunately few tangible clues have been left in the fossil record. One problem with these ideas is that none of them is testable, meaning that supporting data cannot be found, either as fossils or experimentally. Nevertheless, they are thought provoking, which is not necessarily a bad thing (contrary to what Steve Gould, 1984, may think).

Origin Models

The Romer model of egg development is named after the late Alfred Romer, a vertebrate paleontologist and comparative anatomist at the Museum of Comparative Zoology at Harvard University. His specialty was early reptiles because, he felt, they were the key to understanding the great reptile diversification seen in the Late Paleozoic and Mesozoic. Romer's (1957) hypothesis was that some aquatic amphibians called anthracosaurs began to lay their eggs on land at about the time that they were evolving reptilian skeletal features (Fig. 3.1). Indeed, some of these early amphibians and earliest reptiles are so similar in their skeletons that the exact transition point from one to the other is still difficult to determine (Fig. 3.2). Eventually, though, the transition was made, but these early reptiles remained aquatic. The advantage for laying eggs on land was primarily to avoid the aquatic larval stage with its inherent dangers of predators and drying of ponds. However, the land has its own set of dangers, not least of which is the drying effect of the atmosphere. To cope with these problems a series of protective membranes developed around the egg, including a hard shell.

Fig. 3.1. The Romer model for the origin of the terrestrial egg: A female aquatic reptile Limnoscelis *guards her nest along a stream bank of western Texas, 280 m.y.a.* Limnoscelis *is now thought to be a fully terrestrial amphibian.*

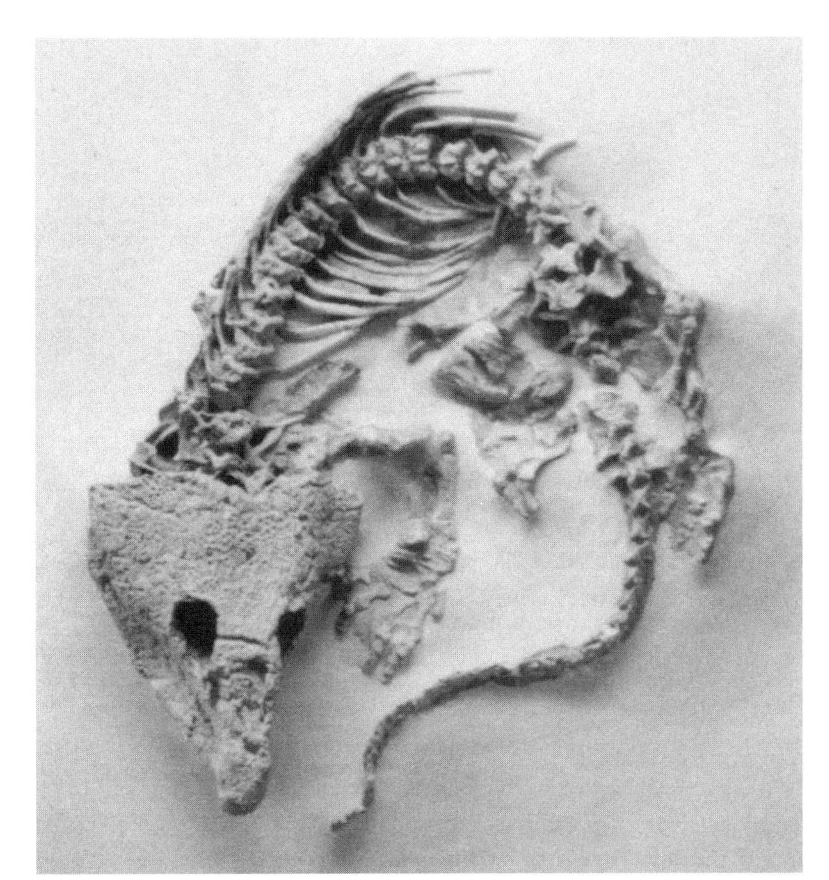

Fig. 3.2. Labidosaurus, *an early reptile from the Permian of Texas.*

Only later did the reptiles completely abandon an aquatic lifestyle. In the Romer model then, the egg came first, then the lizard.

R. Elinson (1989) proposed an unusual hypothesis for the origins of the terrestrial egg that is connected to the very early stages of embryo development (Fig. 3.3). Elinson proposed that the size of an egg is limited by the initial complete division of the germ cell (the cell attached to the yolk that grows into an embryo). This complete division, called holoblastic, is seen in amphibians, which have small eggs (less than 10 mm [3/8"] in diameter). In reptiles and birds, the initial division of the germ cell is incomplete and is called meroblastic. This type of cleavage, Elinson contends, allows the egg to become larger than 10 mm, mostly by increasing the amount of yolk for the embryo (why this should occur is not clear to me). Not surprisingly, the largest known eggs are those of birds and dinosaurs. Once the initial partial cleavage appeared during the evolutionary transition between amphibians and reptiles, the egg became larger. But the larger egg required a way of getting food and oxygen to the embryo, as well as dealing with waste. This was accomplished by development of several membranes, including a vast network of blood vessels. Development of these membranes eventually led to the production of the amniote egg (more about this type of egg below). Addition of a hard shell ensured that everything was held together on land in absence of water buoyancy. Elinson's model requires the lizard to come before the egg.

*Fig. 3.3. Elinson's model for the origin of the terrestrial egg: A, primitive anthracosaur amphibian (*Proterogyrinus*) and an amphibian germ cell showing holoblastic cleavage; B, primitive captorhinomorph reptile (*Hylonomus*) and a reptile germ cell showing meroblastic cleavage.*

Another hypothesis was proposed by German paleontologist Rolf Kohring (1995), whose specialty is fossil eggs. In Kohring's model, amphibians during the Mississippian (360–320 m.y.a.) spread into nutrient-poor or cooler water (Fig. 3.4). Because of the harsher conditions, eggs were produced with larger yolks, i.e., more nutrients for the embryo. With larger yolks, the eggs were bigger and fewer of them could be produced by the female. Rather than thousands of eggs, only hundreds were produced. To keep the larger egg intact, one or more membranes were developed, including one that surrounded and protected the egg. This outer membrane provided a place for the accumulation of calcium ions (Ca^{2+}), which are toxic, deposited from a modified gland in the reproductive tract. Organizing the calcium into a hard shell then made it possible for the egg to be laid on land (it was preadapted to be laid there). In this model, the egg came before the lizard.

Fig. 3.4. Kohring's model for the amniote egg: ~320 m.y.a., a small (20 cm long) primitive reptile lays shelled eggs in the murky shallow water among the horsetails of a small pond.

One other model we should consider is the anti-predator hypothesis proposed by Gary and Mary Packard (1980) to explain the evolution of the hard-shelled egg. Their model was not concerned with the origin of the amniote egg, but continues the story after the amniotic egg appeared. The Packards assume that the earliest reptiles laid leathery shelled eggs on very wet ground where they could absorb water during the embryos' growth. But life on the ground is not without hazards, based on studies of modern reptiles with leathery shelled eggs. Predatory insects and microbes can be a major cause of egg mortality (Fig. 3.5). To counter this loss of eggs, some of the early reptiles began secreting a thin calcareous layer. This hard layer gave the embryos a better chance of surviving until hatching. And these survivors in turn would probably leave more progeny once a few of them reached reproductive age. In time, a thicker and more resistant shell was selected for. However, a thicker eggshell meant that less water could be absorbed for the needs of the embryo. To compensate, larger eggs were produced, containing a great deal more albumen (i.e., egg white, a water storage gel). At this point, the rigid eggshell had reached the bird egg level of complexity.

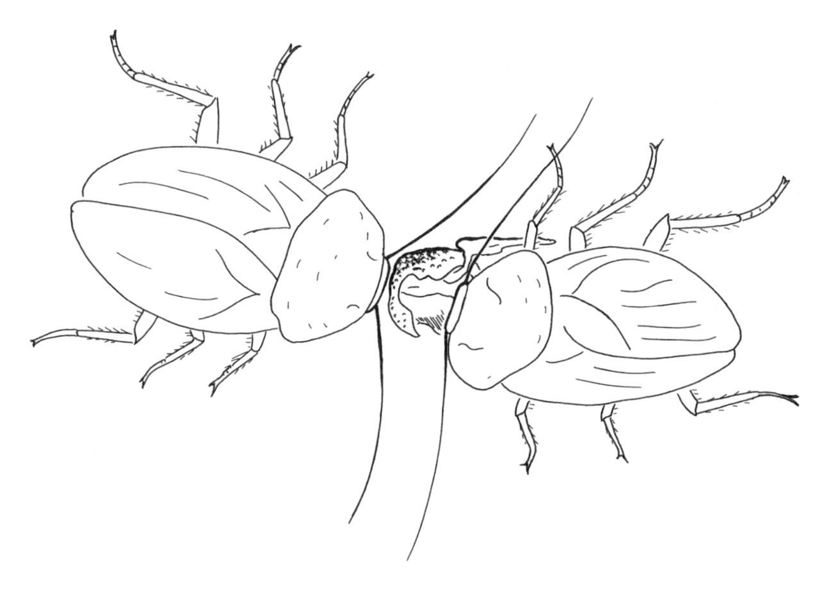

Fig. 3.5. The Packard and Packard model of the hazards of laying eggs on land: Giant cockroaches of the Pennsylvanian feed on a terrestrially laid amphibian egg.

Mary Packard presented yet another model with her colleague Roger Seymour (1997). They note that amphibian eggs can never get very large because the gelatin coat surrounding the developing larva is not very good at transmitting oxygen. Because of this restriction, we will never see frog eggs the size of a chicken's. For Packard and Seymour, the major evolutionary breakthrough was the elimination of the thick gelatin coat and replacing part of it with a fibrous membrane. This change allowed larger eggs to be developed, along with loss of the tadpole stage because the embryos continued development until they became miniature adults. The elimination of the tadpole stage coincides with the appearance of the meroblastic cleavage of the egg (see above). Eventually, the other membranes that made up the amniotic egg appeared, although their timing is unknown.

Critiques of the Models

The Romer model for the origins of the hard-shelled egg was developed in 1957 at a time when much less was known than today about ancient and modern amphibians and reptiles. Admittedly there are problems with this model because the supposed aquatic reptiles, such as *Limnoscelis,* are now considered to be terrestrial amphibians (Berman and Sumida, 1990). The Kohring model, on the other hand, does not explain how the embryo can get enough oxygen through the shell while underwater. Indeed, drowning is a major cause of mortality in crocodile and turtle eggs laid near water (Magnusson, 1982). Also, if the calcium shell was developed for getting rid of excess toxic calcium ions, only the female producing the eggs would benefit; the males would die. The Elinson model, on the other hand, requires the formation of new tissue in order for the egg to become large, but it doesn't explain why reptiles even need eggs larger than those found in amphibians of comparable size. The Packards' model for the origin of the rigid eggshell does resolve the vulnerability problem of leathery eggs. However, it does not explain why many reptiles still find it advantageous to retain a leathery eggshell, let alone consider that not all rigid shells are the same. Finally, the Packard and Seymour model doesn't explain why ancient amphibians needed larger eggs in the first place. After all, amphibians have been a successful group for hundreds of millions of years with small eggs like those of fish. With shortcomings in all the models, perhaps we should reinvestigate the problem.

First Eggs

About 510 million years ago, the first fishes appeared in the oceans of the Late Cambrian. These were not notable as organisms go, being small (probably a few centimeters long), heavily armored, and not very common. Unlike fishes today, which wear their skeletons on the inside, these early fishes wore their skeletons on the outside much like insects. But whereas the insect skeleton is chitin and must be shed as the insect grows, the outer skeleton of these early fishes was calcium phosphate, or bone, and was not shed.

We don't know much about the reproduction of these (Fig. 3.6) or other primitive fish, and fossil fish eggs in general are rare. We do know that some Paleozoic sharks laid eggs, although some may have given birth to live young. The shark eggs were housed in a tough casing much like their modern relatives today (Fig. 3.7). Still other fish eggs from the Pennsylvanian were apparently gelatinous spheres that were attached to the stems of plants (Fig. 3.8; Mamay, 1994).

A typical modern fish egg is a gelatinous sphere within which is an even smaller sphere of nutritious yolk for the developing embryo. Gas exchange between the embryo and water occurs through the gelatin, with oxygen diffusing in and carbon dioxide diffusing out (Fig. 3.9). Because they are so tiny and contain such a high concentration of nutrients, fish eggs are a major food source. They are tasty morsels best served on tiny slivers of French bread with a nice light wine—unless you happen to be one of the thousand small predators that habitually feed on fish eggs, in which case you'll gulp them down one by one. To counter this loss of eggs to predators, the female fish typically lays thousands of eggs at a time. The turbot (*Scophthalmus*), for example, lays about 1.2 million eggs for each kilogram of body weight (Kohring, 1995), so a 9 kg adult would lay an astounding 11 million eggs—that's a lot of caviar! (Fig. 3.10). Other fish such as the

Fig. 3.6. Some of the first vertebrate egg-layers included the small (20 cm long) Pteraspis. *Here, amid a cloud of mud, a wiggling male squirts milt containing sperm as the female squirts out her eggs.*

A B C D E

Fig. 3.7. Fossil (A-C) and modern (D-E) shark egg cases.

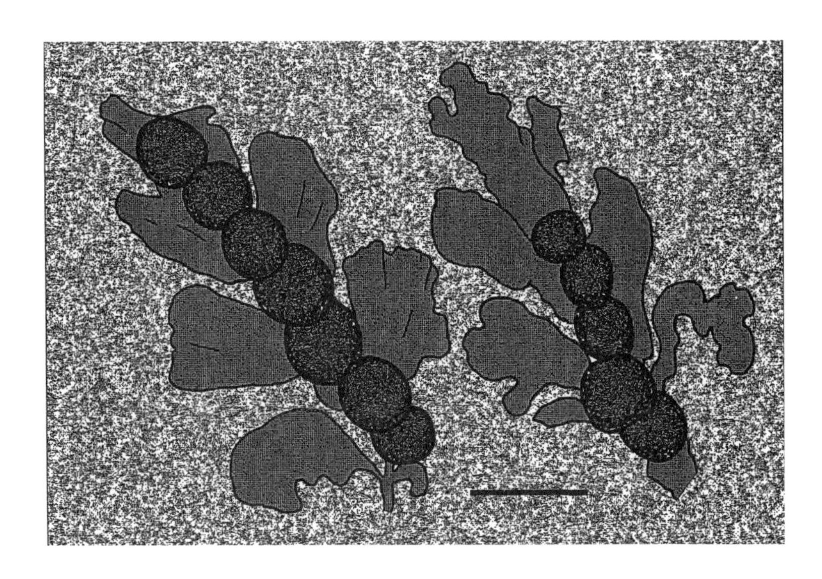

Fig. 3.8. Sketch of possible fossil Pennsylvanian fish eggs adhering to the bottom of fern fronds (adapted from Mamay, 1994).

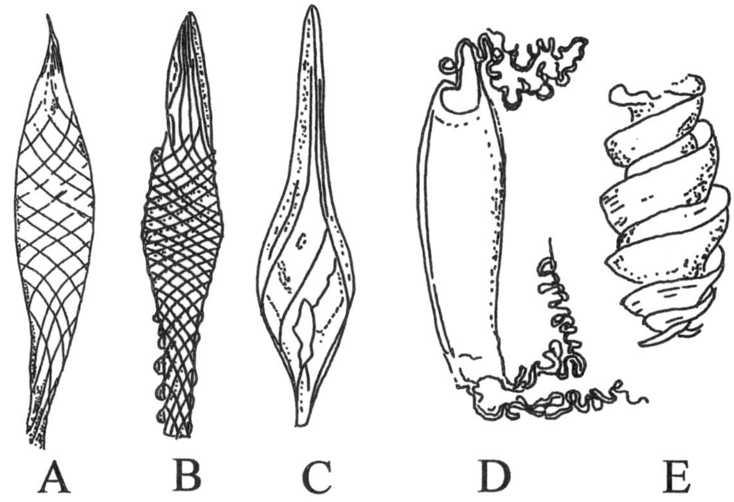

cichlid, however, lay a hundred or fewer eggs, but then vigorously protect them from predators—the mother cichlid carries the eggs in her mouth until the larvae hatch. And even then, the young may seek shelter in the parental mouth when danger approaches.

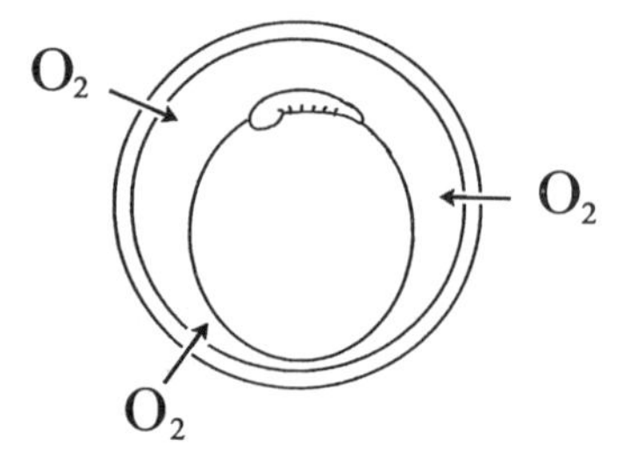

Fig. 3.9. Oxygen passes directly through the outer parts of a fish egg to the embryo.

Fig. 3.10. The champion egg-layer, the turbot Scophthalmus *and her 11 million eggs.*

Producing millions of eggs is what ecologists call an r-strategy of reproduction, whereas producing fewer eggs that are protected is called a K-strategy (there is more to r- and K-strategy, but this portion of the definition will suffice). From our point of view, K-strategy seems more economical for the female because she needs to devote less of her body's resources to producing eggs. However, the r-strategy female, once she has laid her eggs, does not need to expend energy guarding them; she is free to get on with her life. The sheer numbers of eggs laid by the r-strategy female increases the probability that some of them will survive to hatch, and that some of the larvae will mature to reproduce. Although r-strategy and K-strategy are described here as distinct, there is in fact a continuum of reproductive strategies between them.

Sometime around 367 million years ago, during the Late Devonian, the first amphibians appeared (Fig. 3.11A), most likely from a rhipidistian ancestor (Fig. 3.11B). More fishlike than froglike, these early amphibians apparently retained the gelatinous egg of their fish ancestor because that is the type of egg laid by most amphibians today. Quite probably these amphibians employed an r-strategy much like the bullfrog today, which lays 10,000 or more eggs at a time (Kohring, 1995). Modern amphibians generally mate in water and lay eggs in long chains that float freely, or in globular masses attached to vegetation or floating on the surface, and any of these may have been true for some ancient amphibians as well (Fig. 3.12, 3.13). Regardless, the eggs quickly become fish food. Actually, the statement that amphibians need water in which to mate and lay their eggs is overly simplistic. In reality there are amphibians today, frogs mostly but also some salamanders, that mate and lay eggs on land. For example, the male dusky salamander, *Desmognathus fuscus,* deposits a gelatinous packet of sperm on the ground. He then leads the female forward and the packet is taken up into her cloaca (an all-purpose chamber into which the genitals and intestinal and urinary canals open, Fig. 3.14). The female later lays her eggs in a small bowl-shaped depression that she has dug near a stream bank, inside a decaying log, or under stones. Many species of tree frogs mate on the ground or in vegetation, and lay their eggs in masses of foam that dry with a hard crust to prevent desiccation. Still other frogs and toads lay their eggs in moist places such as burrows. The African bush-frog, *Leptopelis karissimbensis,* lays its eggs in a moist, vegetation-lined nest (Porter, 1972).

These diverse examples of reproduction in modern amphibians provide us with clues to the possible means by which some of the ancient

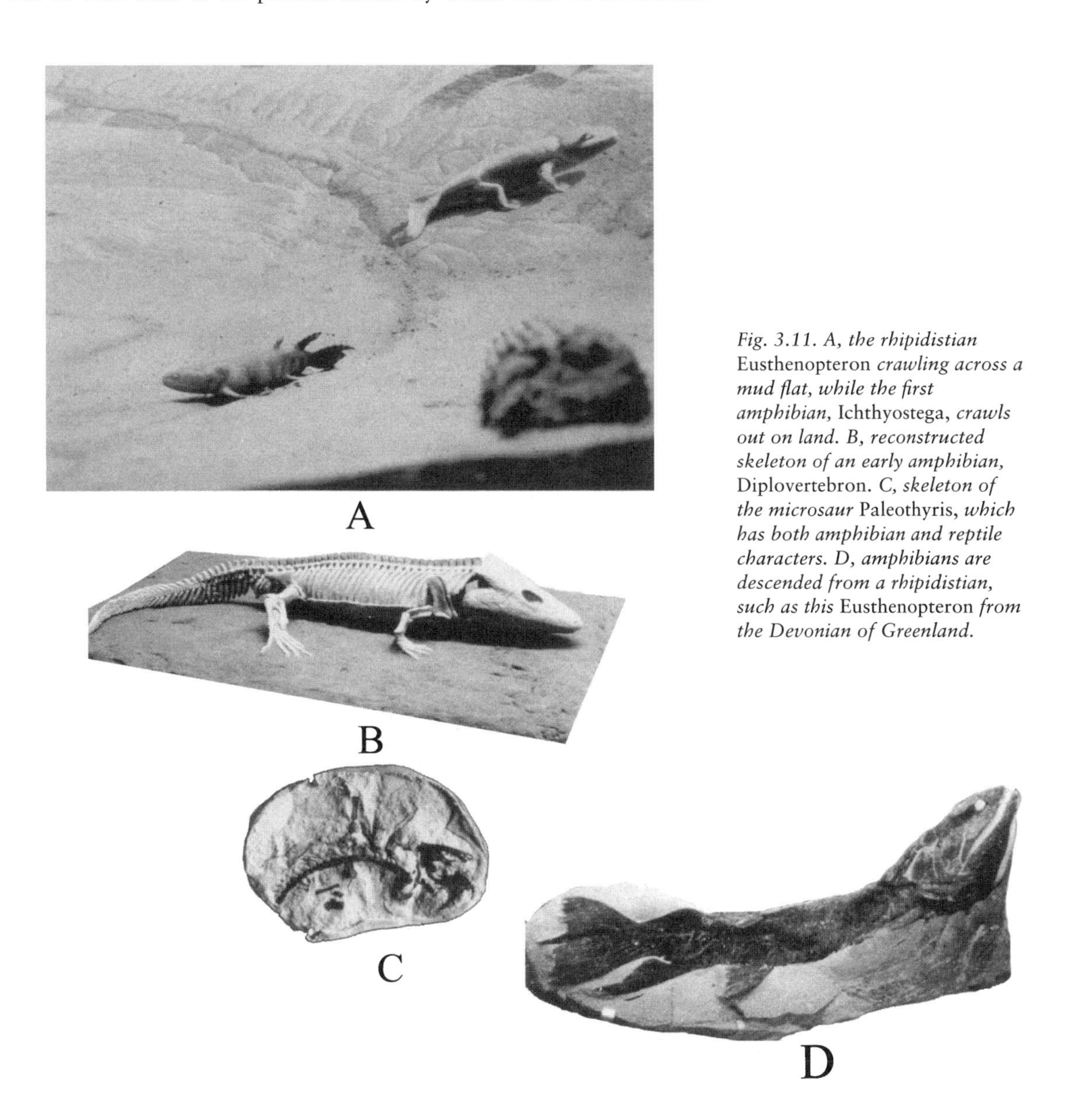

Fig. 3.11. A, the rhipidistian Eusthenopteron *crawling across a mud flat, while the first amphibian,* Ichthyostega, *crawls out on land. B, reconstructed skeleton of an early amphibian,* Diplovertebron. *C, skeleton of the microsaur* Paleothyris, *which has both amphibian and reptile characters. D, amphibians are descended from a rhipidistian, such as this* Eusthenopteron *from the Devonian of Greenland.*

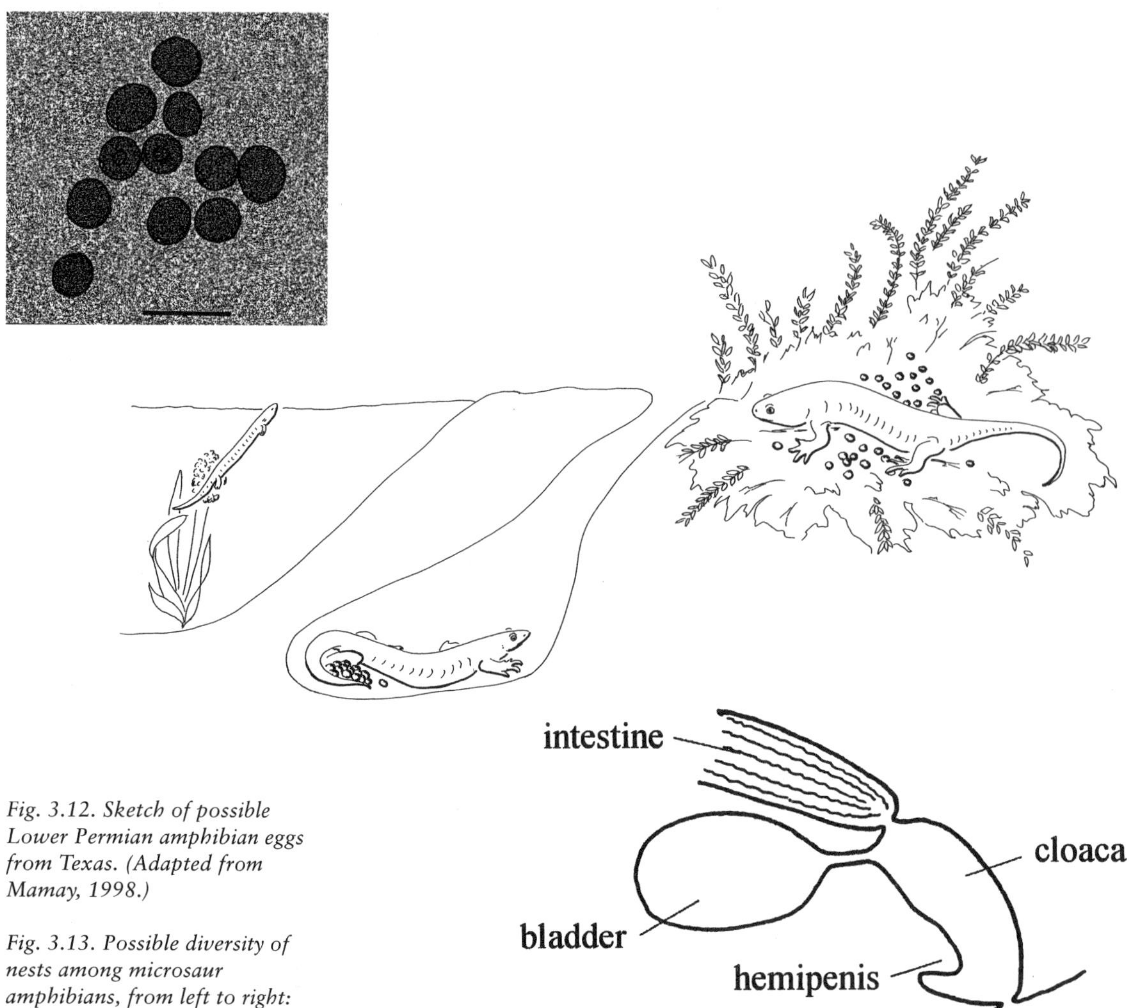

Fig. 3.12. Sketch of possible Lower Permian amphibian eggs from Texas. (Adapted from Mamay, 1998.)

Fig. 3.13. Possible diversity of nests among microsaur amphibians, from left to right: eggs adhering to aquatic plants by Odonterpeton, *eggs laid in a burrow by* Pelodosotis, *and eggs laid in a vegetation nest by* Asphestera. *These images are purely hypothetical, but reasonable.*

Fig. 3.14. Generalized cloaca into which the digestive tract and urinary bladder open. In addition, the cloaca houses the hemipenis in males and the opening to the uterus in females.

amphibians may have broken free of their need for water to reproduce. Some of these amphibians may be "reptilomorph" amphibians, i.e., amphibians with reptilian characters. One example is the microsaurs (Fig. 3.11C), which were small, salamander-like amphibians common during the late Pennsylvanian and early Permian (298–268 m.y.a.). Some were fully terrestrial, while others apparently burrowed in the ground. It seems reasonable then that some microsaurs may have laid their eggs in foam masses, others in piles of moist, decaying vegetation, and still others in damp burrows (Fig. 3.13). Another group, the diadectimorphs, included cow-sized herbivores. The diadectimorphs are problematical because there is still much debate as to whether to classify them as amphibians or as reptiles. Perhaps they had truly crossed that line of reptilian reproduction and laid shelled eggs. If they did, they probably used internal fertilization so that the eggs were fertilized before the shell was added (Fig. 3.15); otherwise the shell of the egg would act as a barrier preventing fertilization. Internal fertilization occurs today in a few amphibians, mostly terrestrial species such as the dusky salamander mentioned above, and also in reptiles, birds, and of course mammals. To help ensure fertilization, a penis or penis-like organ has developed to deliver the sperm. Many male reptiles have a two-headed penislike organ called a hemipenis (double your pleasure, double your fun?). Reptile reproduction is discussed more in chapter 5.

Fig. 3.15. The enigmatic amphibians or reptiles Diadectes *mating. As with many lower vertebrates, the male is smaller than the female.*

What are the advantages for an amphibian in laying eggs on land today? There probably are some because several different groups evolved to do so independently of others. For example, tree frogs are very distantly related to the dusky salamander (their common ancestor probably lived about 200 million years ago), yet both have adapted to mating and laying eggs on land. Their styles differ considerably, indicating that one did not acquire the technique from the other. Determining why some lineages of amphibians found it advantageous to abandon water to reproduce may hold the key to understanding why reptiles evolved in the first place. The most obvious answer is that it is safer for the eggs to be laid in some secretive cavity or burrow than to be left floating in the water where fish and other amphibians might eat them. But this may not be the answer in its entirety because land has its own set of hazards, including predators. Today, these predators include large insects, small mammals (especially rodents), and even some birds. During the mid-Paleozoic, the most common land predators were large carnivorous amphibians and very large arthropods. Another problem faced by creatures laying eggs on land is, believe it or not, lethal fungal infections (Porter, 1972). To counter this problem, the female dusky salamander must periodically roll the eggs in her nest because a "rolling egg gathers no fungus." Still, experimental work with modern amphibians has shown that a greater proportion of eggs survive on land than do in water because of all the secretive places the eggs may be laid. This means that the female need not lay as many as her aquatic counterparts nor spend as much energy in egg production.

Another advantage of laying eggs on land is that the embryo may skip all or part of the larval or tadpole stage. Typically, the tadpole looks more fishlike than amphibian, with gills, no legs, and a tail used for swimming (Fig. 3.16). But as the larva grows, metamorphosis replaces the gills with lungs, legs appear, and, in frogs, the tail is absorbed. In some eggs laid on land, the hatched young can be miniature versions of the adults complete with legs and lungs. However, reaching this level of development requires that the embryo have adequate food supply as it is growing. Not surprisingly, then, the yolk in these eggs is proportionally much larger than in eggs laid in water.

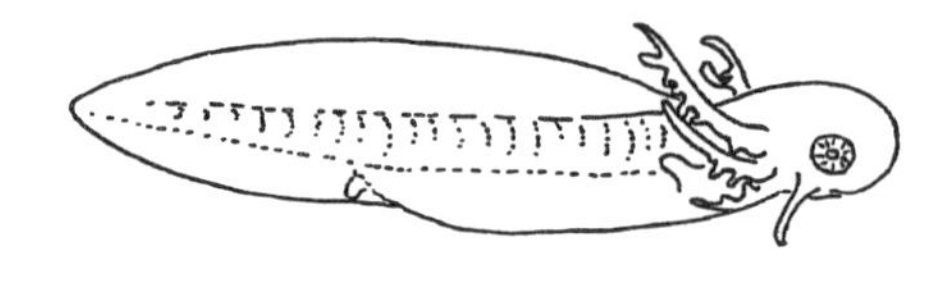

Fig. 3.16. Salamander "tadpole" or larva.

Evolution of the Reptile Egg

Around 315 million years ago, or about 52 million years after amphibians first crawled out on land, the reptiles appeared, signaling a major change in how terrestrial vertebrates reproduced. Most likely reptiles evolved from some amphibian that had already made the breakthrough to life fully on land. Mating occurred on land (Fig. 3.17), but more important was the appearance of new membranes surrounding the egg (Fig. 3.18), the amnion, chorion, and the allantois. These membranes characterize not only reptile eggs but also those of birds and mammals, hence these three groups of vertebrates are also referred to as amniotes. The amnion membrane is present when the egg is formed. It envelops the embryo and supports it in a bag of water (this is the water bag of a woman that breaks during birth). The allantois, on the other hand, is a tiny bag of tissue that grows only as the embryo develops. Initially, it is a small sack, but it soon grows to line the inside of the shell. Its function is to hold waste products produced by the embryo and to exchange oxygen for carbon dioxide.

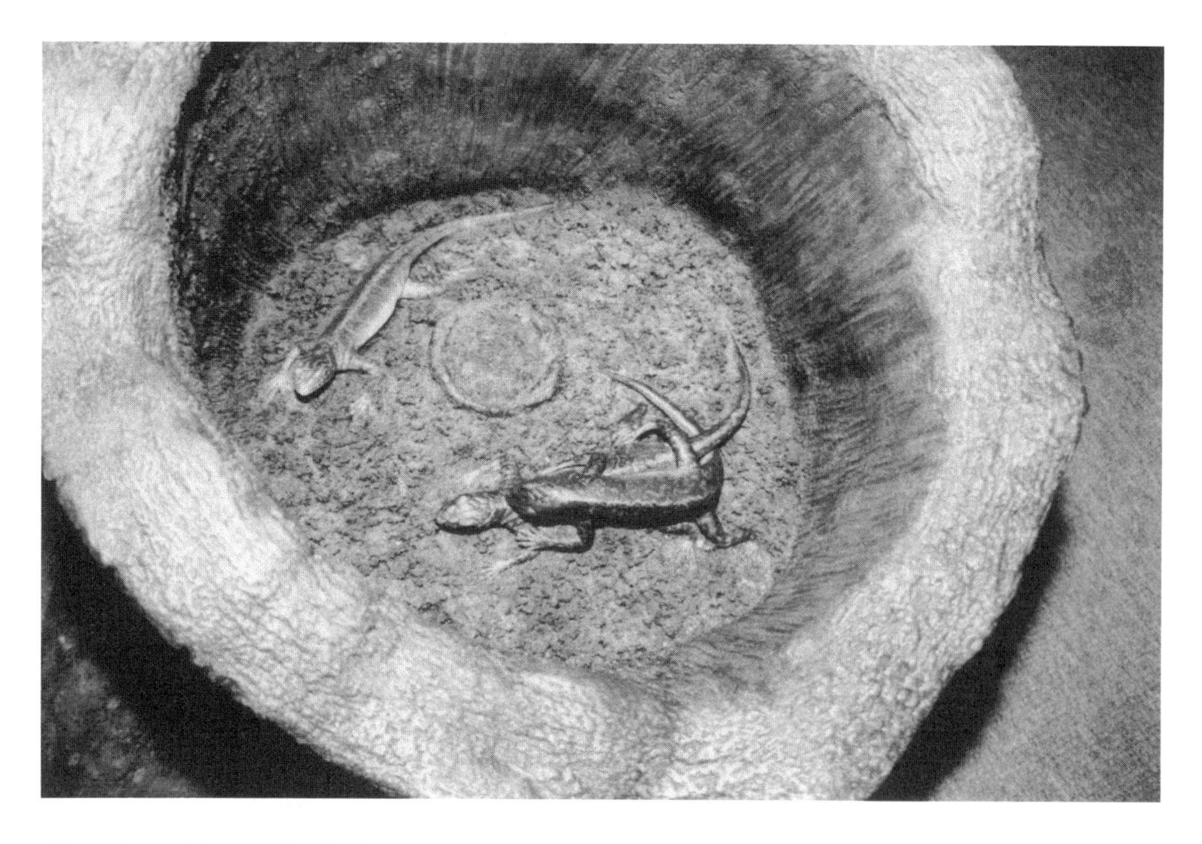

Fig. 3.17. The primitive, lizardlike Hylonomus *mating in a tree stump. Some of the oldest reptiles have been found in what are believed to be fossilized tree stumps at Joggins, Nova Scotia. Tree stumps would afford refuge for these small, secretive reptiles.*

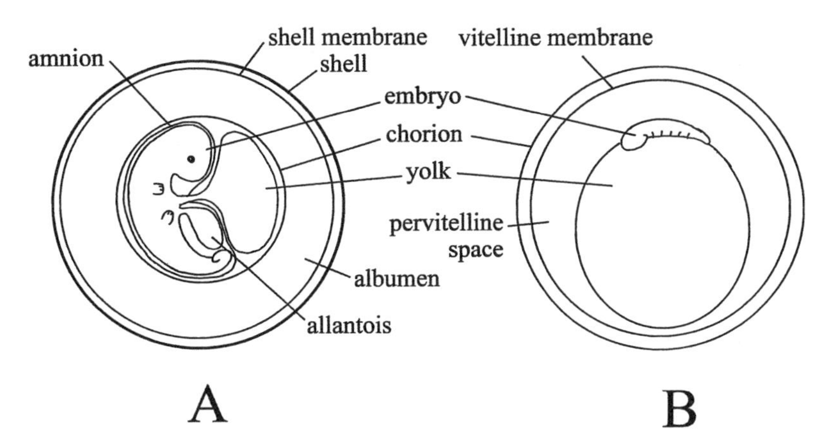

Fig. 3.18. A generalized amniote egg (A) compared with a generalized non-amniote egg (B). The amniote egg gets its name from the amnion sack that envelops the embryo. The appearance of the amnion, chorion, and allantois was probably a gradual process.

The covering of the earliest reptile shell was most likely a tough, leathery membrane (Fig. 3.20A). Such a "shell" is seen around the eggs of many turtles, the most primitive of living reptiles. Unfortunately for us, such eggs have a very low potential for fossilization because the leathery membrane quickly decays. Nevertheless, a possible soft-shelled egg has been described from the Lower Permian. The object, 59 mm (2⅓") long and 37.9 mm (1½") wide, was originally described by Alfred Romer and Llewellyn Price (1939) as the oldest hard-shelled egg. A study by Karl Hirsch (1979), however, failed to demonstrate the presence of a calcite shell. But rather than completely dismiss the object as an egg, Karl suggests that the high phosphorus content of the outer layer might indicate that it is the remains of a soft-shelled egg. If this suggestion is correct, it is indeed the oldest egg known.

Shortly after the earliest or basal reptile group (the captorhinomorphs) appeared, several different lineages evolved (Fig. 3.19). The evolution of the eggshell in these subsequent groups is not known for the most part because most, such as the aquatic plesiosaurs, have no modern descendants. Nevertheless, for a few extinct forms, some spectacular fossils show that the porpoiselike ichthyosaurs gave live birth, much like some lizards and snakes today (more below). Probably this means that they retained the primitive leathery eggshell of the captorhinomorphs. Most of what we know about the evolution of the amniote egg comes from the study of eggs of modern turtles, lizards, snakes, crocodiles (including alligators), the lizardlike tuatara, and birds. From these studies we suspect that at some point during the split of the various reptile groups from the captorhinomorphs, glands in the uterus began secreting a calcium carbonate "sweat" that crystallized as calcite on the leathery membrane (Fig. 3.20). Such a sporadic coating is seen today on the eggshells of some snakes and lizards. The process of adding calcite to the surface of the leathery shell is not as straightforward as dipping a strawberry in chocolate fondue. Instead, as we will see in more detail in chapter 6, the shell usually grows from seed points called nucleation centers on the outer, leathery membrane of the egg. The calcite grows not as a shapeless coating but as well-defined crystals that meet but do not fuse together. Consequently, the eggshell remains somewhat pliable, or semi-rigid (Fig. 3.21C). Such shell is found on the eggs of the snapping turtle, a very advanced turtle. Having a semi-rigid shell allows the developing embryo to still obtain water from the soil the eggs are laid in. In fact, both the leathery and semi-rigid shelled eggs almost double their volume with water absorbed from the ground. Without this water, the embryo would die, so nests are carefully chosen to be in damp ground. Usually this means that the eggs are laid during the rainy season when the ground is wettest. (Standing water, however, drowns the eggs.)

Fig. 3.19. Skeleton of Captorhinus, *a basal reptile that forms part of the group that gave rise to all later reptiles. Skeleton is about 20 cm long.*

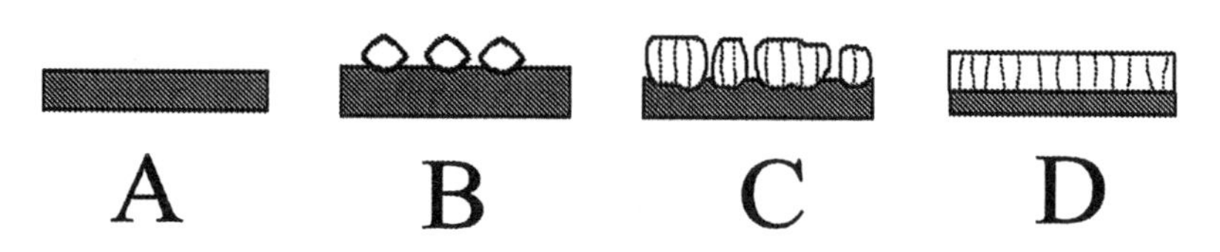

Fig. 3.20. Schematic representation of various types of amniote eggshells: A, leathery, formed by the fibrous shell membrane (seen in some turtles); B, leathery, with a calcite crust (seen in some lizards); C, semi-rigid, in which not all of the calcite crystals interlock, allowing some flexibility of the shell (seen in some turtles); D, rigid, in which the calcite crystals interlock (seen in crocodiles, birds, and dinosaurs).

It was probably inevitable once eggs began to develop a calcite covering that the shell would eventually become rigid (Fig. 3.20D) with the calcite coating the egg entirely. It is not known when this happened, but it most probably occurred well before the end of the Triassic. The oldest rigid eggshells we know of are crocodilian eggshells from the Upper Triassic or Lowermost Jurassic of South Africa. Most likely, however, the rigid eggshell is much older—the rigid shells of various modern reptile groups differ in their structure, suggesting a long separate history of eggshell evolution. In fact, the rigid eggshell of turtles differs enough from the rigid eggshell of crocodiles to imply that the rigid eggshell may have evolved separately in the two groups after the split from the captorhinomorphs (Fig. 3.21, 3.22).

With turtles having a full range of eggshell types, from leathery to rigid, it seems probable that the oldest turtles, *Proterochersis* and *Proganochelys* from the Upper Triassic of Southeast Asia and Europe, retained the leathery shelled eggs of their captorhinomorph ancestors. Otherwise, the presence of semi-rigid eggs in the modern snapping turtle and leathery eggshells in modern sea turtles would indicate a loss in the capacity to produce a rigid eggshell. In other words, the eggshells of turtles would have evolved from leathery to semi-rigid to rigid, then at a latter date some of the turtles lost the capacity of laying rigid shelled eggs; such a change is called a secondary loss. Although possible, it seems more likely that the rigid shell is a more recent acquisition, and that some of the more primitive turtles living today retain the leathery or semi-rigid shells. Support for this hypothesis is seen in the fossil record of turtle eggshells. The oldest semi-rigid

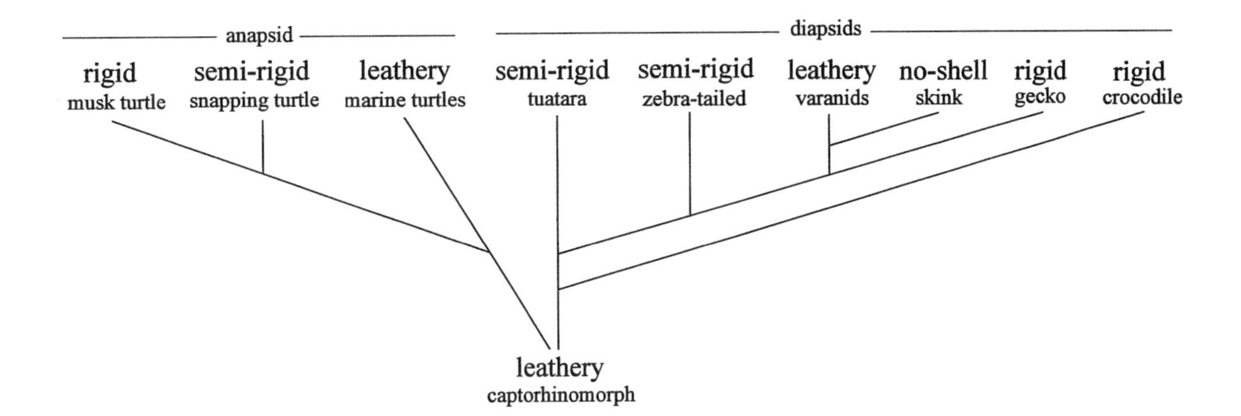

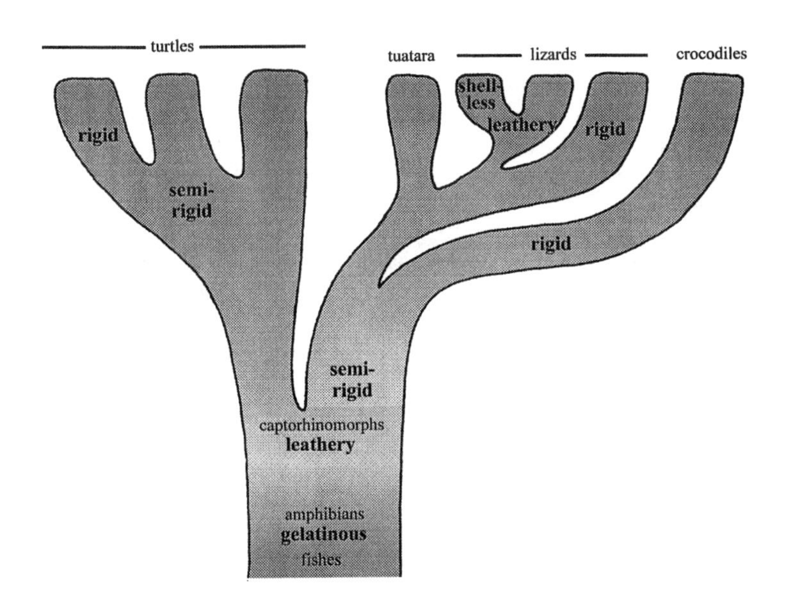

Fig. 3.21. A cladistic relationship (an analysis that uses advanced characters to show relationship) of eggshell type among various modern reptiles. The leathery eggshell is believed to be the most primitive shell type because it is structurally so simple. For this reason, it is thought to have been present in the basal reptiles, the captorhinomorphs. Among turtles (evolutionarily more advanced than captorhinomorphs), the leathery eggshell is primitive for the group, with semi-rigid and rigid shells being progressively more advanced. Among diapsids, the semi-rigid eggshell is the primitive type, based on its presence in the tuatara. It remains the primitive type of eggshell among lizards, with some reverting back to the leathery type (secondary loss) and some becoming rigid (more advanced). Among those reverting to the leathery shell, a few have eliminated the shell altogether (more advanced). Among crocodiles, the rigid eggshell probably appeared among the pre-crocodiles. See Figure 3.22 for another presentation of this information.

Fig. 3.22. A more traditional evolutionary tree showing the evolution of shell types.

turtle eggshells have been found in the Middle Jurassic of England (the egg named *Oolithes bathonicae,* renamed *Testudoflexoolithus bathonicae* by Hirsch, 1996) and in the Upper Jurassic Morrison Formation of North America (Hirsch, personal communication). Hard-shelled turtle eggs are known from the Late Jurassic of Portugal (Kohring, 1990b). Admittedly, the nonexistence of older examples of these eggs may simply be due to no one having yet discovered these fossils, but the absence of older eggshell is tantalizing.

Finally, one other point supporting the idea that turtles acquired their rigid eggshells independently from other reptiles is the composition and structure of the shell. The shell is composed of a form of calcium carbonate called aragonite arranged in radiating needlelike crystals into groups called shell units (more about these in chapter 6). All other reptiles (and birds) with rigid eggshells use a different form of calcium carbonate, calcite. None of these crystals are arranged in the same radiating pattern seen in turtles.

Thus, making the shell from a different form of calcium carbonate arranged in radiating needles seems peculiar and suggests that turtles evolved the rigid shell separately.

Among other modern reptiles is the lizardlike tuatara, which lives today on New Zealand; its eggs have a semi-rigid shell. The tuatara represents an old reptile group, the sphenodontids, that first appeared in the Late Triassic. The ancestor of the sphenodontids was one of several closely related reptiles known collectively as diapsids because of their unique skull features. At some point before the Late Triassic, the early diapsids probably evolved the semi-rigid eggshell seen in the tuatara. Then, during the Jurassic, sphenodontids gave rise to the lizards, which in turn gave rise to snakes by the Early Cretaceous. Most likely the first lizards retained the semi-rigid egg of their sphenodontid ancestor. If true, this would mean those lizards today that lay leathery shelled eggs lost the capacity of depositing calcite on the shell. Why would this have occurred? Many lizards (and snakes) with leathery shelled eggs are ovoviviparous, meaning that the eggs are retained in the mother's body and the young are born live. Actually the process is not like that of mammals because the embryo uses the egg yolk as a food source and does not get nourishment from the mother. Eventually, the lizards hatch inside the female and the young are expelled live. Still other lizards (such as some skinks) have eliminated the shelled egg altogether and are viviparous like mammals. The yolk sac is virtually nonexistent and the developing embryos get their nourishment from the mother's bloodstream through a placenta-like structure (Porter, 1972); the young are born live. Quite possibly, ichthyosaurs—which are not related to any living reptile group—had independently eliminated the shelled egg as well. We'll discuss viviparity more below.

While some lizards have eliminated the semi-rigid eggshell for a leathery shell or have no shell at all, some geckos developed a hard eggshell. The shell differs from that in turtles and crocodiles in that the individual calcium carbonate crystals are not well defined. Furthermore, the shell has a great deal more organic material within and on its surface. These differences suggest that the gecko hard shell was acquired independently of that in turtles or crocodiles.

The oldest crocodile we know of is *Gracilisuchus* from the Middle Triassic of Argentina. We don't know what kind of shell its eggs had, but suspect that it was rigid. The reason is that the oldest crocodile eggshells are from the Lower Cretaceous of Galve, Spain. The microstructure of these shells is identical to that of modern crocodiles, implying that the crocodile egg was quickly established and that no major changes occurred in the eggshell once the hard shell appeared (Fig. 3.23). The structure of crocodile eggshell is primitive in that it is composed of only one form of calcite crystals (more in chapter 6), but these are arranged in tight columns in contrast with those of gecko eggs. It seems possible that the rigid eggshell appeared among some of the early archosaurs, but as yet no proof has been found.

Dinosaurs appeared by the Late Triassic (about 230 m.y.a.). They quickly evolved and eventually became the dominant large-bodied animals on land. We know a lot about their eggs, as we shall see in subsequent chapters, but not those of the earliest forms—yet. The hard shell of dinosaur eggs is remarkably durable, which is why these are the most common of fossil eggs (more in chapter 7). Important information about dinosaur eggs has come from the study of modern bird eggs. These studies reveal an eggshell microscopic structure that is nearly identical to those of some carnivorous dinosaurs, or theropods (more in chapter 8). In fact, if it

Fig. 3.23. A box of rigid crocodile eggs. (Circles are holes in the eggshells.)

weren't for their large size and age, some of these eggs might easily be identified as belonging to birds. And indeed small eggs from Mongolia called *Laevoolithus* could just as easily belong to a bird as to a small theropod. As yet, no embryonic remains have been found to resolve the matter one way or the other. What we do know is that birds retain a complex shell structure that is more similar to that of theropod dinosaurs than to the shell of a crocodile egg. Such similarities, as we shall see, add support to the hypothesis that birds evolved from dinosaurs.

Why Hard-Shelled Eggs?

We said that dinosaurs had hard-shelled eggs, but why did the calcareous or rigid eggshell develop in the first place? Why did evolution not stop at the leathery egg? Perhaps we can better understand these questions by first looking at what the eggshell does for a bird. An egg contains all the genetic information for producing another animal, plus all the nutrients for the developing embryo, all in a hard protective shelter. Thus the egg may be thought of as a house within which the embryo develops (Fig. 3.24). Among all birds and some reptiles, the rigid walls and roof of the house are formed by calcite or its variant, aragonite. Although an egg seems to be self-contained in that the embryo has a yolk to feed upon, in fact, it requires some input from the outside environment. Most crucial are water, oxygen, and carbon dioxide, all which pass through pores in the shell (Fig. 3.25). During incubation, water loss in bird eggs through the pores is about 10 percent to 20 percent of the egg mass. On the other hand, leathery and semi-rigid reptilian eggs absorb water, and the eggs may double their mass (more on water loss and absorption below). Oxygen in all eggs is pulled in for the embryo to use ("breathed in"), and carbon dioxide is expelled ("breathed out"). Rigid shelled eggs do this "breathing" through microscopic pores, whereas leathery and semi-rigid eggs exchange the gases through the leathery membrane or through the many cracks and fissures in the calcite crust.

The rigid shell is also impenetrable to bacteria, fungi, molds, and insects that might eat the egg. Birds supplement this defense with a thin

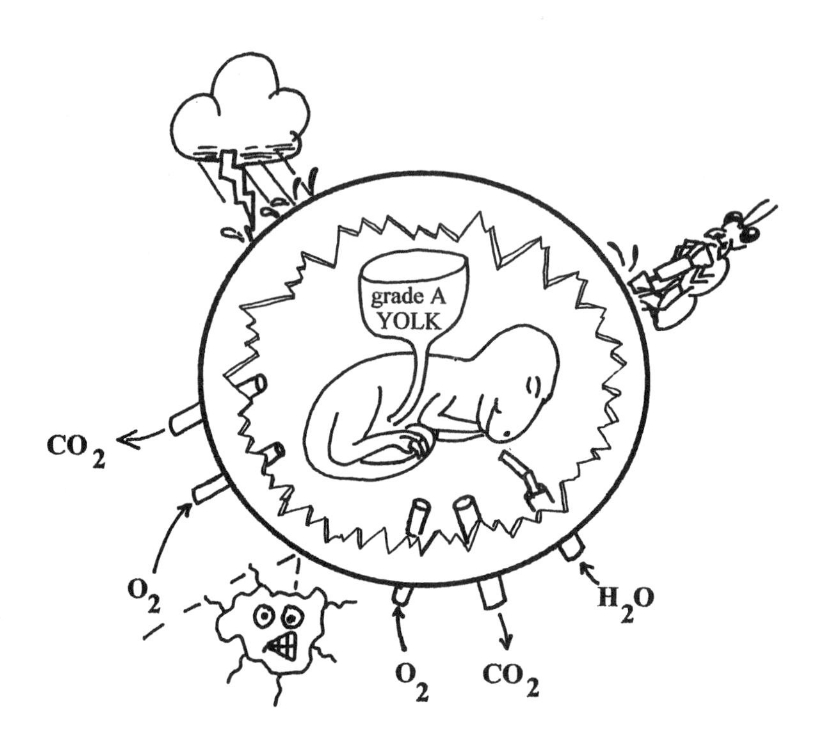

Fig. 3.24. The egg as a "house." The role of the hard-shelled egg is to protect the developing embryo from extremes of the weather, microbes, and insects, while also allowing gas exchange and some water to the embryo.

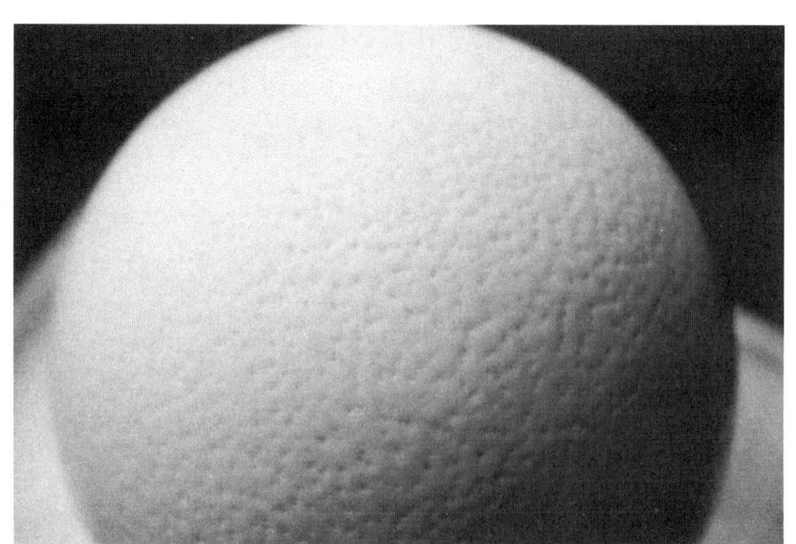

Fig. 3.25. Hard-shelled eggs "breathe" through pores, the pits that cover this ostrich egg.

organic cuticle covering the surface of the shell. Those microbes that do find the pores may be blocked from entry by small plugs of organic material or calcite, organic fibers within the pore canals, or by the shell membrane. In leathery or semi-rigid eggs, only the feltlike fibers of the shell membrane protect the insides of the egg. For ground-nesting birds (and dinosaurs), the rigid shell may have a bumpy or rough surface to keep the dirt or nesting material from plugging the pores, because it just wouldn't do to expend all that energy in laying eggs only to smother the embryo! For most birds, the eggshell must be strong enough to allow the parent to sit on the eggs to brood them without smashing them, as would happen if the eggshell were leathery or semi-rigid. The shell of rigid shelled eggs is also the major source of the calcium the embryo needs for bone growth. In leathery or semi-rigid shells, the calcium is stored in the yolk. Clearly then, the calcareous eggshell serves a lot of functions for the embryo and it is easy to see why it may have evolved. The transition from leathery or semi-rigid to rigid eggshell is not that difficult.

Before closing this section, we should briefly examine the flip side of the rigid shell question: Why did some reptiles retain or reacquire the leathery shell? As mentioned above, the leathery shell occurs today in some turtles, lizards, and snakes (Fig. 3.26). In a few of the more primitive turtles (such as the side-neck turtles of South America), the leathery shell may be a retention of a primitive character that works well enough (sort of a biological "it ain't broke, so why fix it?"). On the other hand, the leathery eggshell of the sea turtles may go hand in hand with the large numbers (upwards of two hundred) of eggs laid in each clutch. Leathery shelled eggs are much smaller when laid than rigid shelled eggs because they will eventually absorb water from the ground. In addition, leathery shells are sometimes dimpled and creased when first laid because large numbers of eggs are tightly packed in the cloaca prior to laying. By laying many eggs, sea turtles make sure that some of their eggs will hatch and that some of the hatchlings will reach reproductive maturity. With oviparous snakes, the leathery shelled eggs are long and slender (Fig. 3.26), and are less hindering to locomotion than rigid eggs would be as the female slithers from side to side. It is difficult to explain why oviparous lizards have a leathery eggshell because some of them, such as the anoles, may lay only a single egg. In these instances, being able to lay large numbers of eggs does not seem to be the reason for the leathery shell. Porter (1972) notes that there is a very strong correlation between the number of eggs laid in these lizards and the size of the female's body. This correlation, however, does not explain why small eggs are leathery shelled when the small gecko lays one or at most two rigid shelled eggs. Somewhere there is an answer.

Why Not Live Birth?

We should discuss the flip side of the rigid eggshell issue: Why didn't viviparity (live birth) evolve in turtles, crocodiles, birds, and dinosaurs? After all, as noted above, viviparity evolved independently several times among lizards and snakes, so it must not be too difficult to achieve. The simplest answer about the lack of viviparity in these animals is that we really don't know. As Richard Shine (1991) has commented, a lot of ideas have been suggested, but very little experimental work has been conducted. Let's look at some of the suggestions.

Viviparous snakes and lizards tend to live at high altitude or in high latitudes where temperatures tend to stay cool or cold most of the year.

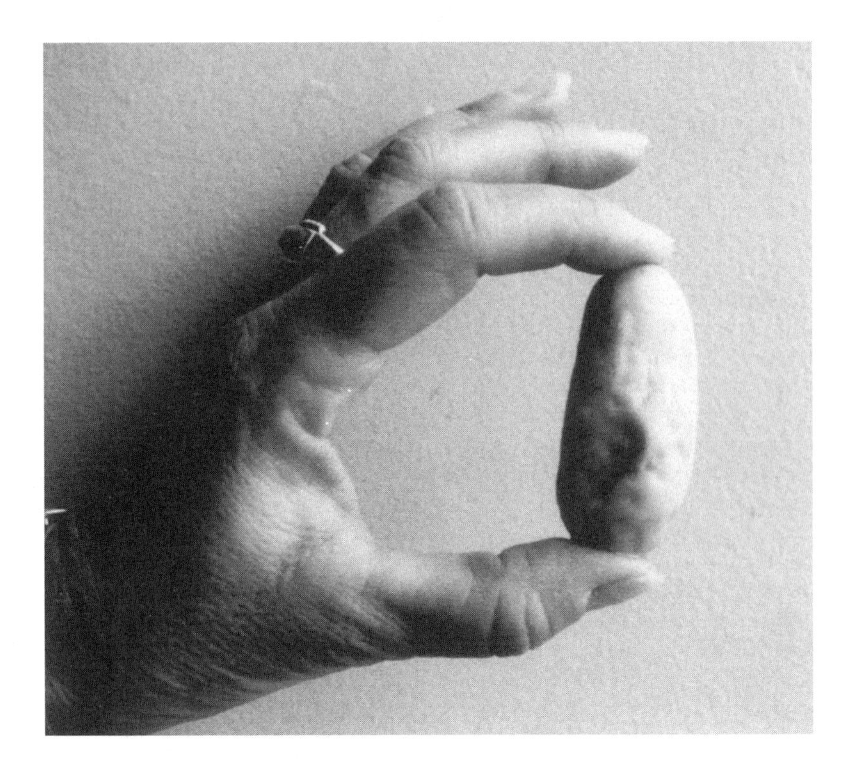

Fig. 3.26. A soft-shelled snake egg.

Retaining the eggs in the body apparently buffers the developing embryo from temperature extremes of night and day. In addition, the embryo is protected from predators. But viviparity is not without its hazards. As the young grow, occupying a large portion of the female abdomen, she may not be able to eat. With turtles—encased in a rigid body shell—the abdomen would not be able to expand to accommodate the growing embryo. But more crucial to all pregnant females is the restriction of movement and speed. If the female cannot outrun a predator, two generations are lost in one bite. As a result, many females become inactive just prior to the birth of the young so as not to draw attention of the predator. It's no wonder, then, that there is a correlation among viviparity, body shape, and speed. Viviparous lizards tend to be slow moving or secretive, whereas oviparous lizards are speedy and agile. Ichthyosaurs, being built like porpoises and probably almost as fast, are the exception among viviparous reptiles; they did bear live young.

One problem yet to be resolved is how the embryo can develop in a moving female. In oviparous reptiles, the embryo must attach itself to the membranes of the shell in order to develop properly. How viviparous lizards manage to get around this requirement is one of those areas Shine (1991) noted as needing research. At any rate, an embryo that needs to attach itself to the membrane can only do so if it is not jostled around, which would most certainly happen inside the moving female. This might not be too much of a problem for a bird embryo; bird eggs are periodically rotated by the parent, so embryos can cope with jostling motions. This periodic rotation is to keep the embryo from attaching itself to the egg membranes lining the shell, in contrast to lizards which apparently do need this attachment.

As for birds, there are at least two reasons why viviparity did not evolve. First, there is the matter of how birds produce eggs. In reptiles, a mass of eggs is produced and laid (or carried) at the same time. In birds, the eggs are produced and laid one at a time until the nest is filled. For a viviparous bird, it would mean that the female could only produce one fledgling at a time, i.e., only one young per breeding season. Second, there is the problem of added weight as the embryo develops and matures within the mother. This added weight might not be crucial for nonflying birds, but it would be the death of fliers.

There may also be physical demands of the embryo that prevent viviparity. For example, the environment of the uterus is different from that of the nest, and the embryo may not develop properly, if at all. For crocodiles and some turtles, which have temperature-sensitive embryos, having a temperature that is too constant might mean that all of the young would be of one sex (more on this in chapter 11).

Finally, the danger of viviparity is that all of the eggs are put in one basket, so to speak. If a pregnant female gets eaten, then two generations are lost. If the female lays eggs, the eggs may get eaten causing a loss of one generation, but the female will survive to lay more eggs. On the other hand, if the female lays her eggs, then gets eaten, the hatchlings will still continue the next generation if they are precocial and don't depend on the mother.

The problem with all of these hypotheses about the lack of viviparity in crocodiles, turtles, and birds is that there is no way to test them to determine their likelihood or probability of occurring. The bug in the ointment is, of course, the viviparous lizards. How do they manage to be the exception to most of the hypotheses? How do they keep from having single-sex young? As for dinosaurs, all the fossil evidence to date seems to indicate that they were all oviparous (we'll discuss the suggestion that some sauropod dinosaurs had live birth in chapter 10).

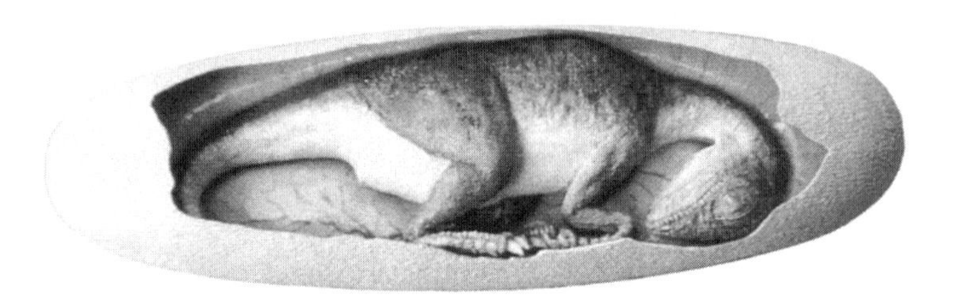

4 • The Mating Game

Before eggs can be laid, they need to be fertilized, and to get them fertilized, a female needs a cooperative male. Usually, however, it is the male who must drive away rivals while attracting the female and convincing her to mate with him. This can involve a great deal of courtship effort (flowers, dinner, maybe a movie . . .). The males are often showy and colorful because they must catch the female's eyes, so to speak. The classic example is the peacock male, which has a large, iridescent green tail, while the female's is a drab color. During courtship the male spreads a showy tail fan while strutting in front of the female (sort of like a body builder on Malibu Beach). The male ostrich, although not as bright as the peacock, nevertheless has contrasting black and white feathers (the female is brown). Instead of showing off a brilliant tail fan, the male ostrich engages in a dance called kantling (Bertram, 1992). This dance starts with the male in a crouched position and alternately lifting one wing over the body then the other while rocking from side to side. Other birds engage in head bobbing, or keep the head lowered as a sign of appeasement (it's hard to imagine a bull *Tyrannosaurus rex* appeasing, but maybe it did). Some male birds also engage in singing, although what makes a female select one singing male over another is one of those mysteries yet to be resolved. We know that some dinosaurs had hollow crests on their heads (e.g., *Corythosaurus*) that may have acted as amplifying chambers for their songs (Fig. 4.1).

Among lizards such as the large Komodo monitor, courtship consists of chases, mutual scratching, licking or flicking of the tongue on various parts of the body, and aborted mounting (Auffenberg, 1981). Sometimes there is a fine line separating courtship and aggressive display. For example, aggressive male behavior in the anole consists of head bobbing, extending the dewlap under the chin, extending a frill along the back to look bigger, and developing a dark patch behind the eyes (Fig. 4.2A). In courtship, the male bobs its head and extends the dewlap, but does not extend the back frill nor form the intimidating eye patch (Fig. 4.2B). Despite the courtship behavior of the male, there is no guarantee that the female will respond by mating. With ostriches, kantling may be only 68 percent successful (Bertram, 1992).

Fig. 4.1. Tall, hollow crest of Corythosaurus, *possibly used as a sound resonator.*

Fig. 4.2. The fine line between aggression and courtship can be seen in the anole lizard. A, male aggression consists of head bobbing, extending the dewlap, extending a frill along the back, and forming a dark patch behind the eyes. B, courtship, on the other hand, consists of head bobbing and extending the dewlap only.

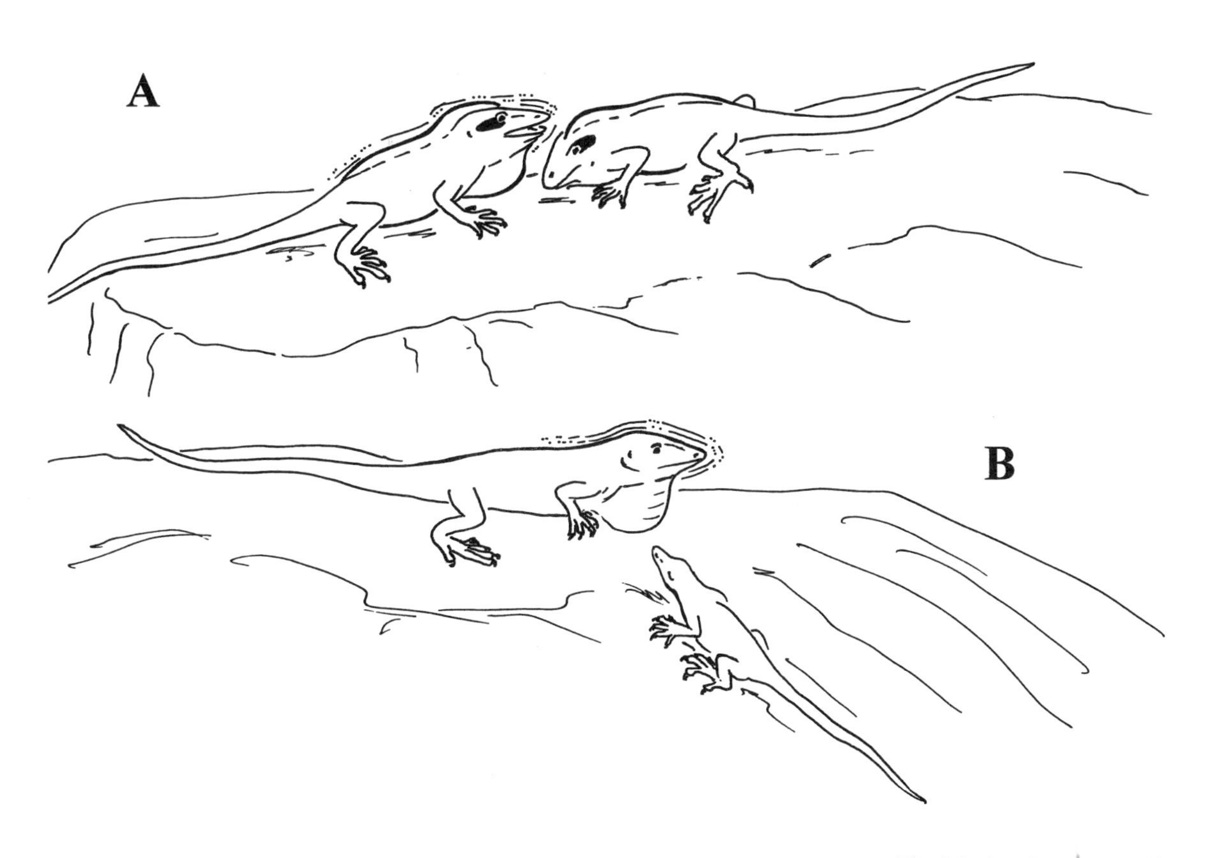

What does this courtship behavior in living animals have to do with dinosaurs? First, the wide range of courtship behavior among living animals suggests that dinosaurs may also have had a wide range of courtship styles. There is certainly no reason to assume that all dinosaurs behaved the same. Second, with dinosaurs that had ornate head or body structures, it may be possible to interpret behavior by analogy with living animals.

Male or Female?

The idea that some dinosaur structures might have no other purpose than sexual display was suggested a long time ago by Hungarian paleontologist Francis Nopcsa. In 1929, he suggested that hadrosaurs with crests were males, while flat-headed forms were females. The crest, he reasoned, was used for display much like a cock's comb. Although an interesting hypothesis, this would mean that for the last 3 million years of the Late Cretaceous in North America, only female hadrosaurs were alive. That is because the only hadrosaurs known are the flat-headed *Edmontosaurus* and *Anatotitan.* No wonder they became extinct! I doubt that these dinosaurs were parthenogenic like some lizards, where the females are able to breed without males. It seems, then, more probable that it takes more than a crest to make a male. Indeed, dinosaur paleontologist Peter Dodson (1975c) discovered just that from a statistical analysis of crested hadrosaur (lambeosaur) skulls. He found that some individuals had large crests, which he interpreted to be males, and some had smaller crests, which he interpreted to be females (Fig. 4.3). He also discovered that juveniles had even smaller crests that, we now know from the work of Jack Horner and Phil Currie, developed from a nearly flat skull roof in hatchlings (Fig. 4.4).

Why should both males and females have crests if only the male uses the crest in courtship? For one, having a crest quickly identifies the animals as being the same species regardless of sex. This is especially important for animals traveling in herds (Leuthold, 1977) the way lambeosaurs are thought to have done. Also, there is evidence that the inside of the crests increased the surface area for olfactory, or smell, tissue, and also the hollow of the crest might have acted as a resonating chamber for calls of both sexes. Who knows what strange cries, bellows, and songs might have been heard in the night 75 million years ago, especially during the mating season.

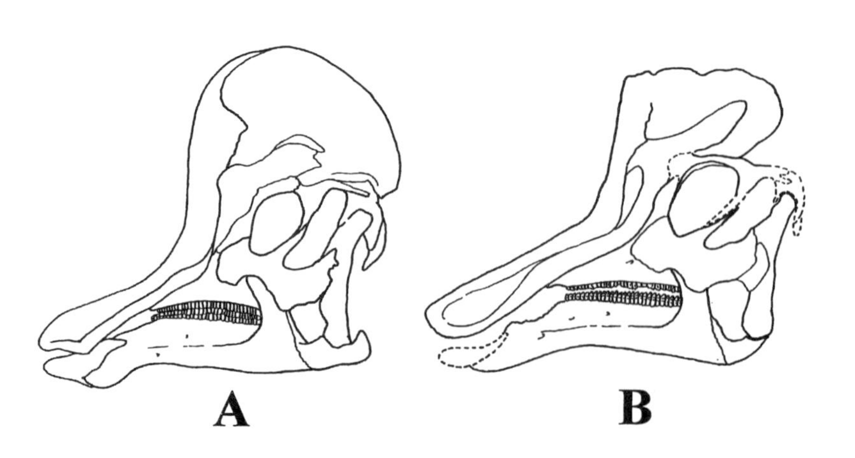

Fig. 4.3. Sexual dimorphism in Corythosaurus *based on the study of Peter Dodson (1975c). The male* Corythosaurus *is believed to have had the larger crest for display (A), and the female had the smaller crest (B). (Adapted from Hopson, 1975.)*

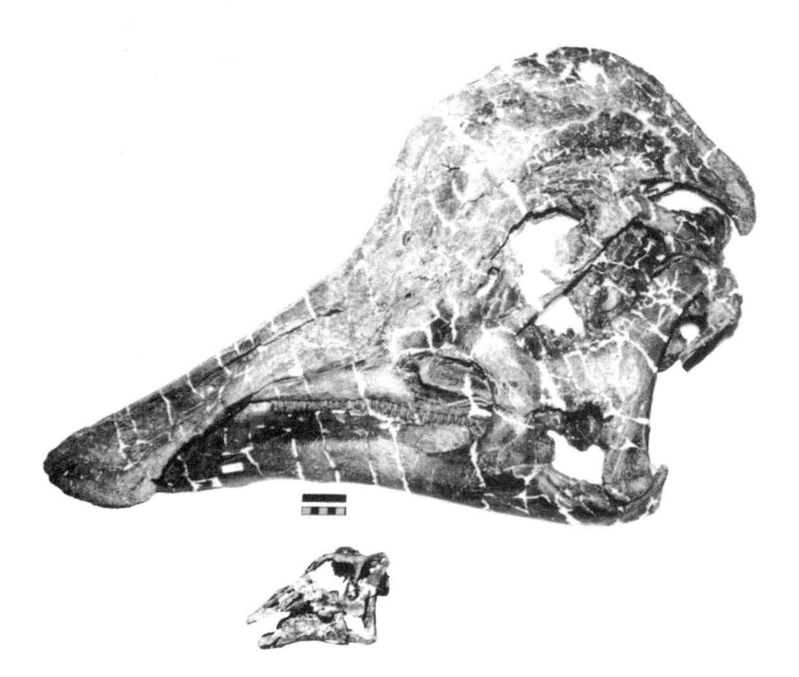

Fig. 4.4. Development of the crest in lambeosaurine hadrosaurs as illustrated by Hypacrosaurus *is shown by the flat head of a juvenile (bottom) and high crest of the adult (top).*

Before we look at possible display structures in other dinosaurs, we need to first discuss how to tell the sexes apart from skeletons. With mammals, it is easy to separate females from males, because of the wider pelvic canal that accommodates the large head of the baby. Part of what makes a baby cute is the proportionally large head relative to body size (Steve Gould, 1980, has discussed the phenomenon in the "evolution" of Mickey Mouse; the original small-headed Mickey is not as cute as the modern big-headed Mickey). Actually, babies don't have enlarged heads just to look cute, but to accommodate a large brain and eyes. Postnatal growth of these structures is much less than that of other parts of the body. Hence, growing up often means the rest of the body catching up with the head.

In dinosaurs, live birth does not appear to have occurred (more on that later). Instead, the female dinosaurs laid clutches of eggs much like birds and most reptiles (Fig. 4.5). Although development of the embryos in birds and reptiles is well under way at the time the eggs are laid, little of the brain or eyes has formed (see chapter 11). This allows the eggs to be small enough that major modifications of the pelvis are not necessary. A minor modification seen in one dinosaur, *Tyrannosaurus rex,* is an increase in the angle between the tail vertebrae and the rearmost downward projecting bone, the ischium (Fig. 4.6). This increase is slight, but may have better accommodated the passage of eggs through the cloaca. As yet, this difference has not been seen in other dinosaurs because no one has looked.

A more reliable method of sexing dinosaurs is the robustness of the skeletons (Fig. 4.6, 4.7). Belgian paleontologist Louis Dollo noted that some of the *Iguanodon* skeletons he was studying were more stoutly built, or robust, than others (Fig. 4.8). David Norman (1980, 1986) has identified these as different species, but I remain unconvinced. The difference could

Fig. 4.5. Dinosaurs laid hard-shelled eggs like birds or crocodiles.

Fig. 4.6. Comparison of a female Tyrannosaurus rex *pelvis (top) and male (bottom). Although the difference is slight, the greater angle in the female between the ischium and bottom of the tail vertebrae (represented by the heavy line) may accommodate the egg. This idea is only a suggestion and has yet to be tested rigorously. Front is toward the left.*

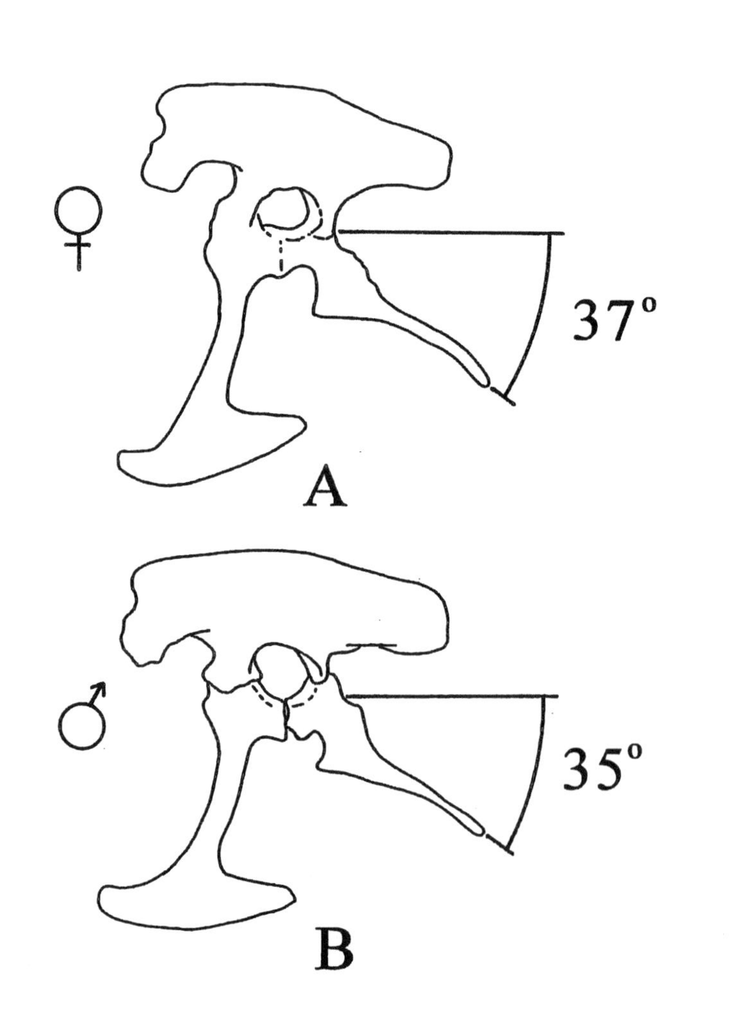

just as easily be sexual. My reasoning is that both robust and gracile forms are typically seen in bone beds where many individuals of a single species of dinosaur are preserved. This difference is best explained as one form being the males and the other the females. But which is male and which is female? The best evidence suggests that the robust form is the female (Raath, 1990; Carpenter, 1990a). This may come as a surprise, because among humans it is the male that is the robust form. However, several lines of reasoning suggest that it is the female that is robust among dinosaurs. First, among living animals, it is the female who is typically the robust form (e.g., Fitch, 1981). The reasons for this are not well understood, but suggestions include the need to defend the nest or young and to have a greater reserve of bone calcium to draw upon for eggshell production or for the development of the fetus' skeleton. Indeed, Anusuya Chinsamy (1990) found erosional cavities within the bones of robust specimens of *Syntarsus*, not in gracile specimens. These cavities formed when the bone mineral was removed, probably during the formation of eggshell. Second, the increased angle between the ischium and tail in *T. rex* mentioned above occurs in the robust form (Fig. 4.6). But it should not be assumed that the female is always the larger, because among crocodiles it is the male which is larger (Cott, 1961; Fitch, 1981). Tom Lehman (1990) suggested that the male of the ceratopsian *Chasmosaurus mariscalensis* had the taller, more erect, and more narrowly spaced horncores than did the female (Fig. 4.9). He argues that the size and position of the horns would made them more effective weapons for shoving contests between rival males (more below).

Crests, Frills, and the Dance of the Titans

Now let's return to courtship. Many dinosaurs have structures on their heads or bodies, the function of which is not always apparent. Among theropods, a double crest occurs on the head in *Dilophosaurus* (of *Jurassic Park* fame). It is doubtful that this dinosaur could spit poison like the spitting cobra of India as the one in the movie does, and it certainly shows

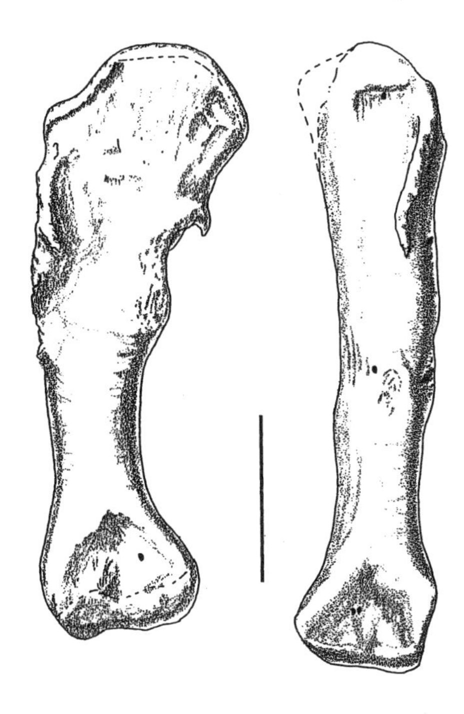

Fig. 4.7. An example of robust vs. gracile in T. rex *is illustrated by the upper arm bone, the humerus. The female is assumed to have been the robust individual (left) because of the greater calcium reserves in the bones for producing eggshell.*

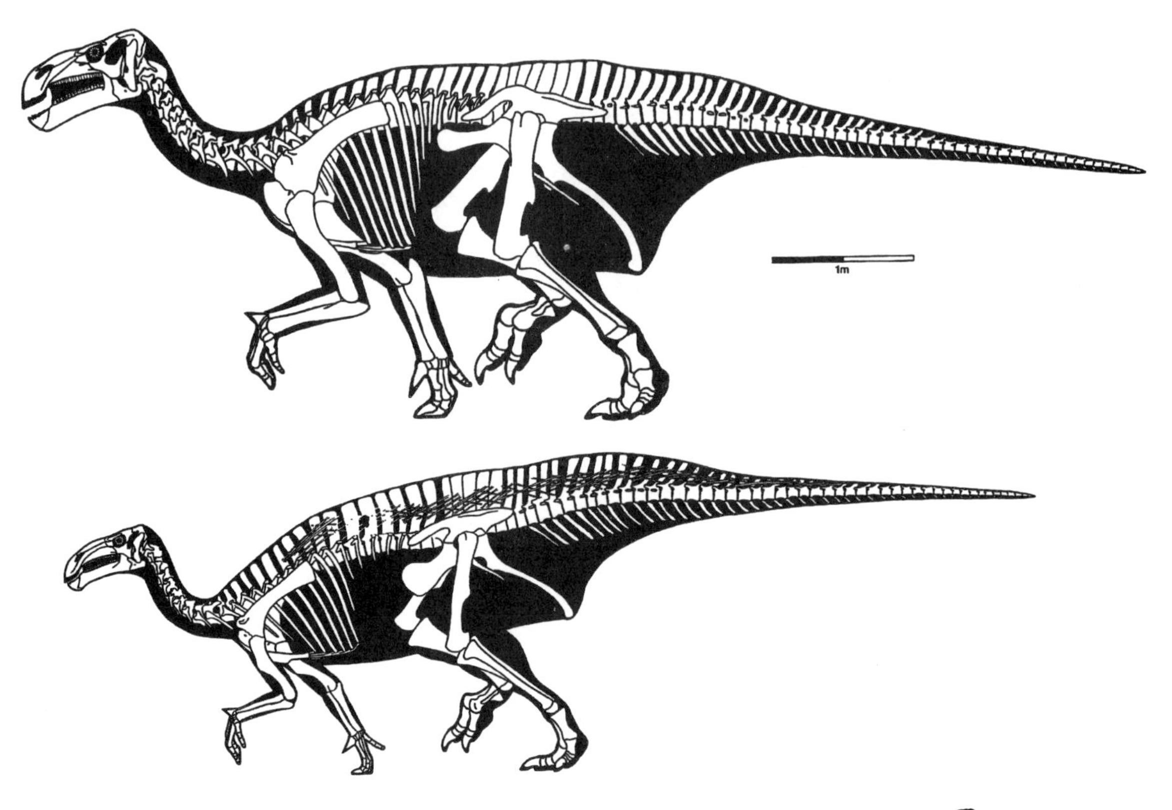

Fig. 4.8. The two forms of Iguanodon *from Bernissart, Belgium. The upper form has been named* Iguanodon bernissartensis *and the lower* Iguanodon atherfieldensis. *Many of the differences, such as in the shoulder blade, may be size related. These differences may not have much significance beyond separating females (larger, robust form) from the males (smaller, gracile form). Other differences, such as the larger blade on the pubis in the "male," are harder to explain and may indeed separate* Iguanodon *into two species. (Courtesy of Greg Paul.)*

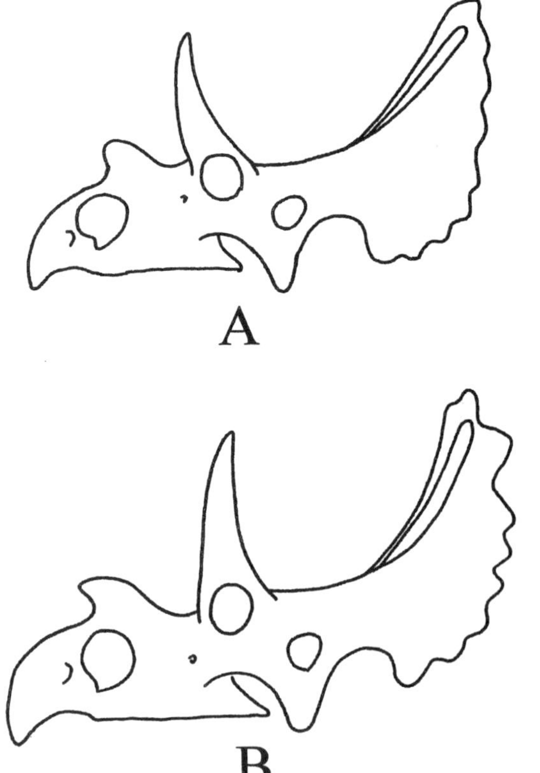

Fig. 4.9. Female (top) and male (bottom) Chasmosaurus mariscalensis *based on the work of Tom Lehman (1990). The males are thought to have the larger frill and more upright horns.*

no sign of an expandable display collar similar to that in the frilled lizard. It does, however, have a pair of thin bone crests atop the head (Fig. 4.10A). These crests are developed as outgrowths from the nasal and frontal bones. They are so thin walled that they must have been used primarily for display. Because they are best seen in profile, *Dilophosaurus* must have presented a profile of the head in display. Among living lizards with display features on the head, such as brightly colored dewlaps, attention is drawn to the head by head bobbing (Fig. 4.2A). So it is conceivable that *Dilophosaurus* bobbed a head with crests covered with bright or iridescent skin. What female dilophosaur could resist such a come-on?

In other theropods, such as *Allosaurus,* a pair of thin hornlike bumps project in front of the eyes (Fig. 4.10B). Considering the variation in the

Fig. 4.10. Display structures in theropods: A, double crest on snout in Dilophosaurus; *B, upright brow horn in* Allosaurus; *C, outwardly projecting brow horn in* Carnotaurus; *and D, tall crest of* Oviraptor. *(Courtesy of Gregory Paul.)*

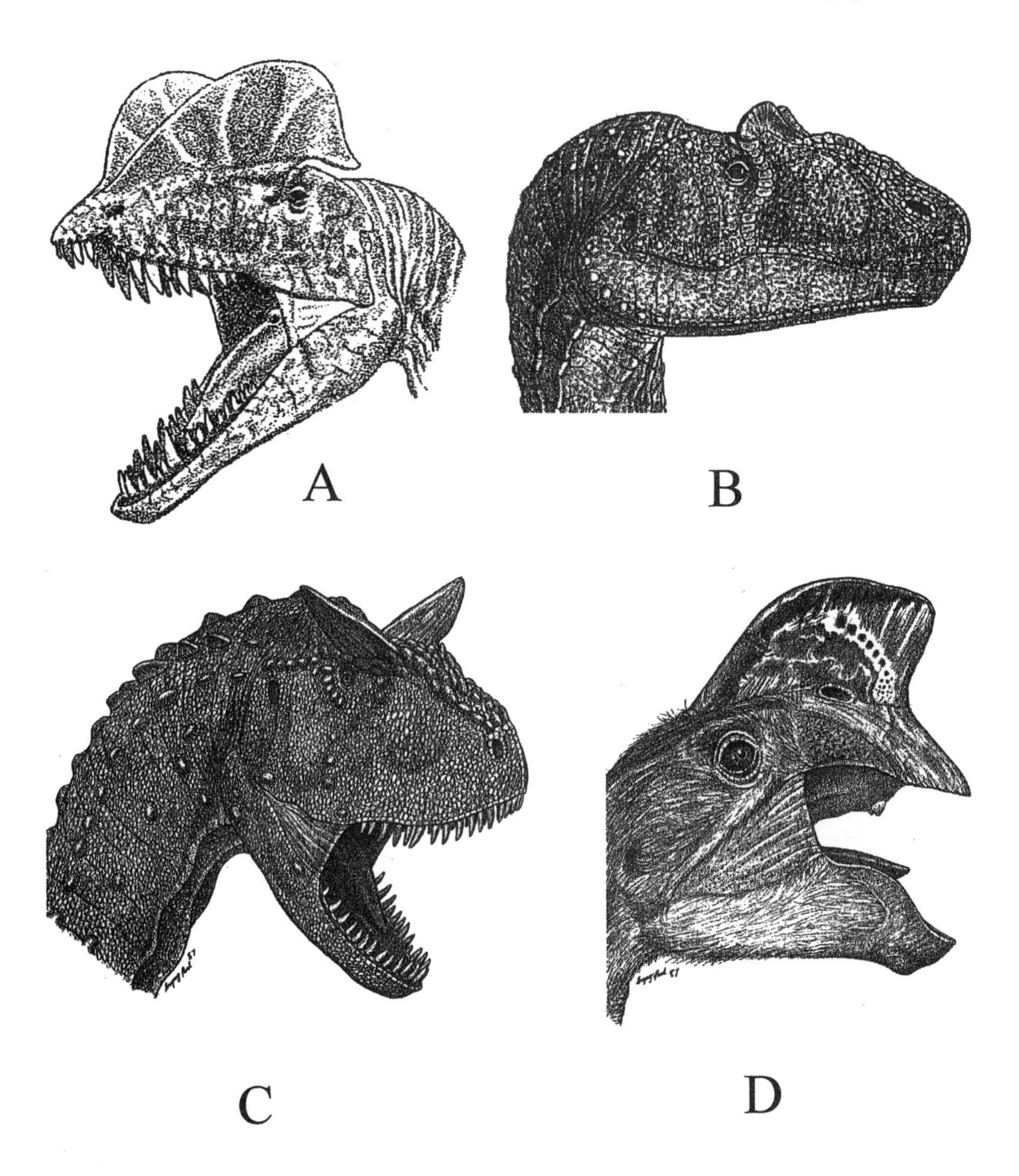

size of these brow horns, it seems probable (although unproven) that the size of the horns is related to the sex of the individual. A variation of the orbital horn is seen in the weird theropod *Carnotaurus* from Argentina (Fig. 4.10C), where the horns project outward and could best be seen at an angle. *Oviraptor* (Fig. 4.10D) has a single, prominent crest on the top of the head. The bones are so delicate that the crest must have been for display only.

Another method of display among theropods involved elongated spines on the back vertebrae. The tallest structure is seen in the aptly named *Spinosaurus* from the Late Cretaceous of Egypt. The spines were probably covered with skin forming a sail-like structure reminiscent of *Dimetrodon.* Perhaps two rival spinosaurs would stand side by side showing off their sails prominently (Fig. 4.12). Perhaps the skin could also change colors like a lizard's. This use of the sail for display seems more likely than for thermoregulation, or control of the body temperature—a related form, *Suchomimus,* that lived about the same time in North Africa did not have a sail at all. If the sail of *Spinosaurus* was for thermoregulation, then *Suchomimus* would have died from the heat.

Tyrannosaurus and related large carnivores lack any display structure of bone on the head or body. *Tyrannosaurus* did, however, apparently have

Fig. 4.11. Nasal horn in Ceratosaurus *used as an antagonistic weapon. (Courtesy of Gregory Paul.)*

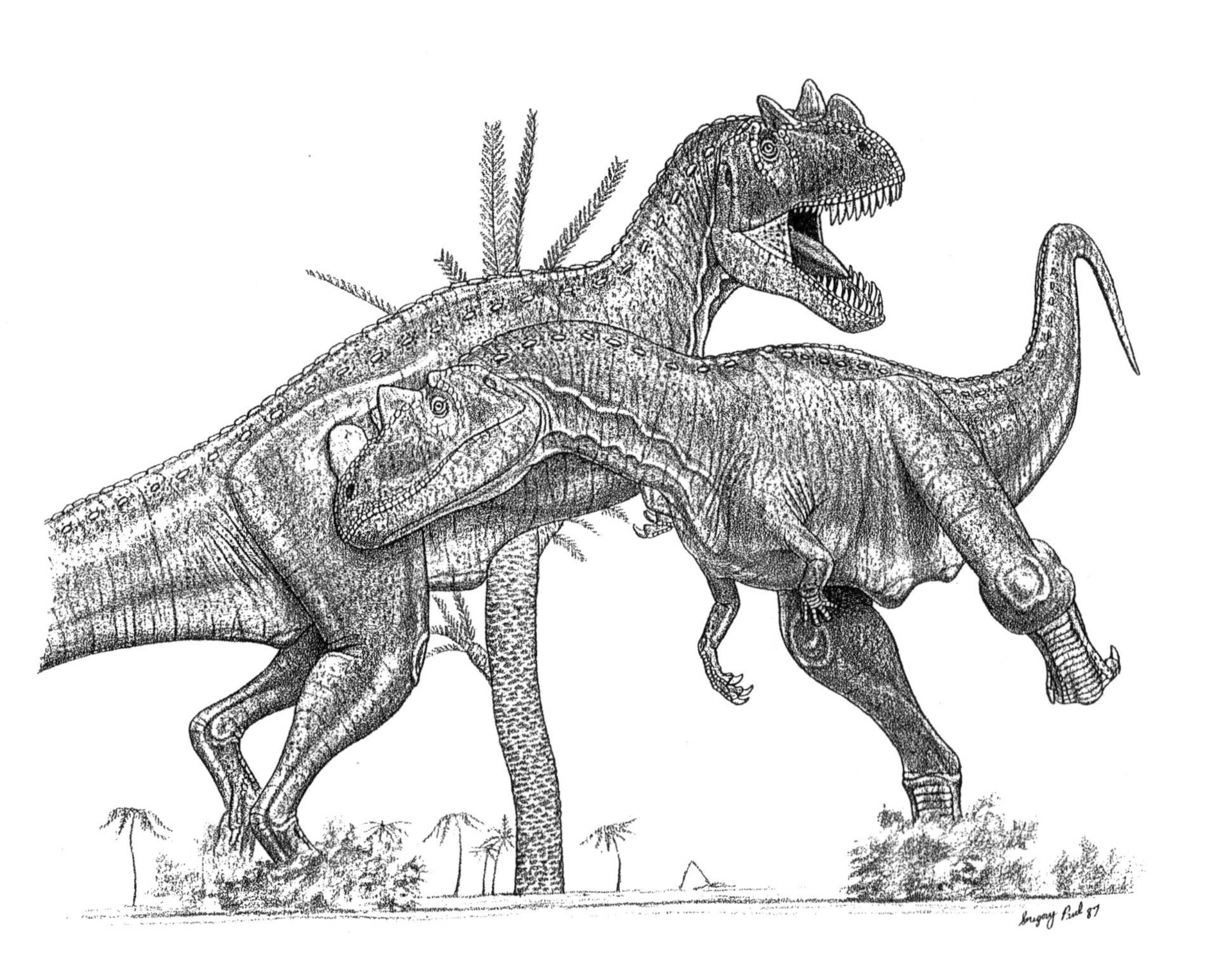

either a pelican-like pouch or dewlap based on an impression of the skin found below a skull in Mongolia (Mikhailov, personal communication). This, like other remarkable traces of dinosaur skin, shows that dinosaurs had display structures that rarely fossilized. If the impression is a dewlap, then *Tyrannosaurus* might have courted by displaying his dewlap. If the impression was a pelican-like pouch, then possibly a brightly colored pouch was inflated and displayed with the head tilted back somewhat like a frigate bird (Fig. 4.13). Some small theropods are known to have feathers and it is possible that these may have used their feathers in courtship display and dance (Fig. 4.14)

Sometimes intimidating the mating rival is not very successful and a fight might break out. Possibly *Ceratosaurus* used its horn on the snout in

Fig. 4.12. Two male Spinosaurus *display their tall sail-backs at each other amid growls and roars.*

Fig. 4.13. Expanded throat pouch in Tyrannosaurus *(A) as compared with a frigate bird (B). The pouch in* Tyrannosaurus *is based on a specimen seen in Mongolia by Konstantin Mikhailov.*

such fights, ramming the opponent in the side (Fig. 4.11). Most theropods, however, probably resorted to what they were good at—biting and kicking (Fig. 4.15).

How and when courtship occurred among theropods is unknown. Possibly they were like Komodo lizards, which assemble near the carcasses

Fig. 4.14. Display in a hypothetically feathered Velociraptor. *The assumption is that* Velociraptor *may have had feathers because of recent discoveries of small theropods with feathers in China and because of the avianlike structure of the* Velociraptor *skeleton. Are these two males threatening each other or is this a courtship dance? (Courtesy of Luis Rey.)*

Fig. 4.15. If intimidation failed, two theropods probably went at each other with tooth and claw. (Courtesy of Gregory Paul.)

they feed upon to court and mate. Perhaps congregating in this manner may explain the numerous individuals of *Allosaurus* found at a site called Cleveland-Lloyd in Utah. The site has been referred to as a death trap for allosaurs attracted to dying dinosaurs trapped in a sticky mud. Might the allosaurs also have been drawn there to mate?

Courtship display among prosauropods and sauropods, the big brontosaur-like dinosaurs, is somewhat less clear because they apparently lacked display structures on the head. Recent discoveries in Wyoming, however, do show that one form, *Diplodocus,* had a spiked frill of skin along the tail in a manner similar to an iguana lizard. The frill may have extended as far as the head, but no evidence for this has yet been found. How this frill was used in courtship is not known and it is possible that it had no role. Having long necks, rival bull sauropods may have used head-neck bobbing or neck weaving. It is also possible that male sauropods engaged in neck combat in a manner similar to giraffes (Fig. 4.16), or nipped one another like stallions. However, considering how diverse sauropods are, it is difficult to imagine that they all used the same form of courtship or male rivalry. Maybe some species used their tails in a non-threatening way, perhaps wagging them like dogs. Another idea for the tail was that it functioned like a bullwhip (Myhrvold and Currie, 1997). However, the spinal cord of *Apatosaurus* and its cousin *Diplodocus* was exposed for almost the last quarter of the tail. Cracking the tail would certainly have been painful (sort of like hitting your "funny bone"). Myhrvold and Currie also suggested that the end of the tail had a "popper," a flap of skin that produced the crack. Unfortunately, if the flap was dead skin, it would be dry and stiff, hardly a good popper. If living skin, popping the tail would fray the end (as they admitted), which would then scab over. As anyone who has picked a scab knows, scabs are very stiff. In addition, once the scab healed, the scar tissue would also be stiff. It therefore seems doubtful that the tail was used as a whip. It is possible that sauropods, like the Komodo lizard, did not engage in display courtship. Lacking display structures, Komodo courtship consists of chases, licks, and nudges. Pity the poor vegetation when sauropods started to engage in sexual frolics!

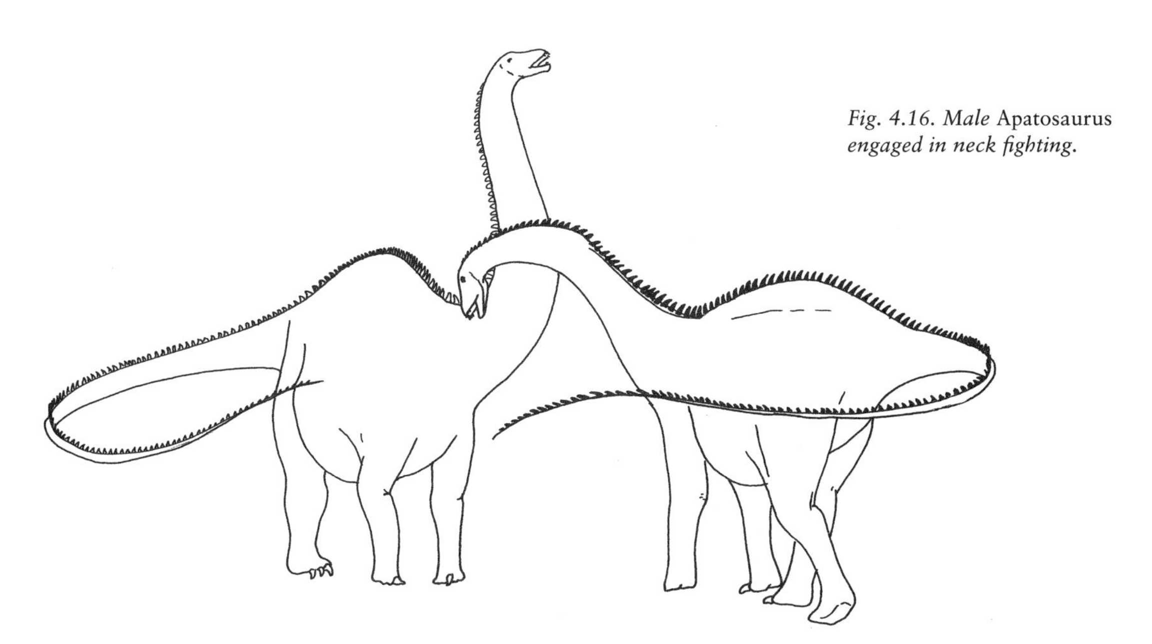

Fig. 4.16. Male Apatosaurus *engaged in neck fighting.*

Prosauropods, which include the distant relatives of the sauropods, resemble sauropods in that they also have long necks and tails. Whether or not they had flaps or frills of skin is unknown. They do have a large claw on the thumbs (as do sauropods), and it is possible that these were used by rival males. Perhaps an old bull prosauropod might be recognized by the numerous scars on its thick hide (Fig. 4.17).

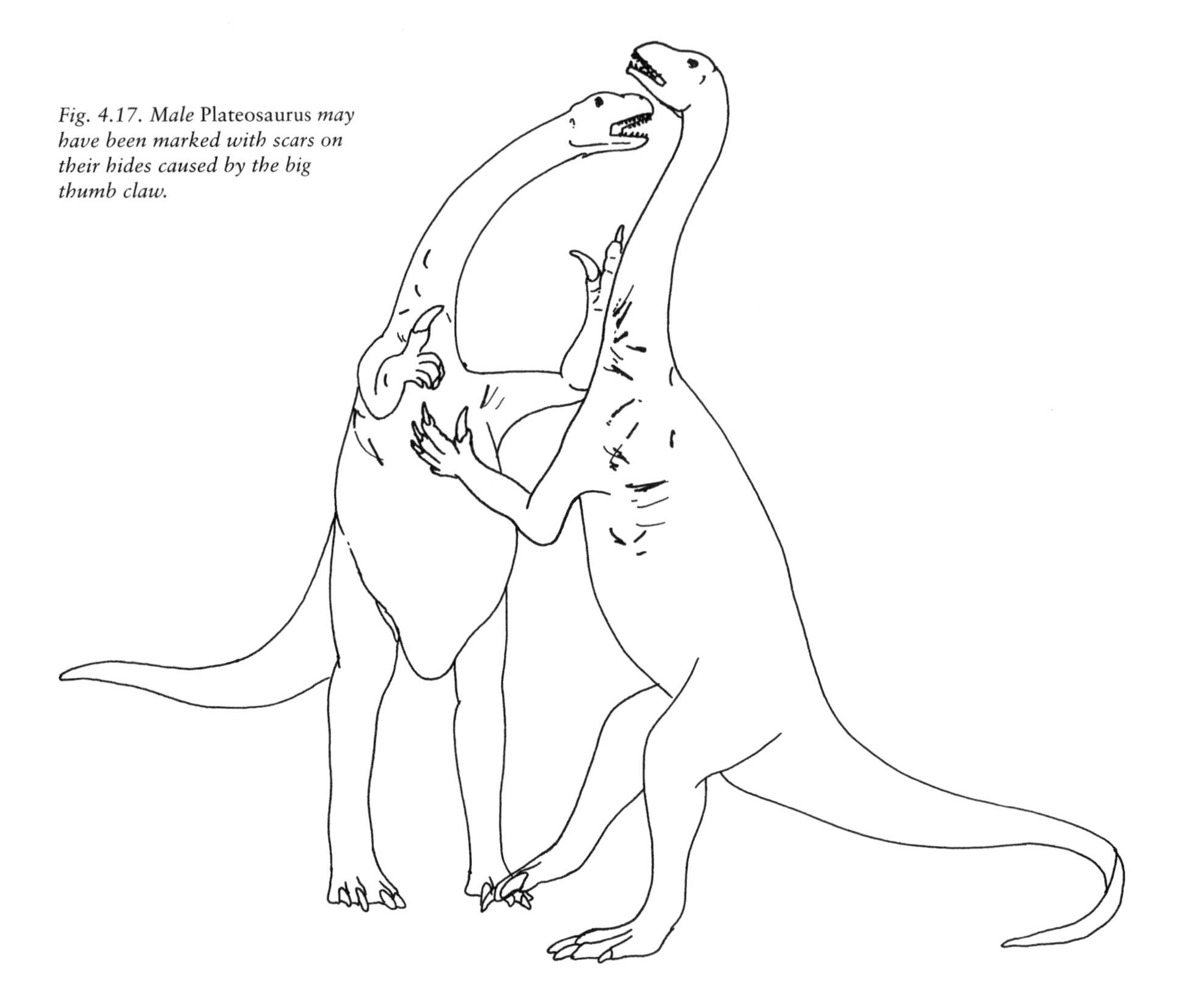

Fig. 4.17. Male Plateosaurus *may have been marked with scars on their hides caused by the big thumb claw.*

Among ornithischian dinosaurs, display structures reach their zenith in crests, horns, and frills. One of the better known early forms, *Heterodontosaurus,* is characterized by a pair of canines in the males, much like those of the mouse deer (*Traugulus*) (Fig. 4.18). These canines might have been displayed with mouth open and only rarely used to bite the opponent to drive him away. Might *Heterodontosaurus* have traveled in small herds overseen by a dominant male? As yet we have no evidence for this, such as a *Heterodontosaurus* bone bed formed by the death of a herd. Slightly more advanced ornithischians, such as the bipedal *Hypsilophodon* from the Lower Cretaceous of England, do not show any display features that we can recognize. Again, this may mean that they had dewlaps or other structures that did not fossilize. One paleontologist-artist, Gregory Paul, has illustrated a related form, *Dryosaurus,* with dewlaps and feathers on its head that might have been used for display (Fig. 4.19). It is difficult to know, however, how far we can go with the bird analogy in illustrations of dinosaurs.

Display structures are more abundant and diverse among the iguanodontids. One of these, *Iguanodon* from the Lower Cretaceous of Europe, had a thumb modified into a spike that might have been used in defense against predators, in fights with rival males (Fig. 4.20), or possibly in courtship. The North African *Ouranosaurus* had a tall fin on its back formed by the elongation of the vertebral spines. Fastovsky and Weishampel (1996) suggested that this fin functioned to control body heat by turning the fin toward the sun when the animal was cool and away from the sun when the animal was too hot. Although such a suggestion is possible, it seems more likely that the fin was a display structure. After all, *Ouranosaurus* lived a little north of the Early Cretaceous equator, where it was warm year round. Both male and female *Ouranosaurus* might have used

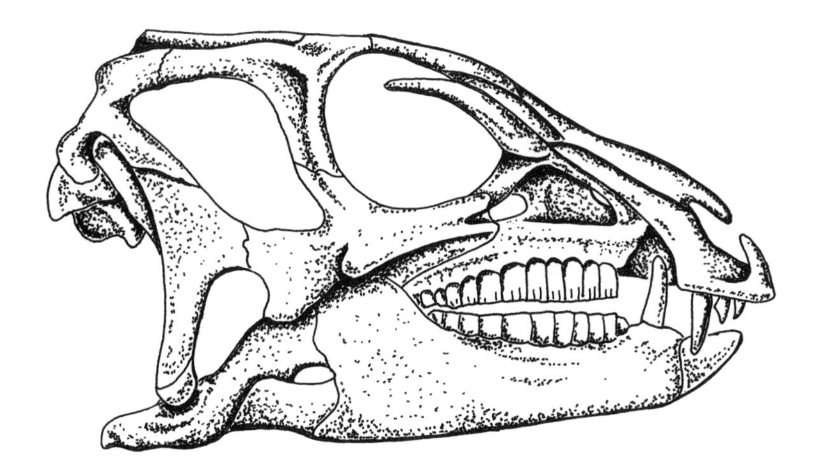

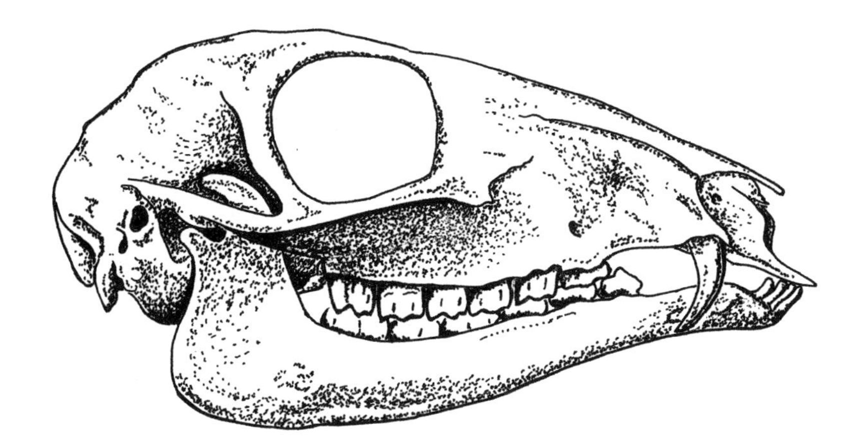

Fig. 4.18. The male Heterodontosaurus *(top) has enlarged "canines" much like the male mouse deer (below) from Southeast Asia.*

A

B

Fig. 4.19. The small ornithischian Dryosaurus *(A) may have had body adornments (B) that have not fossilized. Without a time machine, we may never know. (B drawn by Gregory Paul.)*

Fig. 4.20. The large ornithischian Iguanodon *has a spiked "thumb" that may have been used in fights (either that or they are giving each other a secret handshake). (Courtesy of Gregory Paul.)*

the fin to make themselves look bigger to keep predators at bay, or to keep rivals away from their territory. During courtship the male might have displayed his fin by turning sideways, possibly even blushing, causing the fin to change color.

The flat-headed hadrosaurs, such as *Kritosaurus* or *Edmontosaurus,* have enlarged nasal openings that were covered by skin pierced with a small nostril opening. We know this from a hadrosaur "mummy" found in the Late Cretaceous of Wyoming, and presently on display at the American Museum of Natural History in New York City. The enlargement of the nostril region made for a large display structure on the head. Perhaps the skin covering could inflate like a sack. Admittedly this would look strange to us today, but an inflated nostril sack might send a very clear message to a potential mate (Fig. 4.21). Skin impressions of hadrosaurs show that the hide of the body was covered with a pattern of large disks surrounded by smaller disks. Henry Osborn (1912) suggested that the disks might reflect the color pattern of the living animal.

Hadrosaurs are known to have had a frill down the back. The frill of *Edmontosaurus* had a jagged look (Fig. 4.22A), resembling the parapet of a medieval castle, while in *Kritosaurus,* the frill looked like a row of triangles. In the crest-headed hadrosaur or lambeosaur, *Corythosaurus,* the frill was a flap of skin (Fig. 4.22B) that extended from the neck to the tail. We do not know if these frills played any role in display, or whether they served to help the various species recognize one another. I do, however, suspect that the crests of the lambeosaurs did play some role in display. Peter Dodson noted that the crests were larger in what he identified as male individuals of *Corythosaurus* and *Lambeosaurus* (Fig. 4.23). Perhaps crest size determined which male would dominate a herd or harem. Because these crests were most visible in profile, the antagonists must have stood sideways or at least turned their heads to display the crests. This display

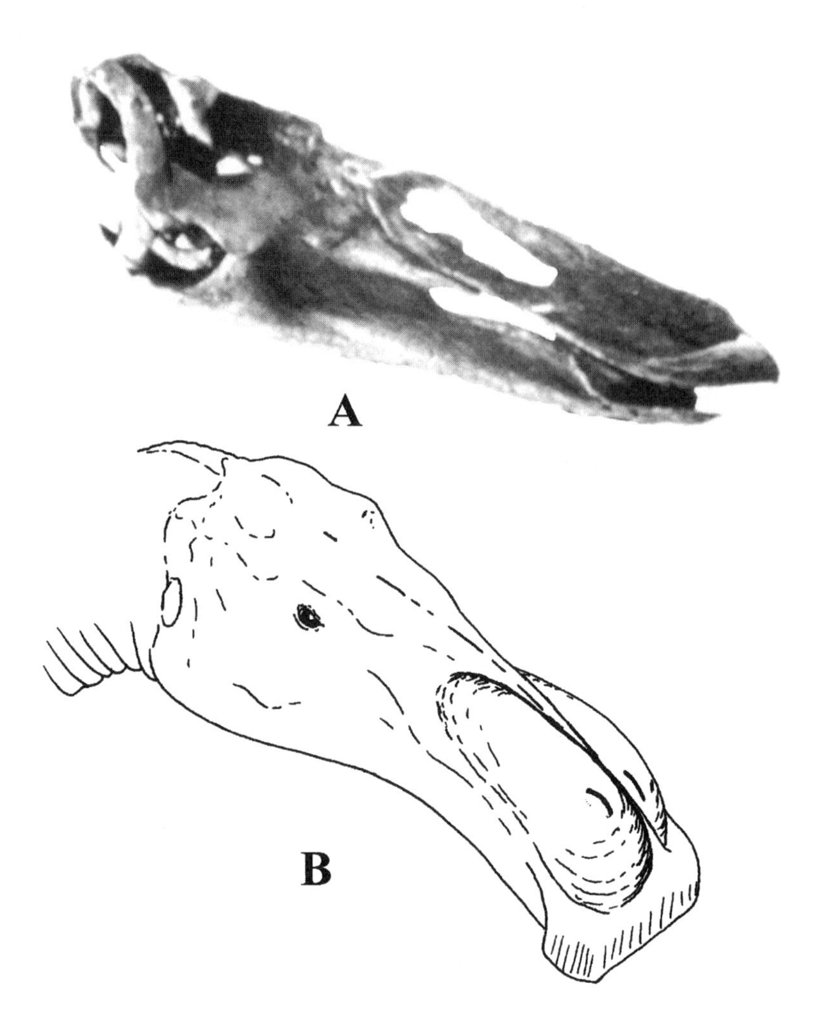

*Fig. 4.21. The flat-headed hadrosaurs have an enlarged nostril area (*Anatotitan *shown here) that may have housed a large inflatable nose sack. Other suggestions for the nostril area include increased tissue for detecting odors, or a salt gland to rid the body of excess salt.*

may have been accompanied by an occasional bellow or bugle using the hollowed crest as a resonating chamber. If intimidation failed, physical confrontation might have consisted of nips with the beak or kicks with the feet. It is doubtful that the thin-walled crests were used as battering rams.

The dome-heads, or pachycephalosaurs, were at one time thought to have used their heads to butt head to head like bighorn sheep. The problem with this hypothesis is that the impact area on the domes is so small that unless the pachycephalosaurs hit exactly, they would have grazed past each other. So what were the domes for? Male pachycephalosaurs may have used them to flank butt, butting each other in the sides (Fig. 4.24). This behavior probably explains why the body is unusual among dinosaurs in being so broad. Such a broad body would protect the vital organs, such as the heart or spleen, from rupture. One pachycephalosaur, *Stygimoloch,* is unusual in that the dome is a tall peak with clusters of spikes on the backside. These spikes may have been for show, perhaps indicating that *Stygimoloch* bobbed their heads, or raised their heads high and profiled the spikes.

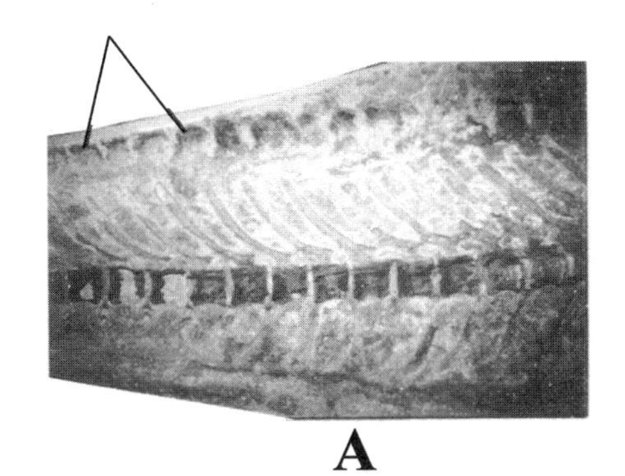

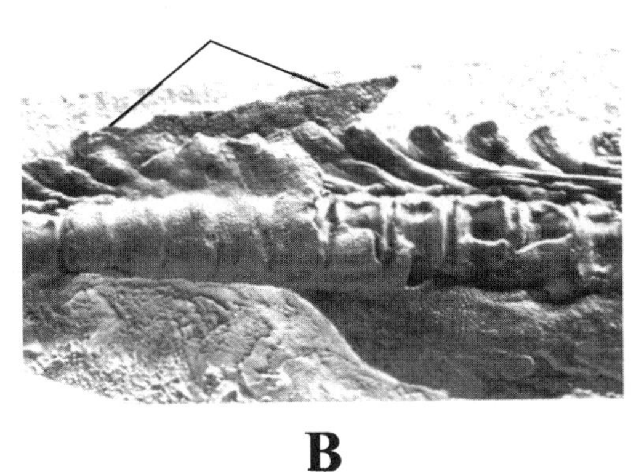

Fig. 4.22. Examples of fossilized frill on the tail of the hadrosaur Edmontosaurus *(A) and* Corythosaurus *(B).*

Fig. 4.23. Comparison of male and female lambeosaurine hadrosaurs: A, male Corthyosaurus *adult; B, male juvenile; C, D, female adults; E, female subadult; F, male* Lambeosaurus *adult; G, female adult; H, I, subadults; J, juvenile, sex unknown. Note the variation in crest shapes as well as changes that accompany growth. (Adapted from Dodson, 1975c.)*

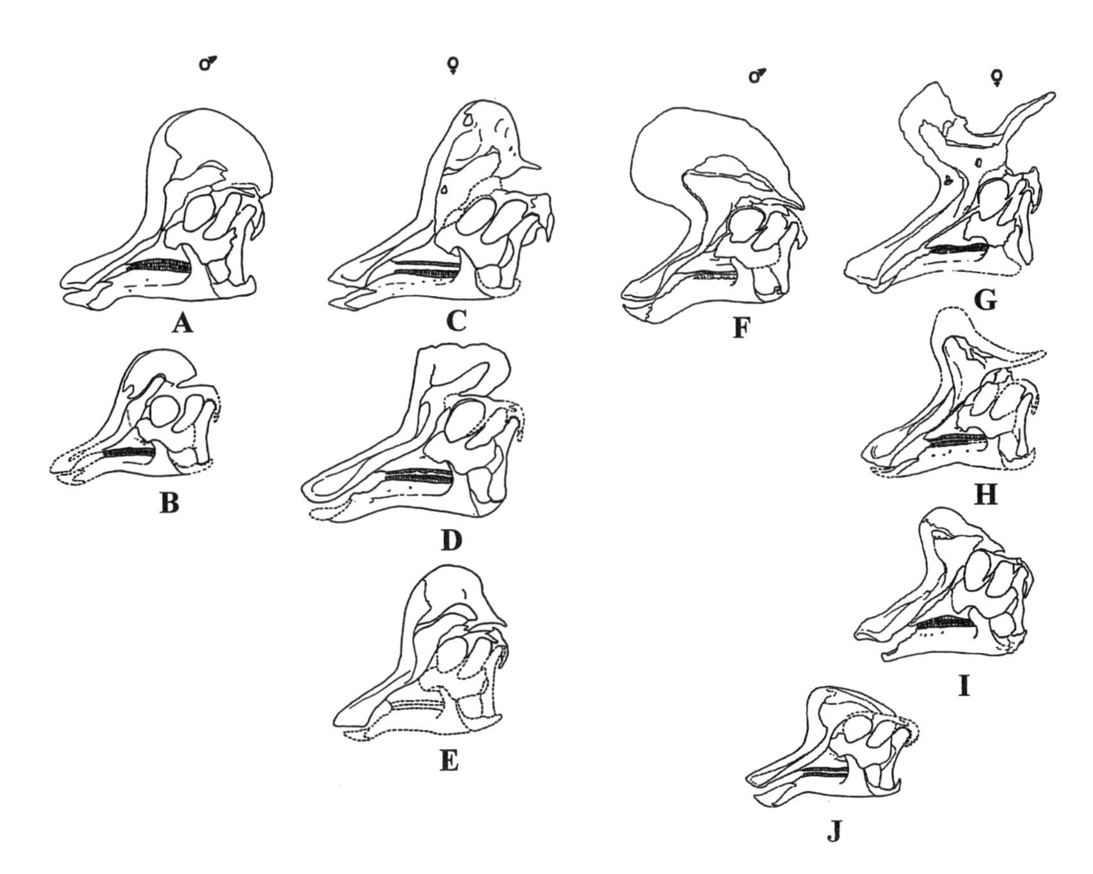

*Fig. 4.24. Two male dome-headed pachycephalosaurs (*Stygimoloch*) fighting for a female. (Life-size diorama at the Denver Museum of Natural History.)*

Among ceratopsians, the most prominent display structures are located on the head. The frill around the back of the head is seen in protoceratopsians as well as ceratopsians (Fig. 4.25). The size of the frill differs considerably, being most developed in *Torosaurus* where it makes up over half the seven-foot skull length. Most ceratopsian frills have a pair of large openings possibly to reduce the frill weight. This opening was undoubtedly covered by skin, making the structure appear solid. The frill is especially noticeable when the head is lowered (Fig. 4.26). It certainly makes the individual look bigger and more intimidating, which might be important to rival males, as well as for a female protecting her nest. Courtship may have consisted of impressing the female with the frill perhaps by shaking the head side to side. It is also possible that the frill could change color or blush, because the bone surface is extensively vascularized, i.e., there is a network of grooves for blood vessels. This vascularization indicates that the frill could receive a large supply of blood.

The other prominent skull features are the horns. These are best developed in the ceratopsians, such as *Triceratops,* but a small nose horn is also present in the protoceratopsians, especially the males (Fig. 4.27). The horns are known to have been used among rivals because several specimens of ceratopsians show healed punctures in the frill. These are especially common among specimens of *Torosaurus.* One living reptile that may provide a clue to horn use is the horned chameleon (Fig. 4.25E). It resembles a ceratopsian in that it has three horns on the face (one on the nose and two over the eyes). Rival chameleons lock horns and try to push one another out of trees. I don't mean to imply that an animal like *Triceratops* pushed

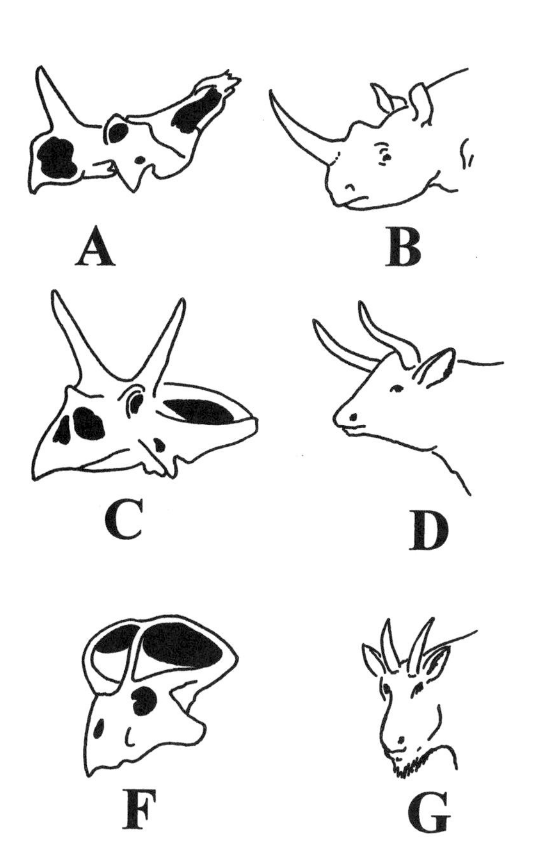

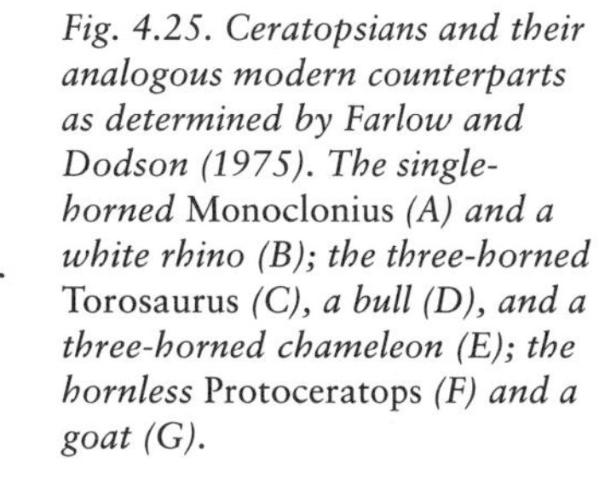

Fig. 4.25. Ceratopsians and their analogous modern counterparts as determined by Farlow and Dodson (1975). The single-horned Monoclonius *(A) and a white rhino (B); the three-horned* Torosaurus *(C), a bull (D), and a three-horned chameleon (E); the hornless* Protoceratops *(F) and a goat (G).*

Fig. 4.26. Front view of Triceratops *with the head lowered to make the individual look bigger. (Drawn by Gregory Paul.)*

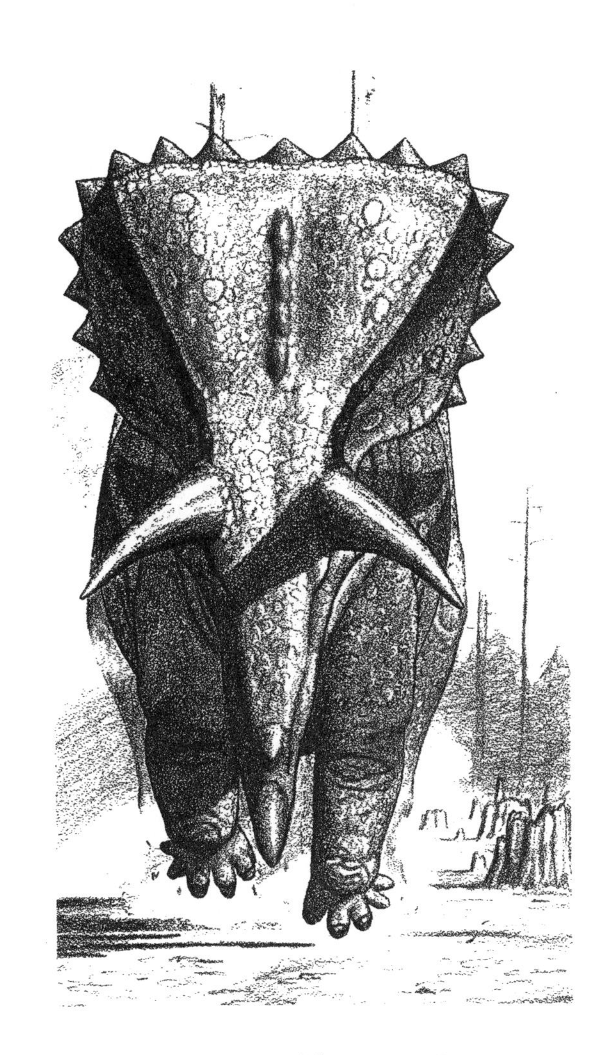

its rivals out of trees, but that the ceratopsians may have engaged in pushing contests of strength like the chameleon (Fig. 4.27).

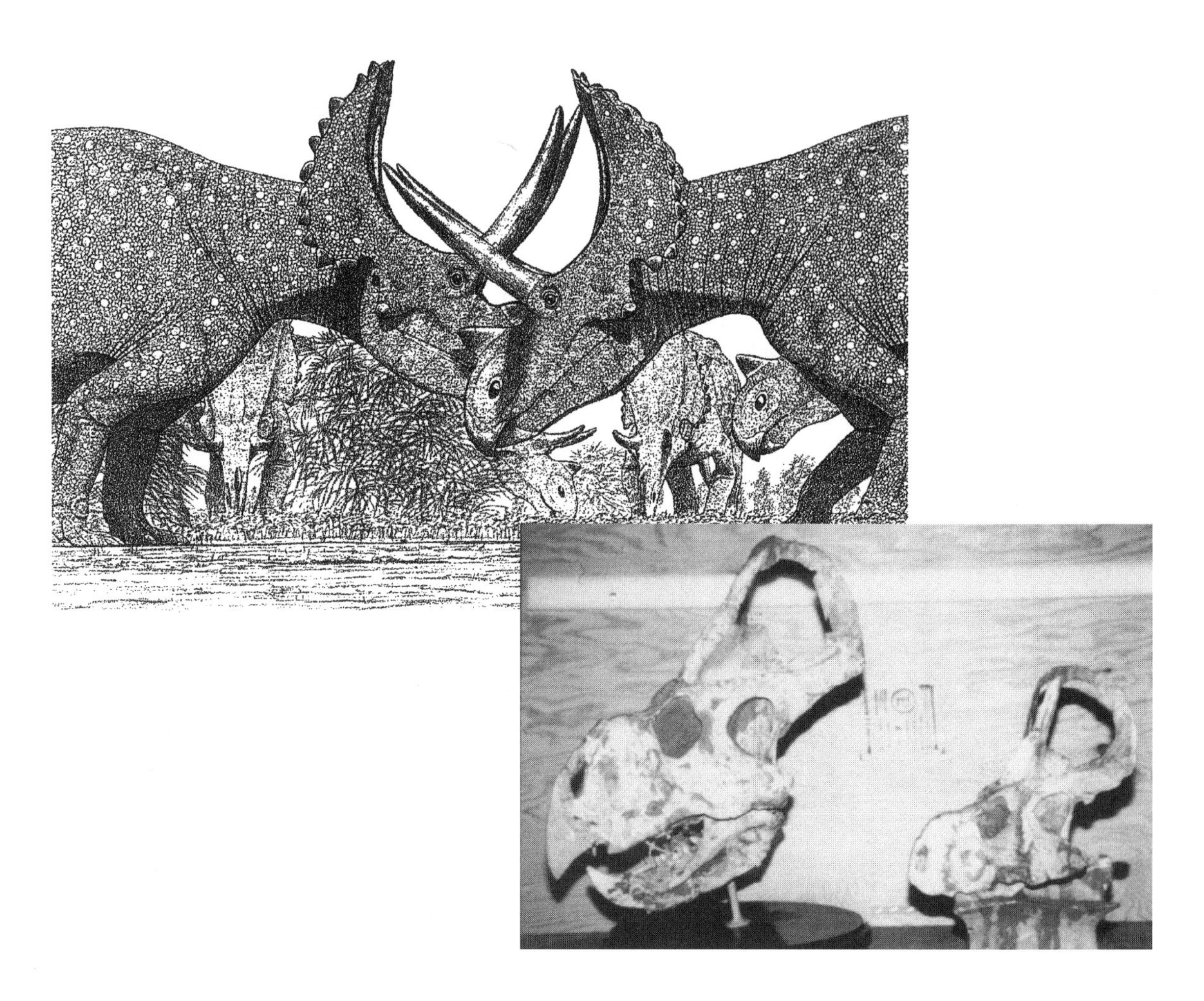

Fig. 4.27. Upper, two male Triceratops *in a shoving contest. (Drawn by Gregory Paul.) Lower, a male* Protoceratops *(left) and subadult female (right).*

Among some horned and antlered mammals, shoving contests are a last resort. For example, a male moose thrashes vegetation with his antlers to let other males know he is present. Sometimes that alone is enough to drive a rival away. Interestingly, the same message helps the female locate the male in the dense vegetation. Might some ceratopsians such as *Triceratops* have sent similar dual messages with their horns? When shoving contests do occur among antelopes, the antler is seldom used as a weapon to injure, because if an individual is able to gore his rival, then it also means that the rival can gore the individual (Leuthold, 1977). Being dead is not a very good strategy for leaving progeny each year. Animals sometimes do get gored, but this is usually (but not always) by accident. It seems most likely, then, that the puncture injuries in the *Torosaurus* frills may also have been accidental, perhaps caused by the horns slipping past one another.

Although most ceratopsians are characterized by frills and horns, the most primitive form, *Psittacosaurus,* lacks these structures. Instead, it has a spike formed by an outgrowth of the cheek or jugal bone (Fig. 4.28). Rival males may have stood side by side and jabbed one another in the sides with the cheek spike until one of them yielded. This type of flank butting is also seen in some small antelopes, such as the dik-dik. Walter Leuthold (1977) observed that the dik-dik horns are sharp enough to maximize the amount of pain without causing serious injury. The same appears to have been true of the cheek spike of *Psittacosaurus* because it is pointed but not long and sharp like a *Triceratops* horn.

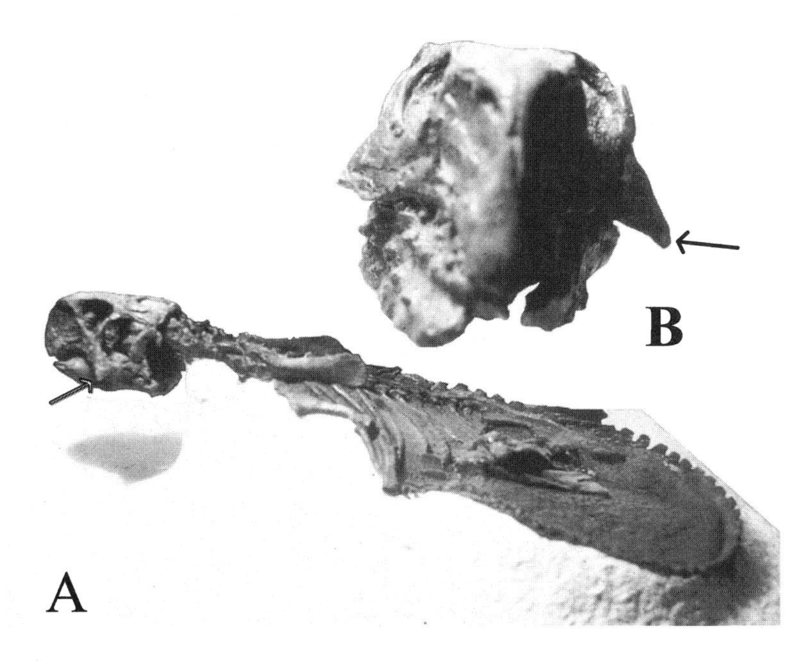

Fig. 4.28. A, Psittacosaurus *in side view showing the cheek spike (arrow). B, skull in front view emphasizing the cheek spike.*

Stegosaurs differ considerably from ceratopsians in their method of display. The most characteristic features about them are the plates and spikes on the neck, back, and tail. These plates can be enormous, as in *Stegosaurus stenops,* where the plate straddling the hip and tail is over 1.3 m (4') tall. In other stegosaurs, such as *Kentrosaurus,* some of the plates are almost spikes. I do not think the plates were used to control body temperature, as has been suggested by Jim Farlow and his colleagues (1976). One reason I disagree is that Farlow and crew based their work on *Stegosaurus stenops,* which has the largest plates of all the stegosaurs. Yet another species, *Stegosaurus armatus,* co-occurs with *Stegosaurus stenops* and is just as big, but has much smaller plates. According to the temperature control model, *Stegosaurus armatus* should have overheated. On the other hand, the wide variety of plates and spikes makes the various species of stegosaurs easily identifiable, which might be important if you are a male in search of a female. A female *Stegosaurus stenops* might not take too kindly to being courted by a male *Stegosaurus armatus.*

Having any kind of structure on the back is bound to make the animal look bigger, and the same was certainly true of the plates of stegosaurs (Fig. 4.29). Any stegosaur presenting itself sideways will look larger than if seen head on. This behavior is similar to a cat arching its back and turning sideways—it makes the cat look bigger and more ferocious. The appearance of the stegosaur might be all the more alluring or intimidating if the plate could blush or change color. Blushing may explain the high vascularization of the bone surface noted by Farlow and others (1976).

Fig. 4.29. Two Stegosaurus *males trying to intimidate one another by turning in profile so as to look bigger.*

The spikes of stegosaurs are obviously an offensive weapon, but they may have also functioned as a nonverbal threat display. Male moose draw attention to their antlers by slowly rocking the head side to side. Perhaps stegosaurs drew attention to their spikes by wagging their tail slowly from side to side. However, it would not look good to intimidate a female by wagging the tail, so perhaps in courtship the tail was kept to one side, away from the female. This would signal a nonaggressive message. Interestingly, several stegosaurs, such as *Huayangosaurus, Tuojiangosaurus,* and possibly *Kentrosaurus,* had a backward-projecting spike on each shoulder. The rearward projection seems to preclude their use as weapons, so perhaps they were important for display. Conceivably these spikes might have been a different color from the body so that they would stand out.

The last group of dinosaurs we'll examine are the ankylosaurs. The nodosaurids are characterized by shoulder or neck spines and no tail club. Besides defense, the spines may also have been used for display. For example, *Sauropelta* has a paired row of neck spines that project up and out from the neck. Viewed from the front, the spines draw the eyes upward, making the animal appear larger. In *Edmontonia* the spines project outward from the body, making the animal look wide and squat in front view. Furthermore, a spike on the shoulder is apparently larger in the males than in the females (Carpenter, 1990b). This spike has a small tine on it, much like on a deer's antler. A male may have locked a spike into the tine of a rival and engaged in pushing contests (Fig. 4.30). The other group of ankylosaurs, ankylosaurids, have no spines on the neck or body, but do have a large club of bone on the tail. Although clearly for defense, the tail club might also have been used by rival males against one another. Finally, both ankylosaurids and nodosaurids have large keeled disks of bone armor

Fig. 4.30. A pair of Edmontonia *engaged in a pushing contest of strength.*

along the neck, back, and tail. These are surrounded by smaller disks in a rosette pattern. It seems possible that the skin covering the keeled armor was colored brightly so that the armor stood out against the background of smaller disks.

Timing of the Dance

What triggered courtship among dinosaurs? The only clue we have for this is what we can learn from living animals. For example, the Komodo lizard engages in courtship almost year round, while actual mating only occurs during three months (Auffenberg, 1981). The lengthy courtship may be due to the male and female mating for life, or because the near lack of seasons in the tropics allows many animals to have extended courtship and breeding periods (Auffenberg, 1981). Perhaps dinosaurs that lived near the equator also had very long courtship or breeding cycles. For animals today that live in mid to high latitudes, courtship and breeding occur within more confined time limits. These animals must mate and rear their young before temperatures drop too low for the young or, for herbivores, before highly nutritious vegetation becomes unavailable. In other animals the lengthening of daylight, or photoperiod, during the spring stimulates the urge to mate. In some birds, the lengthening of the photoperiod coupled with male courtship stimulates hormones in the female, triggering nest building, copulation, and egg laying.

The photoperiod may also have been important for dinosaurs that lived above the Arctic Circle during the Late Cretaceous. At that time, high-latitude temperatures were much warmer than they are now, so that there were no ice caps on either pole. Nevertheless, the poles had six months of darkness during the "winter" months. During the times of darkness, plants could not photosynthesize and hatchling plant eaters would have had little to eat. It was important then for egg-layers to time the hatching in the "spring" when new vegetation was available for the hatchlings. This requires that the parents court, mate, and lay eggs while it was still dark or at least early dawn.

With wider and higher temperature zones during the Mesozoic, perhaps the best model for dinosaur breeding cycles is that of modern animals living in low latitudes, but well above the equator. In many of these animals, the urge to breed is controlled primarily by rainfall (Duvall and others, 1982). For example, alligators and crocodiles engage in courtship just before the rainy season gets under way, with incubation and hatching occurring at the height of the rains (Fig. 4.31). Hatching at the height of the rainy season ensures that the young have abundant food—insects for crocodiles and birds, and nutritious new plant growth for herbivores. If the rains are delayed, such as during a drought, some lizards don't reproduce. It seems very likely that dinosaur reproduction in the middle latitudes was also controlled by rainfall. One thing I am certain of is that considering the diversity of animal courtship and aggressive behavior today, we can't imagine all the different behavioral ways of the dinosaurs.

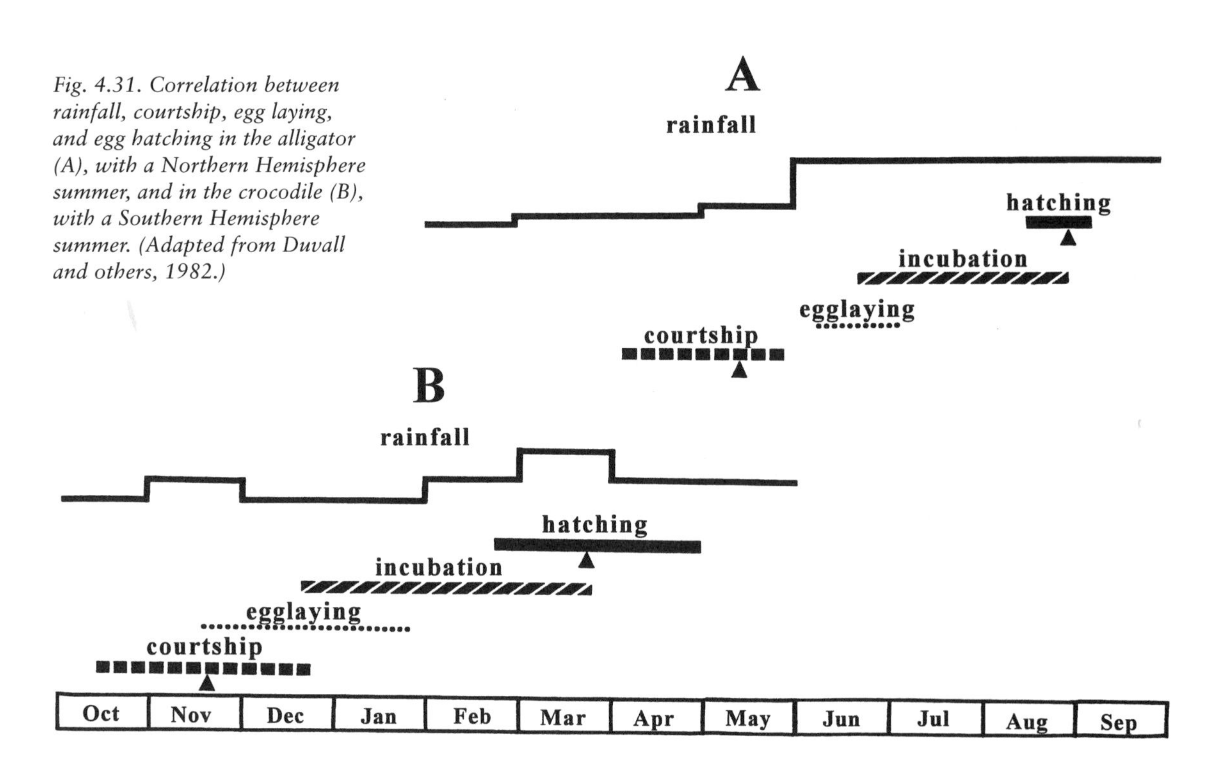

Fig. 4.31. Correlation between rainfall, courtship, egg laying, and egg hatching in the alligator (A), with a Northern Hemisphere summer, and in the crocodile (B), with a Southern Hemisphere summer. (Adapted from Duvall and others, 1982.)

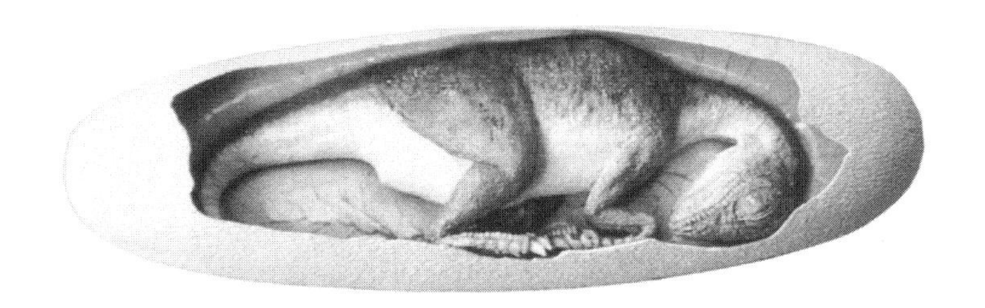

5 • How Dinosaurs Did It

Face it, having gotten this far, you want to know how a pair of three-ton *Stegosaurus* did it—probably like porcupines: very carefully (Fig. 5.1). Actually, we can reconstruct in general terms the how of dinosaur mating, although we do not have any of the genital tissue. Our clues come from comparing the reproduction of living reptiles and birds, and from defective or pathological dinosaur eggs (bizarre, but true).

The Gonads

The gonads (where sperm or eggs are produced) of both birds and reptiles are confined within the body (Fig. 5.2) so it is safe to assume the

Fig. 5.1. Stegosaurus *mate like porcupines—very carefully. (Drawing by Luis Rey.)*

same was true of dinosaurs. With nothing hangin' in the breeze, the male of birds and reptiles can be difficult to identify when there are no outward physical features such as showy plumage (Fig. 5.3). The only certain way is to look inside the cloaca, but only during certain times of the year. Because reproduction is usually seasonal, the gonads change size considerably and typically enlarge just before sperm or eggs develop (Fig. 5.4). So, the most obvious way of sexing reptiles and dinosaurs (assuming you had a time machine) would be to pull or pry open the cloaca. Now, assuming you were stupid enough to sneak up under a *T. rex* and pull the cloaca open, the last thing you would ever see during the last moments of your life would be a penis if it was a male, probably similar to that seen in a crocodile. It is doubtful that it had a hemipenis, which is a paired or forked penis (Fig. 5.5) seen in lizards and snakes. (If you should ever find out which *T. rex* has and live to tell about it, please let me know.) The penis and hemipenis have a groove along the top side for the sperm to travel along.

Fig. 5.2. Gonads in birds and reptiles are internal (female top, male bottom).

Fig. 5.3. Sexual dimorphism can be expressed in size difference (which potentially can be seen in fossils) or outgrowths of the skin, such as scales and plumage (which may or may not fossilize). Here, the female chicken (left) is smaller and has less ornate plumage and comb than a male (right).

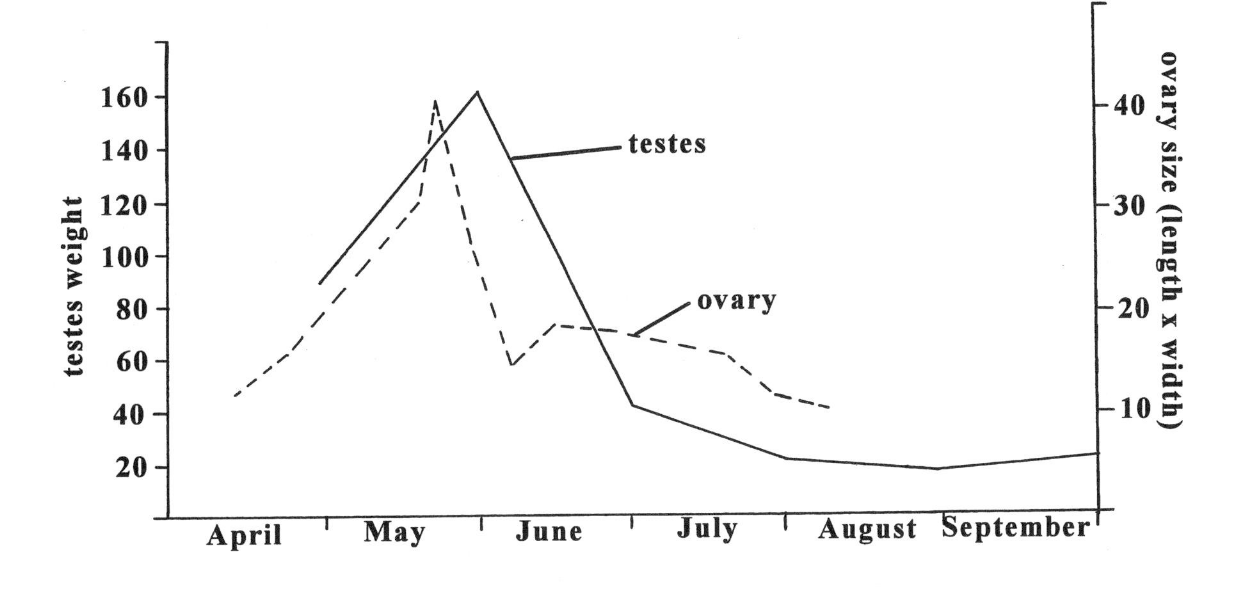

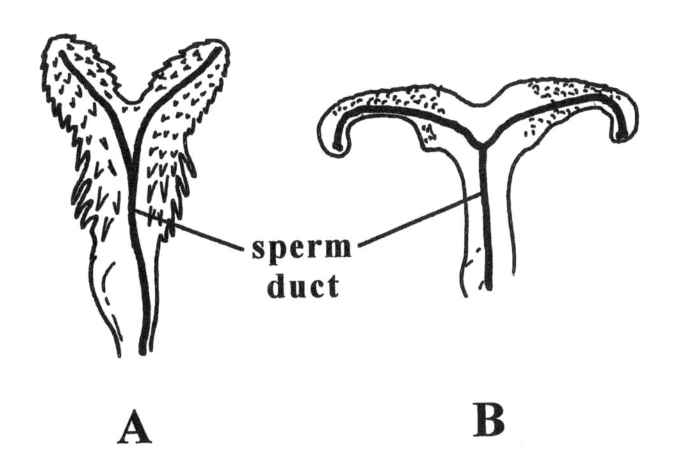

Fig. 5.4. Changes in the size and weight of the alligator gonads. The maximum size and weight peak almost at the same time. (Modified from Joanen and McNease, 1989.)

Fig. 5.5. Hemipenis of the swamp snake Bothrolycus *(A) and arboreal lizard* Plica *(B). The penis of reptiles and birds is different from that of mammals in that the sperm does not travel down an enclosed urethra tube, but along a deep groove. (Adapted from Dowling and Duellman, 1978.)*

Most male birds, however, lack a penis (although some do have the organ). Instead, the male and female invert their cloacae and bring them together and sperm is transferred—sort of like kissing cloacae. At present, we do not know whether the male dinosaurs had a single penis (most probable), a hemipenis (doubtful), or no penis at all. But using what we know about the evolutionary placement of dinosaurs between crocodiles and birds, it seems very probable that advanced theropods, such as the feathered *Sinosauropteryx,* had avianlike gonads, whereas the less avianlike dinosaurs (most dinosaurs) had more crocodilian-like gonads.

Although there is no direct evidence to support my ideas about the gonad types in dinosaurs, there may be indirect evidence based on pathological eggshell. By pathological, I mean eggshell that is not normal, usually because there are multiple layers of eggshell. Birds produce and lay one egg at a time (more on this in the next chapter). Sometimes, the egg may be forced by muscle contractions back up the oviduct where eggs are formed and shelled. This returning egg may meet another egg that has not yet had the shell deposited around it. These two eggs move down the oviduct to where the shell-producing glands are located. A shell is formed around both eggs, resulting in an egg within an egg, a condition called ovum in ovo (Fig. 5.6A). Sometimes the resulting egg within an egg is so large that it causes the death of the female. Usually, however, the egg is laid, but the embryos die (Romanoff and Romanoff, 1949). On rarer occasions, the egg may stall in the shell gland and get a second shell laid over the first, resulting in a double-shelled egg (Fig. 5.6B). The embryo within dies because the pores of the two layers do not match up, suffocating the poor thing.

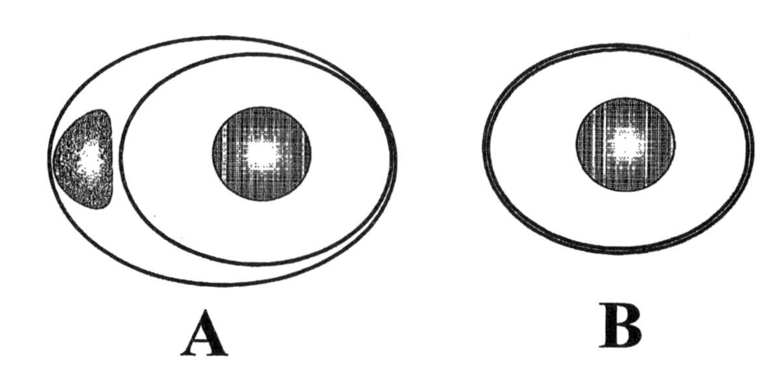

*Fig. 5.6. Examples of pathological eggs: A, ovum in ovo (egg within an egg) seen in birds, but not reptiles; B, multi-shelled egg commonly seen in reptiles, and less commonly in birds; C, multiple, onion-layered turtle (*Geochelone*) eggshell as seen under magnification (courtesy of Darla Zelenitsky and Karl Hirsch).*

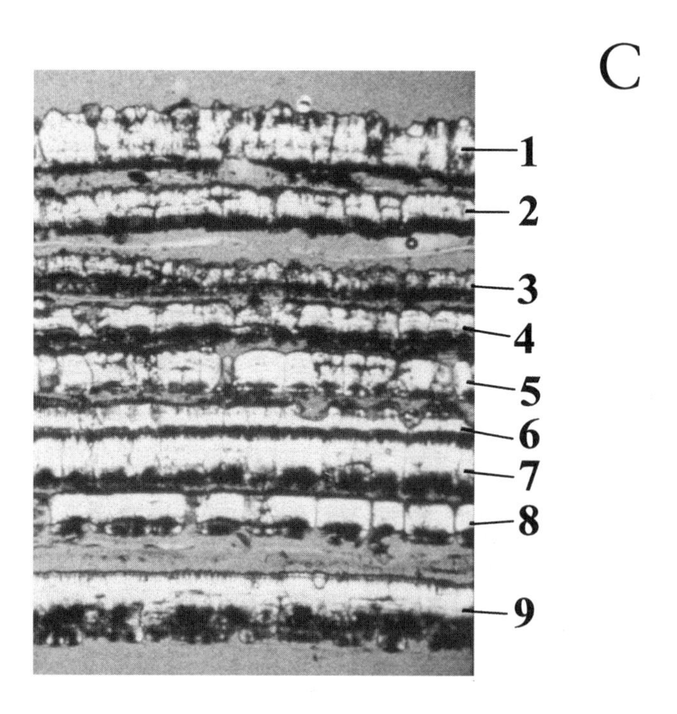

Double-layered eggshell is more common in reptiles than in birds because the eggs are produced in rapid succession but are held in the oviduct until they are laid together in a clutch. Sometimes the last egg or two formed may be forced back up the oviduct while the clutch is being held in the body. However, instead of being enclosed in a shell with another egg, the egg picks up one or more layers of shell that make it resemble an onion (Fig. 5.6C). The embryo invariably dies because the pores of the different layers seldom line up and no gas exchange can occur (Hirsch, unpublished ms).

Among dinosaur eggs, the multiple shell condition has been found, but not ovum in ovo—yet. Multiple-shelled eggs are known for a wide variety of dinosaur egg types (Fig. 5.7; Hirsch, unpublished ms), especially the big, spherical *Megaloolithus*-type eggshell. The multilayers indicate that either the egg stalled in its descent through the shell gland or was pushed back into the shell gland by contractions of the muscles lining the oviduct. Regardless, another whole eggshell was deposited over the egg, and sometimes more. Because multiple eggshell is the most common pathology seen in reptile and many nonavian dinosaur eggshells, we can infer that many nontheropod dinosaurs had a reptilian-type reproductive system, rather than avian. If true, then the male dinosaur probably had a single penis like the crocodile.

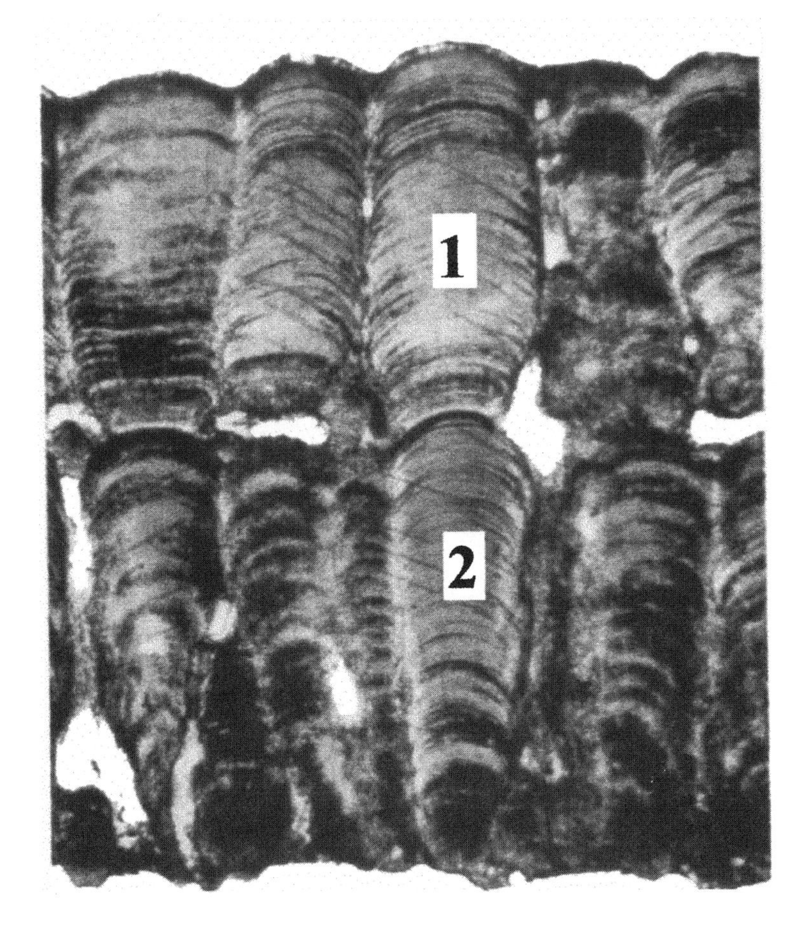

Fig. 5.7. Example of a multilayered dinosaur eggshell, Megaloolithus. *(Courtesy of Darla Zelenitsky.)*

Doin' It

Although many dinosaurs probably had gonads similar to crocodiles, we cannot use crocodile copulation to understand dinosaur copulation because crocodiles do it in water and dinosaurs on land. In water, crocodile bodies are buoyed and the male is able to depress his tail to insert his penis into the female's cloaca. It seems doubtful that dinosaurs mated in water, especially since some of them, such as *Protoceratops,* lived among sand dunes. Perhaps the large, land-dwelling Komodo lizard is a better model. The male of this lizard twists his tail beneath the female until their cloacae meet and the hemipenis is inserted.

This lizard model of sex in dinosaurs was advocated by the late Beverly Halstead (Fritz, 1988). Unfortunately, many male dinosaurs would have had a difficult time with lizard- or snake-style tail twisting (Fig. 5.8). This would be especially true of bipedal dinosaurs that used the stiff tail as a counterbalance in walking. Among most ornithopods, rodlike structures called ossified tendons occur along the spines of the back and tail. These effectively made the back and tail rigid and kept the tail from drooping. They would also have kept the male from depressing or twisting his tail to copulate.

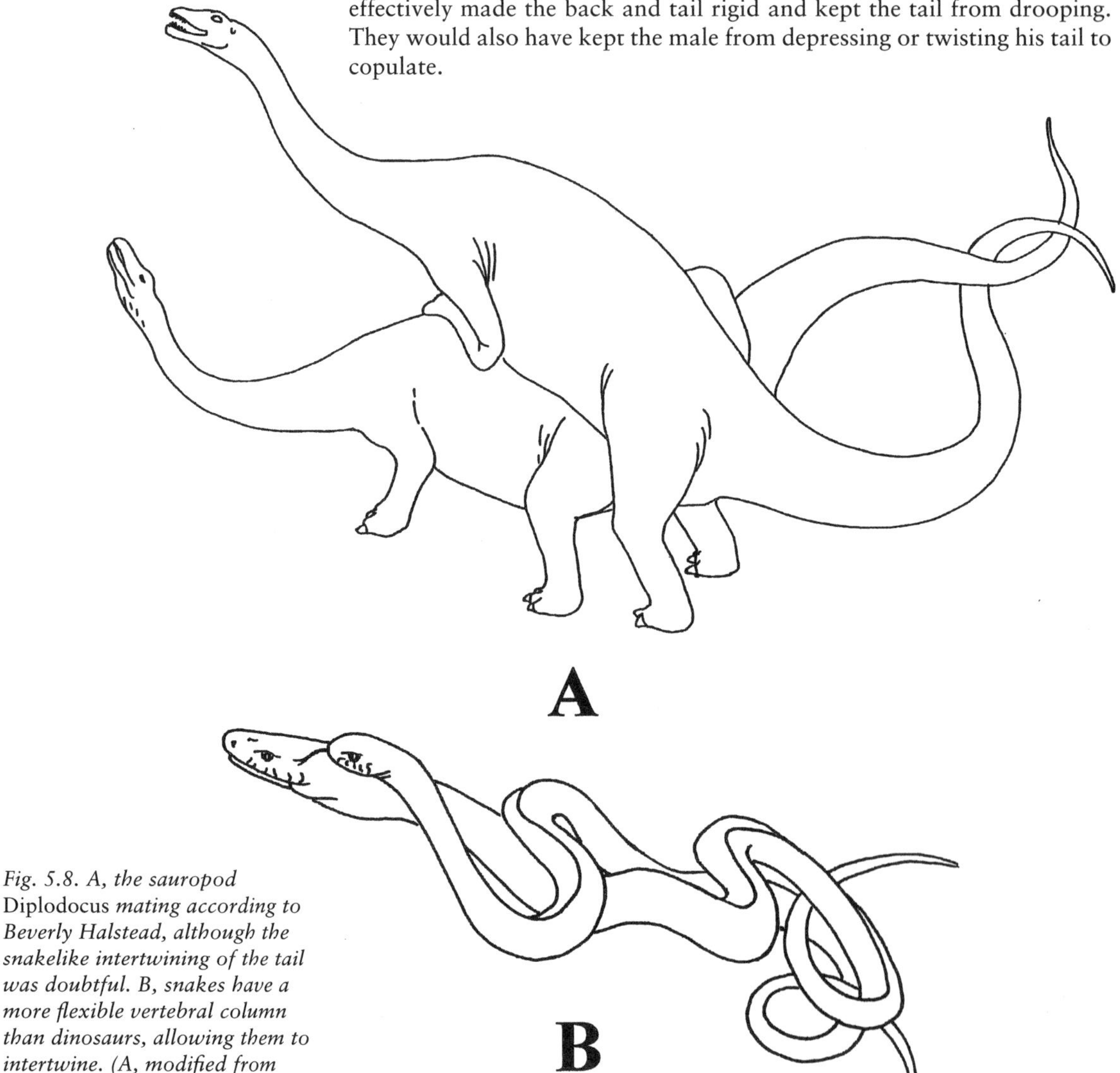

Fig. 5.8. A, the sauropod Diplodocus *mating according to Beverly Halstead, although the snakelike intertwining of the tail was doubtful. B, snakes have a more flexible vertebral column than dinosaurs, allowing them to intertwine. (A, modified from Fritz, 1988.)*

How, then, did dinosaurs do it? It probably varied among the different dinosaurs, although a common method might be for the female to squat on her forelimbs, raising her rear into the air (sort of like a house cat). In this position, with the tail off to one side, the cloaca is well exposed. The male could mount her from behind to one side and support himself with his forelimbs on her back (Fig. 5.9). The male's tail could easily be moved to one side so as to not press into the ground. I certainly don't think it was possible for them to get into the weird contortions of twisted bodies and twisted tails envisioned by Halstead. The anatomy of the skeletons, especially of the back and tail, won't allow for it.

Fig. 5.9. Most quadrupedal dinosaurs, like these two Euoplocephalus, *probably mated like cats. (Sketch by Luis Rey.)*

How *Stegosaurus,* with the big plates on the back, managed to have sex is really not that difficult. Again, with the female squatting in the front while standing on her hind legs, the male could easily rest his forelimbs on one side of her broad pelvis. More difficult to imagine is how the 150' long, 100+ ton *Diplodocus*-like sauropod *Amphicoelias* mated. Perhaps when the animals grew to that size, they had already passed their reproductive years.

Although the dinosaur tail does not seem capable of bending downward for mating, an upward bend was possible in small theropods. A specimen of *Velociraptor,* complete from tip of nose to tip of tail, shows the tail bent sharply upward. Perhaps then, sex was accomplished by the female's raising her tail to a similar angle. This would expose the cloaca to the male from the rear.

Mating in dinosaurs may not have consisted of "quickies." The male Komodo lizard makes several partial attempts to mount (Auffenberg, 1981). Each time the male grasps the skin of the female's neck and he repeatedly touches the base of the female's tail with his hind foot while mounted until the female raises her tail. If the female is not ready for mating (maybe she has a headache), she won't raise her tail at the male's prompting. As a result, mating actually occurs over a span of several days. We don't know if any species of dinosaur behaved in this fashion, but it's fun to speculate. . . .

When Dinosaurs Did It

When dinosaurs had sex was probably determined by the length of the incubation period of the eggs. As stated before, most dinosaurs probably mated seasonally so that the young hatched at optimal times for survival, usually at the beginning of the rainy season for herbivorous species. If true, then the gonads probably changed size seasonally also. Furthermore, gonad development in dinosaurs may also have been environmentally controlled as it is in crocodiles and many birds. Crocodiles are able to sense environmental conditions and adapt hormone production accordingly (Ferguson, 1985). Thus sperm and egg production is environmentally controlled, with mating occurring at the peak of sperm production. This adaptation allows the American alligator to delay mating if, for example, a severe drought is rapidly drying the water holes. For the Nile crocodile, nesting also can be delayed, and when the eggs are eventually laid they are in an advanced stage of embryo development. Either adaptation would have been important for those Cretaceous dinosaurs living in the arid environments of central Asia.

How frequently dinosaurs mated probably varied as it does among modern egg-laying animals. Small lizards and birds in the tropics may breed year round, whereas large birds with a long reproductive cycle (e.g., eagle vultures) breed every other year (Immelmann, 1971). Even crocodiles don't breed every year. In a given group, only 60 percent may breed (Lance, 1987), partially because of population pressures (not the same as peer pressure). When local populations are high, fewer females will lay eggs, thereby reducing the number of mouths feeding on the local resources. In birds, competition for the same food resource may cause delay of breeding in some species (Immelmann, 1971). From these observations in modern animals, it seems reasonable to assume that not all female dinosaurs of reproductive age bred every year. This was probably especially true of hadrosaurs and other herding species.

Dinosaurs of India faced the greatest reproductive challenge. Whereas most continents drifted east-west, India drifted northward toward the equator. For 100 million years India was attached to the northern coast of Antarctica (this was long before there was ice) between 30° and 45° latitude (Smith and others, 1994). The dinosaurs were therefore adapted for reproduction in the mid-latitudes. But beginning around 130 m.y.a., India split off and began drifting northward. By the end of the Cretaceous, 65 m.y.a., India was approaching the equator. During the entire time of the drift Indian dinosaurs were having to slowly adjust to more tropical conditions, and to adjust their reproductive cycle. Were they successful in making the changes, or might their failure to adjust have led to their extinction? Considering that all dinosaurs apparently became extinct at around the same time worldwide, it seems likely that reproduction was not the problem. We'll discuss this issue of extinction further in chapter 13.

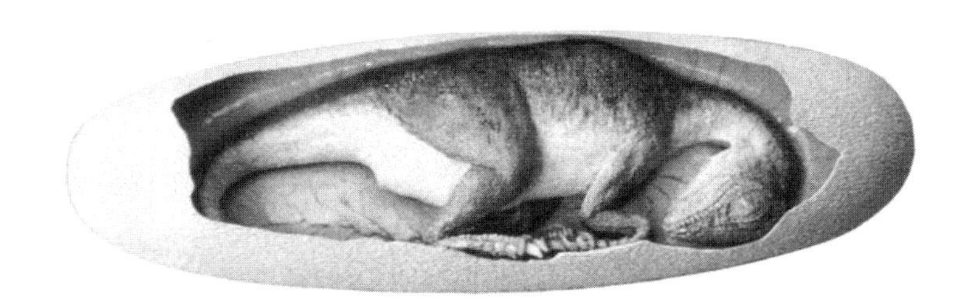

6 • Making an Egg

Before we discuss dinosaur eggs in any detail, we need to understand what eggs are all about. Because dinosaurs are extinct, the only way to understand their eggs is to look at how eggs are formed in two major shelled egg-layers today—birds and reptiles (mammalian egg-layers, monotremes such as the platypus, are not important to our story). Bird eggshells, especially those of chickens, have been so well studied (for economic reasons) that they are used as the standard for comparisons of eggshell. Let's first look at how a chicken egg is produced, the various parts of an egg, and then how a reptile egg is formed. This background will help us understand dinosaur eggs in the next chapter.

Bird Eggs

The formation of an egg takes a long time, involving many steps. All of it occurs in the oviduct, which is actually five organs in one (Fig. 6.1). The inner lining of the oviduct houses glands that secrete various parts of the egg, and the outer surface of the oviduct has muscles to push the egg along in waves of contraction from one set of glands to the next. The raw material for egg production begins with the processing of food by various organs in the body. If the diet lacks an important mineral or vitamin, the result can be unhealthy or pathological eggs. For example, copper deficiency results in abnormally large, weak eggs (Burley and Vadehra, 1989).

The cell called an oocyte that eventually becomes a chicken originates in the ovary. Two ovaries are present in the chick, but only one ever matures. This one ovary is packed with 600–700 microscopic oocytes. In theory, every one of these could become an egg, but this seldom happens before old age stops egg production. When the female reaches maturity, egg formation begins when a group of these oocytes start to mature. For the first sixty days, maturation is slow, but then it speeds up for the last six to ten days. During this time, the ovary begins to look like a bunch of grapes of various sizes (Fig. 6.1). The first structure of the egg, the yolk, is deposited as the oocyte matures. The yolk provides nutrition for the embryo and antibodies for protection against disease. The yolk may appear to be a homogeneous mass, but in reality is composed of alternating white and yellow layers (Fig. 6.2). Microscopically, the yolk is made up of large and

Fig. 6.1. *The left oviduct of a bird. The ovary is where the egg starts. It then descends through the magnum where the egg white is deposited, then on to the isthmus for the feltlike shell membrane. Over this membrane the shell is deposited in the uterus. The egg then exits through the cloaca.*

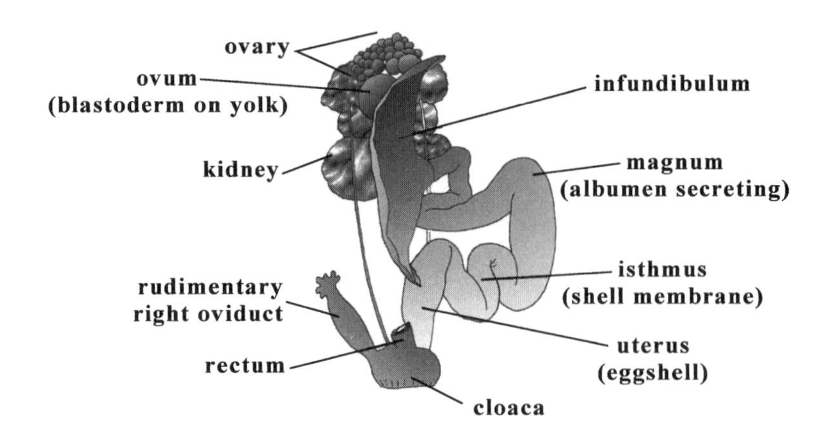

small particles, and particles within particles. These particles are protein, fats, and phospholipids (Burley and Vadehra, 1989). Yolk size relative to egg size depends upon whether the chick is altricial (hatched naked and underdeveloped) or precocial (hatched with feathers and well developed). Generally, altricial birds take less time to incubate before hatching than precocial birds, and so don't need large yolks to feed on. This allows the females to lay proportionally smaller eggs. The trade-off, however, is that the chicks hatch blind, naked, and helpless (more on this in chapter 11).

Once the yolk is formed, the egg moves to the infundibulum, a funnel-

Fig. 6.2. *Detailed structure of the bird egg and eggshell.*

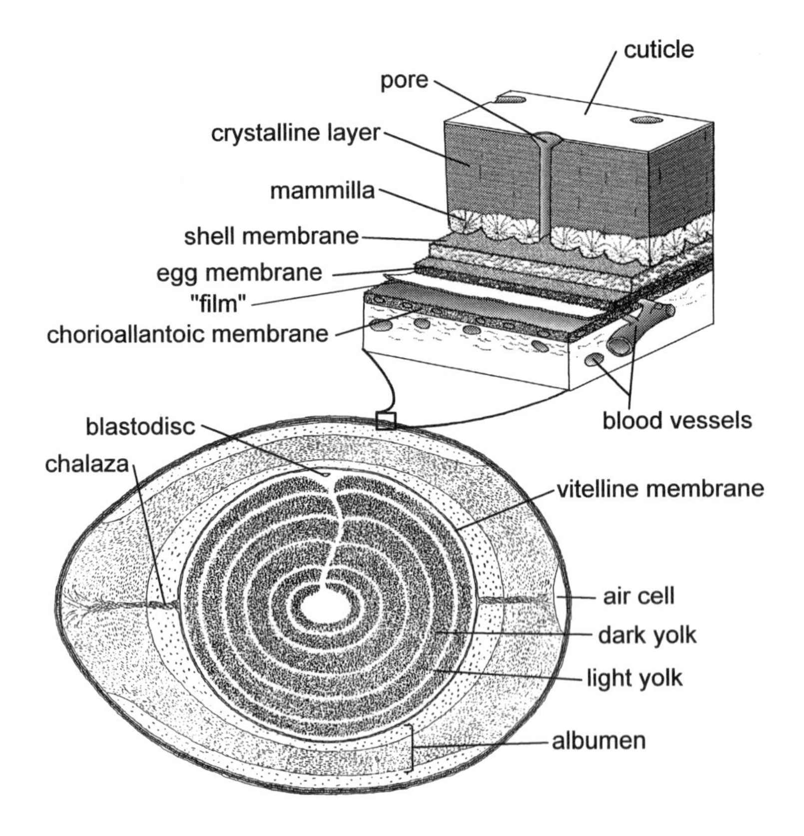

like structure, where fertilization occurs. After fertilization, the oocyte begins to divide until a small whitish disk of cells is produced. (We'll discuss embryo development in chapter 11.) This disk, called the blastoderm, begins to differentiate into the embryo. In the infundibulum, the egg also gets the first of several membranes (Fig. 6.3). The vitelline membrane forms an outer coat to the yolk, thereby keeping it intact (the membrane is what makes it possible to separate the yolk from the egg white when you're making hollandaise sauce). The vitelline membrane is primitively the outer layer of fish and amphibian eggs. Also added in the infundibulum is the chalazal layer of the albumen. The chalaza—the white ropelike structure on each side of the yolk—isn't distinctive until the egg has moved farther down the reproductive duct. Its purpose is to keep the yolk from drifting around inside the egg and to keep the embryo right-side-up. This entire process of fertilization and membrane production takes about fifteen to thirty minutes.

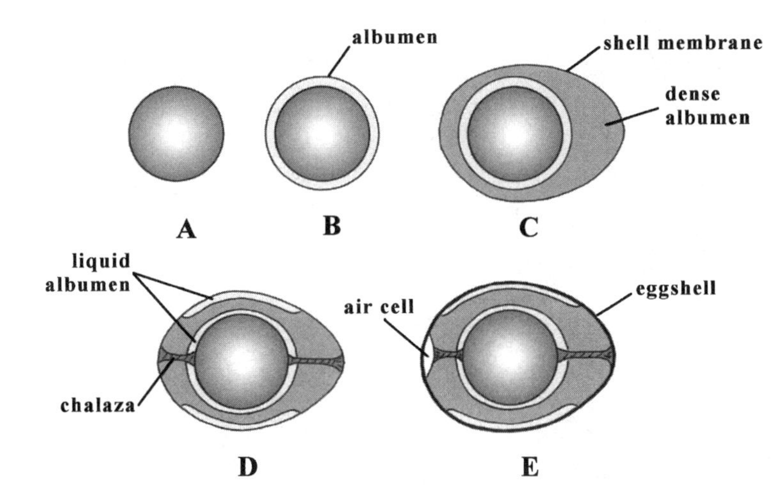

Fig. 6.3. Stages in the formation of the bird egg: A, yolk with blastodisc; B, albumen is added; C, dense albumen and shell membrane added; D, liquid albumen added and chalaza begins to differentiate; E, air cell and eggshell.

The egg is next pushed into the magnum, or albumen-secreting region. The albumen is an elastic, insulating gel with a high water content (Fig. 6.3). It acts like a shock absorber cushioning the embryo. It also has various antimicrobial chemicals to protect the embryo from disease organisms. The albumen is secreted in concentric layers of thick and thin consistency around the yolk in about two to three hours. Several membranes within and around the albumen keep the fluids of the albumen intact. Once the last membrane is in place around the albumen, the egg moves to the isthmus, where the shell membrane is produced. In the isthmus, the albumen absorbs a great deal of water, a process called plumping, thereby increasing the volume of albumen. As the egg continues through the isthmus, it stimulates secretion of protein fibers that form two thin, felt-like coats; this takes about an hour. The inner or egg membrane surrounds the albumen, while the coarser outer or shell membrane lines the inside of the shell (Fig. 6.4). Together, the inner and outer membranes strengthen the shell, as well as determining the eventual shape of the egg (round, tapered, or elongated—Fig. 6.5). The weave of both membranes is loose enough to allow respiratory gases to pass to and from the embryo, yet the weave of the inner membrane is tight enough so that the albumen doesn't seep out.

Fig. 6.4. SEM showing the feltlike shell membrane. The membrane has many functions, including keeping the contents of the egg from leaking out the pores, providing growth sites for the shell units, and keeping bacteria out. (Courtesy of Karl Hirsch.)

Fig. 6.5. A comparison of a spherical turtle egg, elongated crocodile egg, and tapered bird egg.

The last part of the egg to be formed is the shell. Shell production occurs in the uterus and takes eighteen to twenty-four hours. Although composed predominately of calcium in the form of calcium carbonate, the shell also contains calcium phosphate and magnesium carbonate. Prior to the formation of the shell, there is a rise in the dissolved calcium level of the mother's blood. This calcium is obtained partially from the mother's bones (30–40%), and the rest from her diet. In the uterus, calcium is removed from the blood by secretion glands. Shell production isn't a one-step process; it requires several sequential steps, each producing a distinct zone within the shell.

Shell production starts slowly, with numerous small granules of calcium first appearing within the outer part of the shell membrane forming the zone of basal plates (Fig. 6.6). These granules, called eisospherites or basal plate groups, are composed of irregular plates of calcite (Fig. 6.7). A small sphere of organic material, called the primary spherite or organic core, is often deposited above the eisospherite. Its purpose may be to serve as a focal or nucleation point for the start of shell deposition. Initially, calcite crystals grow as tiny spikes radiating in all directions from the organic core. These spikes fuse together into plates of calcite that resemble flower petals. Several of these petals are stacked within one another like a partially open bud, forming the secondary spherite. The secondary spherite is seen with a polarizing microscope as an indistinct dark mass. The secondary spherites, located on the surface of the shell membrane, collectively form the zone of radiating calcite crystals.

The next layer that forms is the wedge zone, which incorporates the secondary spherites (or zone of radiating calcite crystals) at its base. Under the scanning electron microscope (SEM), each wedge is actually made of tabular crystals that resemble flagging stones. Because of their appearance, they are referred to as having a tabular ultrastructure (ultrastructure is the molecular level and is seen under the SEM, the most powerful of all microscopes). Groups of radiating wedges grow mostly in an outward direction and form cuplike structures that look like breasts (well, with a lot of

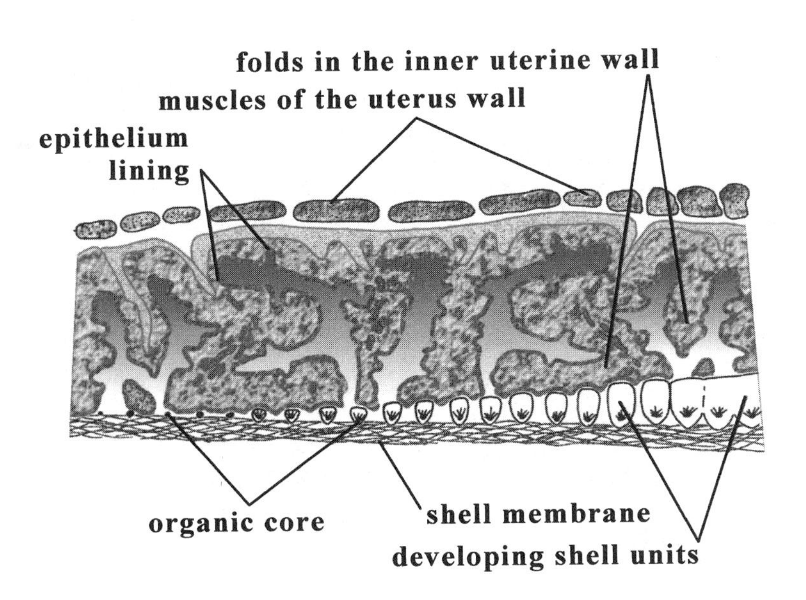

Fig. 6.6. Formation of the eggshell by the uterus, site of the shell glands. The organic core is deposited on the shell membrane where it acts as the nucleation site for the shell. The shell is shown schematically as forming assembly line fashion, but in reality probably forms sequentially—first organic cores form on the shell surface, then the shell units form.

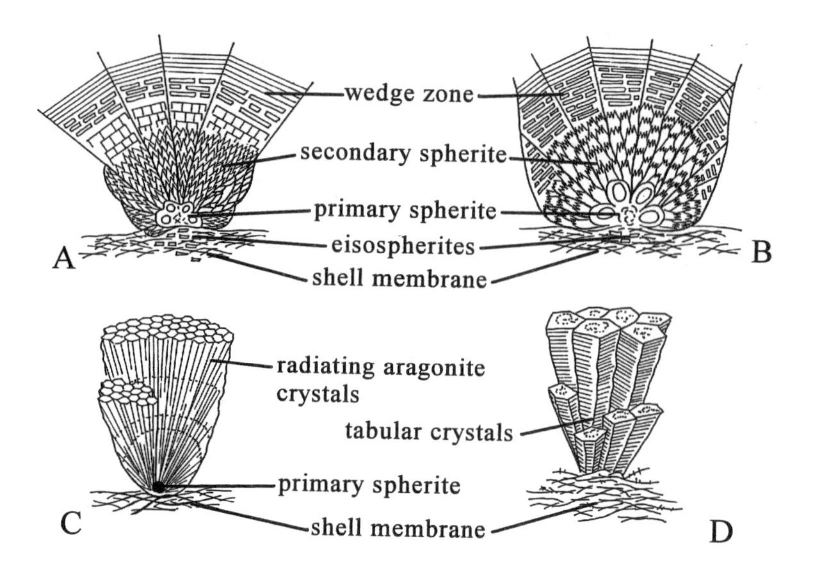

Fig. 6.7. Structure at the base of eggshell: A, emperor penguin (neognathous bird); B, ostrich (paleognathous bird); C, hard-shelled turtle; D, crocodile. (A and B modified from Mikhailov, 1987b; C and D modified from Mikhailov, 1992.)

imagination they do), hence the name mammilla (singular) or mammillae (plural) (Fig. 6.2); the mammillae are sometimes called basal caps (guess because it sounds polite). As each mammilla grows, it eventually meets its neighbors. Growth from this point on is restricted to an outward direction, perpendicular to the shell membrane, because of the closely abutting, adjacent shell units (Fig. 6.6). The shell units form a layer variously called the continuous layer because of its uniform appearance, or the crystalline layer because the faces of the calcite crystals can be seen on the broken surface. However, in turtles, crocodiles, and many dinosaurs, the shell units remain distinct and this layer is called the column layer or prismatic zone, depending on the author (Fig. 6.7). In birds, the outermost portion of the continuous layer may have very distinct, vertically oriented calcite crystals. This layer is called the external zone and is missing in some bird species.

The shell units are the basic building blocks of the eggshell. The shell unit consists of a column or prism of calcite with a mammilla at its base (Fig. 6.2). The column is in turn composed of tall, polygonal calcite prisms. Each prism is composed of small calcite units, 10–15 microns across, surrounded by organic material (Fig. 6.8). The SEM shows the appearance of the prisms to resemble the skin of a lizard, hence the name squamatic layer (Fig. 6.9). If the calcite is dissolved away, the remaining organic material looks like a sponge, hence the alternative name, the spongy layer. The term spongy layer has been applied to the shell above the mammillae, although, as Chao and Chiang (1974) have noted, the term actually refers to an organic residue. Long organic fibers may also be deposited during the formation of the columns, especially near mammillae. These fibers help strengthen the shell by connecting calcite crystals. The development of the shell units varies considerably, being best developed in large ground birds, or ratites, such as the ostrich, and less developed in the chicken. In some eggs, the shell units cluster and the suture between adjacent units is obliterated through fusion (Fig. 6.10).

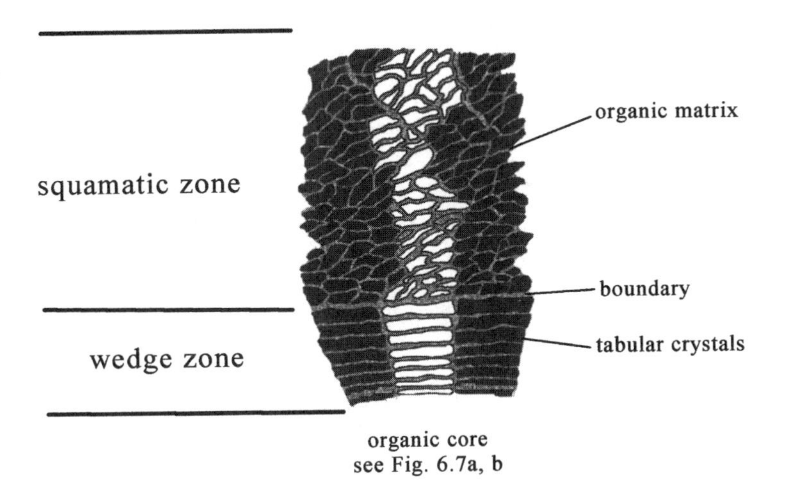

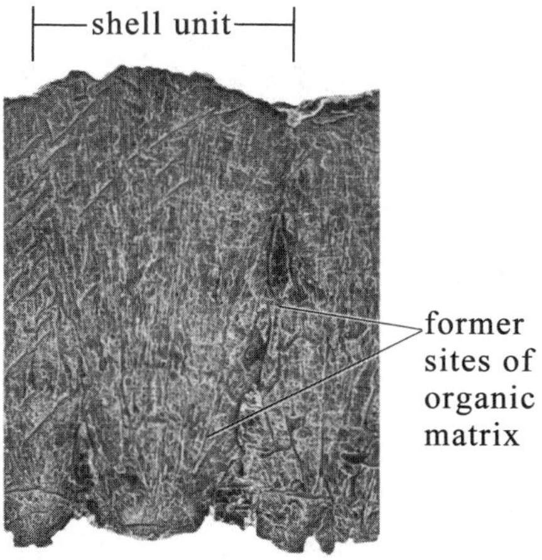

Fig. 6.8. *Schematic representation of the eggshell ultrastructure seen through polarized light resulting in light and dark bands of extinction. At the base, the calcite crystals are tabular and form the wedge zone. Above this, the calcite crystals are less organized and are slightly overlapping (imbricating), forming the squamatic zone. The calcite crystals might be thought of as bricks that are held in place by a cement of organic matrix formed by proteins. (Modified from Mikhailov, 1988.)*

shell unit

former sites of organic matrix

Fig. 6.9. *SEM of an eggshell (*Maiasaura*) showing the squamatic ultrastructure. The numerous pits and streaks are former sites of organic matrix. (Courtesy of Karl Hirsch.)*

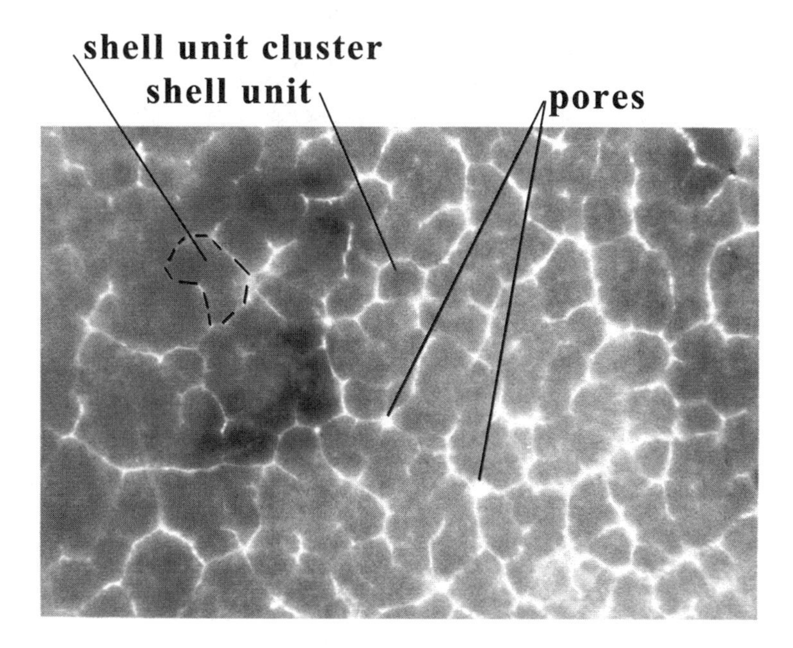

Fig. 6.10. *Thin section parallel to the surface of a crane eggshell showing the shell units just above the mammillae. Note how some shell units cluster. (Courtesy of Karl Hirsch.)*

Randomly interspersed among the shell units are the pore canals, which connect the surface of the egg with the network of air spaces between mammillae. The pore canals allow the embryo to breathe (Fig. 6.11). Gas exchange (oxygen in, carbon dioxide out) is by diffusion, meaning that the lower oxygen level within the egg pulls the higher oxygen level outside the egg to the embryo. Even so, the amount of oxygen that eventually reaches the embryo is less than at the surface of the egg (Fig. 6.12). The oxygen supply declines because the narrow diameter of the long pore canal restricts air flow, because the inner shell membrane acts as a barrier, and because the oxygen is diluted with more and more carbon dioxide and water vapor with depth.

The rate of gas diffusion is controlled by the shape of the pore canal (straight or curved), as well as its narrowest point (Fig. 6.13). Because the oxygen needs of the embryo differ among bird species, the number, size, and shape of the pores on the surface of the shell vary considerably (Fig. 6.13), as they do in dinosaurs (more in chapter 8).

Fig. 6.11. Simple experiment showing the porosity of the eggshell. Air is injected into an empty chicken egg.

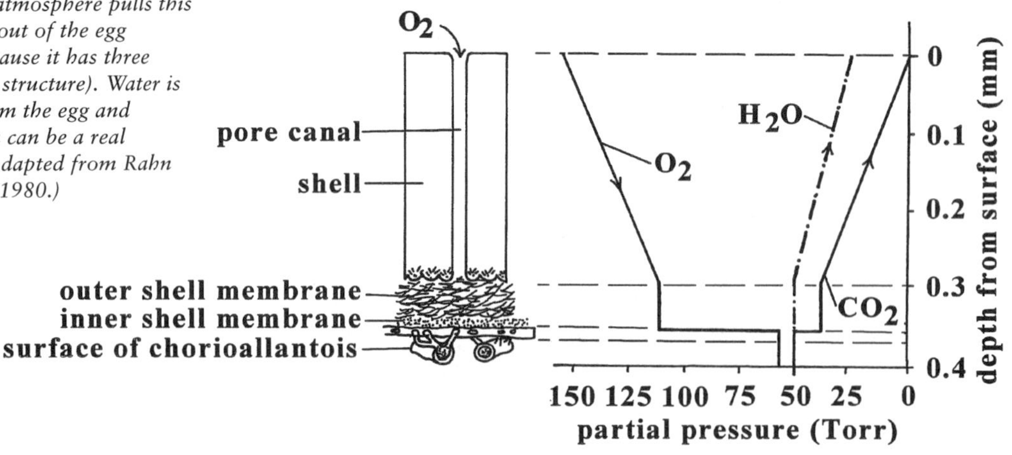

Fig. 6.12. The amount of oxygen available to the embryo is less than the amount of oxygen in the air because the various parts of the shell act as barriers to diffusion. Oxygen is passively pulled into the egg because of its lower pressure within the egg. As can be seen on the right, the amount of oxygen (O_2) slowly decreases from around 150 torr to around 110 torr as it travels down the pore canal (left). The oxygen passes unhindered through the shell membrane, but its pressure drastically drops to about 55 torr as it is pulled into the blood capillaries at the surface of the chorioallantois. The lower partial pressure of CO_2 in the atmosphere pulls this heavier gas out of the egg (heavier because it has three atoms in its structure). Water is also lost from the egg and dehydration can be a real problem. (Adapted from Rahn and others, 1980.)

Fig. 6.13. The size and shape of pores vary considerably in birds (and dinosaurs, as discussed in chapter 8). Many are simple straight tubes, but others are forked. A, ostrich; B, rhea; C, swan; D, goose; E, kiwi; F, chicken; G, tinamou; H, robin; I, hummingbird. Scale along left side is shell thickness in mm. (Adapted from Rahn and others, 1980.)

How pores form is a mystery that has yet to be solved (Fig. 6.14). One idea is that threads of protein fiber extend upward from the shell membrane at the site for each pore. Once the shell is deposited, the thread shrivels or dissolves leaving the pore behind (Stewart, 1935, in Board and Sparks, 1991). It is difficult to explain pores that fork or are interconnected within the shell by this method, and furthermore the existence of such threads has not been substantiated by detailed microscopic work. Another idea involves the presence of fluids between the growing shell columns that keep shell from being deposited (Tyler and Simkiss, 1959). This idea was elaborated upon by Board (1982), who suggested that the fluid is magnesium-rich. Russian egg paleontologist Konstantin Mikhailov (1987a) advocates the magnesium-rich fluid model, but conceded he does not understand how the fluids would prevent shell deposition. Yet another suggestion for pore formation (Richardson, 1985) is that pores form where tiny folds of the shell gland region of the uterus occur between the developing shell columns. Although this might explain elongated pores, it is difficult to explain the circular cross-section of some pores. At present no adequate explanation is forthcoming. We'll return to pores again in the next chapter.

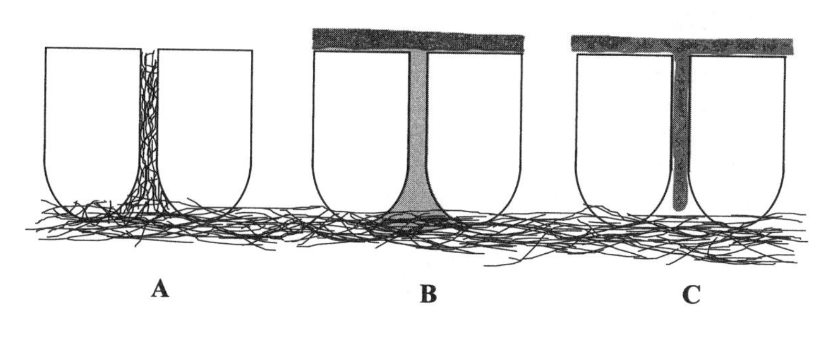

Fig. 6.14. The three models for pore formation. A, Fiber Model: bundles of protein fibers extend upward from the shell membrane. The shell is deposited, and then the fibers dissolve, leaving the pore canals. The problem with this model is that no traces or impressions of the fibers have been found in the pore canals as would be expected. B, Fluid Model: fluids from the uterus (above) keep pore canals open as the shell is formed. Once the egg is laid, the fluids evaporate, leaving the pore canals. The problem with this model is that it is difficult for fluid channels to produce the complicated ostrich shell (see Fig. 6.13A). C, Fold Model: folds in the uterus keep what will become the pore canal open. The most obvious problem with this model is that it also cannot produce the complicated pore system of the ostrich egg (Fig. 6.13A).

On the shell surface, the pores appear as minute oval to circular openings (Fig. 6.15). Generally there is a rough correlation between the number of pores and the number of mammillae. The greater the number of pores, the smaller and more numerous the mammillae, but this relationship is not rigid. Although most eggs laid by a single individual have almost the same number of pores per cm^2 of shell surface, eggs of a flock typically show variation in these numbers. In contrast, there is up to 10 percent variation in the size of the pores among eggs laid by a single female. Even the distribution of pores on an egg varies, with fewer pores on the pointed end. On average, the egg of a domestic chicken has about 150 pore openings per cm^2.

One problem of pores is that they can allow microbes to enter. To block these, the pores often contain a filter of protein fibers. Some of these fibers have allegedly been found in the pores of dinosaur eggs. Oddly, among some species of birds, some of the pores may be partially or completely plugged with inorganic material, including calcite crystals, or be capped with organic or inorganic spheres. Naturally, this reduces the effectiveness of these pores in respiration.

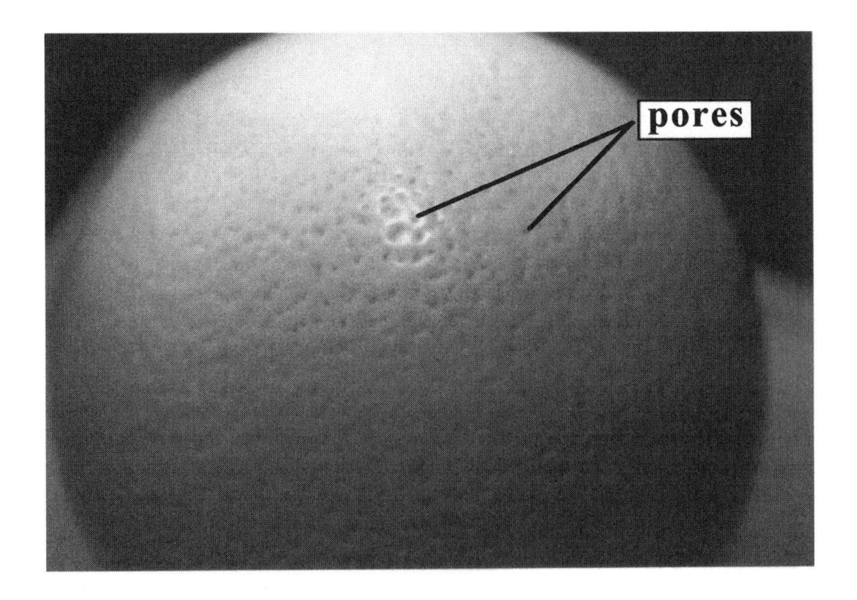

Fig. 6.15. Pore openings on the surface of an ostrich egg. Note that many of them form rows, which interconnect within the shell (Fig. 6.13A).

The surface of the eggshell can vary considerably among different birds, with the appearance sometimes controlled by diet. Trace elements from food, such as strontium, iron, manganese, and magnesium, may be incorporated into the calcium as it is being deposited in the outer layer. These minerals can give the shell a distinctive color, such as the pale blue color of a robin's egg or the black of an emu's egg. Some pigmentation camouflages the egg in the nest, but other pigmentation helps emit excess heat or absorb heat from the sun. The surface of the egg may also be glossy or dull, smooth or rough, glassy or chalky, pitted or granulated. The chalky appearance is sometimes due to a thin layer of calcite. This calcite generally lacks any internal structure and sometimes cannot be distinguished from the top of the underlying columnar layer. The glossy texture is often due to a very thin, transparent protein coat on the surface called the cuticle. It is absent on the eggs of some birds (such as the gull) and thick on others (such as the cassowary). The purpose of the cuticle is not well understood because it is so variable in appearance and because it is not present in all species. Most suggestions for the cuticle lean toward some sort of protection, such as preventing dehydration of the egg in an arid environment, sealing out disease microbes, or resisting dissolution of the shell by soil acids. The problem with these suggestions is that the cuticle is not restricted to birds that lay their eggs in an arid environment, or on the ground where soil acids are present. If the cuticle keeps out microbes, then why don't all eggs have it? With some eggs, the cuticle also contains granules of color pigment. These granules give the eggs color spots, flecks, blotches, streaks, or fine lines. Sometimes, there are granules of calcium in the cuticle causing the shell to have a rough texture.

Once the outer surface of the egg has formed, the egg passes into the cloaca. From here the eggs pass out and are laid one at a time. Usually only a single egg is laid per day (roughly the time it takes an egg to be formed). The timing of when the first egg is laid can be controlled by the hen. This keeps the eggs from being laid when conditions are not favorable for survival, such as during a drought.

Reptile Eggs

Let's now look at eggs in reptiles. As with bird eggs, there is a tiny embryo and a large mass of yolk at the time of laying. The egg provides the developing embryo with its nutrients, whereas the nest environment determines the incubation temperature, and how much water and gas there is for exchange. The nesting temperature is very important because it often determines the sex of the embryo, as we shall see in chapter 11. Reptile eggs have the same amniotic membranes, the amnion, chorion, and allantois, as do those of birds. These membranes differ in thickness among the various types of reptiles. Reptile eggs do not, however, have a chalaza to anchor the yolk. Instead, the yolk is suspended in a thicker albumen and the embryo is attached to the shell membrane.

Once the yolks are formed, the eggs in a clutch pass into the infundibulum, where each acquires albumen (Fig. 6.16). The proportion of albumen in turtle, crocodile, and bird eggs is roughly proportional to egg size, but is much less in lizards. This may be due to the lizard embryo being more developed at the time of egg laying. Once the albumen is formed, the eggs move into the uterus, where they are wrapped first in the shell membrane, then in the shell. The formation of the shell membrane and shell differs considerably among various living reptiles. In the tuatara, the shell membrane and shell are formed almost at the same time in the uterus. In lizards with a semi-rigid eggshell, there is a slight lag, with the shell membrane first being formed and then the shell over that; all of this occurs in the same part of the uterus. In crocodiles, the shell membrane is laid down first in the anterior part of the uterus, after which the egg is moved along to the posterior part where the shell is deposited (Palmer and Guillette, 1992). The anterior uterus in crocodiles is the precursor to the isthmus of birds.

Slight differences in where and when the membrane and shell are produced in the tuatara, lizard, and crocodile seem to reflect different evolutionary stages of the uterus. If this is true, then the multipurpose uterus of the tuatara evolved from a single-purpose uterus of the earliest amniotes, in which only the shell membrane was produced. The next stage in evolution may be seen in lizards, where a slight time delay occurs be-

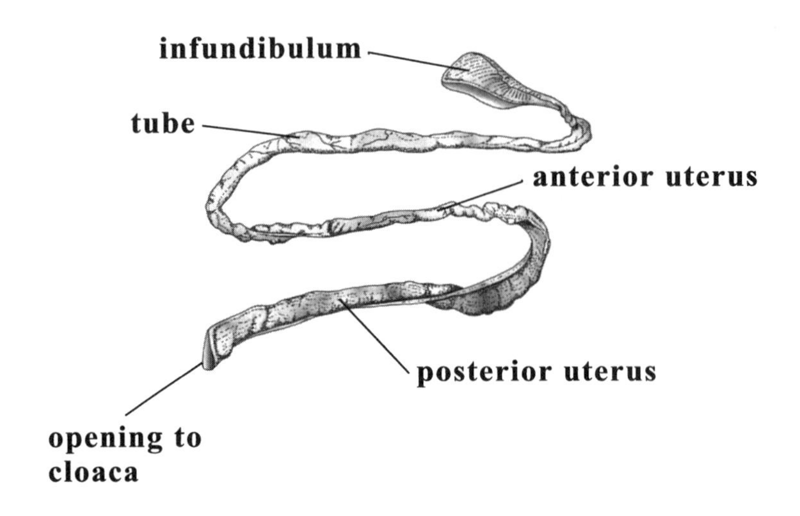

Fig. 6.16. Oviduct of the crocodile—a less complicated structure than that of the bird oviduct (Fig. 6.1).

tween when the shell membrane is formed and when the shell is deposited. It is difficult to know which extinct group of reptiles had such a uterus. Eventually, the glands for the production of the shell membrane and the shell were separated to opposite ends of the uterus. The advantage of this separation is that each egg of a clutch could be treated rapidly in assembly-line fashion. (And you thought Henry Ford was the first to use the assembly line!) We don't know when this innovation first appeared or how widespread this type of uterus was among extinct reptiles. We do know that birds retain an assembly-line uterus, with the upper part of the uterus called the isthmus (this change actually reflects the terminology of a structure, rather than what the structure does). Furthermore, rather than shell a clutch of eggs in rapid succession, birds send only one egg at a time through the system, presumably as an adaptation for flight (more eggs, more weight). Because dinosaurs bridge the gap between crocodiles and birds, it seems a safe bet that they had a differentiated uterus like a crocodile or bird, rather than a multipurpose uterus like a lizard or tuatara. The big question is whether the eggs were shelled a clutch at a time like a crocodile, or singly like a bird. We'll come back to this question later.

The thickness of the shell relative to egg size is roughly similar in crocodiles, turtles that have a hard-shelled egg, and precocial birds. Eggshell thickness is greater than that seen in altricial birds or turtles with a parchment shelled egg. The multiple layers seen in bird eggshell (mammillae, continuous layer) do not occur in non-dinosaurian reptiles. For example, in hard-shelled turtle eggs, the shell is composed of what are comparable to the secondary spherites of the bird egg. The shell also differs from bird eggs in that it is made of aragonite, a different form of calcium carbonate. In some lizards, the shell consists of only a thin crust of irregular-shaped grains of calcite overlying the shell membrane. Sometimes the grains are arranged into a flower or rosette pattern, and other times are randomly scattered on the shell membrane. Shell columns may be present, but these are irregular in shape and distribution. They appear more like pillars supporting a roof of crusty shell. There are no mammillae at the base of the columns. Instead, the columns intermingle with the fibers of the shell membrane. The gecko is unusual among lizards in that the eggshell structure is organized into columns (see Fig. 8.16A). The columns are irregular in shape and are made of tabular calcite crystals. They have no mammillae; rather, the ends of the shell unit are jagged. The pores are scattered and irregular in size.

In the tuatara, the eggshell consists of circular columns that extend to varying depths into the underlying shell membrane. Above the columns is a layer of granular calcite of different sizes, shapes, and orientation. The surface is very rough and irregular, giving the shell the appearance of being deeply weathered. In many ways, the columns and granular calcite layer are similar to that seen in some lizards, which is not surprising given their close evolutionary relationship. The eggs of crocodiles (Fig. 6.7D) are composed of what is thought to correspond to the wedge zone in the mammillae in a bird shell. The calcite crystals are tabular and arranged horizontally into fan-shaped wedges that are tightly packed against one another.

Dinosaur Eggs

With dinosaur eggs, we are limited in what we know because about all that remains is the hard shell. We assume that they had the major amniotic membranes, the amnion, chorion, and allantois, found in both birds and reptiles. As yet these structures have not been found fossilized, but that is not to say that no soft tissues are known from fossilized eggs. As long ago as 1954, Chinese paleontologist Minchen Chow reported on traces of cuticle on the surface of dinosaur eggs from Laiyang, China. A similar report was later made by Sochava (1971) for dinosaur eggshell from Mongolia. Zhao and Li (1988) illustrate what they believe to be fossilized shell membrane adhering to the bottom of *Dendroolithus* eggshell, while Terry Manning has found what he believes to be globules of yolk in an egg from China (Fig. 6.17). The spheres are of different sizes and suggest that they may indeed be fat and/or lipid globules. As yet no chemical analysis has been done on these remarkable structures to verify their identity. There was actually one previous report of yolk, by Chung-Chien Young in 1954. He noted that an uncrushed egg cut open had a yellowish-tint calcite infilling (Fig. 6.18) and speculated that the yellow was from the yolk. It now seems more likely that the tint was due to barite,

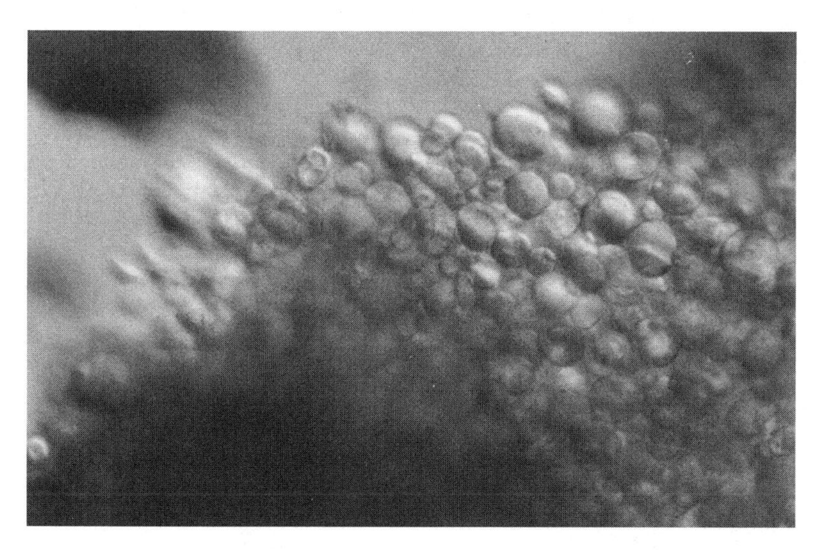

Fig. 6.17. Microscopic view of possible fossilized yolk from an egg from China. The spheres resemble the lipid and fat globules seen in modern eggs. (Courtesy of Terry Manning.)

a mineral precipitate. Still, the fossil record has produced many amazing examples of soft tissue, including the impression of embryonic sauropod skin (Chiappe and others, 1998), so who knows what awaits discovery in an uncrushed dinosaur egg?

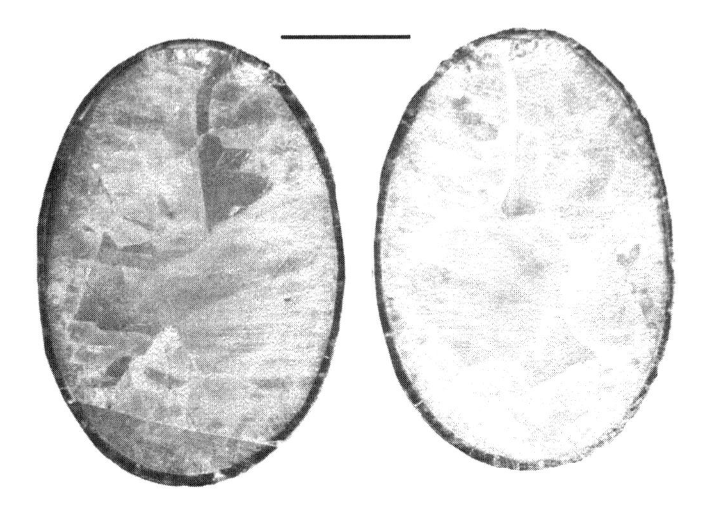

*Fig. 6.18. Calcite-filled dinosaur egg (*Ovaloolithus*). The calcite has a yellowish tint, which led Chinese paleontologist C. C. Young (1954) to suggest that the infilling* might *represent fossilized yolk. Such an interpretation is now considered doubtful.*

As we shall see in later chapters there is a considerable amount of variation in the structure of dinosaur eggshell—more than in any living group today. Shell units are visible in non-theropod dinosaur eggs (Fig. 6.19A, B), but not in those of theropod dinosaurs (Fig. 6.19C), which are like bird eggshell. The shell units are somewhat parallel-sided and taper toward the bottom into cones or mammilla-like structures. Mikhailov (1997) argues that these structures are not true mammillae because they don't have a differentiated ultrastructure. Instead, the ultrastructure is a continuation of the ultrastructure making up the shell unit. This difference can be illustrated by comparing Figure 6.19A with 6.19C. Note the distinct layer (mammillary zone) along the bottom of Figure 6.19C (theropod eggshell) and its absence in Figure 6.19A (sauropod? eggshell). Figure 6.19B (ornithopod) has faint boundaries separating the shell units and no distinct mammillary layer, so is more like Figure 6.19A than Figure 6.19C. We'll look at dinosaur eggshell in much greater detail in chapter 8.

Egg Size and Shape

Size

Reptile and bird eggs come in a variety of sizes and shapes. On average, the egg is between 1 percent and 10 percent of the female's body weight in birds. Egg size in birds also depends on whether the chick is precocial or altricial (although remember that there is a continuous spectrum between these two end categories). The eggs are largest in precocial

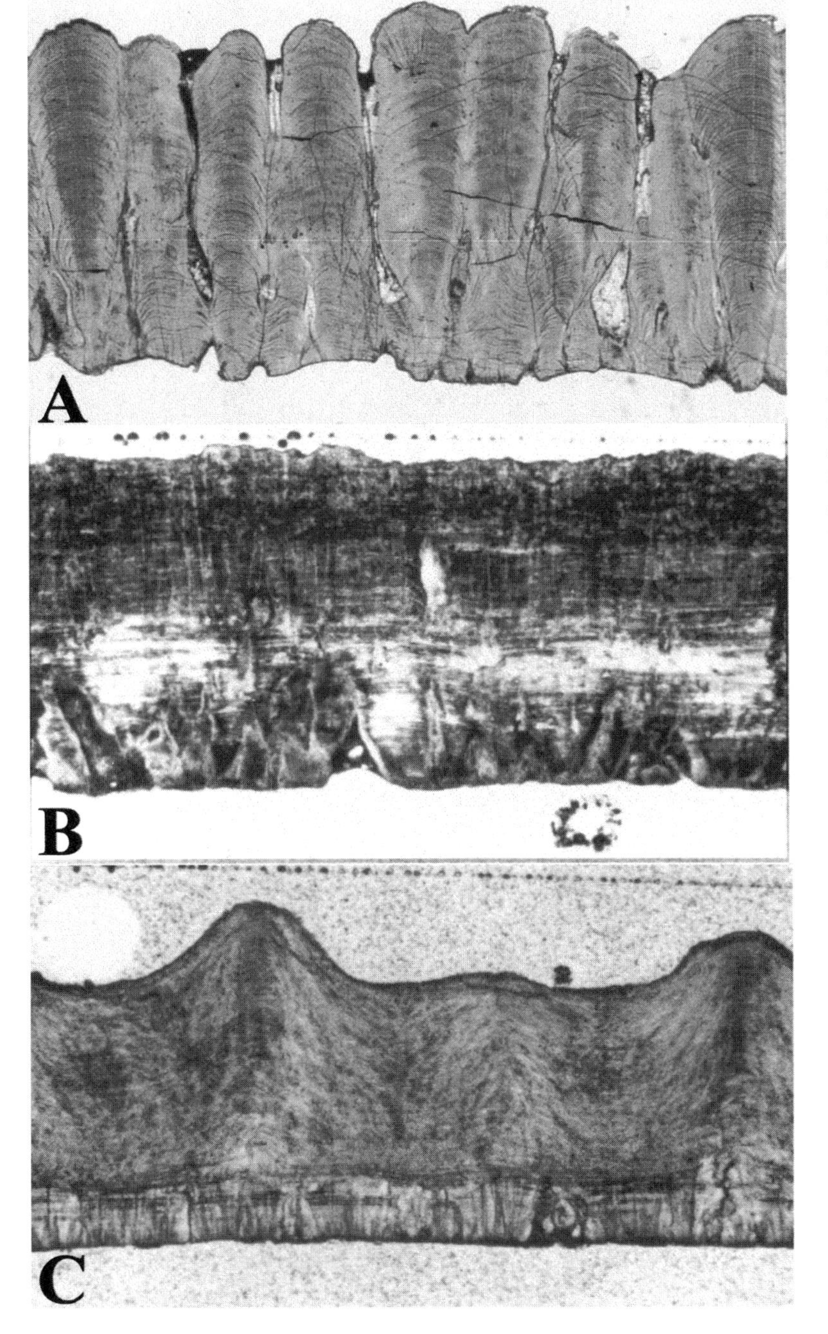

Fig. 6.19. Examples of dinosaur eggshell: A, Megaloolithus *eggshell has very distinct shell units that do not have a differentiated zone at their base; B,* Spheroolithus *eggshell with faint shell units, with poorly differentiated bases; C,* Macroelongatoolithus *eggshell with a continuous layer lacking any traces of the shell units, and with a distinct zone at the base formed by the mammillae. (A, courtesy of M. Vianey-Liaud; B, C, courtesy of K. Mikhailov.)*

birds because the young remain in the egg longer prior to hatching and so need more yolk to feed them. As a result, the number of eggs laid by precocial birds is less than altricial birds. The oddball among birds is the kiwi, which produces a single huge egg weighing about 25 percent of the female's weight; a similar quirk does not occur among living reptiles, nor has it been found as yet among dinosaurs. The maximum size of an egg ultimately depends upon the maximum diameters of the uterus and pelvic opening and the age of the female. Even so, small birds tend to have proportionally larger eggs than large birds. For example, a 3 g (1 oz) hummingbird has a 0.3 g egg, which is 10 percent of body weight. The ostrich, on the other hand, weighs 150 kg (330 lb), and has a 1.6 kg (3.5 lb) egg, which is 1 percent of the body weight (Fig. 6.20, 6.21). The largest bird egg is that of the extinct elephant bird (*Aepyornis*) from Madagascar (Fig. 6.21, 6.22). At 7,300 cm^3, the egg has a larger volume than the biggest known dinosaur egg, *Megaloolithus,* which is about 4,189 cm^3.

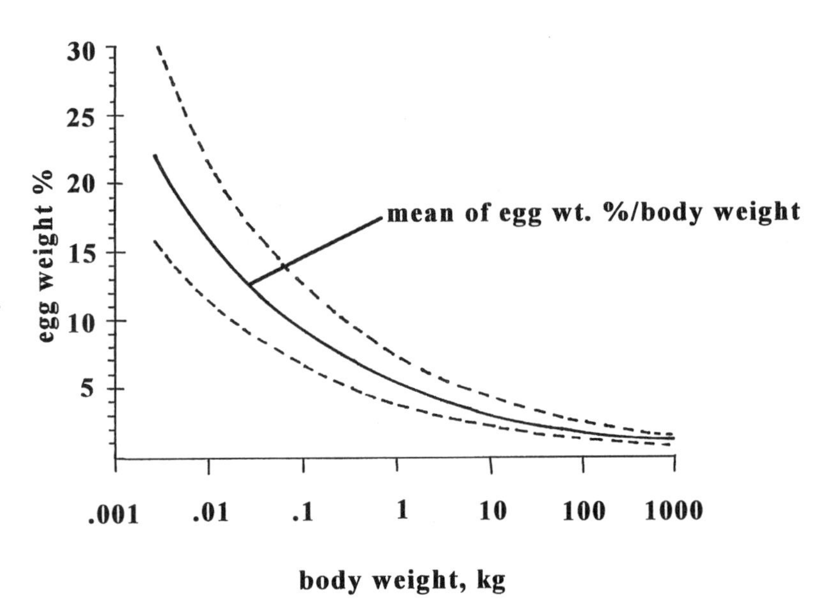

Fig. 6.20. Correlation between body weight and egg weight as a percentage of the body weight in birds. Note that the range of variation (between the dashed lines) narrows, the larger the bird. Part of that is due to there being fewer and fewer species, the larger the bird. (Modified from Rahn and others, 1975.)

Not surprisingly, there is also a correlation between egg size and shell thickness, with larger eggs having a thicker and stronger shell. The reason for this increase is that with larger eggs, there is an even larger increase in volume, meaning that there is more "stuff" for the eggshell to hold in. How much stronger the eggshell is can be measured by gradually increasing the load or force until the egg breaks. The results of such experiments by Ar and his colleagues (1979) resulted in a formula:

$$F = 50.86\ M^{0.915}$$

where F is the crushing force and M is the mass of the egg in grams. Once this formula was determined, the correlation between shell thickness and the breaking point of eggshell became clear, resulting in another formula:

$$BP = T^2(1718)$$

where BP is the breaking point in kg, T is the shell thickness in cm, and 1718 a constant. The advantage of this formula is that the strength of eggshell pieces can be determined. Assuming that unfossilized dinosaur eggshell had the same mechanical properties as modern bird eggshell, we can then use the breaking point formula to predict the breaking point of dinosaur eggshell (Table 6.1; Fig. 6.23). Remarkably, both the ostrich and the extinct elephant bird have eggshell that is thicker and stronger than 90

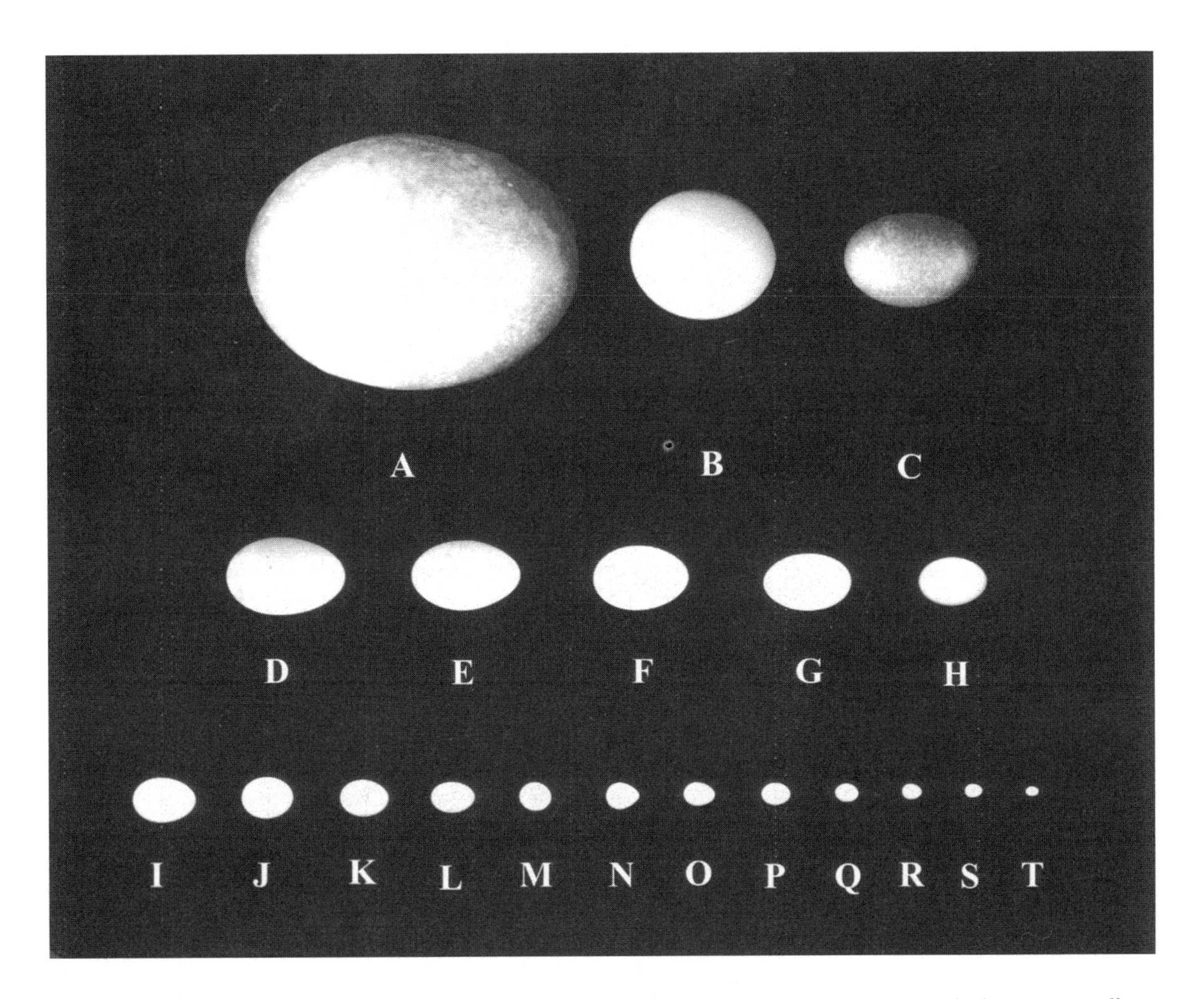

Fig. 6.21. The largest to smallest bird eggs: A, elephant bird; B, ostrich; C, emu; D, albatross; E, swan; F, goose; G, H, mallard; I, hen; J, marsh hawk; K, teal; L, road runner; M, burrowing owl; N, bobwhite; O, flicker; P, woodpecker; Q, lark bunting; R, bank swallow; S, goldfinch; T, hummingbird. (Courtesy Photo Archives, Denver Museum of Natural History.)

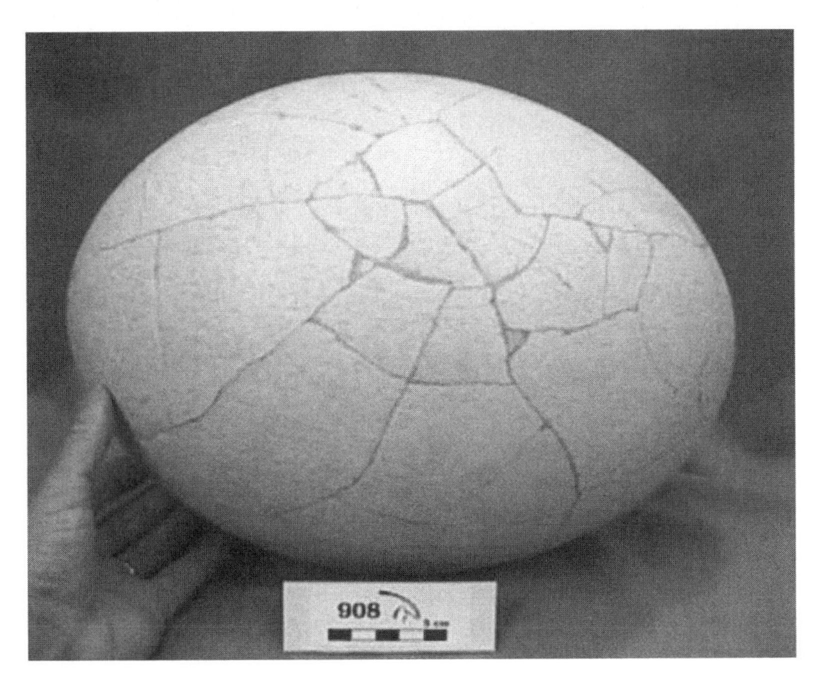

Fig. 6.22. The largest egg known is that of the extinct elephant bird, Aepyornis, *from Madagascar. Scale in cm. (Courtesy of the Stone Company.)*

percent of all dinosaur eggs. Only the largest megaloolithid eggs are stronger (although not larger) than the elephant bird egg.

Shape

Both reptile and bird eggs are either spherical, elongated, or somewhere in between (but no corners, which would be painful to lay). Elongated reptile eggs can be fifteen times longer than wide (in one species of legless lizard), but bird eggs are rarely twice as long as wide (Iverson and Ewert, 1991). Elongated dinosaur eggs are up to three times longer than wide, hence are more elongated than bird eggs, but are much less elongated than some reptile eggs. Elongated reptile eggs tend to be symmetrical in that both ends are the same size and shape. This is not true of elongated bird eggs, where one end tends to be much larger and rounder, and the other end smaller and more pointed (Fig. 6.5). Elongated dinosaur eggs show greater variation, with some being symmetrical like reptile eggs, and some being asymmetrical like bird eggs.

Why the different shapes? Among reptiles, spherical eggs (Fig. 6.5) occur almost exclusively in turtles that lay large numbers of eggs in a clutch (up to two hundred in the green sea turtle). Because eggs are held in the cloaca before being laid, spherical eggs can be more closely packed together than they could if elongated. In contrast, the very elongated eggs occur in long-bodied reptiles, such as snakes and legless lizards. Crocodiles tend to have eggs that are only about twice as long as wide (Fig. 6.5). Egg shape in reptiles is symmetrical, with each end of the egg the same size and shape, because waves of muscle contraction move the eggs along one after another. With each egg separated the distance of one wave of contraction, each contraction is the same, hence each end of the egg is distorted the same amount.

The shape of eggs in birds is due in part to nesting styles, and partly to the uterus' moving only one egg through at a time. Ostriches and other ground birds tend to have almost spherical eggs that are frequently rearranged by rolling. The plover lays pear-shaped eggs that are laid in clutches of four, with the pointed end inward, and then covered. Although some dinosaur eggs are asymmetrical, none reached the extreme seen in the guillemot (Fig. 6.5). This bird lays tapered eggs on narrow rock ledges. If the egg gets away, it tends to move in a tight circle rather than roll off the ledge (O'Connor, 1991a). Elongated eggs in both reptiles and birds also allow for an egg with a greater volume, i.e., a bigger egg, within the constraints of the oviduct and diameter of the pelvis. Why a bigger egg? Because the embryo can develop longer before hatching, thereby increasing its chances of survival. The asymmetrical egg of birds may reflect the smaller, more compact abdominal cavity (Iverson and Ewert, 1991). As each egg is slowly pushed down the oviduct from behind by muscle contraction, the front of the egg is distorted as it pushes the oviduct aside. This shape is retained once the shell membrane is laid over the egg.

But what about egg shape in dinosaurs? Dinosaur eggs may be elongated or spherical, although the length-width ratio of elongated eggs is less than seen in reptiles. Interestingly, most (but not all) elongated eggs have avian-like eggshells and are mostly confined to theropod dinosaurs, whereas the more spherical eggs belong to non-theropod dinosaurs.

Why are some dinosaur eggs elongated in the first place? Polish egg paleontologist Karol Sabath (1991) has examined this question and reached two conclusions. First, that the shape helped the eggs to be deposited in a stable, spiraled configuration. Second, that the shape might also be "better

TABLE 6.1.

Correlation between shell thickness and shell breaking point in dinosaurs and birds (see Fig. 6.23). A, average bird eggshell thickness and B, their breaking point; C, thinnest dinosaur eggshells and D, their breaking point; E, thickest dinosaur eggshell and F, their breaking point. Data for birds from Ar and others, 1979.

bird genus	A (cm)	B (kg)	dinosaur egg genus	C (cm)	D (kg)	E (cm)	F (kg)
Nectarinia	0.0052	0.0464	*Dendroolithus*	0.17	49.6502	0.38	248.0792
Prinia	0.0069	0.0817	*Dictyoolithus*	0.15	38.655	0.28	134.6912
Carduelis	0.007	0.0841	*Faveoloolithus*	0.18	55.6632	0.26	116.1368
Passer	0.0088	0.1330	*Youngoolithus*	0.138	32.7175	0.175	52.61375
Muscicapa	0.0076	0.0992	*Cairanoolithus*	0.15	38.655	0.24	98.9568
Carduelis	0.0068	0.0794	*Megaloolithus*	0.08	10.9952	0.5	429.5
Melopsittacus	0.0116	0.2311	*Sphaerovum*	0.25	107.375	0.5	429.5
Erythropygia	0.008	0.1099	*Ovaloolithus*	0.06	6.1848	0.33	187.0902
Lanius	0.0095	0.1550	*Shixingoolithus*	0.23	90.8822	0.26	116.1368
Passer	0.0102	0.1787	*Spheroolithus*	0.08	10.9952	0.58	577.9352
Galerida	0.095	15.504	*Preprismatoolithus*	0.07	8.4182	0.1	17.18
Pycnonotus	0.0083	0.1183	*Prismatoolithus*	0.024	0.9895	0.12	24.7392
Turdus	0.0122	0.2557	*Protoceratopsidovum*	0.03	1.5462	0.07	8.4182
Streptopelia	0.012	0.2473	*Pseudogeckoolithus*	0.03	1.5462	0.03	1.5462
Streptopelia	0.0132	0.2993	*Ellipsoolithus*	0.12	24.7392	0.16	43.9808
Streptopelia	0.0119	0.2432	*Elongatoolithus*	0.03	1.5462	0.15	38.655
Streptopelia	0.0137	0.3224	*Macroelongatoolithus*	0.2	68.72	0.32	175.9232
Glareola	0.0151	0.3917	*Macroolithus*	0.08	10.9952	0.2	68.72
Falco	0.0194	0.6465	*Nanhsiungoolithus*	0.08	10.9952	—	—
Athene	0.0187	0.6007	*Trachoolithus*	0.08	10.9952	0.12	24.7392
Chlidonias	0.0148	0.3763	*Phaceloolithus*	0.05	4.295	0.07	8.4182
Gallinula	0.0229	0.9009	*Laevisoolithus*	0.1	17.18	—	—
Corvus	0.0177	0.5382	*Subtiliolithus*	0.03	1.5462	0.045	3.47895
Falco	0.0242	1.0061	*Continuoolithus*	0.094	15.1802	0.124	26.4159
Alectoris	0.0278	1.3277	*Dispersituberoolithus*	0.026	1.1613	0.028	1.3469
Himantopus	0.0185	0.5879	*Porituberoolithus*	0.05	4.295	0.065	7.2585
Tyto	0.0241	0.9978	*Tristraguloolithus*	0.032	1.7592	0.036	2.2265
Sterna	0.0171	0.5023					
Nycticorax	0.0205	0.7219					
Bubulcus	0.0204	0.7149					
Egretta	0.0218	0.8164					
Phasianus	0.0308	1.6297					
Burhinus	0.0266	1.2155					
Gallus	0.0295	1.4950					
Strix	0.0268	1.2339					
Larus	0.0231	0.9167					
Ardea	0.0242	1.0061					
Geronticus	0.0394	2.6669					
Gallus	0.0359	2.2141					
Anas	0.0315	1.7046					
Buteo	0.0371	2.3646					
Bubo	0.0349	2.0925					
Ciconia	0.0502	4.3294					
Aquila	0.052	4.6454					
Anser	0.0741	9.4332					
Gyps	0.0067	0.0785					
Struthio	0.2245	86.5876					
Aepyornis	0.38	248.0792					

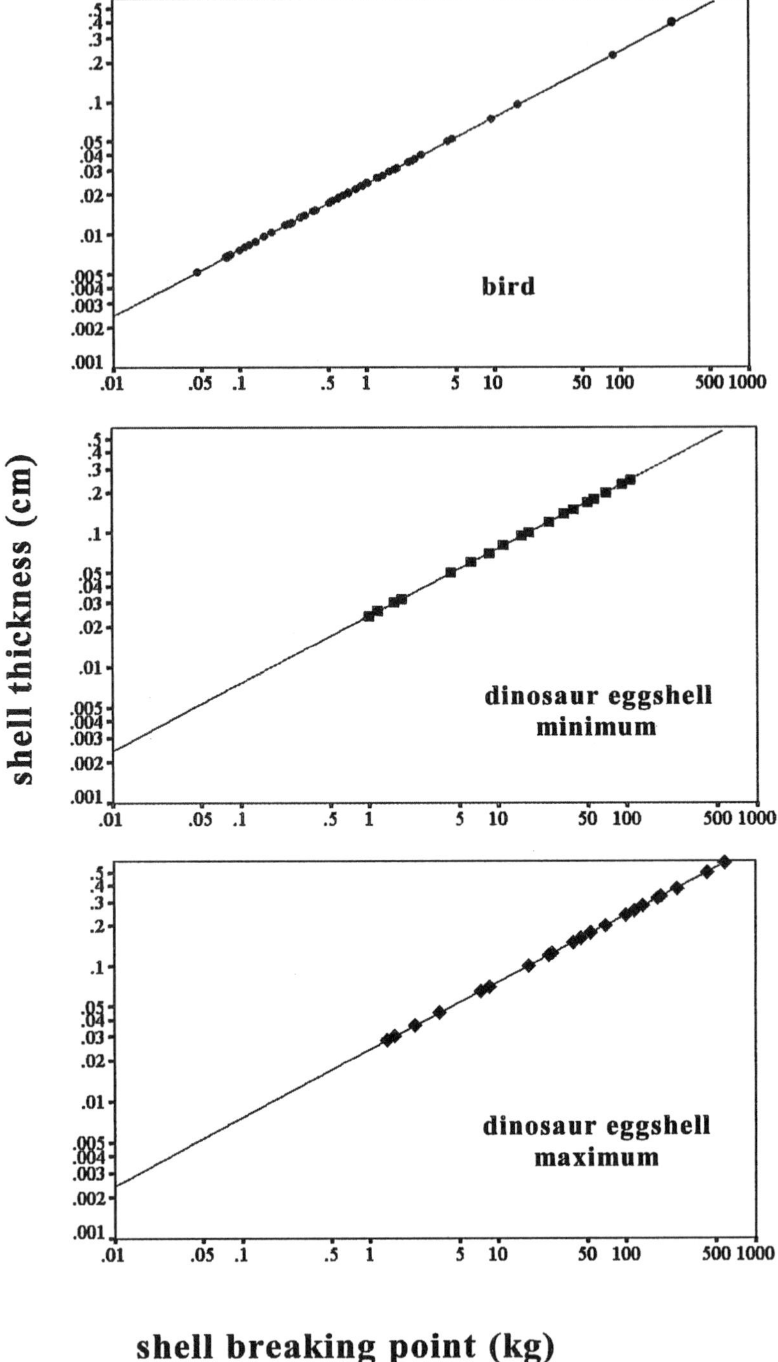

Fig. 6.23. Correlation between shell thickness and shell breaking point for modern birds (top) and dinosaurs (for the thinnest and thickest known shell, data in Table 6.1). Note that the log-log plots produce the same straight line in all three graphs, showing a strong correlation.

for the elongated hatchling" (whatever that means). However, Smart (1991) concluded that the blunt end of the bird egg accommodates the air cell, a feature not seen in reptile eggs, but necessary for the survival of the developing chick embryo (more in chapter 11). Because theropod eggs are the most avianlike in their shell, the elongated shape is to accommodate a similar air cell. The implication is that the modern bird egg was already present in small theropods. Another suggestion about the elongated shape of eggs is from the mathematical breaking point analyses by Zhao Zi-Kui and his colleagues (they were apparently unaware of the study by Ar and others, 1979). One of their studies (Zhao and Ma, 1997) concluded that elongated eggs (e.g., *Elongatoolithus*) were better at resisting crushing the more vertical they stood. But as we shall see in chapter 10, few theropod eggs were ever laid standing upright.

Abnormal Eggs

Thus far, we have looked at the way eggs should be, but sometimes things go wrong. As was briefly mentioned earlier, defective eggs provide us with important clues about dinosaur reproduction. The study of defective dinosaur eggs is still new, but this is not true of modern eggs, especially those of birds. Understanding defects in modern bird eggs, especially those of chickens, is economically important. What these studies have shown is that some defects are congenital, some are nutritional, and others are environmental. Not all egg defects occur in dinosaur eggs, such as DDT thinning of shell, but others can and do occur. It is these that we will look at. It is important that these defects be recognized for what they are, otherwise the fossil eggs might be misidentified or misinterpreted.

A common pathology is ovum in ovo, meaning an egg within an egg. Such eggs have long attracted attention, at least as far back as the days of Aristotle. Romanoff and Romanoff (1949) described an egg within an egg and several variations of one or more of the eggs missing yolks. The ovum in ovo pathology occurs when an egg is pushed back up the oviduct by contraction of the oviduct muscles. The egg may get pushed back as far as the magnum before it encounters the next egg coming down the system. There it gets surrounded by albumen along with the other egg and the two begin the descent down the oviduct. Eventually a shell membrane and shell are deposited around both eggs. Why was the first egg sent back up the oviduct in the first place? Stress may be one important factor. The female may have been under stress when she started forming eggs (that is why some eggs have no yolk). At some point during the descent the oviduct muscles contracted in reversed waves pushing the egg back up the system. Although ovum in ovo eggs have not yet been reported for dinosaur eggs, they could potentially occur in small theropod eggs with an avian structure, such as those found with *Oviraptor* (Norell and others, 1994). The best way to detect such eggs may be to break a fossil egg open (unless it is found broken), but few people would resort to that. The other option is to CAT scan the egg, although there is no certainty that this would work (see chapter 8).

One pathology that may be easier to detect is the occurrence of abnormally large or small eggs in a nest. For example, the average chicken egg is 5.8 cm ($2\frac{1}{4}$") long and 4.4 cm ($1\frac{3}{4}$") in diameter. However, an occasional egg can be a whopping 9.2 cm ($3\frac{2}{3}$") by 6.8 cm ($2\frac{2}{3}$") or even a minuscule 1.25 cm ($\frac{1}{2}$") by 0.9 cm ($\frac{1}{3}$"). An undersized chicken egg is the most common defect, and it often has little or no yolk (Romanoff and Romanoff, 1949). Abnormally large or small dinosaur eggs might not be recognized as such if they occurred alone, but would definitely stand out in a nest. As yet, however, no one has reported such pathological dinosaur eggs, but perhaps no one has looked.

Some chicken eggs can have odd shapes, being pear-shaped or arranged in odd spirals. Such eggs are very rare and have not been reported for birds in the wild. Other times, a normally elongated or oval egg may be spherical, or a spherical or oval egg might be elongated. Again, such eggs would be obvious if they occurred in a nest among normal dinosaur eggs, but might be misidentified if found alone unless really weird looking.

Defects also occur in the shell and these fall into two categories: those that affect the shell structure and those that occur on the surface of the shell. Shell structure defects include multiple layers of shell, abnormal growth of the shell, and shells abnormally thick or thin. Multiple layers generally occur in reptile eggs and are caused by the egg being retained or sent back to the shell-producing region of the uterus. Multiple-layered eggshell is especially common in turtle eggs, but also occurs in crocodile and gecko eggs (Schmidt, 1943; Erben and others, 1979; Hirsch, unpublished ms). Sometimes, layering can be extreme, as in a case reported by Hirsch (unpublished ms) in which a Galapagos turtle at the San Diego Zoo produced an egg with nine shell layers (Fig. 5.6). Although such a high number has not been found among dinosaur eggs, many specimens with two or three layers are known. These occur among eggs referred to *Megaloolithus* (sauropod?) from France, Spain, Mongolia, India, and Argentina (Hirsch, unpublished ms), *Spheroolithus* eggshell from Montana and Alberta, and *Preprismatoolithus* from Utah (we'll cover eggshell names in chapter 9). Although the most common pathology in dinosaur eggs, multiple-layered eggshell is still rare. Good thing too, since the pathology is mortal to the embryo—the pores of the various layers do not line up and gas exchange cannot occur. Basically, then, the embryo asphyxiates.

Identifying pathological multiple layers of eggshell can be tricky and mistakes have been made. If an egg gets buried and crushed so that the opposite sides of the shell are in contact, the result can produce fossil eggshell that looks doubled—or more if several layers of eggs were involved. The key, however, is to look at the relationship of the eggshell parts microscopically in cross-section. An egg that gets crushed will show mammillae of the opposite sides in contact with one another. If several layers of eggs are involved (crushed nest), eggshell will show outer surface–outer surface contact, as well as mammillae-mammillae contact. Why do I raise this seemingly obvious point? Because some reports of pathological eggshell are actually crushed eggshell—that is how tricky it can be sometimes.

The formation of shell in the shell gland may not always go smoothly for a variety of reasons, including stress (sudden appearance of a predator, drought, etc.), illness, overcrowding, or poor nutrition. The result may include abnormally shaped shell columns or extra spherites (Fig. 6.24). These are known in modern shell as well as in dinosaur eggshell (Hirsch, unpublished ms; Vianey-Liaud and others, 1987; Zhao, 1994). Zhao and others (1991) also report that sometimes the mammillary layer may be abnormally thin or thick in relation to the crystalline layer in *Macroolithus* eggs from China. Sometimes the mammillary layer in *Macroolithus* eggshell may also be doubled or more, probably because the egg got moved back and forth in the same region of the oviduct where the shell-producing gland is housed.

Another pathology seen in modern eggshell is an abnormal thickness. If the shell is too thick, the hatchling cannot get out, or if abnormally thin, the egg is crushed by the brooding parent or overlying nesting material. Abnormally thin eggshell has been reported for *Megaloolithus* eggshells from France and Spain by Erben and his colleagues in 1979. They concluded that the thinning was due to hormonal imbalances of vasotocin and estrogen in the local population of sauropods following environmental changes. The resultant increase in egg breakage and dehydration of the embryo, they believed, was a contributing factor to the extinction of the dinosaurs. Sounds plausible, doesn't it? But recently, French paleontologist Vianey-Liaud and her colleagues (1994) have shown that different species of eggshells were used by Erben and his colleagues. This can be shown by comparing a histogram of eggshell thickness, where the greatest number of shells of a particular size form discrete clusters (Fig. 6.25). The importance of these graphs is that they also show the distribution curve of the samples, with most of the shells clustering near the middle of the groups and the number of specimens more or less than the average thickness tapering off on either side. The range of thickness for some species was over a millimeter, part of which is due to shell thickness not being uniform around an egg. Nevertheless, the histograms do show several specimens well separated from the curves that may possibly identify abnormally thin and abnormally thick specimens.

Surface defects of the shell are perhaps the most common defect in birds (Romanoff and Romanoff, 1949). If the shell surface is normally smooth, defects may cause it to be partially or entirely wrinkled or ridged. There may be calcareous granules over part or all of the egg, or a thin coat of calcite may cover all or part of an egg. Transparent spots may occur, and even the shell pigments may vary, with splotches or uneven colors. As yet, none of these surface pathologies have been noted among dinosaur eggshell, so keep your eyes open.

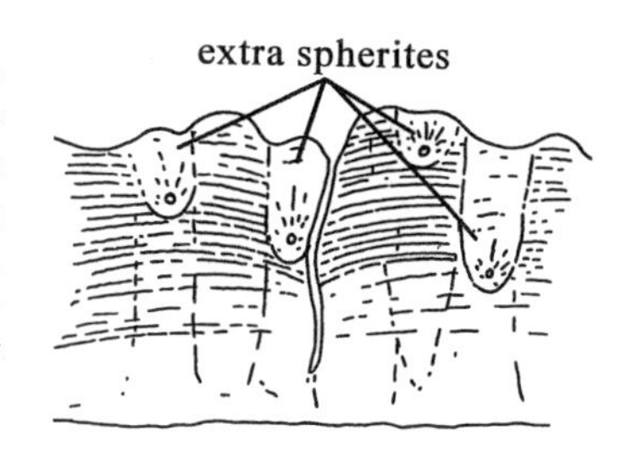

Fig. 6.24. Example of pathology, extra spherites, near the surface of Maiasaura *eggshell. Spherites normally form before the shell units, and the occurrence of a second group of these on formed shell units suggests the egg was pushed back into the portion of the uterus where spherites are formed on the shell membrane.*

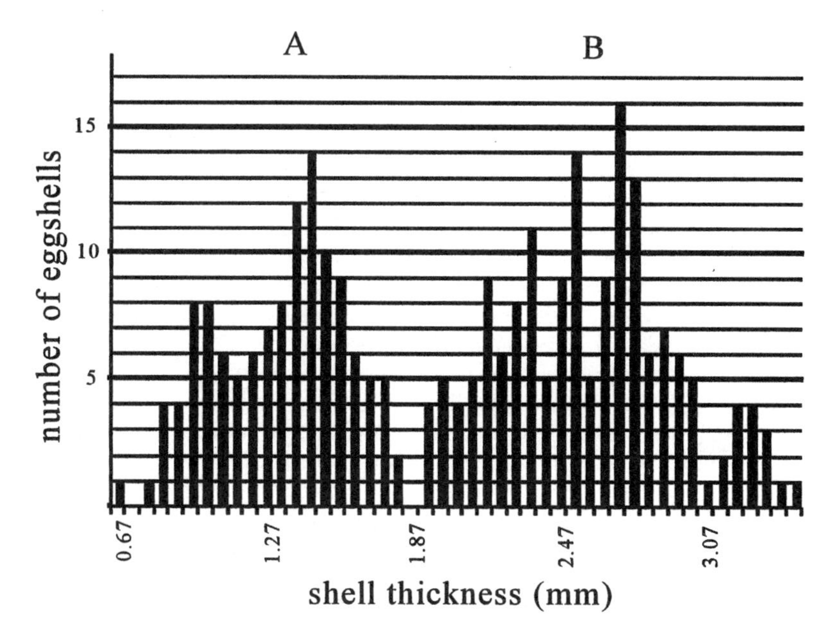

Fig. 6.25. Bar diagram showing shell thickness by frequency for eggshells from the Cretaceous of France. Note the two distinct clusters, with A now called Megaloolithus aureniensis *and B* Megaloolithus siruguei. *At one time the difference in thickness was used to argue that eggshell became thinner during the Cretaceous resulting in the extinction of the dinosaurs. (Data from Vianey-Liaud and others, 1994.)*

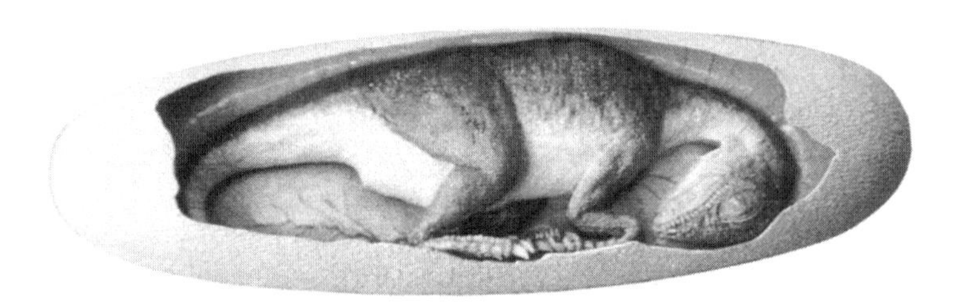

7 • Eggs as Fossils

How to Fossilize an Egg

We generally think of eggs as delicate structures because they are so easily broken. While that is true of whole eggs, it isn't necessarily true of the shell fragments. Try breaking a chicken egg shell into smaller and smaller pieces. You'll find that the pieces are stronger than you might think. The toughness is due to the organic material deposited within the shell as it is formed (see chapter 6). The organic material might be thought of as the cement binding calcite crystal bricks together thereby making the shell more resistant to breakage than if the bricks were simply stacked.

In flowing water, eggshell pieces can be carried for quite a distance. Just how far was demonstrated experimentally by Canadian paleontologists Tim Tokaryk and John Storer (1991). They placed pieces of chicken eggshell in a rock tumbler along with a little sand and water to simulate transportation in a stream. After rotating for seventy hours, equivalent to the shells' traveling sixty-eight kilometers (forty-two miles), the pieces showed remarkably little change in size. Admittedly a host of additional factors could influence the distance traveled in the real world: size of the sediment particles (sand vs. gravel), acidity of the water, etc. Nevertheless, the experiment did demonstrate that eggshell is a lot more rugged than we usually think. Water transport explains the occurrence of dinosaur eggshell fragments far from the nest where they originated.

Some of the best sites for dinosaur eggs, eggshells, and baby dinosaur bones occur in the mudstones of ancient floodplains (Fig. 7.1, 7.2). Today these rocks may be hard, but originally they were soft sediments deposited periodically by floods. Between flood episodes, which in some places and times may have been annual, soil processes modified the sediments—plants established themselves and sent down roots, insects burrowed, earthworms ingested and expelled the sediment, the water table rose and fell seasonally, rainwater seeped downward, and the sun baked and cracked the surface. Because the ground was dry most of the year, the dinosaurs would lay their eggs in hopes that they would hatch before the next flood buried the landscape.

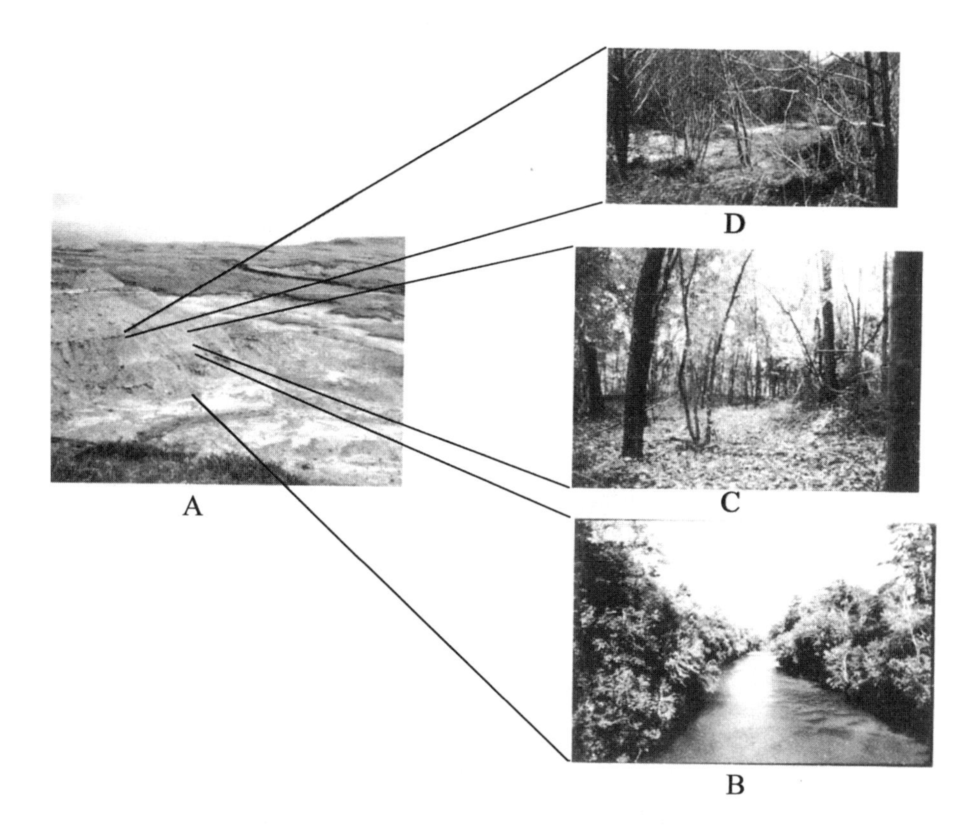

Fig. 7.1. The present terrain where dinosaur eggs have been found are typically exposed as badlands (A, near Egg Mountain). But the landscape wasn't always like this and it is a matter of reading the rock record. Near the base of the exposure are sandstones of an ancient river channel, much like the Black River in Mississippi (B). Above the channel sandstones are mudstones of the ancient floodplain similar to that adjacent to the Black River today (C). Thin layers of sandstone mark intervals of flooding when sand spilled out onto the floodplain (D, sand seen as light color zones among the vegetation).

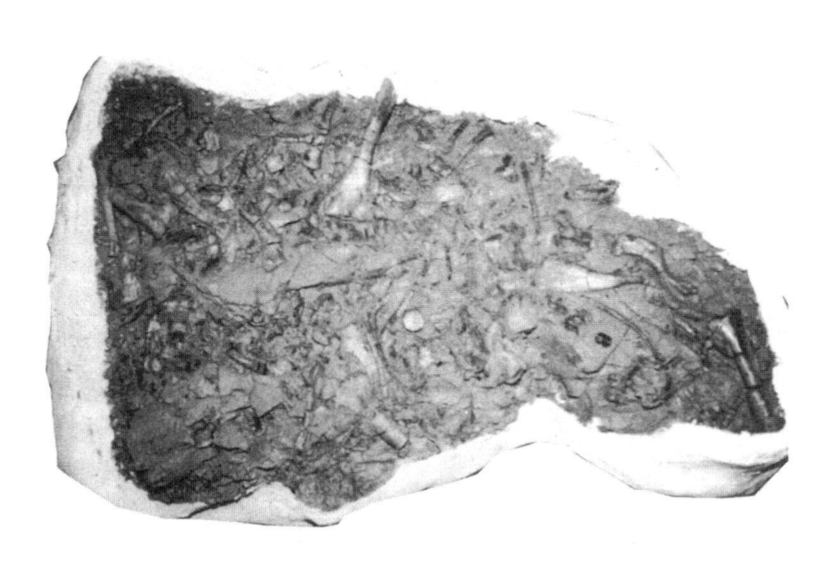

Fig. 7.2. Mass of fossilized baby dinosaur bones, Hypacrosaurus, *from the Two Medicine Formation of Montana. The rock containing the bones is a flood plain mudstone.*

All of these soil processes leave their imprint, which can remain long after the soil has been deeply buried and hardened to rock. Millions of years later, erosion may reveal these ancient soils, called paleosols, within which clusters of dinosaur eggs are often found. Sedimentologists can reconstruct the ancient environment by looking at the rocks containing the eggs (Retallack, 1988). Ancient plant roots typically occur as twisting, branching, and tapering cylinders of discolored rock, which often fizz when a drop of acid is put on them because of the calcium cement. Insect burrows may also twist and branch, but are of uniform thickness, and usually a centimeter or less in diameter. Sometimes excavation marks can be seen on the burrow wall. Traces of earthworm activity are seen as tiny ricelike fecal pellets. The rise and fall of groundwater is marked by horizons of calcium carbonate nodules. Often the most obvious evidence for the presence of paleosols are color bandings of the outcrop (Fig. 7.3).

Another important place where dinosaur eggs and eggshells have been found is in sandstones in southern Mongolia and northern China (Fig. 7.4). We know that some of these sandstones were once active sand dunes because the sand consists of uniform-sized grains of fine quartz that show a characteristic frosting of the surface due to bouncing off each other dur-

Fig. 7.3. Paleosols, or ancient soil profiles, are often expressed as color banding, such as seen here in the Morrison Formation near Arches National Monument, Utah.

ing Late Cretaceous windstorms (Jerzykiewicz and others, 1993). Over the millennia the sand has compacted into sandstone. It's hard to imagine sand dunes having life among them, let alone dinosaurs nesting there. Nevertheless, whole and partial eggs and eggshells occur there in abundance. In fact, the dinosaur eggs found by the Central Asiatic Expedition at the Flaming Cliffs came from ancient sand dunes. Today, the remains of lizards, turtles, birds, small mammals, small crocodiles (there were small ponds and rivers among the dunes), dinosaurs, and of course, eggs are known from sand dune sandstones. The discovery of *Oviraptor* skeletons sitting on their nest of eggs indicates that sandstorms may have been the chief method of egg burial (Norell and others, 1995; Dong and Currie, 1996).

Fig. 7.4. Polish paleontologist Karol Sabath looking for dinosaur eggs among the ancient sand dune deposits of the Djadokhta Formation in southern Mongolia. The frosted and uniform-sized sand grains establish these sediments as aeolian, or windblown. The steep angle of the sandstone beds is the leeward side of a sand dune. (Courtesy of K. Sabath.)

Dinosaur eggs have been found in Cretaceous beach sands in northeastern Spain (Sanz and others, 1995). The area must have been a popular resort—an estimated 300,000 eggs are believed to be present. Evidently dinosaurs (actually, sauropods) liked to visit the seashore as well. Actually, the eggs were probably deposited in the sands because of how easy it was to dig. Once buried, heating by the sun would ensure incubation. The sand subsequently lithified into rock.

Perhaps the oddest discoveries of eggs are those that occur in marine deposits. From England, Buckman (1859) described some eggs from the White Limestone, Carruthers (1871) from the Stonesfield Slate, and van Straelen (1928) from the Oxford and Gault Clays; and Dobie (1978) described some from the Mooreville Chalk of the southern United States. (The specimen described by Melmore, 1930, as a possible reptile egg from the Lias Formation of Whitby, England, is probably not an egg because the "shell" lacks any structure.) All of these finds are from marine sediments, as indicated by invertebrate fossils, as well as plesiosaurs, ichthyosaurs, and fishes. The eggs described by Buckman and Carruthers have been studied by Karl Hirsch (1996) and identified as belonging to turtles (Fig.

2.9). The other eggs have yet to be identified taxonomically. How did these eggs get to the bottom of the sea? From what we know about modern reptiles and about modern eggs, the female did not lay her eggs on the seafloor. Were the eggs expelled or dropped from drifting carcasses? Might eggs have been transported to sea in drifting mats of vegetation, perhaps vegetation nests washed to sea by floods? No one knows, and without a time machine we may never know.

Eggs have also been discovered in volcanic deposits. The specimens were not found in lava, but in the debris thrown out of a volcano. The specimens consist of hard-shelled turtle eggs found on one of the Canary Islands in the Atlantic (Hirsch and Lopez-Jurado, 1987). At least two clutches were found, suggesting that very large land tortoises had dug nests in the old volcanic debris. But not all volcanic deposits are suitable for the preservation of eggs. Jim Haywood and his colleagues (1991) studied the aftermath of volcanic ash from Mount Saint Helens' burying the nests of ground-nesting gulls. They found that after only two years, the shells had mostly dissolved because of the acidic quality of the ash.

We have considered the various environments in which fossil eggs have been found, but this does not tell us how the eggs came to be fossilized. To understand that we need to turn to another field of paleontology called taphonomy. We know that many eggs are hatched or were eaten because the tops of the eggshells are missing (Fig. 7.5). The opening provides a way for sediments to fill the egg (brought in by water or wind). But some eggs apparently never hatched, especially those under study by Terry Manning. Many of his dinosaur eggs have insect fecal pellets within them, indicating that some insect (fly?) laid its eggs within the egg where the larva fed on the dead hatchling and other tissue (Fig. 7.6). Eventually the larvae entered into the pupal stage, metamorphosed into adults, and flew off to repeat the cycle elsewhere.

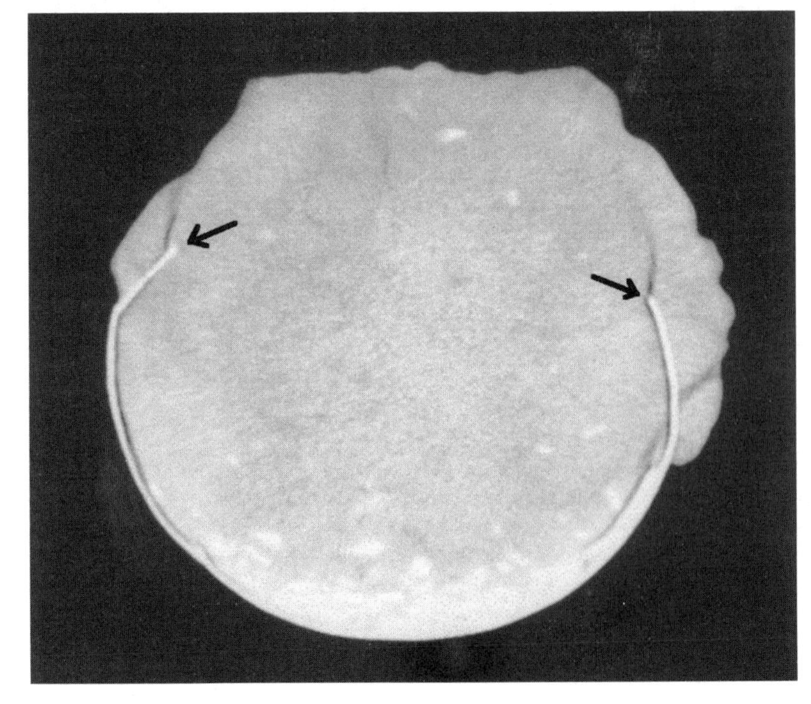

Fig. 7.5. CAT scan of an egg from China showing the end of the eggshell still buried within the rock (arrows). Note the fragments of eggshell in the bottom of the egg (white mass at bottom), which might be an indication of predation (see Fig. 10.5).

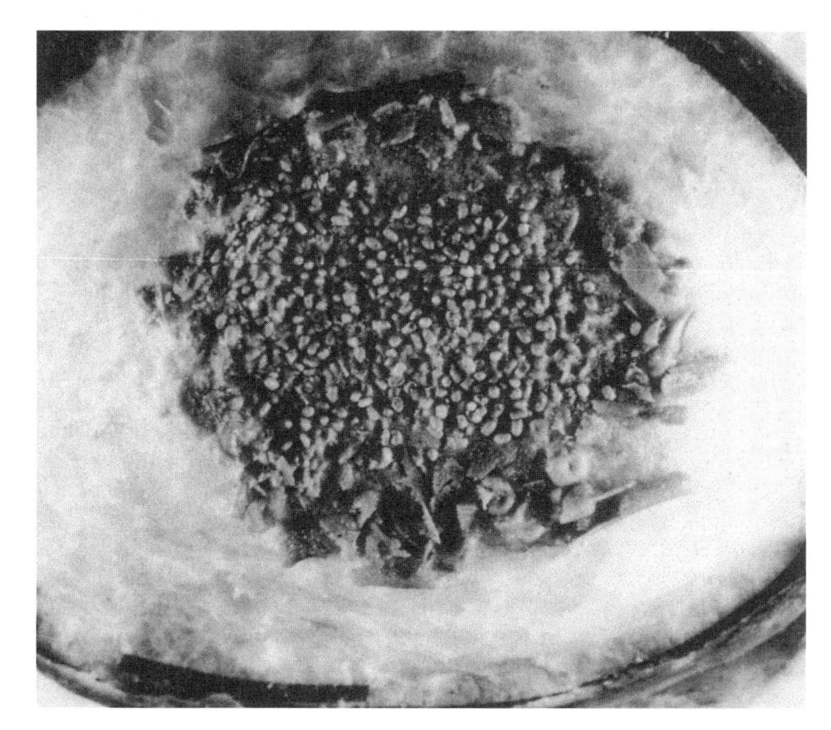

Fig. 7.6. Fecal pellets of maggots preserved among embryonic dinosaur bones inside an egg. The pellets and bone were uncovered by Terry Manning using weak acid. (Courtesy of Terry Manning.)

What killed these and other dinosaur hatchlings in the first place? We may never know, but based on studies of modern reptile and bird eggs it could have been overheating, drowning from too much water, dehydration from not enough water, asphyxiation from being buried too deep (by flood sediments, blowing sand, or volcanic ash), disease, or some congenital condition. Of course if the egg is buried too deep, then insects may not be able to get to the egg. Nevertheless, bacteria are present, as well as fungi and other microbes, and these begin the process of decay.

Once the eggs are buried, additional sediments may bury them deeper until the shells crack, allowing sediments to enter. Sometimes, though, the water table may rise and the process of fossilization begins before the eggs crack. If fossilization is rapid enough, the inside of the egg might become filled with minerals, especially calcite, protecting the egg from being crushed by the overlying sediments (Fig. 7.7). How does the calcite get into the egg? Remember that the eggshell is covered with microscopic pores. Dissolved calcite in the water is carried through the pores and is then deposited inside the egg. This calcite gets into the groundwater in the first place because rain combines with atmospheric carbon dioxide to produce carbonic acid (H_2CO_3), a very weak acid. This acid, plus organic acids from decaying plant material in the soil, seeps downward with the rainwater, dissolving any calcite present, which is usually in the form of calcium carbonate (you can see this effect on limestone by the caves that are formed). Eventually the rainwater carrying the dissolved calcium (as well as other dissolved minerals) as ions reaches the water table and seeps through the earth and through any eggs (and even bones) that happen to be there (Fig. 7.8). These dissolved minerals can be selectively precipitated by decay bacteria within the egg. Sealed from contact with atmospheric oxygen, these bacteria use alternative energy sources for their metabolism. For example, some bacteria use organic sulfur for cellular metabolism, which is what produces that rotten egg smell. Bacteria also use ions to rid themselves of metabolic waste. One of the most common methods is to attach carbon dioxide molecules (CO_2) to calcium ions (Ca^{2+}) that are dissolved in the groundwater. The result is the instant precipitation of calcium carbonate ($CaCO_3$).

Some decay bacteria use nitrogen as their energy source, converting it to ammonia. This ammonia can leach outward from the egg, setting up an alkaline microenvironment that leads to the precipitation of other minerals. Even the organic material within an eggshell fragment may be enough to set up a microenvironment suitable for calcite deposition. The resulting thin coat of calcite deposited over the shell or egg fragments sometimes makes it difficult to recognize eggshell in the field. Because all eggshell is calcite (or aragonite in the case of turtle eggs), most fossil eggshell is the original eggshell produced millions or hundreds of millions of years ago. (Isn't that neat?) That is what makes the study of fossilized eggshell possible as you will see in the next chapter.

Sometimes, however, the eggshell can be slightly or completely altered, a process called diagenesis. Aragonite of the rigid turtle eggshell is not a very stable form of calcium carbonate; it generally changes under heat when buried deeply to your basic, run-of-the-mill calcite. But even the calcite of eggs can be slightly modified because of pressure from being deeply buried. The result can be a herringbone pattern that looks like cross-

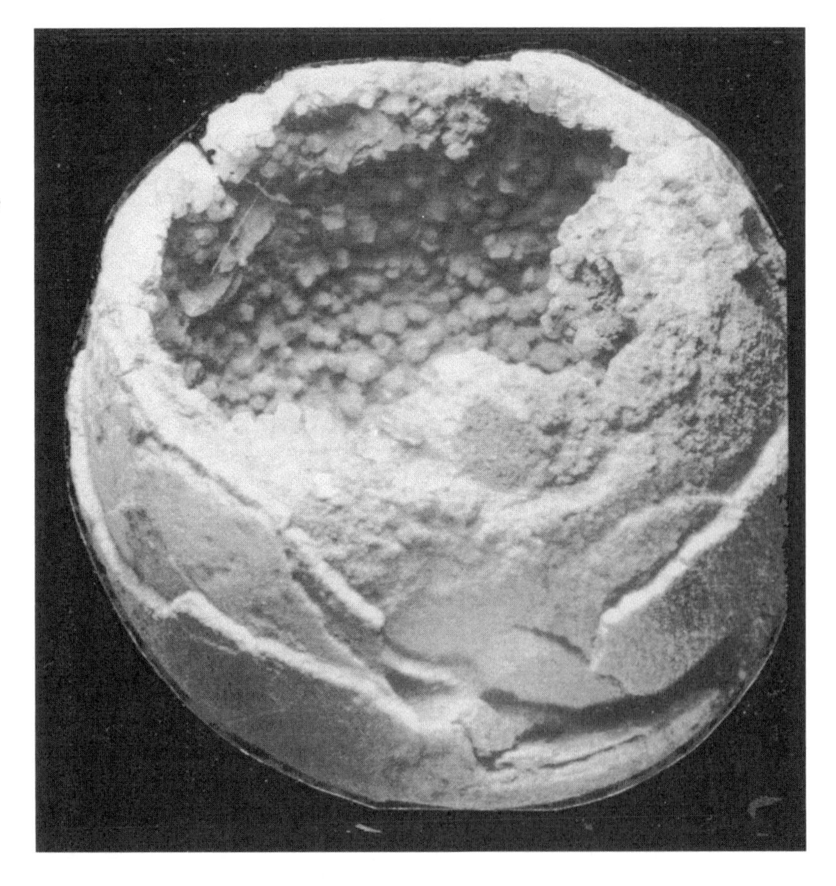

*Fig. 7.7. Calcite crystals lining the inside of a dinosaur egg (*Dendroolithus microporosus*) from Mongolia. Such hollow eggs are rare because the weight of sediments usually crushes them before calcite can reinforce the inside. (Courtesy of K. Mikhailov.)*

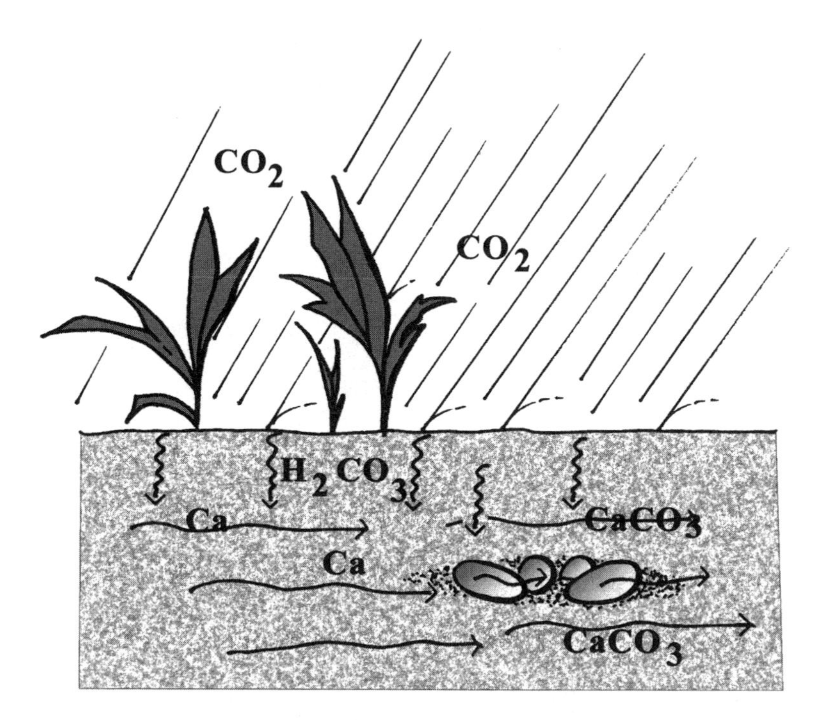

Fig. 7.8. Once floods have deeply buried a nest of dinosaur eggs, the process of fossilization can begin. First, carbonic acid (H_2CO_3) is formed by the absorption of carbon dioxide (CO_2) by rainwater. Percolating downward through the soil, calcium is dissolved from the soil by the acidic water. Calcium enriched, the water is added to the groundwater. Seeping through the ground, the water passes through the decaying eggs, where calcium carbonate precipitates.

hatching under the microscope, or even a total loss of all internal structure in more extreme cases.

Silica is occasionally found in eggshell because it also precipitates out in alkaline conditions. Because the silica molecules are of a different size than the original calcite molecules of the shell, most of the internal structure is obliterated. Other minerals can also be present in trace amount, the most important of which is iron. Iron is important because it can impart color to a fossil, even when present in minute amounts. It is iron, possibly as iron sulfide (pyrite), that gives some fossil eggshell its black color, and iron oxide (hematite) that gives other eggshell a reddish tint.

Collecting Eggs

Okay, so tens of millions of years have gone by, the eggshell is now fossilized and the surrounding sediments lithified. Collecting these eggs cannot begin until some evidence of their presence is made known. This usually means erosion has cut down through sedimentary rocks exposing the eggs or eggshells. Usually the slopes of hills or gullies (a.k.a. ravines, arroyos, wadis, or coulees) are littered with the eroded-out fragments of eggshell (Fig. 7.9A, B). By tracing these fragments back to their source, a little careful digging may reveal a clutch of eggs. The shell fragments are recognized by three major criteria: (1) the uniform thickness of the fragments, (2) their slight curvature (usually), and (3) a surface peppered with tiny pores amid a pattern of bumps or ridges or on a smooth surface. A fourth criterion is sometimes visible as well, the presence of tiny bumps, the mammillae, on the underside (concave side). Sometimes, however, the embryo may have removed so much calcium from the mammillae that they may be difficult to detect without a powerful magnifying lens or microscope.

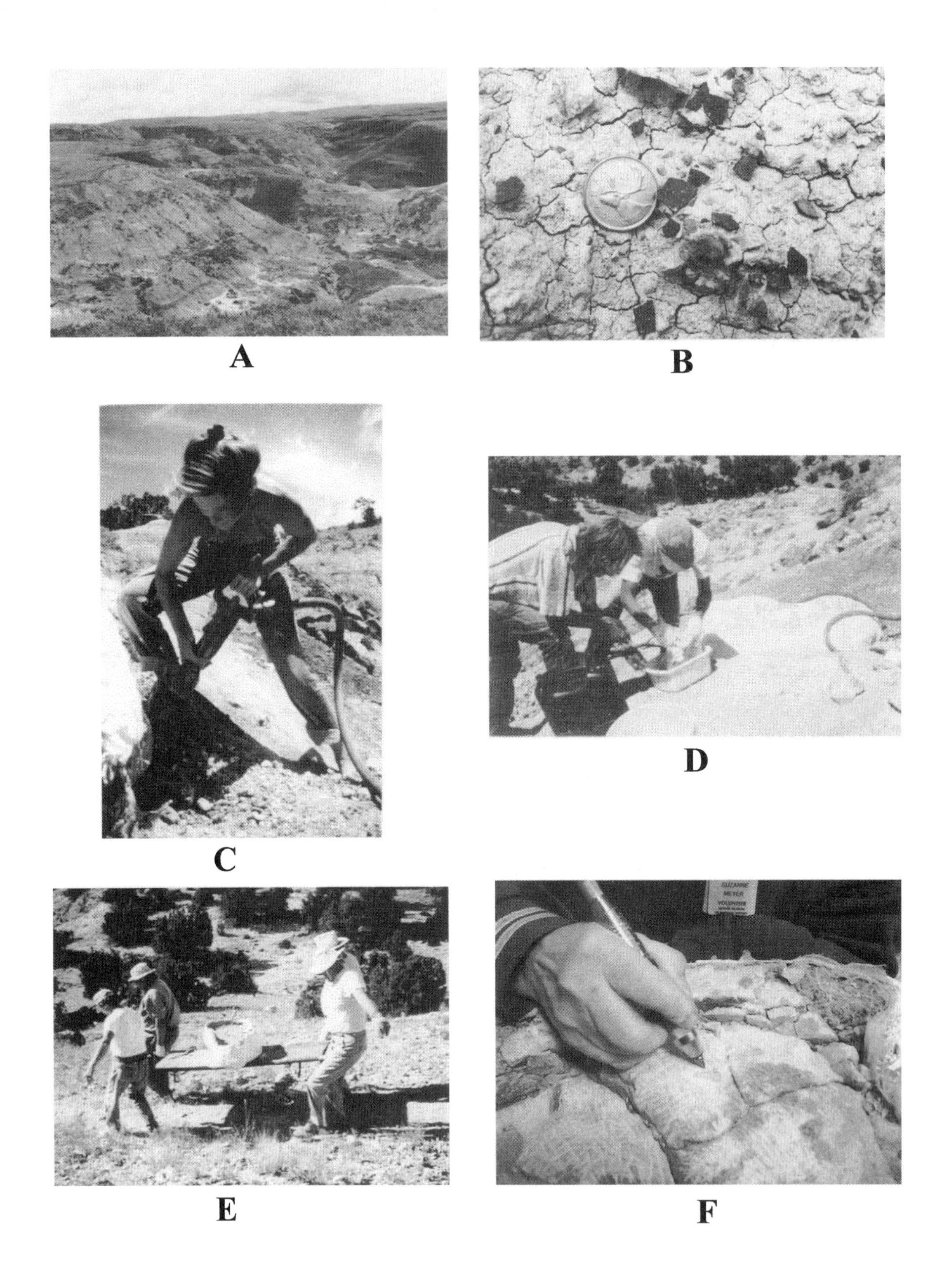
A
B
C
D
SUZANNE
MEYER
VOLUNTEER
E
F

Once an egg has been located, the surrounding rock is scraped, chiseled, or dug away to locate other still buried eggs. Once the clutch of eggs has been delineated, then the rock around the sides is excavated down to well below the estimated bottom level of the clutch (Fig. 7.9C). This can be especially tricky to determine because some dinosaurs are known to have laid several layers of eggs in a single nest (more in chapter 11). The block containing the eggs is partially undercut, then covered with newspaper, aluminum foil, or tissue, after which the entire block is encased in multiple layers of burlap strips soaked in plaster of Paris (Fig. 7.9D). Once the plaster has dried, the block is undercut the rest of the way and the block rolled upside down and transported back to the lab (Fig. 7.9E). In the lab, the plaster and burlap are usually removed from what was originally the bottom side because the bottom of the clutch is usually in better shape than the top (Fig. 7.10). The rock is carefully removed (a process called preparation) by small pneumatic engraving tools, dental picks, needles, X-Acto blades, or any other tools that do the job (Fig. 7.9F). The work can be

Fig. 7.9. (facing page) Steps to collecting eggs. Random digging is a poor way of finding eggs. It is more expedient to let Mother Nature do the work. A, badlands at Devil's Coulee with eggs at or near the surface. B, as the rock is eroded, the slightly more resistant eggshell accumulates on the surface. C, once a high concentration is found, the rock is carefully excavated until the eggs are delineated. D, the eggs are encased in layers of burlap and plaster of Paris making a hard jacket or shell to protect the delicate fossils. E, the jacket is then transported to a vehicle and back to the museum. F, there, technicians carefully scrape or scribe away the rock.

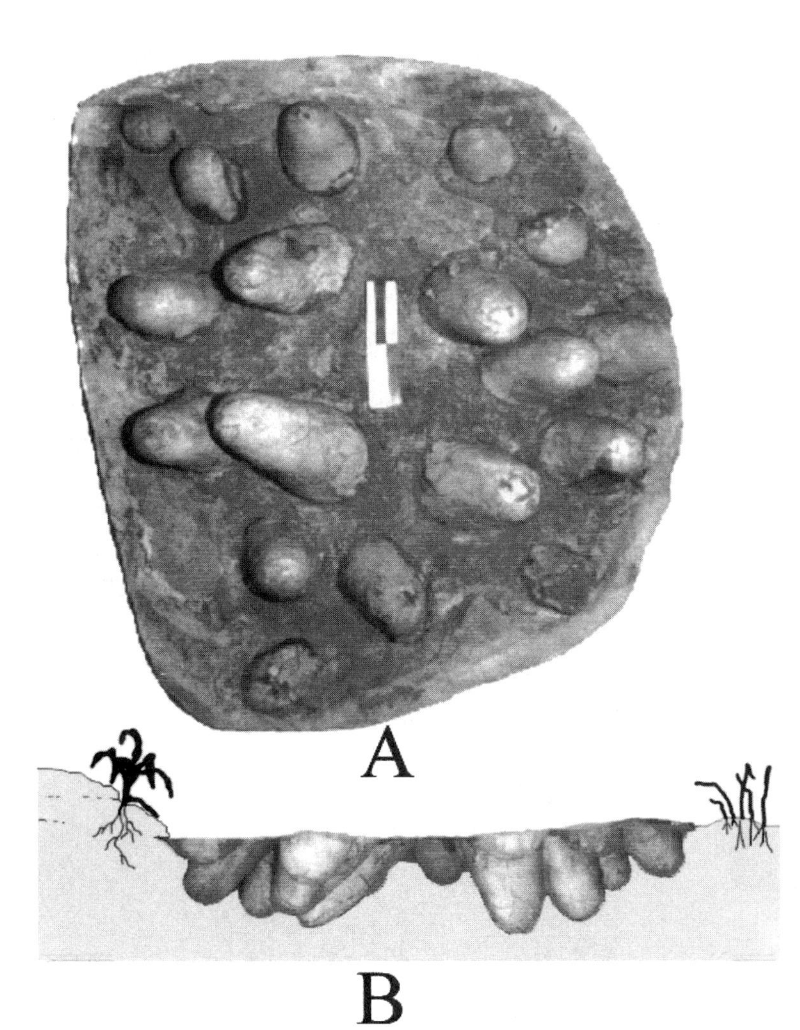

Fig. 7.10. Clutches of fossil eggs can seem to be intact (A), but in reality are the bottoms of hatched or eroded eggs (B).

tedious, and a great deal of patience is required in order to clean the eggs without damaging them. The surface of eggs is often ornamented with bumps or ridges that are diagnostic for different egg species (more in chapter 8), hence it is important that these features not be damaged or destroyed. Because of this, it is generally not a good idea to use a miniature sandblaster, called an airbrade, nor an air eraser (used in art). Although the airbrade does make the eggshell look pretty, it does so at the expense of the shell surface.

When to stop removing rock is a judgment call. The eggs can be completely freed, in which case it may be difficult to place them back in their original position relative to one another, or left partially embedded in rock. Leaving them in rock does have the advantage of concealing the fact that the eggs may be hatched. Many of the eggs exported from China during the mid-1990s remain on a small rock pedestal (Fig. 7.11A). X-ray and CAT-scans of these eggs show that the shell does not continue through the rock. A distinct gap is present on what was originally the upper side of the eggs, through which the hatchling emerged (Fig. 7.11B). By preparing the eggs from the bottom side, the vendors make the eggs look nicer and intact, thus ensuring their sale.

Fake Eggs

It looks like an egg, therefore it must be an egg, right? Not necessarily. Nature makes egglike structures that can even fool professional paleontologists. (Pay attention, I'll give you the secret how to avoid misidentifications.)

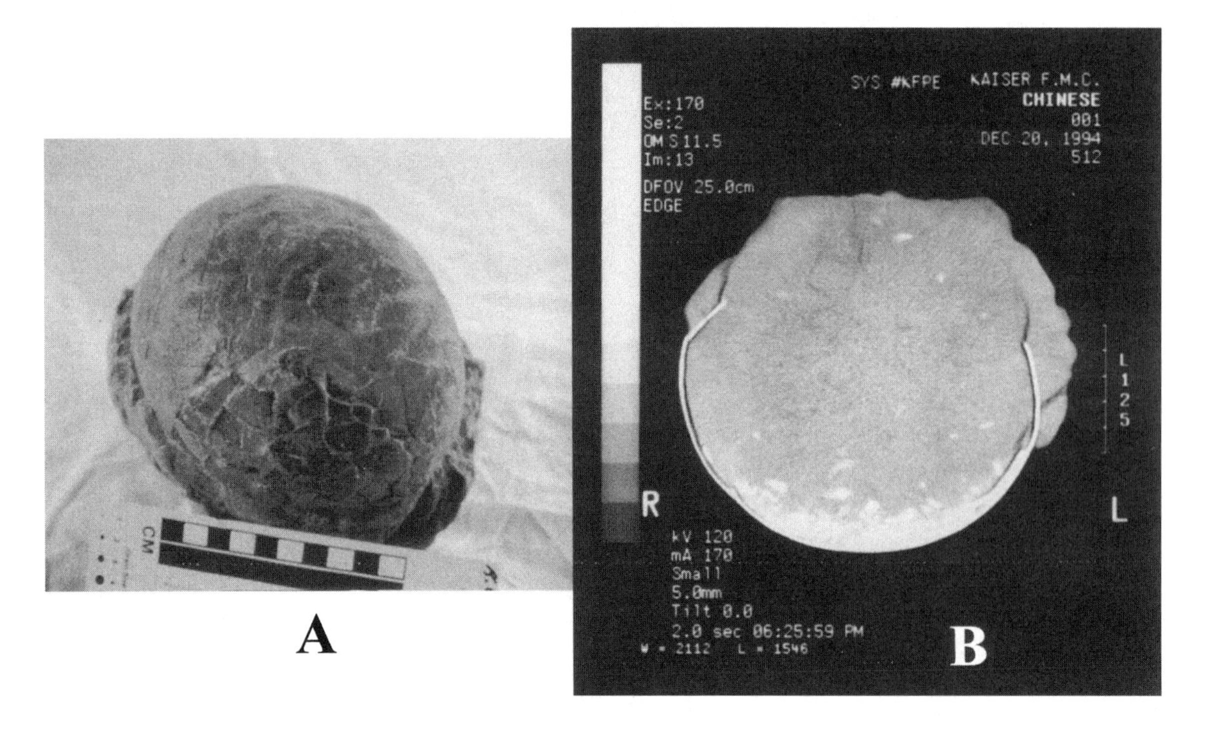

Fig. 7.11. Many eggs from China appear to be complete when purchased (A) with a rock pedestal, but CAT scanning reveals that the pedestal hides the opening of a hatched egg (B). Note shards of eggshell littering the bottom of the egg. Some shards within eggs have been wrongly identified as embryonic dinosaur bones.

The most common objects brought to me as fossil eggs are egg-shaped, water-worn stones and concretions. Concretions can assume almost any shape, and many require only a little imagination to see a stone heart, foot, leg, rabbit, or practically any shape or animal. Naturally, these concretions are often the source of reported dinosaur eggs (Fig. 7.12A). Concretions start as a localized concentration of organic material buried in sediment. This might be a dead fish at the bottom of a lake, a crab on the sea floor, or a pocket of finely fragmented plant material on the bottom of a river. Decay of this organic material in the sediments produces the same bacterial and alkaline conditions discussed above in the fossilization of eggs. Ammonia can leach outward through the sediments, and minerals—especially calcium—dissolved in the groundwater can be precipitated, thus cementing the sediment together. This cementation can occur rapidly, before the surrounding sediments are compacted. That is why concretions produce some of the best—least damaged or crushed—fossils, while similar fossils in the same stratum but not inside concretions are badly crushed.

The shape of the concretion only partially reflects the shape of the organic material contained within. Generally, the farther from the organic material the alkaline environment extends, the less resemblance between concretion and organic material. The most common source of this organic material is very fine, disseminated plant (land and aquatic) fragments, invertebrates, and microbes. You can see some of this organic material in a handful of sediment from the bottom of a pond or slow-flowing river. This organic material in the sediments is not evenly distributed for a variety of reasons (currents, loss through decay, etc.), so where the concentration is highest, concretions are liable to form. Odd-shaped concretions occur because of the way the organic material was dispersed in the sediments in the first place.

For most egg-shaped concretions, breaking them with a hammer will reveal a nearly uniform interior, which, with a magnifying lens, is often cemented sand grains (the cement is calcite if the rock fizzes with vinegar or muriatic acid). Sometimes color banding looking like layers of an onion may be present. This coloration records the growth of the concretion

Fig. 7.12. Pseudo-eggs: A, nonorganic concretions formed by the cementation of sand grains or mud particles; B, organically formed insect burrows from the Morrison Formation of Colorado. (The concretion was brought to me as a fossil egg.)

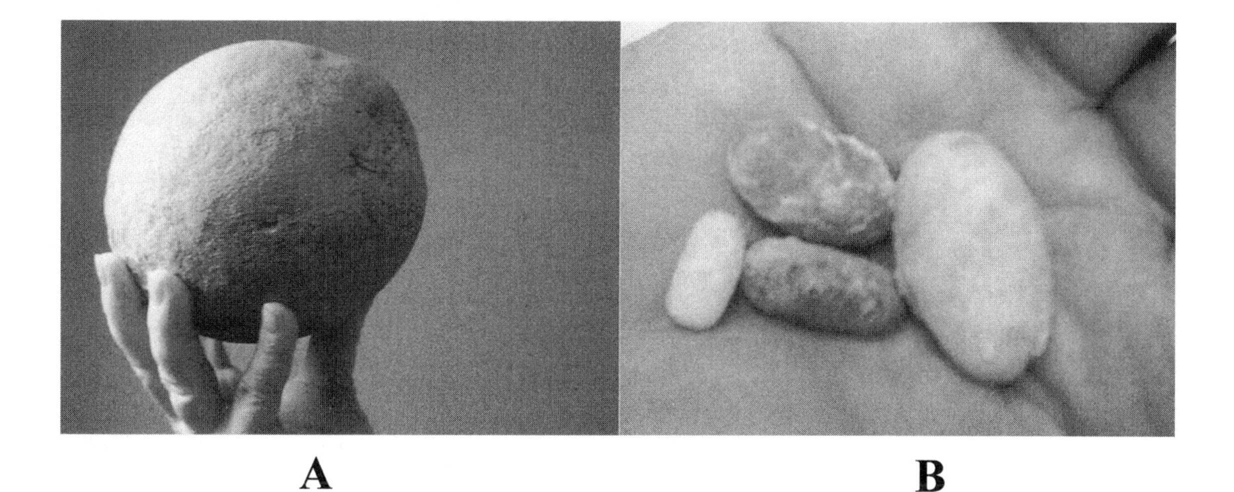

through time; it is not fossilized yolk and egg white as has been suggested to me (Fig. 7.13A). The most common colors are various shades of yellowish-brown (limonite) and brown (siderite), both of which are iron minerals. Sometimes a yellowish core of sulfur may be present, but the presence of sand grains under the magnifying lens will show that the yellow is a superficial coloring and not the remains of the yolk.

The most peculiar of the egglike concretions are those that have a hard rind and soft interior. When emptied, the concretion looks like an empty eggshell. The formation of the rind is called case hardening and is a phenomenon that occurs in the weathering zone in the upper few meters of rock exposed by erosion. When encountered in deeper rock, these concretions are only a little harder than the surrounding rock and are often white and yellow sandy blobs.

Fossilized insect burrows and pupae are sometimes misidentified as eggs, even by professional paleontologists (Fig. 7.12B). Many insects dig egg-shaped chambers in the soil and sand dunes in which they live or breed, and these cavities can sometimes be preserved. In soft soil, insects may coat the insides of the chamber with saliva to bind the soil together. The chamber may eventually get filled with sediment washed or blown in. Millions of years later, the chambers can be eroded out of the surrounding rock and litter the ground. Fossilized pupae sometimes look like small eggs. Millions of years previously, an insect pupa died for some reason (perhaps smothered by the surrounding sediments or drowned by a rising water table). Regardless, once dead, the pupa decayed, leaving a cavity that was later filled by calcite or clay minerals brought in by groundwater. Later, erosion can also leave these pupae littering the ground.

The most egglike structures are objects called calculi or stomach stones—the bane of paleontologists (at least one scientific paper described an "egg" in great detail that later proved to be a stomach stone). These stones are formed in the stomachs of modern ruminants, such as deer, elk, goats, and cattle, and are later regurgitated. Cut open, these stones show multiple layers of calcium phosphate (the material that makes up bone) surrounding a foreign object swallowed by the animal (Fig. 7.13B, C). The stones form to protect the stomach from damage by the alien object as the stomach churns during digestion (Hirsch, 1986). The surface of the stone may even have tiny dimples that resemble pores.

The giveaway with concretions, burrows, pupae, and stomach stones is the lack of a shell on the outer surface. A true fossil egg still retains the shell (otherwise it would not be an egg, right?), hence all of the shell characteristics (see chapter 8), including pores, mammillae, and prismatic or continuous layers are present. Furthermore, fossil eggs are rarely undamaged. Usually the shell is badly cracked or crushed and pieces can be lifted off for study. Concretions are seldom all the same size, so a "clutch" of eggs is suspect if different sizes of "eggs" are present. The largest known egg, that of the extinct bird *Aepyornis*, is about the size of a soccer ball, and dinosaur eggs are smaller. Concretions, on the other hand, can be as large as a house, so again, abnormal size is another reason to suspect a giant dinosaur "egg." Concretions typically have sand grains on their surface that are visible with or without a hand lens, whereas an eggshell will be smooth or have very distinct bumps (called nodes). Insect burrows and pupae tend to be small, about a centimeter or two (half an inch to an inch) long. Again, the absence of eggshell is suspect. Breaking the object will show a uniform structure of calcite or infilled sediment. Sometimes scratch marks produced during digging of the burrow by the insect can be seen on the surface under a strong light.

Fig. 7.13. A, limestone nodules, whole and broken open, sent to me as fossil eggs. The white interior was thought by the finder to be the egg white. B and C, one of the most egglike of pseudo-eggs, a stomach stone, cut open to reveal its true nature. (One such stone was described in great detail by a professional paleontologist as a fossil egg, while others turn up in museum collections as fossil eggs as well.) In this case, a deer apparently swallowed a fishing weight (lead has a sweetish taste, which is why it was added to wine by the Romans), and multiple layers of calcium phosphate formed around the object to protect the delicate tissues of the stomach. Later, the deer regurgitated the stone. Someone found the stomach stone and gave it to Karl Hirsch as a fossil egg. (B and C courtesy of K. Hirsch.)

The trickiest objects to correctly identify are stomach stones. They are usually found lying on the ground and their "perfect" egg shape is suspect because, as stated above, fossil eggs are seldom undamaged. They tend to range in size from a centimeter to almost 6 cm (larger sizes are very rare). The absence or scarcity of pores on the surface is a giveaway as to their true nature, as are the onion layers within. Although multiple-layered eggshell is a known pathology, these layers do not extend to the core of the egg as they do with stomach stones. We'll discuss how fossil eggs are studied in the next chapter.

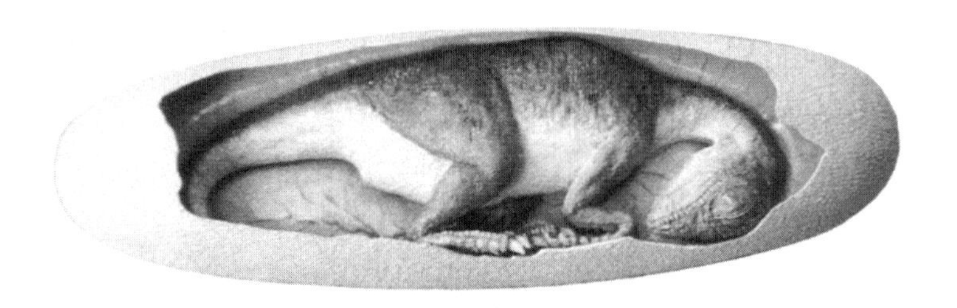

8 • How to Study a Fossil Egg

Dinosaur eggs are treated like any other fossil object—they are collected, studied, categorized, and described. How this work is done has changed considerably over the years as new techniques, new technologies, and even new philosophies have been introduced.

Tools of the Trade

In chapter 1, I mentioned that Paul Gervais conducted microscopic studies of various eggshells during the 1870s in order to determine the affinity of the fossilized eggshells from southern France. The technique Gervais used is little changed today, with the exception that epoxy resins are now used in place of natural resins such as balsam to glue the eggshell to the microscope slide. The steps have been presented in detail by Quinn (1994), but essentially entail embedding a small piece of shell in special epoxy resin (Fig. 8.1A). Once the resin has hardened, one end is cut exposing the shell in cross-section. The cut is made using a very thin, water-lubricated rock saw (the water keeps the resin from melting and gumming up the blade). This cut surface is then lapped using progressively finer emery paper or lapping grit (as fine as 1200 grit) to remove the marks left by the saw. This surface is then epoxied to a frosted microscope slide (Fig. 8.1A) and the block is cut with the thin rock saw close to the slide (Fig. 8.1B). The shell exposed on the cut surface is again ground using progressively finer polishing grit or emery paper until the shell is almost transparent and free of scratches (Fig. 8.1C). The result, called a thin section, may now be studied using a petrographic microscope (Fig. 8.1D). To interpret the eggshell in three dimensions requires several vertical thin sections (called radial sections, Fig. 8.2A), as well as several horizontal sections (called tangential sections, Fig. 8.2B). Because making thin sections destroys so much of the shell, several pieces of eggshell are needed.

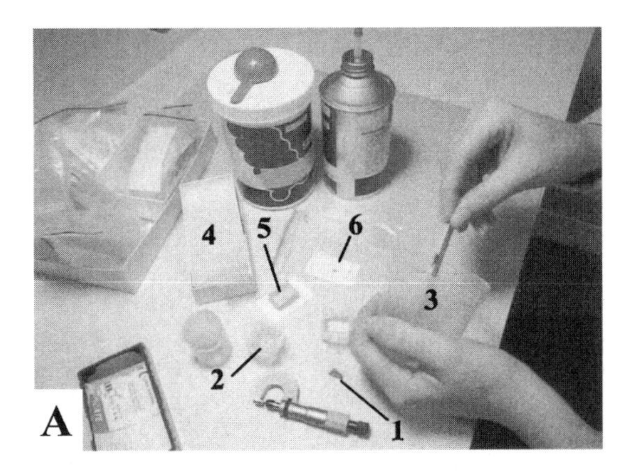

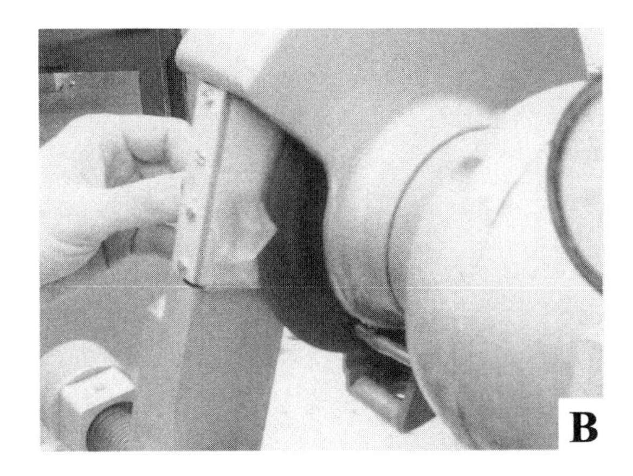

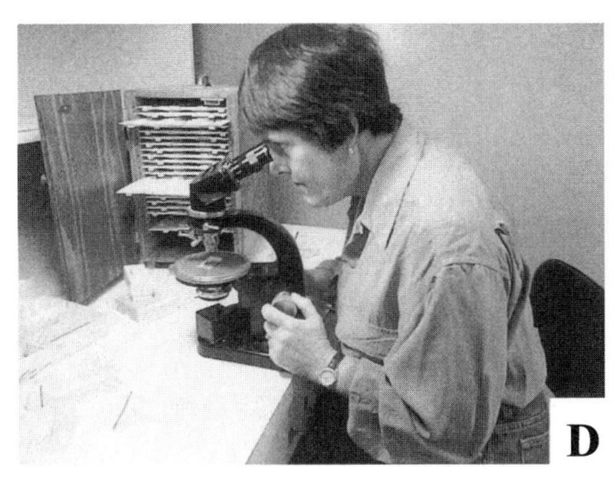

Fig. 8.1. A, embedding an eggshell: (1) The shell itself, (2) container, (3) resin, (4) slides, (5) partially completed slide, (6) finished slide. B, trimming the excess resin and eggshell. C, lapping the eggshell until transparent. D, studying the slide under a microscope.

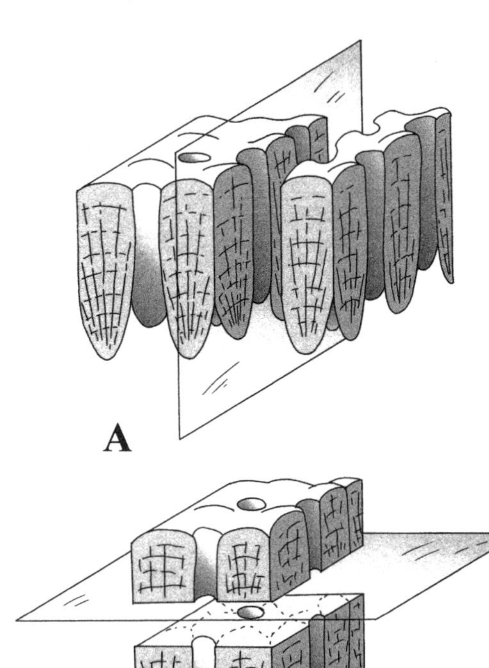

Fig. 8.2. Two types of thin sections used to study fossil eggshell denoted by a two-dimensional plane: A, radial or vertical thin section; B, tangential or horizontal thin section.

Why thin sections? They are the only way to determine the size and shape of the shell units and the shape and size of the pore canals, features that are used to classify fossil eggshell. The relative size and shape of the shell units are best seen when viewed through a polarizing filter on a microscope. A polarizing microscope, also known to geologists as a petrographic microscope, is used to determine the optical properties of crystals. A polarizing filter *polarizes* the light coming through the microscope, meaning that only light waves traveling in a certain direction can pass through (Fig. 8.3). You might have noticed the effect while wearing a pair of polarizing sunglasses. As you look at a distant cloud, it and the sky get lighter in color as you slowly tilt your head to one side. This effect is due to the polarizing lenses' allowing progressively more light to pass through as they become more parallel to the light waves.

Some crystals, especially calcite, can also act as natural polarizing lenses. When the eggshell is ground so thin that light passes through it, the calcite of the shell units polarizes the light. Viewed under a polarizing microscope, the individual shell units can be made to "wink" light and dark by rotating the stage or microscope slide (Fig. 8.4). This occurs because two overlapping polarizing filters oriented perpendicular to one another block all light. Why? Because the first filter within the microscope

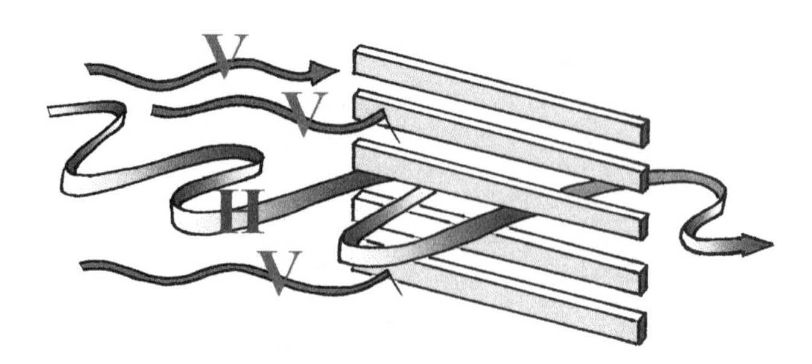

Fig. 8.3. Polarized filters aligned vertically (top) allow light waves in the vertical plane (V) to pass through, but block horizontal waves (H). Rotate the filter 90° (bottom) and horizontal light waves pass through, but not vertical.

polarizes the light in one direction, while the second filter, the shell units, polarize the light 90° in another direction. The result is that the light wave is blocked by the second filter (sunglasses are usually only partially polarized, so some light still passes through if two pairs of glasses are placed over each other at right angles). The blockage of all light is called extinction. With eggshell, the shell units have a different extinction pattern (Fig. 8.4), and this can be important in the study of dinosaur eggshell. Such studies have shown that not all round eggs are the same because they have different extinction patterns (see Appendix II).

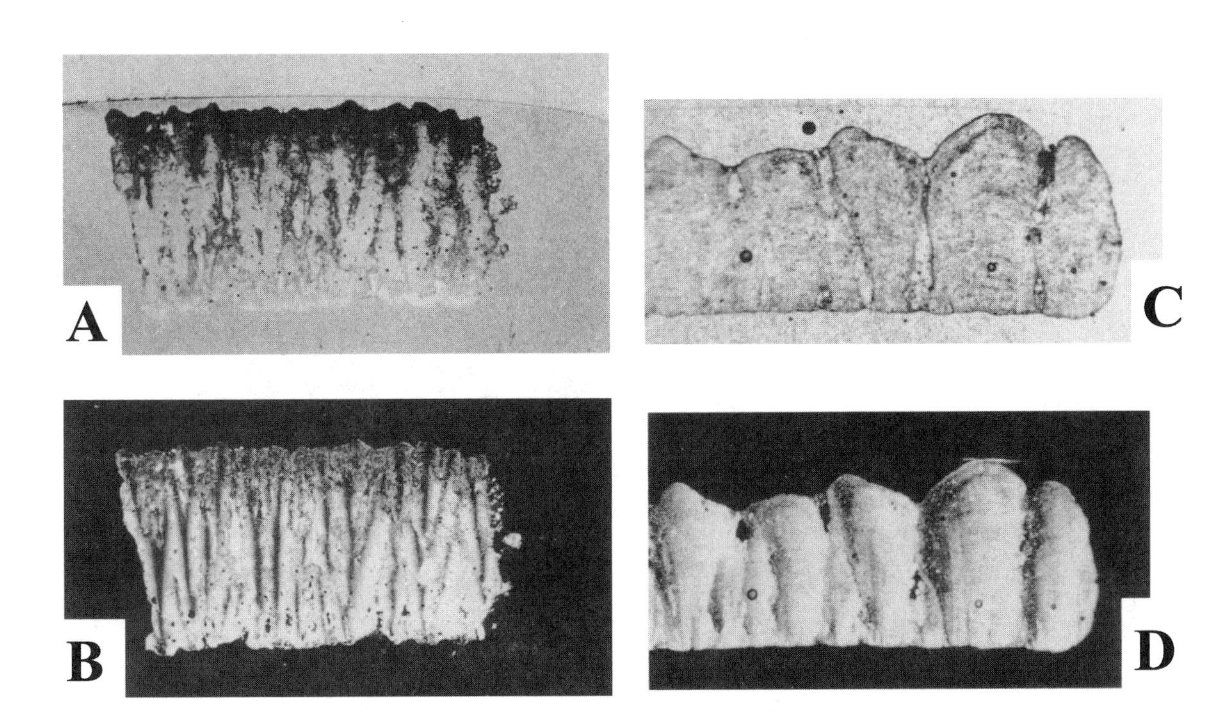

Fig. 8.4. Two examples of fossilized eggshell seen under normal light (A, Faveoloolithus; *C,* Spheroolithus*) and polarized light (B,* Faveoloolithus; *D,* Spheroolithus*). (Courtesy of K. Hirsch and K. Mikhailov.)*

Smaller shell detail can be seen with the scanning electron microscope (SEM) (Fig. 8.5). With the SEM, a small specimen is glued to a metal button, then coated with atomized gold or platinum. A tiny beam of electrons is fired at the specimen in a vacuum and the electrons reflected off the gold or platinum are gathered to produce a picture. Because the electron is so much smaller than the calcite molecule, very small details can be seen. The best detail is seen along a freshly broken surface of the shell because the break occurs along planes in the crystal lattice of the calcite (Fig. 8.6). Thus the detail and information revealed with the SEM are different from that revealed by the standard light microscope. It is because of these differences that both techniques are used to describe fossilized eggshell.

The first person to use the SEM on dinosaur eggshell was German paleontologist H. Erben in 1970. In the process, a whole new world was revealed for the first time. The minute detail was far greater than that seen with the conventional light microscope, and so was referred to by Erben as the ultrastructure. To understand the significance of this level of detail, it is

Fig. 8.5. The scanning electron microscope is a large machine containing a monitor which receives the image produced by the electron detectors within the specimen chamber. Inset, Polish paleontologist Karol Sabath uses the electron microprobe feature of the scanning electron microscope to study the elements in fossilized eggshell. (Courtesy of K. Sabath.)

*Fig. 8.6. Example of dinosaur eggshell (*Faveoloolithus*) seen with scanning electron microscopy (compare with Fig. 8.4A, B).*

important to digress and examine the calcite crystal. Calcite, a form of calcium carbonate ($CaCO_3$), is unusual among minerals in that the crystal occurs in over three hundred forms. These variants are usually versions of prism, rhombohedron, or scalenohedron shapes (Fig. 8.7). Having variants is important for eggshells because different parts of the shell are made up of different shapes of calcite (Fig. 8.8). Why eggshells have these different crystals isn't understood very well, although what little work has been done suggests that organic material affects crystal growth.

The differences in crystal form affect the appearance of the eggshell as seen in thin section and SEM, allowing separate types of dinosaur eggshells

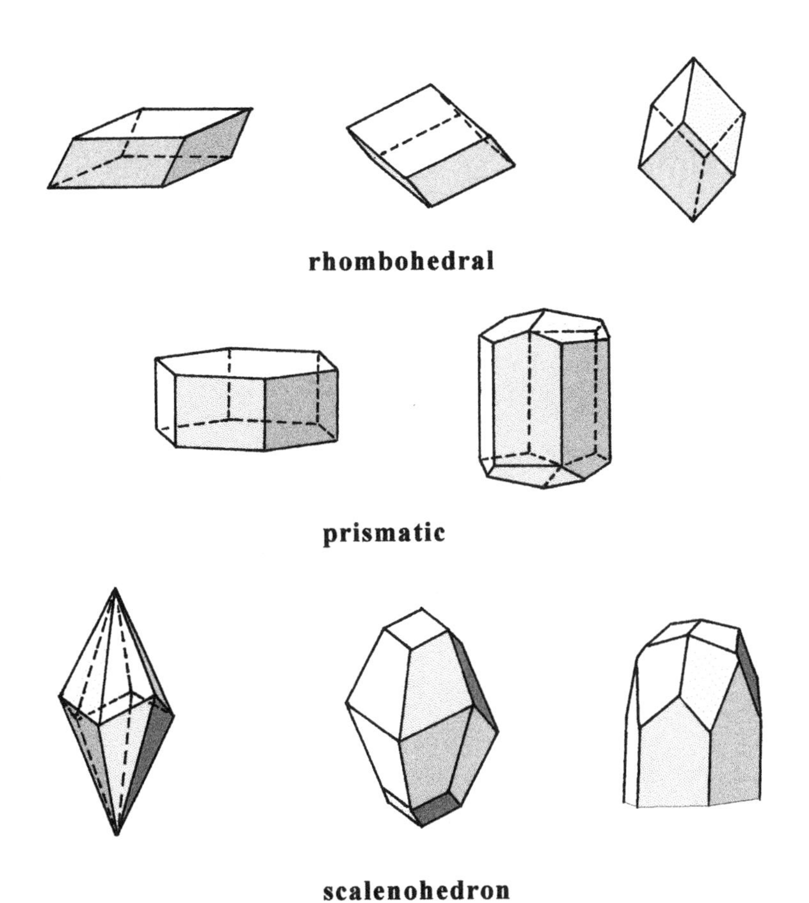

Fig. 8.7. Some of the various shapes of calcite crystals found in eggshell.

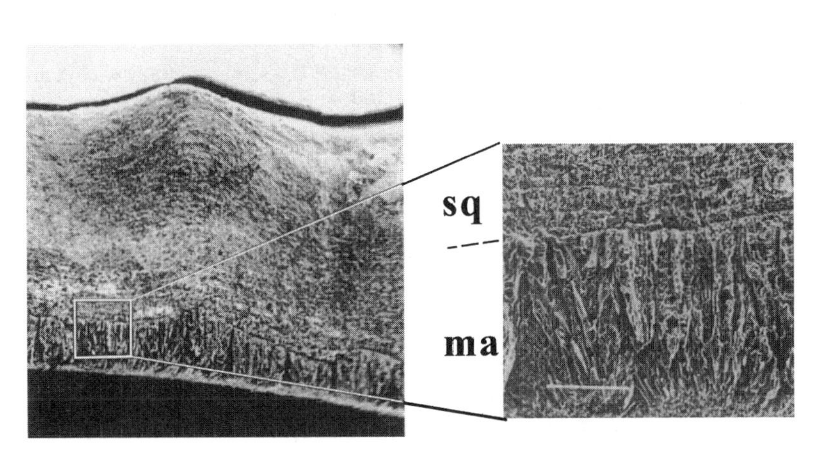

*Fig. 8.8. SEM of an eggshell (*Macroolithus*) with a close-up showing the abrupt change between the mammillary layer (ma) and the squamatic- (sq) appearing continuous layer. The change can be seen as due to the different shapes of the calcite crystals.*

to be identified (Appendix II). Sometimes the crystal faces of fossil eggshell are not very distinct due to weathering or the nature of the eggshell. In this case, it is possible to enhance the individual crystals with either dilute acetic acid or a chemical called EDTA (ethylenediaminetetracetic acid). These chemicals etch the edges of the crystals so that the individual crystal faces can be seen.

The biggest and most expensive tools used to study fossil eggs are standard X-ray photography and computerized axial tomography, or CAT scanning. Both X rays and CAT scans rely on differences in density to "see" inside the egg (Fig. 7.11). The eggshell, being mostly well-arranged calcite crystals, has a different density from that of the infilling matrix or rock. It is this difference that determines the amount of X rays passing through to a photographic film. X rays are a form of electromagnetic radiation, like light, and hence travel as waves. Because the "crests" of the waves are close together (less than 1/1000 the wavelength of visible light), X rays have high energy and great depth of penetration. The less dense the object is, the more X rays pass through and the darker that region of the film when developed. The standard X-ray machine basically takes a "picture" of the entire egg, reducing the entire volume into two dimensions. The CAT scanner, however, uses rotating beams of X rays to build a two-dimensional "slice" of the egg. With multiple slices, a three-dimensional image of the egg emerges.

Because X rays and CAT scanners can peer inside an egg, they have been used mostly to look for embryos. Reports of alleged embryos inside eggs have appeared in the press and even in the scientific literature. But without exception, these images are not those of embryos, but rather of differences in density because of differential mineralization and cementation of the matrix within the egg, or are pieces of eggshell from a hatched egg. Even when bones are known to be present, the results of X-ray and CAT scanning have been poor because the bones are so poorly developed (ossified) that they have the same or nearly the same density as the infilling matrix (Fig. 8.9). The only foolproof method of determining the presence of an embryo is to break or cut the egg in half, or to dissolve a portion of the eggshell away.

The use of the CAT scanner on juvenile dinosaur skulls has produced mixed results, again because of problems associated with differences in density. The problem is apparent in Figure 8.10, where areas known to be bone (e.g., premaxillary region) disappear in the CAT scan, while an unidentifiable mass appears in the rear of the skull. The conclusion from this is that the use of CAT scans in the study of dinosaur eggs and baby bones has to be treated with caution. You certainly must be cautious in your interpretations of the ghostly images.

As mentioned earlier, the best way to study embryonic bones within an egg is to dissolve the shell and matrix away with acid. The technique was pioneered by Terry Manning, a knowledgeable and skilled amateur. The technique involves selecting only those eggs that are known to be intact (i.e., not hatched). The eggs are partially submerged in very dilute (about 5%) phosphoric acid and monitored for several weeks. Every few days the eggs are soaked in distilled water to remove the acid that has seeped into the eggs. The eggs are then air dried before examination under a microscope. Any bone seen is carefully cleaned with a needle and artist's paint brush, then coated with a clear plastic preservative (usually Paraloid B72, Vinac B15, or Acryloid B67). The eggs are then returned to the acid bath and the whole cycle repeated for months until the embryo is revealed (Fig. 8.11A). The technique is very time-consuming and only about 20 percent of the

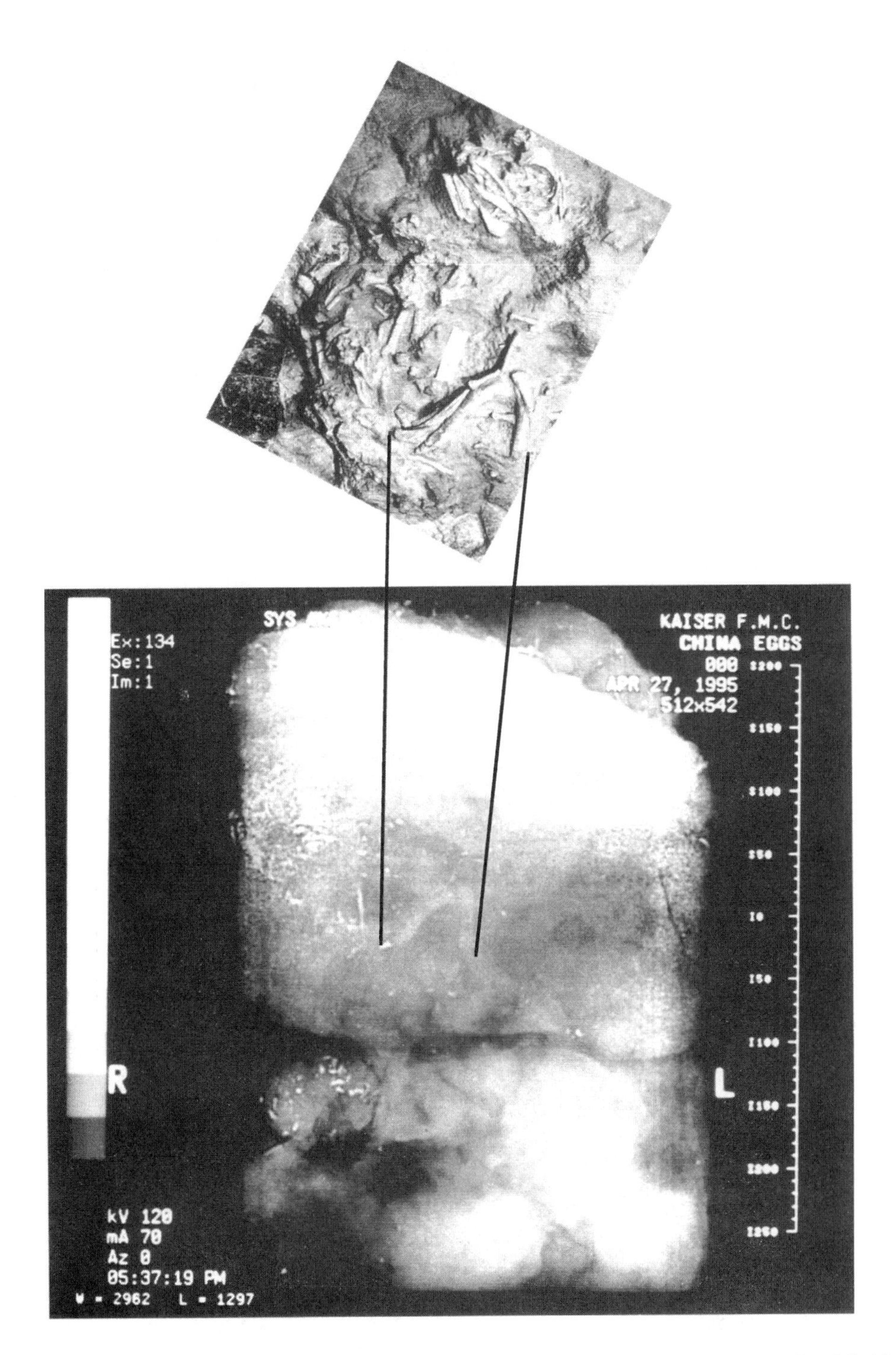

Fig. 8.9. The difficulty of using a CAT scanner to find an embryo within an egg may be illustrated with this hatchling or embryo lying atop some eggs. Below is a CAT-scan image of the specimen. Note that except for the legs, the rest of the skeleton is not visible.

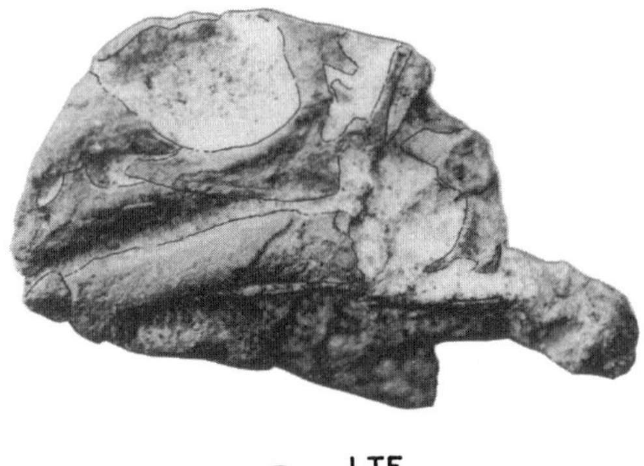

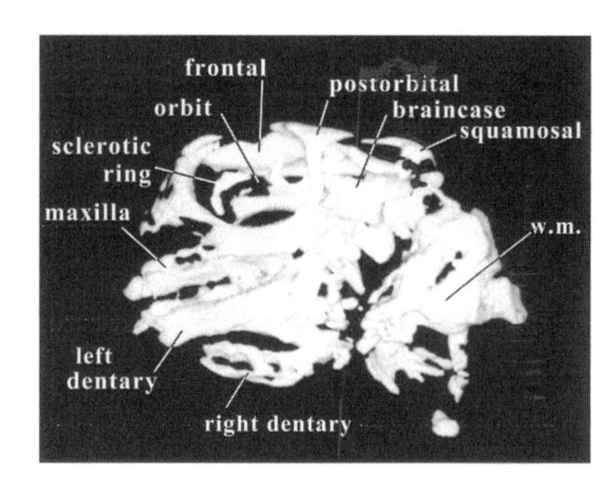

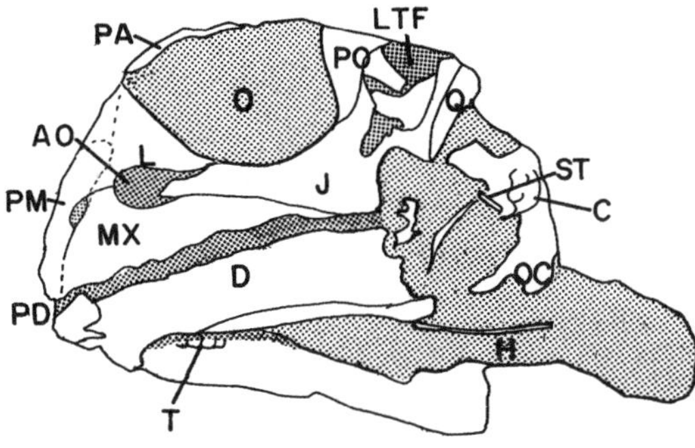

Fig. 8.10. Because X rays rely on differences in density to show what is not visible, bones can "magically" disappear and new ones appear. Note all the missing bones in the snout region of this baby Dryosaurus on the CAT scan compared to the skull in the photograph. Note also the white mass (w.m.) of "bones" in the rear part of the skull. This mass is not bone, but hard, dense rock.

eggs ever reveal any trace of the embryo. Still, Terry has uncovered some of the most remarkable dinosaur skeletons ever, and even more remarkably, what appears to be fossilized soft tissue, including fat globules from the yolk, as well as muscle and cartilage (Fig. 8.11B–D).

Other methods of studying eggshell are much more esoteric than using acid to reveal bones in an egg, and involve the chemical analysis of the eggshell. The most sophisticated technique uses the electron microprobe, which is part of the SEM, while the other two include X-ray diffraction and the mass spectrometer. The eggshell must first be cleaned of all contaminants, usually the rock in which the eggshell was found, and a final cleaning using an ultrasonic bath.

The microprobe fires a narrow beam of electrons at the eggshell within the SEM chamber. The electrons strike the atoms in the eggshell, producing X rays that have a characteristic pattern or signature for each element in the shell. A computer processes the pattern as picked up by the detector and displays the result as a jagged line, the peaks of which correspond to particular elements (see the computer monitor in Fig. 8.5). The pattern of the peaks is compared with those of a database of known elements. When the peaks of the eggshell sample match with the peaks in the database, the various elements in the eggshell can be identified. X-ray diffraction works in a similar manner to the microprobe, although a narrow beam of X rays is used to bombard powdered eggshell. When the X rays strike the atoms of the eggshell, some of the X rays are diffracted. The angle and intensity of the diffraction depends on the elements present in the eggshell. Again, the peaks are compared to a database to identify the elements.

For mass spectrometry, the eggshell is ground to powder and a small sample is placed inside the vacuum chamber of the mass spectrometer.

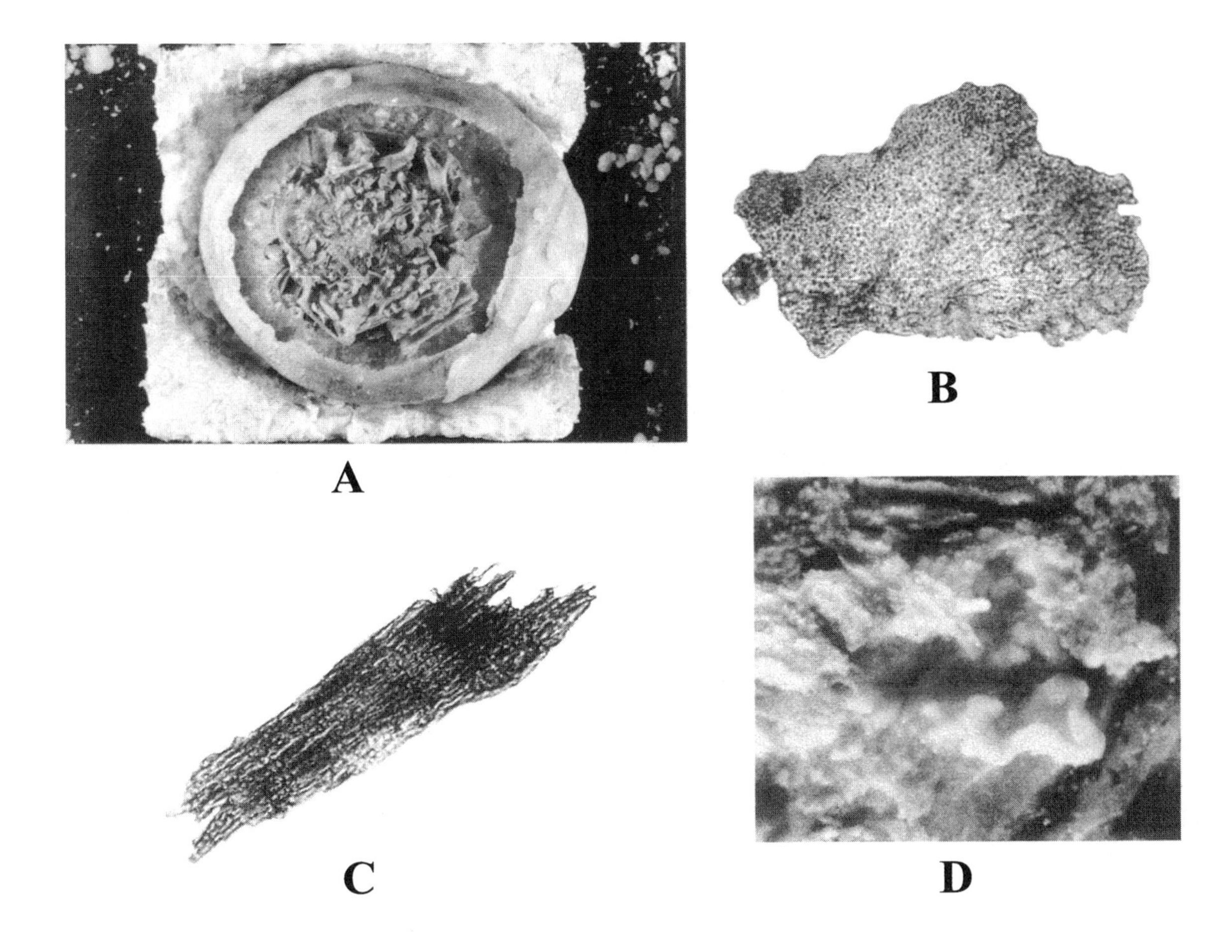

Fig. 8.11. On rare occasions, even soft tissue may become fossilized. Terry Manning has recovered examples from egg specimens such as this one (A). The soft tissue includes cartilage (B), possible muscle fibers (C—note the banding that may be myomeres), and cartilage at the ends of bone (D). (Courtesy of Terry Manning.)

Under vacuum, the powder is vaporized by the high temperature of a laser. The molecules of the eggshell are then bombarded by high-energy electrons, which fragment them and give the atoms a positive charge, i.e., make them into ions. These charged ions are then passed through a powerful magnetic field, separating them according to their mass before they strike the detector where the strikes are tallied. A jagged line graph, or mass spectrum, is then produced showing peaks of the various ions in the sample. The spectrum of peaks is compared with those of a spectra database of known compounds; matches identify the various compounds or elements in the eggshell.

Folinsbee and his colleagues (1970) were among the first to use the mass spectrometer in the study of dinosaur eggshell. They discovered that *Protoceratops* eggs (actually, probably *Oviraptor* eggs) had more of the heavy isotope of oxygen ($\delta^{18}O$—delta oxygen eighteen), rather than the light ($\delta^{16}O$) in their shell (calcium carbonate—$CaCO_3$). This difference must mean that the drinking water had concentrated the heavier isotope through evaporation, hence the environment must have been hot (more on this in chapter 13). That this difference in isotopes is from the drinking water (H_2O) and not from the air was determined by controlling the oxygen isotope of drinking water given to various birds and reptiles.

Folinsbee and his colleagues also found that the carbon in Late Cretaceous eggshell ($CaCO_3$) is mostly the heavier isotope, ^{13}C (i.e., carbon 13), rather than the lighter ^{12}C, implying that herbivorous dinosaurs fed primarily on C3 plants (i.e., leafy plants that use three carbon atoms in photosynthesis products, unlike C4 plants, such as grasses, that use four carbon atoms). Similar results for the oxygen and carbon isotopes of dinosaur eggshell have been obtained by Erben and others (1979) for eggs from France, Sarkar and others (1991) for eggshell from India, and Zhao and others (1991) for eggshell from China.

An example of how the results for $\delta^{18}O$ and $\delta^{13}C$ of dinosaur eggshells compare is presented in Figure 8.12. As may be seen by the scatter of points around the regression line, the correlation between the environmentally controlled ^{18}O and ^{13}C might seem to be only moderately significant because the points don't cluster along the line. Actually, what strong correlation there might have been was probably lost through diagenesis. In other words, groundwater that was active during the process of fossilization may have altered the oxygen isotope ratio, or carbon from bacteria that once fed on the eggshell may have altered the carbon isotope ratio. Which of these processes had the biggest effect is probably something that will be difficult to determine.

The work with the microprobe, X-ray diffraction technique, and mass spectrometer has produced some interesting but controversial results. As we shall see in a later chapter (chapter 13), Chinese, French, and German paleontologists have used the results of their analysis of dinosaur eggshell to show a correlation between changes in eggshell chemistry and environ-

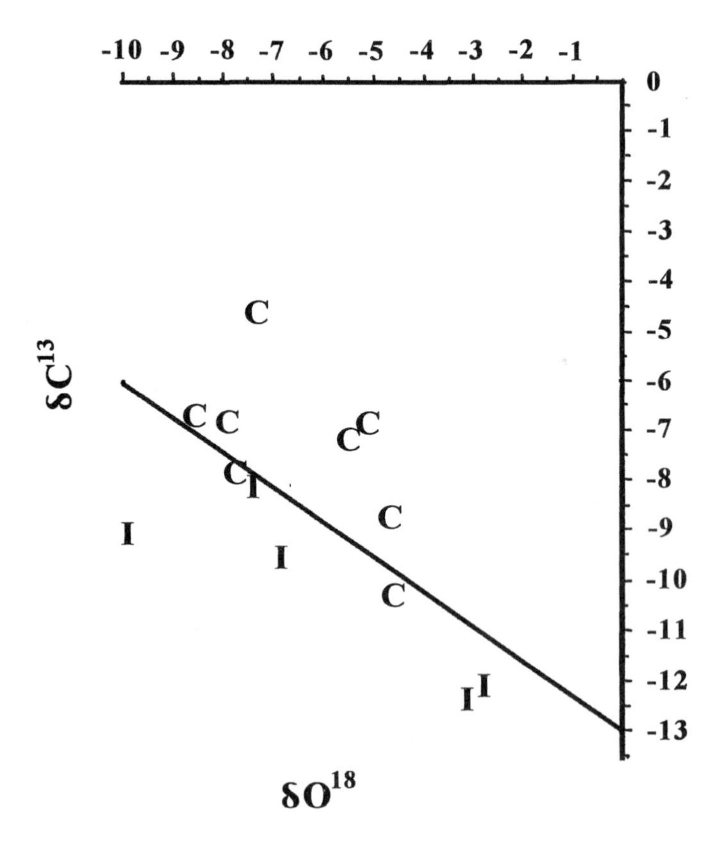

Fig. 8.12. Plot of ^{18}O and ^{13}C isotope values as determined from eggshells from India (I) and China (C). Data from Zhao and others (1993) and Tandon and others (1995).

mental changes that may have contributed to the extinction of the dinosaurs (Erben and others, 1979; Zhao and others, 1991; Dauphin, 1990a).

French and Russian paleontologists have also analyzed the organic part of dinosaur eggshell for amino acids (e.g., Kolesnikov and Sochava, 1972; Vianey-Liaud and others, 1994). The most common technique uses electrophoresis to identify the amino acids (a process also used in criminology labs). To avoid contamination, only samples not previously in contact with human hands are selected. After cleaning, the eggshell pieces are dissolved in EDTA, leaving the organic matrix of the eggshell (see chapter 6). This organic material is homogenized (often in a blender), then embedded in a gel and subjected to an electrical current. The gel acts like a sieve so that the proteins, which make up the various amino acids, can be separated by both electric charge and size. The electrical current causes the various proteins to migrate to specific levels in the gel, producing the characteristic bands when treated with protein silver stain.

The pattern of these bands is compared with previously established standards of amino acids and the unknowns in the eggshell can be identified (Fig. 8.13). The problem with this type of study is that amino acids tend to deteriorate at moderate temperatures, such as occur in deeply buried rock, and can be leached by groundwater under certain conditions; so are we really looking at an accurate profile of the eggshell proteins? Still, the results of such studies on pristine eggshell show that dinosaur eggshell has many of the same amino acids found in bird eggshell (see Table 8.1).

One other analytical technique is used to show that diagenetic change has occurred during fossilization. In chapter 7, I mentioned that the calcium of the eggshell can be altered. The most important tool to reveal these changes is cathodoluminescence. Pure calcium or magnesium-rich calcium

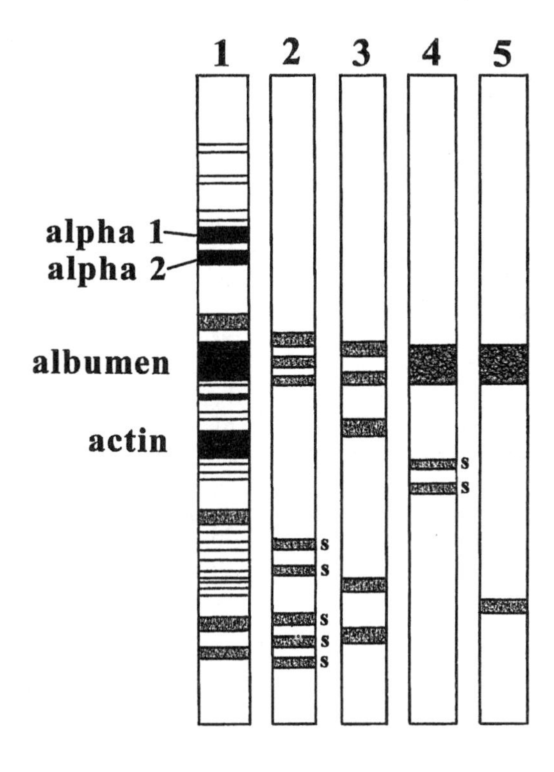

Fig. 8.13. Electrophoresis amino acids from dinosaur eggshells from France produce bandings. Column contents: 1, reference markers; 2, eggshell from Rousset-Erben; 3, eggshell from Rousset; 4, eggshell from La Cairanne; 5, eggshell from La Cairanne. Alpha 1 and 2 are collagen markers. Note that the egg proteins albumin and actin occur in several of the samples. S = shell protein. (Modified from Vianey-Liaud and others, 1994.)

TABLE 8.1.

Percentage of proteins in modern bird and dinosaur eggs (data for *Gallus* averaged from Vianey-Liaud and others, 1994; for *Cairanoolithus* averaged from La Cairanne in Vianey-Liaud and others, 1994; for *Faveoloolithus* from Kolesnikov and Sochava, 1972, identified by Mikhailov, 1994b; for *Macroolithus* averaged from Zhao and others, 1993).

Amino Acid	*Gallus*	*Cairanoolithus*	*Faveoloolithus*	*Macroolithus*
alanine	8.4	14.25	9.5	15.3
arginine	5	1.45	0.1	3.3
aspartic acid	8.3	12.1	10	4.7
cystine	3.3	0.8	3.5	—
glutamic acid	11.4	15.7	11.5	8
glycine	14.2	18.2	15.1	45.9
histidine	2.8	1.3	1.25	—
isoleucine	3.1	2.5	2.5	1.76
leucine	5.9	3.6	5.5	2.4
lysine	3.4	3.7	3.5	3.1
methionine	1.7	1.3	1.5	0.86
phenylalanine	2.3	3.1	5	17
proline	8.5	5.0	2	1.76
serine	8.4	6.6	6.5	4.8
threonine	5.8	8.25	6	8.9
tyrosine	1.6	1.05	3.25	—
valine	6.8	5.4	4.5	2.2

as is found in eggshell will not luminesce, but alteration to manganese-rich calcium will luminesce orange (Fig. 8.14).

In marked contrast with the high-tech scanning electron microscope, mass spectrometer, and electrophoresis procedure is the Geneva Lens Measure. This instrument is used to estimate the size of an egg, a technique developed by Sauer (1968) for fossil ostrich eggshell; it has since been used on other fossil eggshell (e.g., Hirsch, 1983). The Geneva Lens Measure is a gauge with three prongs, the middle one of which is moveable and controls the pointer of a dial. The dial is a scale for measuring a curved surface, specifically that of a lens (so is used mostly by opticians). But the instrument also works to measure the curved surface of eggshell fragments and allows the size of the egg to be estimated. Of course not all eggs are round, so one measurement of the shell curvature may not be enough to convey egg size and shape. The technique works best when a crushed egg is found and measurements can be taken of large fragments (greater than 3 cm; for smaller fragments an Obrig Radius Dial Gauge for measuring contact lenses can be used). The Geneva Lens Measure gives readings in diopters, a unit of measurement related to the focal length of a lens, which must be converted to the radius of the curvature in millimeters. For a perfectly spherical egg, the formula is $r = 530/P$, where r is the radius and P is the diopter measurement. Thus a shell fragment with a diopter of 5 has a radius of 106 mm (4").

But not all eggs are spherical; many are elliptical or elongated. For such eggs, the shell fragments having the shortest radius, also have the same curvature in all directions. The radius corresponds to a value of b^2/a, where a = the length of half of the polar axis and b = the length of half the equa-

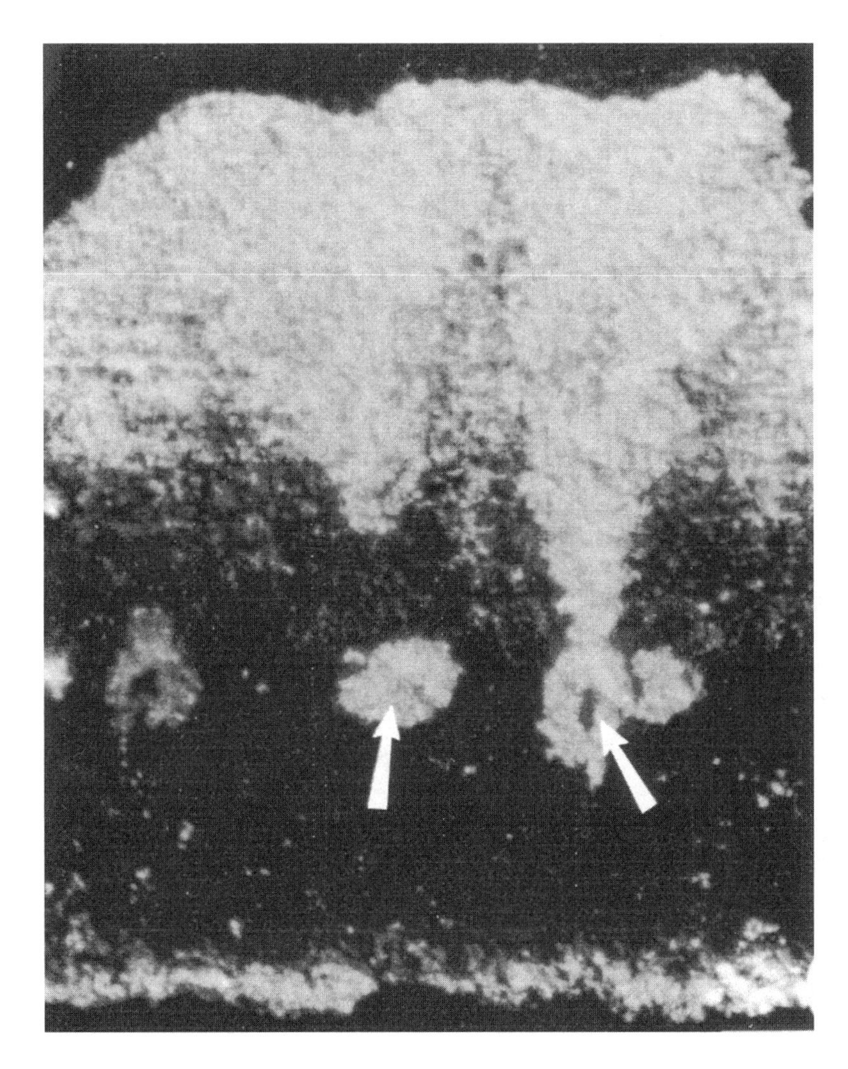

Fig. 8.14. Diagenetically altered calcite eggshell will luminesce with cathodoluminescence. Here, manganese-rich calcium luminesces (arrows). (Courtesy of K. Hirsch.)

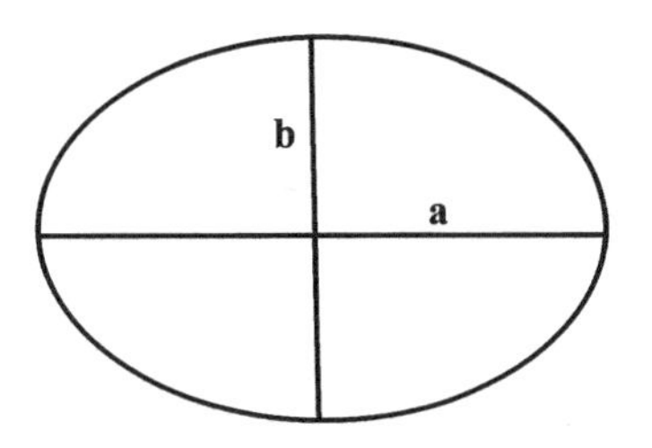

Fig. 8.15. Terminology for measurements of eggs: a, polar axis; b, equatorial axis.

torial axis (see Fig. 8.15). The flatter fragments having the longest radius actually have two very different radii at right angles to each other. The shortest radius is b and the longest is a^2/b. Once the lengths of the radii are known, the thickness of the shell is added to each (because the diopter was measured on the inside of the shell). The radii are then doubled for the diameters, which are the same as the greatest length and width of the egg.

Types of Egg Structures

Early in this chapter I discussed how thin sections are made in order to study eggshell under the microscope. The study of the shell microscopic structure, or microstructure, is more than just a scientific exercise—it is about the only way in which to identify various species of fossil eggshell. It is a human trait to put everything into little boxes, thereby making sense out of the complexity of the world, or in our case, eggshell. In chapter 6, I discussed how the eggshell of the various major egg-layers, turtle, crocodile, and bird, differs. These differences are very specific and consistent, allowing us to recognize six basic types of eggshell, which reflect the egg-layer.

Basic Types of Eggshell

The very essence of the eggshell is at the ultrastructural level, where we can see the individual crystals of calcium carbonate that make up the shell unit. It is at this molecular level that the characteristics of the eggshell are established and six basic types have been recognized.

Geckonoid Basic Type (Fig. 8.16A): The most primitive of the six types of basic shell design is found in the gecko egg. How the eggshell is produced has not yet been studied as it has in the chicken. Surprisingly, no nucleation site has yet been found at the shell–shell membrane interface; consequently, there are no shell units. Instead, the shell consists of homogeneous calcite, except near the top where they show a distinct vertical arrangement. Probably the shell is secreted as a whole, although how this is achieved is unknown.

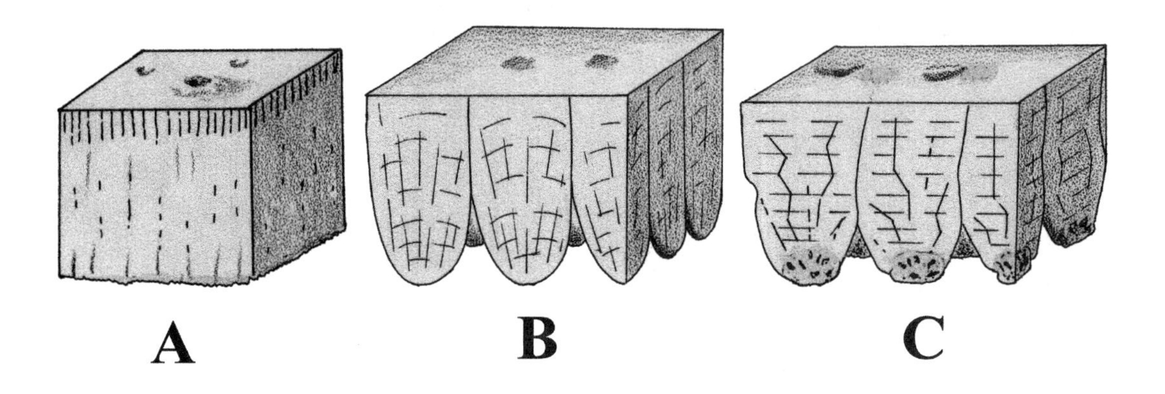

Fig. 8.16. The basic shell types seen in modern reptile eggshell shown as schematic block diagrams: A, geckonoid; B, testudinoid; C, crocodiloid. See text for details.

Fig. 8.17. (facing page) Spherulitic basic type of eggshell shows six variants or morphotypes: A, discretispherulitic; B, angustispherulitic; C, prolatospherulitic; D, filispherulitic; E, dendrospherulitic; F, reticulospherulitic. Almost all non-theropod dinosaur eggshell is of the spherulitic basic type.

Testudinoid Basic Type (Fig. 8.17B): Distinct shell units are present in hard-shelled turtle eggs, with an organic core for the nucleation site. The shell units are composed of radiating, slender crystals of aragonite. Superficially, the crystals resemble the eisospherites of bird eggs, and at one time were thought to be the same. It seems more likely that the similarity is more apparent than real, and is due to the simplicity of the ultrastructure of the turtle shell.

Crocodiloid Basic Type (Fig. 8.16C): The ultrastructure of the shell units of the crocodile egg is made up of tabular plates of calcite and loosely resembles the wedge zone of the bird egg. The shell units are also wedge shaped, being distinctly wider at the top than bottom, thus accentuating the similarity to the wedge zone. Again, this resemblance is incidental and has never been demonstrated as being the same as the wedge zone of a bird egg. There is no organic core at the base of the shell units, but rather an aggregation of calcite plates that serve as the nucleation site.

Spherulitic Basic Type (Fig. 8.17): The shell consists of a single ultrastructure composed of tabular calcite crystals arranged in a radiating pattern into shell units. The boundaries of the shell units can sometimes merge, making them difficult to discern except with polarized light. Each shell unit

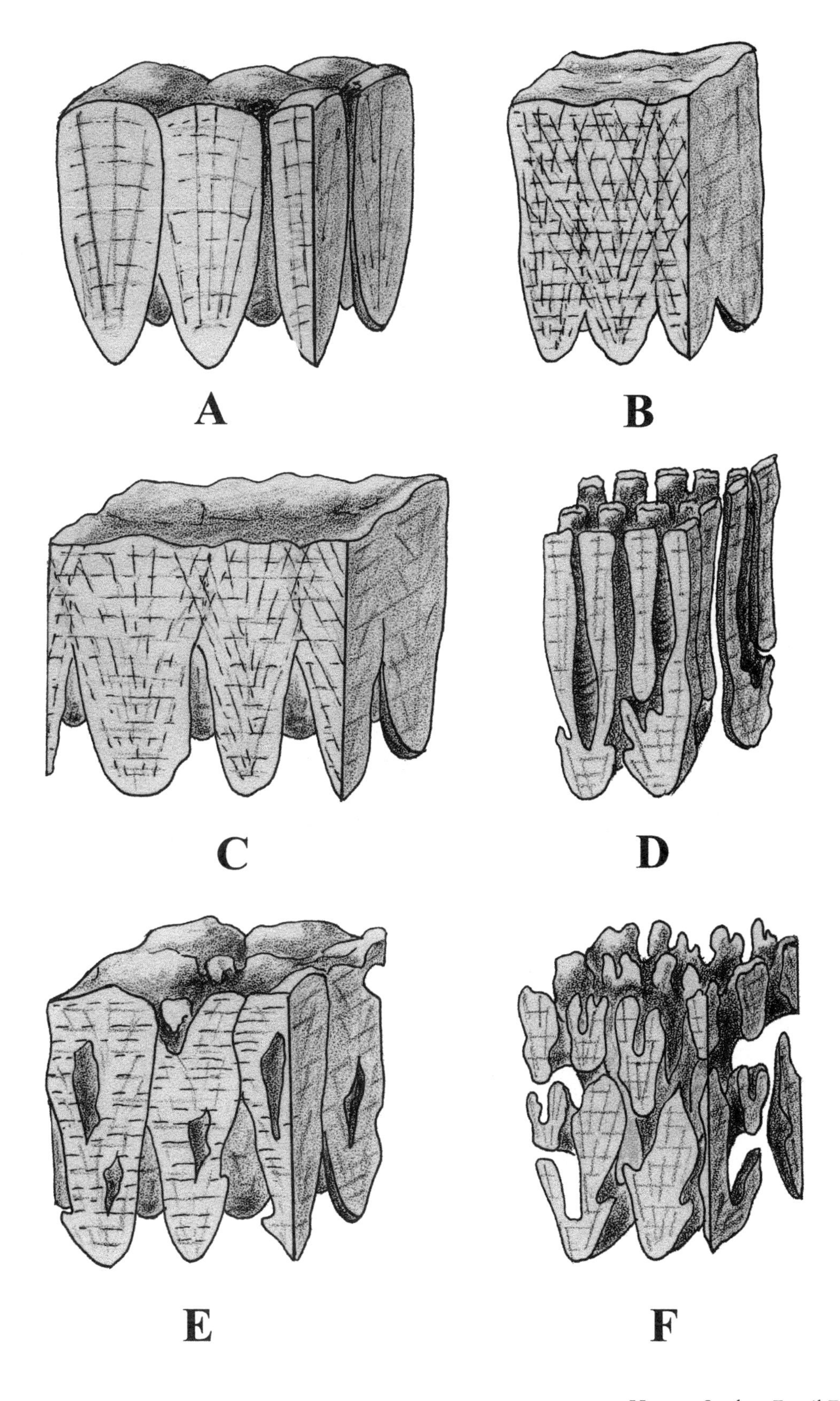
A
B
C
D
E
F

grows upward from an organic core attached to the shell membrane; thus technically, spherulitic eggshells do not have true mammillae at the base (Mikhailov, 1997). The mammilla-like structures on the inside surface of the shell are actually the bases of the shell units (a subtle but important distinction). There are several variations, called morphotypes, and these are discussed below. The spherulitic type of shell occurs in non-theropod dinosaur eggs and is sometimes referred to as dinosaur-spherulitic.

Prismatic Basic Type (Fig. 8.18A): The shell unit has a two-part ultrastructure with no clear boundary separating the two. The lower quarter to half has a radiating tabular ultrastructure much like the spherulitic shell units. The upper portion, however, changes into a more homogeneous tabular ultrastructure. The irregular boundaries of the shell units may or may not be seen in this upper portion; if not, then the upper portion may look like a single homogeneous structure. The individual shell units are usually more prominent when viewed with the polarized light microscope. Prismatic eggshell is present in primitive theropod eggshell and has in the past been referred to, in part, as dinosaur-prismatic.

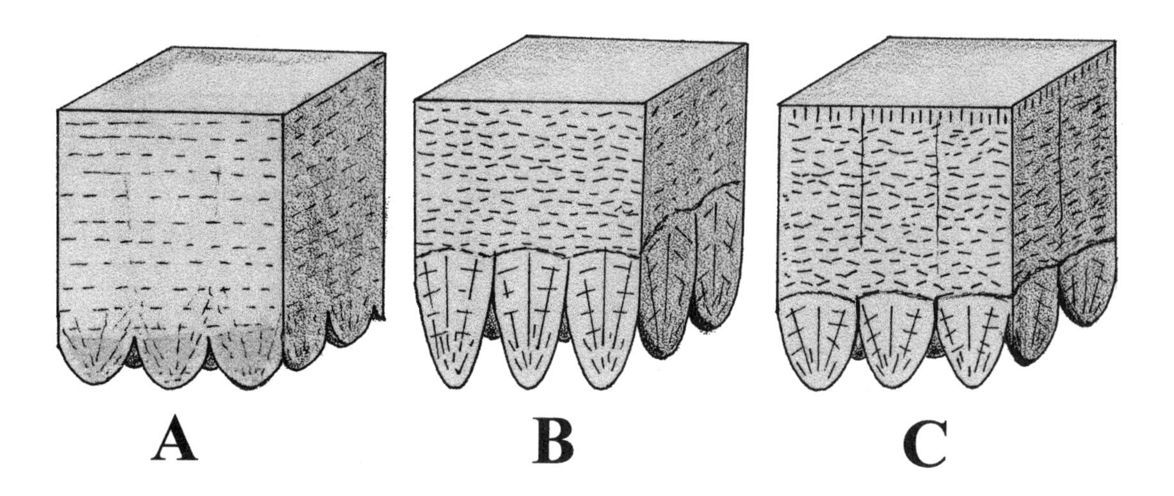

Fig. 8.18. Two other basic types of eggshell: A, prismatic; B, C, ornithoid. Ornithoid eggshell includes two morphotypes: B, ratite; C, neognathe. Most theropod eggshell is of the ornithoid basic type.

Ornithoid Basic Type (Fig. 8.18B): The shell resembles the prismatic type in having two zones, but the boundary between the two is very abrupt. The lower ultrastructural zone forms the mammillary zone, and consists of radiating calcite crystals surrounding an organic core, or tabular calcite crystals arranged into a wedge zone above the organic core. The upper layer has no distinct shell units and is in fact continuous. This portion has a distinct squamatic ultrastructure (Fig. 8.19). The ornithoid basic type is found in bird and theropod dinosaur eggshell.

Structural Morphotypes

There is a considerable amount of variation in the basic plan of dinosaur eggshell and this is called the structural morphotype. These vari-

*Fig. 8.19. Ornithoid eggshell of the ratite morphotype (*Elongatoolithus*) showing the sharp boundary between the mammillary layer and overlying continuous layer. Also note the faint growth lines in the continuous layer that parallel the surface, rising beneath the node. (Courtesy of K. Mikhailov.)*

ations are only seen in the spherulitic and ornithoid basic types of eggshell, and are apparently nonexistent in the geckonoid, testudinoid, crocodiloid, and prismatic types of eggshell. Let's look at those structural morphotypes that are known and have been defined. The first six are all variations of the spherulitic basic type.

1. Discretispherulitic (formerly called tubospherulitic): This may be the most primitive morphotype because the shell units are usually distinct, although tightly appressed against one another (Fig. 8.17A). The shell units have a fan-shape pattern and arch-shaped accretionary or growth lines. On the surface of the shell, each shell unit forms a bump or node. This morphotype is vaguely similar to the testudinoid basic type in that it is very simple, and as a result Gervais originally thought the large eggshells from France were giant turtle eggs (see chapter 1). We now know that this similarity is superficial. The discretispherulitic morphotype occurs in the large spherical eggs referred to as sauropods.

2. Angustispherulitic: The shell units are slender and partially fused

together (Fig. 8.17B). This fusion makes the individual units somewhat difficult to discern except in the lower one-third to one-half of the eggshell, where the fan-shape pattern may be seen. Growth lines in the fans are generally horizontal. In the upper portion, where adjacent shell units fuse, the fan-shaped patterns may overlap one another in a crisscross. Angustispherulitic shell is seen in elliptical eggs thought to belong to ornithopods.

3. Prolatospherulitic: The shell units are fan-shaped and broader than those of the discretispherulitic morphotype (Fig. 8.17C). In addition, they are partially fused together, making it difficult to trace out each shell unit. Under the polarized light microscope, horizontal growth lines can be seen. These lines are almost restricted to the lower two-thirds to three-quarters of the shell unit. On the shell surface, the shell units form large bumps or nodes. This morphotype of shell occurs in hadrosaur eggs.

4. Filispherulitic: The shell units consist of very long, slender, irregular branching prisms of calcite crystals separated by a honeycomb of pore canals (Fig. 8.17D). This morphotype is seen in eggs referred to as *Faveoloolithus* by Zhao and Ding (1976); at present, we do not know what kind of dinosaur laid the eggs.

5. Dendrospherulitic: The shell units are irregular fan-shapes that branch. Scattered among the wider shell units is a network of dead-end canals that may be remnants of pore canals left when some of the shell units grew together during shell formation (Fig. 8.17E). The dendrospherulitic morphotype is only known from eggs called *Dendroolithus* from Mongolia.

6. Reticulospherulitic: The shell units are irregular, branching prisms that appear to form a network in thin section (Fig. 8.17F). The shell is pierced by numerous, interconnected pore canals that give the shell an even more honeycombed appearance than in the dendrospherulitic morphotype.

Two other morphotypes have been identified among the ornithoid eggshells:

1. Ratite: The shell unit is made up of different types of calcite crystals: Radiating prisms in the mammilla, stacks of prisms forming wedges above the mammilla, and an upper zone called the continuous layer because shell units are fused together so that the individual shell units are not visible (Fig. 8.18B). The mammillae are distinct and may occupy up to half of the shell thickness. The name—ratite morphotype—is derived from the ratite bird (e.g., ostrich) egg, but it is also seen in most theropod eggshells.

2. Neognathe: The shell unit is similar to that seen in the ratite morphotype, except that the sides of the shell units are visible in the upper portion of the shell (Fig. 8.18C). This morphotype resembles the prismatic morphotype (see above under Prismatic Basic Type) and distinguishing the two can be difficult, especially if the specimens are less than pristine. Generally, the neognathe morphotype eggshell is thinner, and the eggs smaller than the prismatic morphotype. In addition, the microstructure of the prismatic layer of the eggshell is squamatic, meaning that etching with weak acid causes the calcite crystals to look rough, like lizard skin. There is also often a thin layer of vertical prisms at the surface of the shell. This eggshell morphotype is only known in recent and fossil birds.

Other Features of Eggshell

Besides the calcite structure of the eggshell, there are two other important features of eggshell that are used to identify the various types: the pore canals and the surface ornamentation or texture.

Pore Canals

As stated in chapter 6, the embryo needs to breathe and does so through tiny canals that connect the inside of the egg with the outside. Some pores are simple vertical canals, while others are more complicated, with changes in diameter and branches. These differences led Soviet paleontologist A. Sochava (1969) to develop an eggshell classification scheme based on them because they were easily seen. Unfortunately, Sochava's scheme proved impractical when it was realized that different eggs have similar pore canals. Still, pore canals are an important character of eggshells when used in conjunction with other characters. The names and definitions of pore canals are essentially the same as those proposed by Sochava, with the addition of several new names. The various types are illustrated in Figure 8.20.

1. Angusticanaliculate: The pore canals are long, narrow, and straight (Fig. 8.20A). The canals are small (0.01–0.1 mm) and not very abundant (about 3–20/100 mm^2), so there is relatively less gas exchange than other pore types (e.g., multicanaliculate). Mikhailov (1997) suggested that this type of pore is found primarily in eggs laid in a dry environment so as to limit evaporation of the albumen water.

2. Tubocanaliculate pore canals resemble the simple angusticanaliculate canals, but are larger in diameter (0.5–0.2 mm), and have enlarged, funnel-shaped openings on both the outer and inner surfaces of the shell (Fig. 8.20B). The pores are numerous (400–500/100 mm^2), indicating that gas exchange would be high, as would be water loss. This type of pore occurs in eggs that are buried in humid mounds (Mikhailov, 1997).

3. Multicanaliculate pores have many closely spaced canals that are often branching, hence the name meaning "multiple canals" (Fig. 8.20C). The canals are large (0.1–0.3 mm) and with numerous openings on the surface (600–1000/100 mm^2). Gas exchange and water vapor loss are high, restricting this egg type to humid mounds (Mikhailov, 1997).

4. Prolatocanaliculate canals vary in width along their length (Fig. 8.20D). They may be subdivided into foveocanaliculate, which have enlarged pore openings on the surface of the shell (Fig. 8.20I–K), and lagenocanaliculate, which have narrow pore openings (Fig. 8.20G, H). Generally, the canals are large (0.05–1 mm) and moderately abundant (30–150/100 mm^2). Gas exchange and water loss are moderate to high, suggesting that the eggs were laid in a variety of environments.

5. Rimocanaliculate pores have slitlike canals (0.01–0.03 mm × 2–5 mm). This type of canal co-occurs with angusticanaliculate and prolatocanaliculate in short chains of canal openings on the shell surface (Fig. 8.20E). The ostrich egg has rimocanaliculate pore canals (see Fig. 6.13), suggesting that dinosaurs producing eggs with similar pore canals laid their eggs in the open as ostriches do today.

6. Obliquicanaliculate pores have simple canals that extend diagonally thereby cutting across several shell units rather than extending between them as in all other canal types (Fig. 8.20F). Thus far they are only known from a single type of egg, *Preprismatoolithus*.

To determine the true shape of the pore canals requires multiple thin sections (or a lucky break that exposes the length of the pore canal on the edge of the shell fragment). The simplest method is to make a vertical or radial thin section and hope that a canal is intersected at some time during the grinding or polishing phase. Because a thin section is essentially a two-dimensional view of the eggshell, only the simplest straight canals can

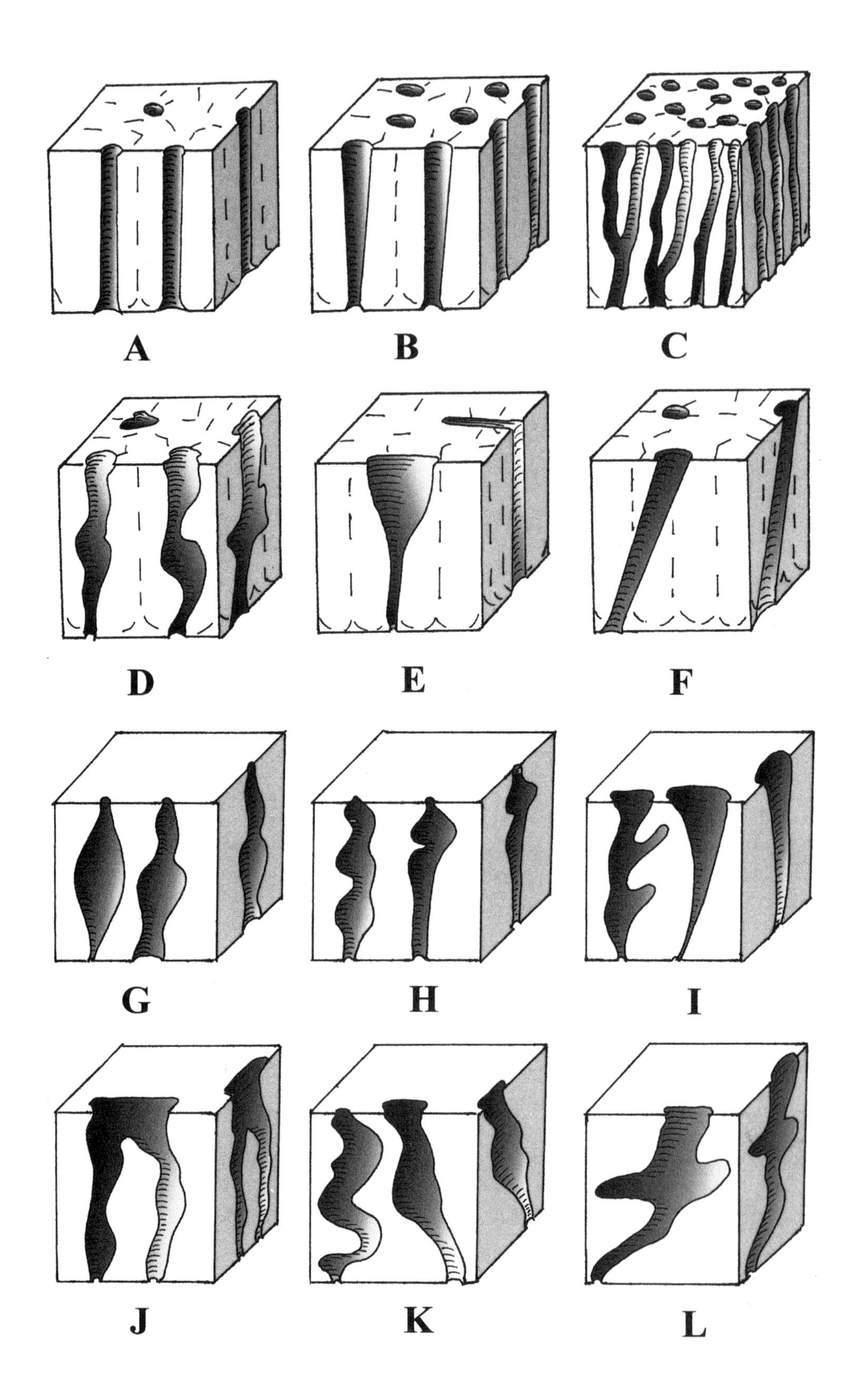
A
B
C
D
E
F
G
H
I
J
K
L

reliably be detected. A better technique, albeit more time-consuming, is to grind the eggshell tangentially (i.e., parallel to the shell surface) starting at the shell surface (Fig. 8.2). Fortunately, the pore canals are easily distinguishable from the surrounding shell because they are filled with mineral of a different color (usually the same as the host rock in which the shell or egg was found, but sometimes calcite). This difference allows the canals to be studied in cross-section while the shell is still too thick for light to pass through. Drawings or photographs of the canals can be made at intervals as the shell is ground down. By overlaying the drawings (on tracing paper), the meandering, branching, and changes in the diameter of the canals can be seen.

Surface Ornamentation

One other feature of eggshell that has proven useful in eggshell classification is surface texture or ornamentation. Although chicken eggs are relatively smooth (except for pores), other eggs can have a rough texture. This ornamentation may consist of bumps (nodes) or ridges and valleys scattered randomly or in patterns (Fig. 8.21). In radial thin section, growth lines usually follow the contour of the surface ornamentation, showing that the ornamentation was deposited as the egg formed and is not an artifact of later alteration of the shell surface (see Fig. 8.19). Several different types of ornamentation are recognized:

Fig. 8.20. (facing page) Variations in pores: A, angusticanaliculate; B, tubocanaliculate; C, multicanaliculate; D, prolatocanaliculate; E, rimocanaliculate; F, obliquicanaliculate. G–L, pore canals can vary considerably from these simple illustrations as demonstrated by variations of the prolatocanaliculate form. (Modified from Mikhailov, 1997.)

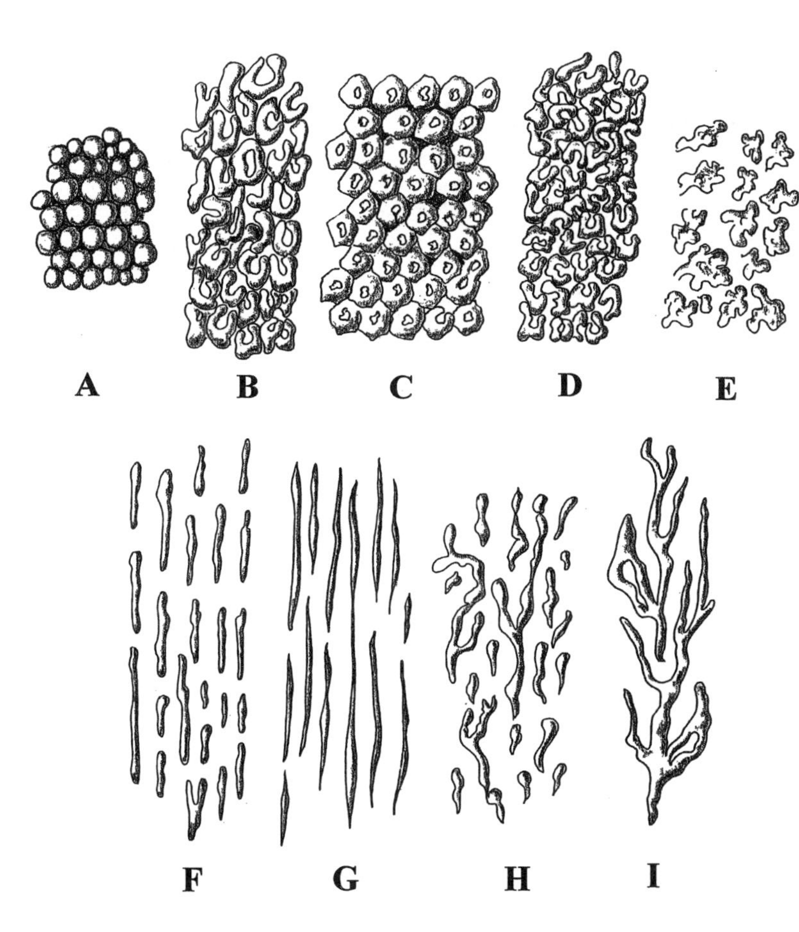

Fig. 8.21. Surface texture or ornamentation of eggshell: A, compactituberculate; B–D, sagenotuberculate; E, dispersituberculate; F–G, lineartuberculate; H, ramotuberculate; I, anastomotuberculate.

1. Compactituberculate ornamentation appears as a dense covering of nodes that are the tops of shell units (Fig. 8.21A).

2. Sagenotuberculate ornamentation has a netlike pattern of nodes and ridges, with pits and grooves between them (Fig. 8.21B–D). The ornamentation is produced on top of the shell units. It occurs on some spherical and ellipsoid eggs.

3. Dispersituberculate ornamentation consists of scattered, or dispersed, nodes (Fig. 8.21E). This ornamentation occurs on the poles of elongated eggs.

4. Linearituberculate ornamentation has linear ridges (hence the name) extending parallel to the long axis of the egg, and is especially well developed around the midsection (Fig. 8.21F, G). Sometimes the ridges are composed of numerous nodes or bumps forming linear chains.

5. Ramotuberculate ornamentation is produced by irregular chains of nodes (Fig. 8.21H). Some of these chains meander and split or join other chains. This type of ornamentation is usually found in the transition zones between the midsection and ends (called the poles) of elongated eggs.

6. Anastomotuberculate ornamentation forms linearituberculate ridges that are wavy and branched, or anastomosed (hence the name) (Fig. 8.21I). They resemble wind or water ripple marks in sand. This ornamentation is common in elongated eggs.

These different types of ornamentation correlate with specific egg shapes. Elongated eggs typically have linearituberculate, ramotuberculate, and dispersituberculate ornamentation that forms a continuum from the midsection (equator) to the ends (poles). Compactituberculate ornamentation occurs on spherical megaloolithid eggs.

In the next chapter, we shall see how all these long-winded names are used in the naming of eggshells.

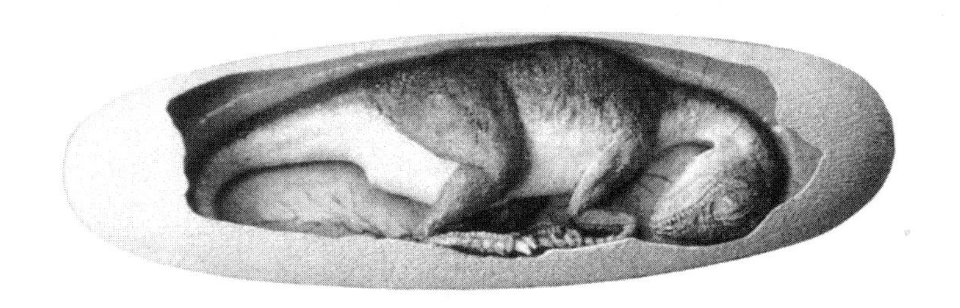

9 • What's in a Name

It is human nature to name things and probably has been since early humans could talk. Names succinctly identify an object or creature in a word or two which would otherwise take many words to describe: "a walking shag rug with a pair of long white things on its head and a tail on both ends" can be more concisely identified as "a woolly mammoth." Likewise, "an elongated egg, 51 cm long, with 3.2 mm thick eggshell with a ratite morphotype, linearituberculate, ramotuberculate, and dispersituberculate ornamentation, and angusticanaliculate pores" is more cumbersome than *Macroelongatoolithus xixianensis.* Because fossil eggs are traces of ancient life, we can treat them as zoological objects and subject them to the same zoological laws governing names, or taxonomy (Table 9.1).

Carolus Linnaeus (1707–1778) developed the hierarchical taxonomic system used today, whereby a plant or animal (living or fossil) is given a two-part name (binomial nomenclature, or "scientific name"). For example, modern humans are *Homo sapiens,* with *Homo* the genus name and *sapiens* the species name. In one sense, the genus name is like your family's last name, whereas the species name is comparable to your first name. You may have several members in your family, each with a different first name, but same last name. Similarly, with cats we have *Felis concolor* (mountain lion), *Felis leo* (African lion), and *Felis domesticus* (house cat). Using the analogy of your name to explain genus and species might seem backward since the last name is given first, but for a great many people in the world emphasis is placed on the family rather than the individual. Thus, Chinese egg paleontologist Zhao Zi-Kui would write his name Zi-Kui Zhao in Western convention. Hopefully this analogy of last name first, first name last helps you understand that the fossil eggs *Spheroolithus irenensis* and *Spheroolithus tenuicorticus* are similar enough to belong to the same genus, but different enough to belong to different species.

Taking this analogy of names one step further, your extended family (aunts, uncles, grandparents, etc.) may not all have the same last name, yet you share blood (actually, you share DNA, but you know what I mean). An extended family of sorts also occurs in taxonomy, although this is simply

TABLE 9.1.
Comparison of the Linnaean and parataxonomic classification schemes

Linnaean	Parataxonomic
Kingdom: Animalia	
Phylum: Chordata	Veterovata
Class: Mammalia	Basic Type: Spherulitic
Order: Primate	Morphotype: Prolatospherulitic
Family: Homonidae	Oofamily: Spheroolithidae
Genus: *Homo*	Oogenus: *Spheroolithus*
species: *sapiens*	oospecies: *irenensis*

called the "family." It comprises several genera (plural of genus), thus *Acinonyx* (cheetah) and *Felis* together make the family of cats, the Felidae (note the family name ends in -idae), and the fossil eggs *Elongatoolithus* and *Macroolithus* form the family Elongatoolithidae. Sometimes a family may have only one genus and species.

Note that the genus and species names are given in italics and the family name is not. Why? Partly convention, but partly to make it clear that a name is referring to the scientific name of a specific animal, not its common or vernacular name.

One other point we need to address is why scientific names are so difficult to pronounce. It's really all about communication. Linnaeus' classification scheme attempted to standardize the names given to plants and animals. He wanted a zoologist in England to understand that when a zoologist from the United States was writing about the dolphin, he was referring to *Delphinus,* the whale, not *Coryphaena,* the fish. The problem gets worse when zoologists don't even speak the same language. Under the standardized Linnaean system, an egg paleontologist from China, one from France, and another from India all understand what an egg paleontologist from Russia means by *Elongatoolithus* (it certainly beats saying, "the long, skinny egg, with bumps and ridges on its surface, having a shell structure similar to an ostrich egg"). The complexity of the names is because Linnaeus decided to use Latin in his system. Why Latin? Because it was one language that the learned gentry throughout Europe at the time Linnaeus lived all shared because of the influence of the Catholic Church. (Mass was still given in Latin.) Greek is also used sometimes as an alternative.

Today, the International Code of Zoological Nomenclature sets forth guidelines for proposing new scientific names, thus ensuring that the names are valid. This code is followed throughout the world and still requires that the scientific names be in Latin or be Latinized. Thus, when Zhao Zi-Kui and Li Zuocong named some eggs found near Wangdian village, the species name became Latinized: *Dendroolithus wangdianensis* (*-ensis* is a common way of Latinizing a non-Latin name).

Early Methods of Egg Classification

Naming fossil eggs actually goes back to the earliest discovery of eggs in England. In 1859 (the same year that Charles Darwin's *The Origin of Species* appeared), J. Buckman described a clutch of eggs under the name *Oolithes bathonicae* ("stone eggs from Bath"). The description was based on obvious exterior features, called macrostructure (meaning large structure), that could be seen with the eyes, including egg shape and size (Table

TABLE 9.2.
Changing emphasis in the study of fossil eggshells

Study	Features	Details
Buckman 1859–*Oolithes bathonicae* Linnaean Classification scheme		
	macroscopic features	
		size
		shape
Pouech 1859		
	macroscopic features	
		curvature to estimate egg size
Gervais 1877		
	microscopic features	
		shape of the shell units
van Straelen 1925		
	macroscopic features	
		surface ornamentation
Young 1954		
	macroscopic features	
		width/length
Sochava 1969		
	microscopic features	
		shape of pore canals
		mammillae and spongy layer (continuous layer)
Erben 1970		
	ultrastructure of shell with the SEM	
Chao and Chiang 1974 (= Zhao Z. and Jiang Y. K.)		
	microscopic features	
		shell thickness
		mammillae thickness
		number of mammillae/cm^2
Zhao 1975–Parataxonomic scheme		
	macroscopic features	
		arrangement in a clutch
		dimensions of egg
		surface ornamentation
	microscopic features	
		shell thickness
		mammillae thickness
		pore dimensions (diameter, length, shape)
		shape of shell units
		proportion of mammillary zone to shell thickness
Mikhailov, Bray, and Hirsch 1996–formalization of parataxonomy		

9.2). Unfortunately, this technique does not allow egg-shaped stones or mineral deposits to be separated from true fossil eggs, which can be a problem, as we saw in chapter 7. A further problem develops when the eggs are represented only by fragments. Size and shape of the egg are unknown, so macrostructure is of limited use. That was the problem facing Paul Gervais, who was introduced in chapter 1. Gervais attempted to identify the egg-layer by microscopically comparing the fossils with eggshells of known living egg-layers. Pouech had earlier used the curvature of the shell fragment to estimate a radius of 18 cm, or 7" (he assumed a nearly spherical egg). This radius, he noted, indicated an egg with four times the volume of an ostrich egg. Gervais also made note of the shell thickness, a point that would be used later by Vianey-Liaud and her colleagues (1994) to differentiate similar-sized eggs from France. In the 1920s, van Straelen added surface texture or ornamentation to the list of physical characters used in describing eggs.

The strictly "eyeball" method of separating egg types was not very exact, a problem that became greater once hundreds of eggs were found in China during the 1950s. To resolve this difficulty, Chinese paleontologist C. C. Young used width/length ratios to quantify the differences among eggs. By this method, the closer the ratio was to 1 the more spherical it was. For example, an egg 8.0 cm long and 7.0 cm in diameter would have a ratio of 0.87 (7.0/8.0 = 0.87). In contrast, an egg 14.3 cm long and 7.0 cm in diameter would have a ratio of 0.49. In these examples, the first egg was more round than the second, and was named *Oolithes spheroides;* the second egg was more elongate and was appropriately named *Oolithes elongatus.* The shortcomings of even the use of width/length ratios became apparent through the studies of Chao and Chiang (Wade-Giles romanization of Zhao Z. and Jiang Y. K.) in 1974. They discovered that eggs referred to *Oolithes spheroides* could be further divided into five types based on the thickness of the shell and microscopic features.

In Russia, little was done with the dinosaur eggs collected in Mongolia in the late 1940s and the 1950s until Sochava looked at them in the 1960s. She looked at their microscopic structure and noted that some were very similar to modern bird eggshell; these she called ornithoid. During the course of her studies, which also included eggshells from Soviet Central Asia, she noted that three distinct types of pore canals were present; these she named angusticanaliculate, prolatocanaliculate, and multicanaliculate (Sochava, 1969). At about the same time, Erben was working with the scanning electron microscope and noting detail finer than what could be seen with the light microscope. This level of detail, at the micron level, he called the ultrastructure.

Growth of the Modern Classification System

Zhao Zi-Kui continued his study of fossilized eggshell, and realized that it was not practical or possible to name eggs after the egg-layer unless the embryo was present. For example, elongated eggs from Mongolia were long assumed to be those of *Protoceratops* because the eggs and *Protoceratops* skeletons were the most common fossils found at the Flaming Cliffs (see chapter 1).

By 1975, however, Zhao began to have doubts about the original identification of these eggs and urged that a more neutral approach be taken in the naming of eggs. This approach, called parataxonomy ("beside taxonomy"), parallels the classification of animals in having families, genera, and species (Table 9.1). However, unlike the animal classification, parataxonomy makes no assumption about the identity of the egg-layer. The classification is based only on a physical description of the egg, including macrostructure, microstructure, and ultrastructure. Hypothetically, therefore, the same egg type could be laid by two different but closely related dinosaurs. For example, two very closely related genera of crested hadrosaurs, *Corythosaurus* and *Lambeosaurus,* are separated primarily by the shape of the crest on the adult skull. Because there are few other distinguishing characters in their skeletons, it is quite probable that these two genera of hadrosaurs laid eggs so identical in structure that a single egg species name could be applied to both of their eggs. But even if an embryo for either of these hadrosaurs was found in an egg, it is doubtful that it could be identified as either genus with certainty because the characteristic crest does not develop until long after hatching (see chapter 12).

The concept of parataxonomy was slow to be adopted by non-Chinese paleontologists because Zhao's work was published in Chinese. Beginning in 1991, Russian paleontologist Konstantin Mikhailov published a series of papers on Mongolian eggs housed in the Paleontological Institute in Moscow that further developed egg parataxonomy (Mikhailov, 1991, 1994a,b, 1997; Mikhailov and others, 1994). These papers were either published in English or were soon translated into English from Russian. Nevertheless, it took several more years until egg parataxonomy began to be used worldwide (e.g., Vianey-Liaud and others, 1994; Khosla and Sahni, 1995; Kohring and Hirsch, 1996). Initially, Mikhailov considered parataxonomy as an informal classification scheme, but later accepted that the principles were as valid as the traditional scheme used to classify plants and animals. The basic principles were recently formalized by Mikhailov and others (1996).

Principles of Egg Parataxonomy

Parataxonomy is a method of naming and classifying fossil eggs based upon physical characters of the egg, not upon who laid the egg. Therefore, the list of characters that define or diagnose an egg must be those that apply to all eggs of the same species. In other words, preservation peculiarities such as crushing are not used, but the type of ornamentation of the shell is used.

Parataxonomy employs the Greek prefix "oo," meaning "egg," to distinguish the egg species, genus, and family: oospecies, oogenus, and oofamily. Although technically and visually distinct, pronouncing the words sounds funny, and it remains to be seen if the words replace the longer but clearer and more easily said "egg species," "egg genus," and "egg family." Parataxonomy also uses the Greek root "*-oolithus*" (meaning stone egg) in the genus and family name (e.g., *Spheroolithus,* Spheroolithidae). This root avoids confusion with animal names: *Elongatoolithus* vs. *Tyrannosaurus.*

But how is an egg species, genus, or family recognized? The criteria for the species include the range of egg size (length, width, and volume), range of eggshell thickness (best determined from a clutch of eggs because all the eggs are of the same species), pattern of the pores on the surface of the shell, and details of the ornamentation. The egg genus is distinguished by egg size, morphotype, type of pore canals, and surface ornamentation. And finally, characters of the egg family include morphotype, pore canal type, and surface ornamentation. Note that the same criterion (e.g., size) may be used to define the egg species as well as the egg genus, although the range of difference is different. For example, the eggs of *Dendroolithus* range from 60 mm to 210 mm (2⅓" to 8¼") in greatest diameter, but its three species have a narrower range: *Dendroolithus microporosus* 60–70 mm, *Dendroolithus verrucarius* 90–100 mm, and *Dendroolithus wangdianensis* 170–210 mm (when the same genus name is repeated, it is common convention to just use its first letter; thus: *Dendroolithus microporosus, D. verrucarius,* and *D. wangdianensis*).

Why are there ranges or variation in the size of these different egg species? Variation occurs because eggs are not stamped out by some organic cookie cutter. Instead, they are formed by organic tissues and subject to the "whims" of those tissues. For example, we know that eggshell thickness varies even among chicken eggs by as much as 20 percent (Romanoff and Romanoff, 1949), but we do not know why this happens. Is the egg held a few minutes longer in the egg gland region of the oviduct so that more shell material is secreted? Or are the shell glands overly active at that particular time? Regardless, the range of variation of the shell is taken into account in the diagnosis of an egg species, but only if it is consistent and well defined. For example, suppose you measured 200 fragments of dinosaur eggshell and most clustered between 1.1 mm and 1.4 mm in thickness, with one fragment having a thickness of 2.3 mm. The range of shell thickness used in the diagnosis would be 1.1–1.4 mm, not 1.1–2.3 mm because the one measurement is so inconsistent with the majority of the measurements. The odd fragment may indicate a fragment of a different species of egg washed into the site, or that the eggshell is pathological.

Let's look at a diagnosis for an egg genus and species taken from Appendix II. The diagnosis is an abbreviated description of the genus or species.

> *Elongatoolithus* Zhao, 1975
>
> The eggs are up to 17 cm (6½") long and have a length/diameter ratio of 2–2.2. The surface is covered with a fine linearituberculate ornamentation. There is a sharp separation of the mammillary and continuous layers. Wavy growth lines are present in the continuous layer and parallel the surface of the shell. The mammillary layer is about ⅙–½ the shell thickness. Eggs are usually paired in a clutch, with their wider end out. Known from the Lower and Upper Cretaceous of China, Mongolia, and Kyrgyzstan. Originally called *Oolithes elongatus* by Young, 1954.

The information given means that *Elongatoolithus* was named by Zhao in 1975 for eggs previously named by Young (1954) as *Oolithes elongatus.* No species name is given, so the diagnosis, or list of distinguishing features, is only for the egg genus. The shell surface has a fine ornamentation described as linearituberculate (see chapter 8), meaning that the ornamentation has the form of long, narrow ridges. The rest of the diagnosis describes the size of the eggs, where the shell is thickest, and how the eggs are arranged in clutches.

> *Elongatoolithus andrewsi* Zhao Zi-Kui, 1975. Eggs 13.8–15.1 cm (5½–6") long, 5.5–7.7 cm (2–3") in diameter. Length of eggs is 2–2.5 times the diameter. The shell is 1.1–1.5 mm thick, and the mammillary layer is about ¼ of the shell thickness (Fig. AII.19). Clutch is about a dozen eggs.

Again, *Elongatoolithus andrewsi* was named by Zhao in 1975. Because the species name is given, the diagnosis is for the species, not the genus, which was given earlier (no need to repeat all that). The rest of the diagnosis gives additional information about what characterizes the egg species.

With this information, you can now translate the diagnosis for the eggs given in Appendix II.

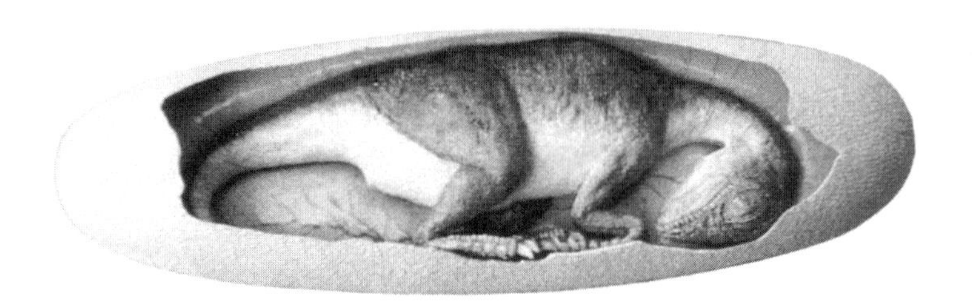

10 • The Nest

The Nest as Home

Before eggs can be laid, a place must be prepared to receive them. Such a place is called a nest. The different styles of nests are almost as varied as the animals that lay eggs. The nest may be as simple as a small clearing in the gravel, as with penguins, the elaborate apartment building–like structure of the Sociable Weaver (*Philetairus socius*) of Africa, or the vegetation mounds of the Mallee Fowl (*Leipoa*) and crocodiles (Fig. 10.1). The nest structure of dinosaurs can be inferred, but not always with a great deal of confidence. The problem is that the sedimentary environments that best preserve dinosaur eggs are not those that preserve the vegetation that may have been incorporated into the nest. Because the structure holding the eggs is generally not preserved, a group of dinosaur eggs is more properly called a clutch.

A nest is supposed to be a safe haven for the eggs, protecting them from rain, cold, heat, and predators. Without a nest, the world can be a very hazardous place for a developing embryo. The embryo lies helpless within an egg, unable to defend or hide itself from predators or protect itself from the environment. Dinosaur eggs undoubtedly faced many of the same hazards as eggs today. It is therefore possible that dinosaurs dealt with such problems in a similar manner as modern egg-layers. In some modern animal species, the parents offer some protection to the nest by keeping predators at bay, unless the predators work in teams where one distracts the guardian, allowing the other to grab an egg. Such behavior has been reported for large monitor lizards preying on crocodile eggs (Cott, 1961).

To some extent environmental conditions of the egg can be controlled by brooding (sitting on the eggs). During brooding, the parent uses its body to keep the eggs at the proper incubation temperatures, usually between 33° and 37° C. This may involve sheltering the eggs from rain or sun, or providing them with body heat. Some birds develop a brood patch, a bald

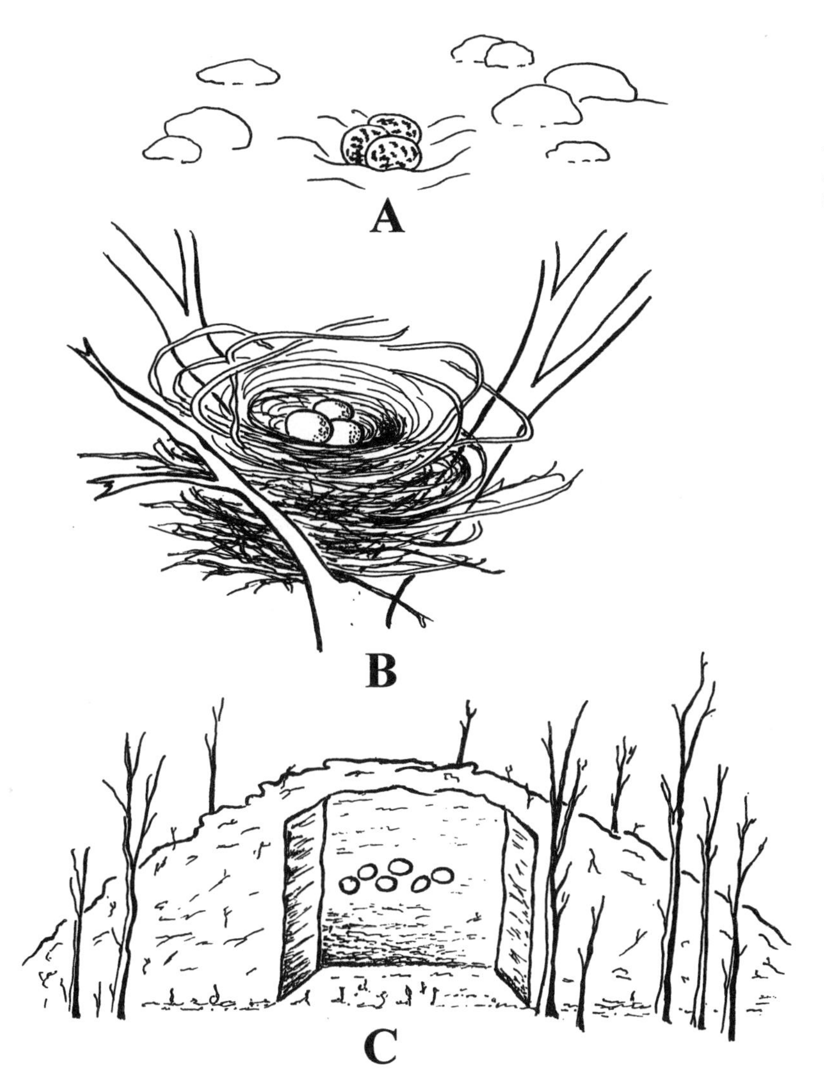

Fig. 10.1. Major nest types: A, ground nest; B, tree nest; C, mound nest.

or nearly bald spot on the chest or belly. The exposed skin ensures that the eggs are not insulated from the body heat of the parent by the feathers. The appearance of the brood patch is usually controlled hormonally and it only appears during the breeding season. Some birds pluck their chest bald to make a brood patch, using the feathers for the nest. The female python may also brood her eggs by producing heat through muscle contractions. The female crocodile does not brood, although she may rest her chin or belly on the nest while on guard. Brooding has been alleged for *Oviraptor* because the adult skeleton was found among the eggs (Norell and others, 1995). But unless the *Oviraptor* was regulating the temperature of the eggs with its body, the behavior may not have been brooding in the strictest sense.

Hazards of the Nest

Predation is the greatest hazard faced by an egg. For a predator, an egg is a very high source of nutrition in a small package. Although eggs may contain up to 87 percent water, the remainder is proteins, fats, carbohydrates, lipids, and minerals (iron, calcium, magnesium). A good deal of these nutrients are in the yolk, which is greater by volume in the eggs of precocial, or ground-nesting birds than among altricial, or tree-nesting birds. For example, the egg of the ground-nesting Brush Turkey of Australia has a yolk that is 62 percent of the egg volume, whereas in tree-nesting passerine birds such as the robin it is only 20 percent of the egg volume. It seems a safe assumption that most dinosaur eggs had large yolks as well. Because of their eggs' high nutritional content and accessibility, egg loss by predation is very high among ground-nesting birds. Ostrich egg losses are approximately 46–58 percent during the incubation period (Bertram, 1992). The loss can be higher for eggs laid early in the season. To offset this loss, some ostriches will lay a second batch of eggs if there is still time for the eggs to hatch and the young to mature.

Evidence of predation on fossil eggs is rare, yet has been documented by Williams (1981) and Hirsch and others (1997). Williams concluded that marks on an egg were made by a large mammalian predator, whereas Hirsch and others concluded that peck marks on bird shell fragments indicated predation by a large bird (Fig. 10.2). Predation of dinosaur eggs is difficult to prove, but must have been quite common. Perhaps some of the large sauropod eggs from France provide this evidence. These eggs show the tops of the eggs collapsed inward, with the pieces lying within the egg. Some French paleontologists have suggested that these eggshell pieces belonged to the "hatching window," that is, where the hatchling pushed its way out of the egg (e.g., Cousin and others, 1994). However, as pointed out by Polish paleontologist Karol Sabath (1991), if this were true, the eggshells should be outside the egg. Their presence inside suggests implosion of the egg as would be expected if a predator smashed in the top.

To lessen the chances of predation, many birds nest in trees, cliffs, or other high areas. It is not known if any of the smaller dinosaurs nested in

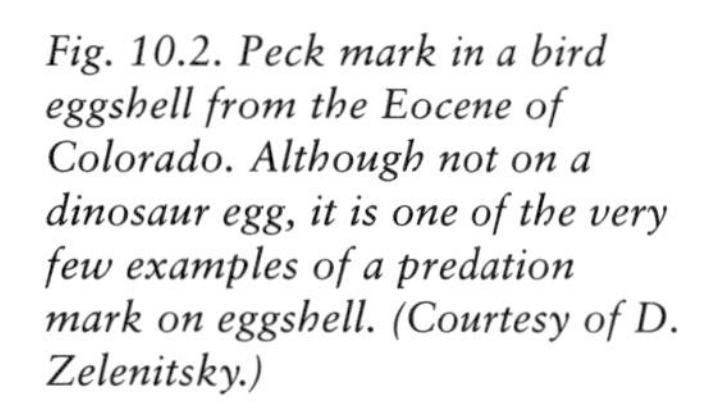

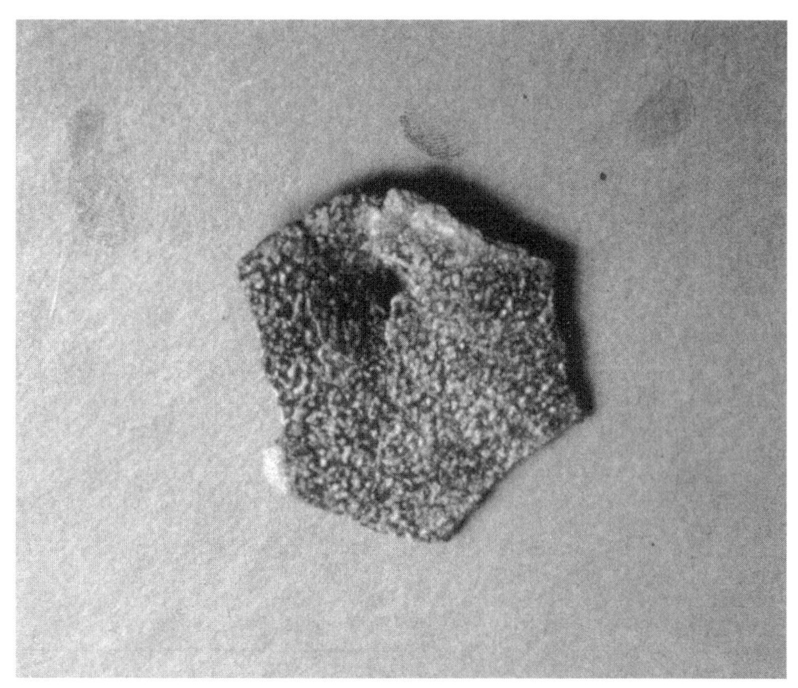

Fig. 10.2. Peck mark in a bird eggshell from the Eocene of Colorado. Although not on a dinosaur egg, it is one of the very few examples of a predation mark on eggshell. (Courtesy of D. Zelenitsky.)

trees, although the possibility cannot be discounted. Some reptiles, notably snakes, decrease the risk of egg predation by retaining the eggs within the body, as discussed in chapter 3. The young are thus born "live." Whether any dinosaurs engaged in this type of behavior is not known, despite allegations to the contrary. As early as 1883, famed paleontologist Othniel Charles Marsh of Yale University speculated on live birth. Marsh gave a very brief description of what he called a fetal sauropod skeleton found in association with an adult skeleton (Fig. 10.3).

The specimens came from the Upper Jurassic Morrison Formation at Como Bluff, Wyoming. The implication of the association was that sauropods were ovoviviparous, with the eggs retained in the body of the mother where they hatched and the young were birthed live.

More recently, Robert Bakker (in Morell, 1987) has claimed that the large pelvic canal in sauropods demonstrates that they gave birth to live babies. Perhaps, but at present we are unable to separate male and female sauropods, and all sauropod skeletons found to date have large pelvic canals. Surely we haven't only found females! It may be just as likely that the adult found with the young sauropod bones is a male, in which case the Marsh-Bakker hypothesis can be dismissed. In addition, there are now specimens of embryonic sauropod bones within the large *Megaloolithus* eggs (Chiappe and others, 1998). So, for now, there is no evidence that dinosaurs retained their eggs in the manner of certain snakes, but there are very good reasons to suspect that they all laid eggs.

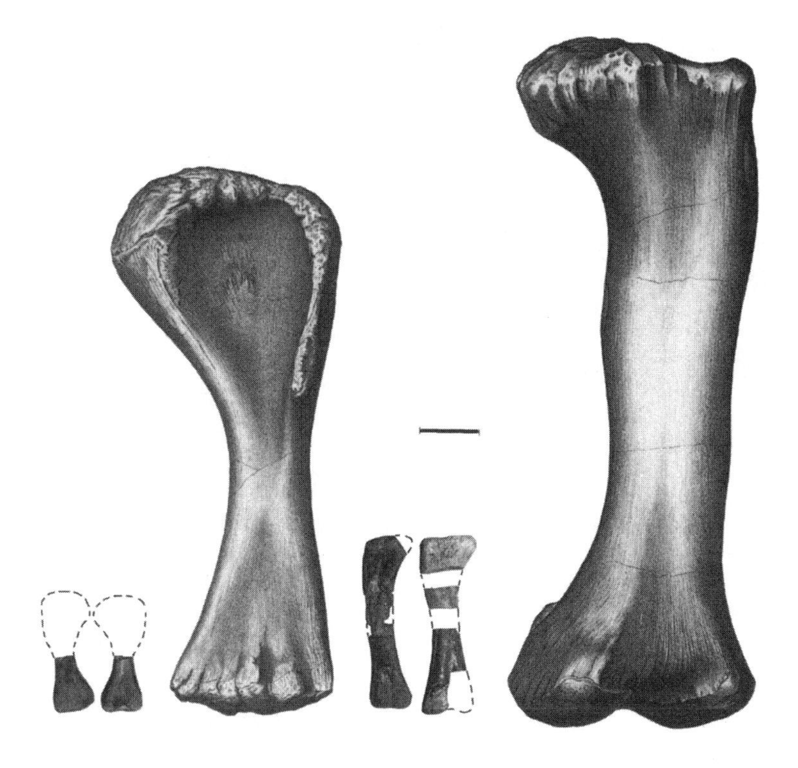

Fig. 10.3. Comparison of the juvenile and adult leg bones of Camarasaurus grandis: *left, humeri; right, femora. O. C. Marsh referred to the small bones as a "foetal" dinosaur, implying that sauropods gave live birth. Considering how few bones there are (there are parts of four vertebrae also), it seems more probable that the occurrence of these bones with an adult sauropod was incidental.*

Although dinosaurs probably did not give "live birth," there is evidence that at least some of them protected their eggs. A good example of such behavior is that of *Oviraptor* (Fig. 10.4, 10.5), several skeletons of which have been found associated with clutches of eggs. Few of the eggs are directly beneath the skeleton, so it seems unlikely that many of them were in direct contact with the body (i.e., brood patch). That being the case, it seems doubtful that the *Oviraptor* was brooding the eggs in the manner of a bird. Instead, it seems more likely that the *Oviraptor* was guarding the eggs from a marauding predator. We'll return to this topic later.

Making a Nest

Although we cannot rule out the possibility that some small, nonavian dinosaurs nested in trees, we can only study the remains of eggs laid on or in the ground, or fallen out of a tree. It is very difficult to determine, however, if the eggs were laid on the ground, buried in the ground or in a

Fig. 10.4. A, skeleton of an oviraptorid associated with eggs. B, scene as reconstructed using a skeleton. C, embryonic oviraptor found in an egg.

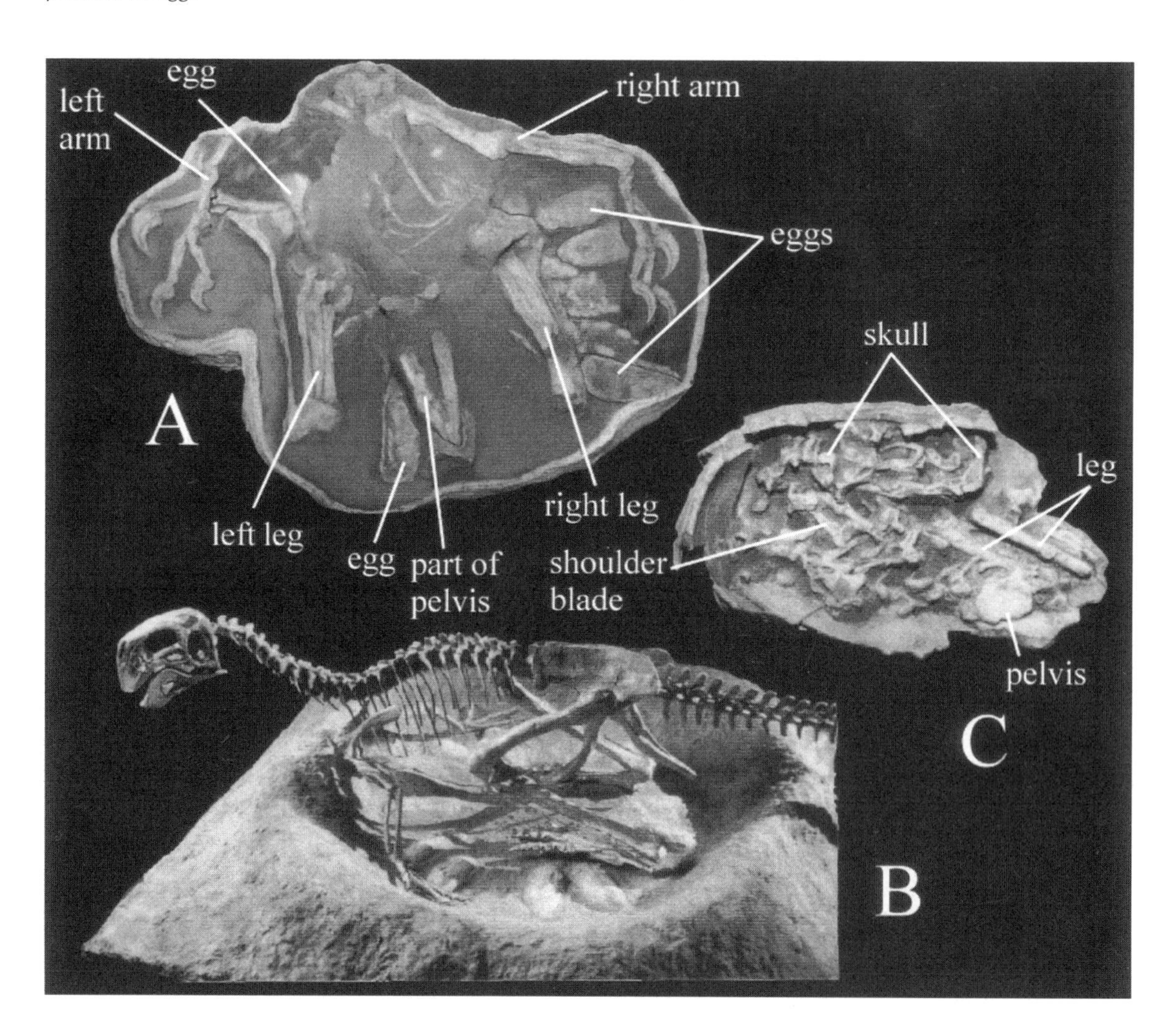

Fig. 10.5. Oviraptor *reconstructed sitting among eggs. Although the adult may be trying to cover all the eggs with its body, it is clear that many eggs were not in contact with the body indicating that true avian brooding did not occur. (Courtesy of G. Paul.)*

mound of vegetation, or something in between. Considering how diverse nests are among birds today (Table 10.1), we may assume diverse nests among dinosaurs as well. Some nests, however, such as the colonial nests of the weaver bird, require flight, so we may rule out these types of nests from nonavian dinosaurs.

Why such a variety of nests in birds today? Each nest type has advantages and disadvantages. The nest used is the one best suited for a particular bird and its lifestyle; there is no "one-size-fits-all" among nests. The simplest nest is no more than a bare patch of ground (Fig. 10.1A). Such a nest is used by such diverse birds as the Grey Gull (*Larus modestus*) and the ratites, such as the ostrich (Collias, 1991). In vegetated areas, this patch may be formed by scraping away the plants, a form of behavior best seen in ratites. Another nest form is actually within the ground. The Egyptian plover simply buries its eggs in the sand where the sun's heat incubates them. The European Bee-eater (*Merops apiaster*) excavates a tunnel a meter or two long in an earthen bank. The digging is done with the bill and the dirt kicked out with the feet. The burrowing owl is lazier and nests in prairie dog burrows.

TABLE 10.1.
Nest types of modern birds and reptiles

Egg-Layer	Nest Type	Gender of Nest Builder	Incubation Type	Incubation by Gender	Average Incubation Time
ostrich (*Struthio*/ratite)	scraped area on ground	male	brooding	male and major hen	35–42 days
robin (*Turdus*/ passerine)	open cup of twigs, grass	female	brooding	female	11–14 days
Brush Turkey (*Alectura*/ galliform)	vegetation and earth mound	male	fermentation of mound	(none)	49 days
European Bee-eater (*Merops*/ coraciiform)	burrow in mud bank	male and female	brooding	male and female	20–22 days
alligator (*Alligator*)	vegetation and earth mound	female	fermentation of mound	(none)	63–65 days
Nile crocodile (*Crocodylus*)	vegetation and earth mound	female	fermentation of mound	(none)	84–98 days
sea turtle (*Caretta*)	bury in sand	female	solar heat	(none)	31–65 days
Komodo lizard (*Varanus*)	bury in hole	female	solar heat	(none)	240–255 days

The most elaborate nests of any ground-nesting bird are the vegetation and earthen mounds of megapode birds, such as the Brush Turkey. It is the male who constructs the nest and guards it until the eggs hatch. About five months are spent building the nest. Once a suitable place is found, a shallow, elongated pit about 1 × 3 m (3¼' × 10') is excavated with the feet. Next, the pit is filled with leaf litter until a 60 cm (2') mound is built (Fig. 10.1C). This mound is then sealed by a thin layer of dirt. As the leaf material decays, heat is produced by fermentation, raising the internal temperature to 29°C (84°F). Into this fermenting mass, eggs are laid at six- to seven-day intervals until there is a clutch of fifteen to twenty-four eggs. The eggs incubate for two months at temperatures between 32° and 36°C (90°–97°F). The male maintains the nest, checking its temperature with prods of his bill during the early morning. If the nest temperature is too hot, he will cool the nest by stirring the mound. This stirring also ensures that there is sufficient oxygen for the eggs. Even so, the oxygen level is often less than 10 percent of atmospheric levels, whereas carbon dioxide may be 10–12 percent higher. As the mound temperature rises during the day, its height is reduced—then is restored in the late afternoon to trap the heat for the night. If the weather is too hot or cold, the mound is left at its nighttime height or made even higher so as to insulate the eggs (Fig. 10.6). Experiments have shown that the mound temperature rises to incubation temperature in about two and a half months through fermentation of the vegetation, whereas it takes six months by the sun alone (Frith, 1957).

The alligator and caiman make a vegetation mound similar to a megapode's, although an initial pit is not dug. Over a span of several days, the female gathers vegetation and mud in her mouth and piles them until a mound 1.8 m (6') in diameter and a meter (3') tall is built. Then the eggs are deposited in the top center of the mound. The female next crawls back and forth over the mound to smooth it into a conical structure. Decay of the vegetation raises the temperature of the nest. The female alligator does not monitor the nest temperature as does the male megapode. The eggs are kept from overheating by being close to the surface where excess heat can escape (Joanen, 1969; Deitz and Hines, 1980).

Other crocodilians nest in pits excavated in sandy soil (Cott, 1961). Here, the female may spend a considerable amount of time digging "test" pits until she finds a suitable spot. What exactly she is looking for is unknown, but like females everywhere, she knows it when she finds it. After digging a hole about 60 cm (2') deep, she deposits her eggs and covers

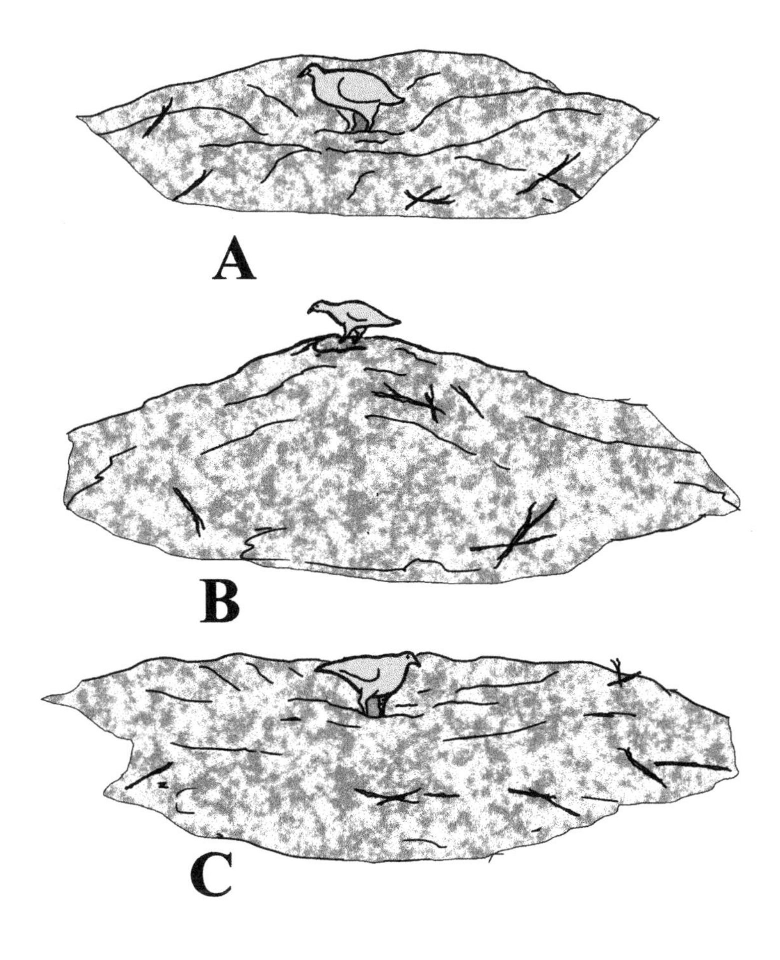

Fig. 10.6. Sketches from photographs showing a male Mallee Fowl adjusting the insulation of the incubation mound. A, mound being opened to release excess heat from the fermentation of vegetation. B, mound built up to its maximum in order to insulate the eggs from the sun. C, the mound is opened to let the sun's warmth reach the eggs. (Drawn from photographs of Frith, 1959.)

them with sand. She then stays nearby to guard her nest against predators, especially from the Nile monitor, a very large lizard. The sun's heat warms the sand, thus incubating the eggs. Because all the heat is obtained from the sun, the nest may undergo greater temperature fluctuations than the vegetation mound nest. That is because, although the earth loses its heat slowly, it is not accumulating any at night or on cloudy days. Other reptiles nest in secretive places among rocks or vegetation (Porter, 1972) as we discussed in chapter 3.

Dinosaur Eggs and Nests

We are hard-pressed, however, when it comes to nests made by dinosaurs. Any vegetation they may have used has rotted away, leaving only clutches of eggs in rock. Polish paleontologist Karol Sabath has taken this into account in his study of dinosaur clutches (1991). He believes that evidence for vegetation mounds can be seen in the preservation of some clutches today (Fig. 10.7). Sabath has noted that elongatoolithid eggs with a linearituberculate ornamentation are found in a radial pattern (Fig. 10.7D) and suggested that the high points of the ornamentation kept contact between the vegetation and the egg at a minimum. The grooves that separate the linear ridges allowed gases (oxygen and carbon dioxide) access to and from the pores. As fermentation continued with the slow decomposition of the vegetation, eventually there would not have been enough

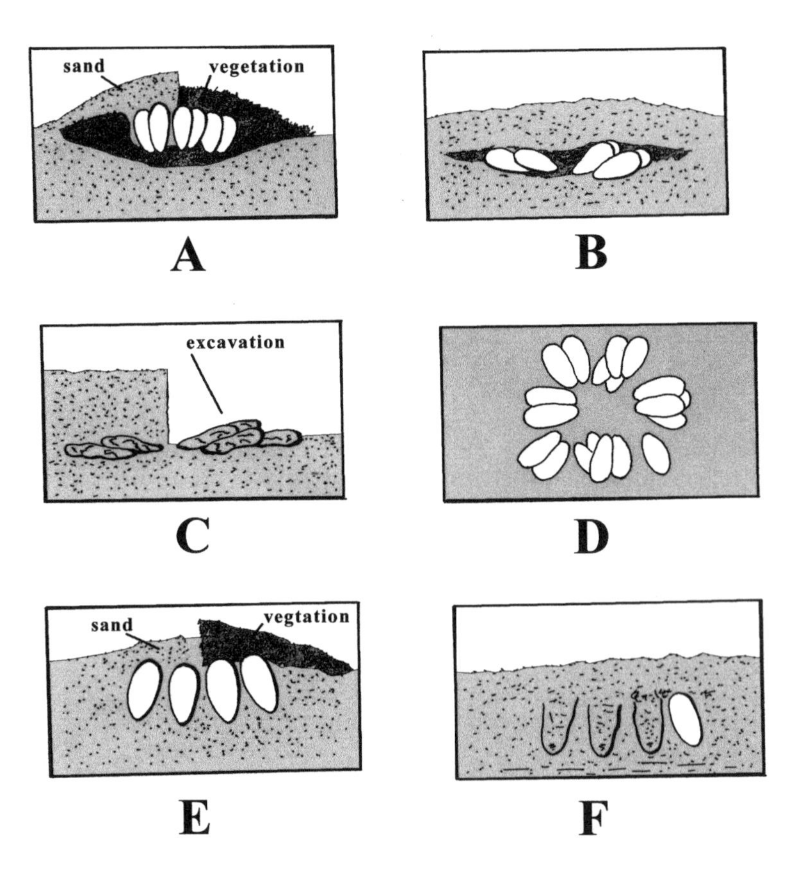

Fig. 10.7. Polish paleontologist Karol Sabath has proposed two hypotheses about the nests of Mongolian dinosaurs. For elongated eggs with linearituberculate ornamentation, he suggests that the eggs were laid upright in a mound of vegetation or in a pit dug into the vegetation (A). The eggs were either buried with more vegetation (right side of A) or with sand (left side of A). Eventually, the supporting vegetation would rot and the eggs would slowly topple outward (B). The eggs would be excavated millions of years later (C), revealing the circle of eggs (D). Smooth eggs, he suggests, were laid upright in the sand (E), with either vegetation piled on top (right side of E) or more sand (left side of E). As more sediments were later deposited over the eggs, the tops would be pushed in and the insides filled with sand; an occasional egg might be infilled with minerals before being crushed by sediments.

vegetation to support the eggs in their upright position. The eggs would slump over from the weight of the overlying sand into their radiating pattern.

Smooth-shelled eggs are also laid upright, but are supported by sand. The eggs may be covered by sand (left side of Fig. 10.7E) or vegetation (right side of Fig. 10.7E). The weight of overlying sediments may eventually collapse the tops of the eggs allowing sand to fill them (Fig. 10.7F). Occasionally, an egg may get filled in with minerals before the weight of the sediments can crush the egg.

How plausible are these two scenarios? The first one assumes that CO_2 will rise along the grooves of the lineartuberculate ornamentation and escape into the atmosphere, and O_2 will descend along the same route. While it is true that these gases pass each other through the pores, the distance traveled is measured in millimeters, not centimeters. Furthermore, even in the short distance across the pores, some mixing does occur (see Fig. 6.12). That mixing would be greater in the linearituberculate grooves and it is doubtful that the embryo would receive enough oxygen. Furthermore, because both crocodile and megapode bird eggs have a smooth shell, linearituberculate ornamentation must have some other purpose. Write me when you figure it out.

Another problem with the hypothesis is the occurrence of *Oviraptor* skeletons among the eggs. These eggs give every indication of having been laid sloped inward, with the adult sitting among them. So rather than the sloped radial pattern of these eggs being the result of decayed vegetation, it seems more probable that the eggs really were laid the way that they are found.

There is less problem with the second scenario involving the upright, smooth-shelled eggs. Some eggs show telescoping of the shell as would be expected from the pressure of the overlying sediment. Some eggs are missing their tops, not because crushing pushed the shell inward, but because the hatchling had escaped.

Dinosaur eggs have been found in a variety of clutch arrangements (Fig. 10.8). Most occur in a single plane, although multiple layers are known for a few taxa. We can infer that the multilayered eggs (Fig. 10.8D, H) must have been laid in a pit dug into the ground and buried, otherwise the tiers would not have remained distinct. By the same reasoning, we can infer that eggs found erect (Fig. 10.8E, F) must have been at least partially buried in order to keep the eggs upright. Some eggs were clearly dumped in pits excavated by the female (Fig. 10.8B). This is especially true of megaloolithid eggs from France (Fig. 10.9A) when viewed in cross-section. Although such piles vaguely resemble the clusters of sea turtle eggs (Fig. 10.9B), the clusters are not as dense and only one layer deep.

These megaloolithid eggs are also said to have been laid in arcs (Fig. 10.10A), although the pattern is weak; other alternative patterns are possible (Fig. 10.10B), which leads me to question whether any deliberate pattern is present. There are distinct clutches that probably represent nests, with a wide scattering of eggs that are not in nests. Assuming that these non-nested eggs were laid by the same species of dinosaur, several possibilities may explain the occurrence of these eggs. These single or paired eggs may have been laid by young sauropods in the first year of breeding. As Romanoff and Romanoff (1949) have noted, birds in their first year of egg laying produce fewer eggs than their older relatives; alternatively, the clutches may be from older females toward the end of their reproductive life.

Another possibility is based on the behavior of the ostrich. The nesting

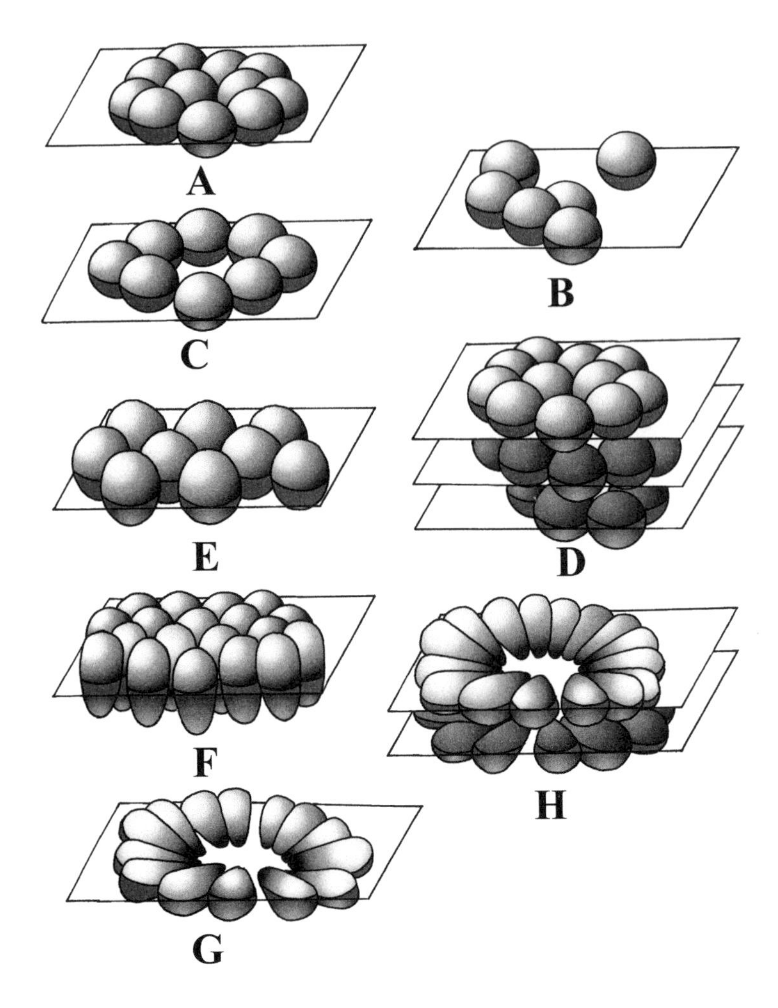

Fig. 10.8. Various patterns of egg clutches: A, close packing clumped (spheroolithid); B, random loose packed (megaloolithid); C, circular (spheroolithid); D, multilayered closely packed (faveoloolithid); E, loose cluster (ovaloolithid); F, multilayered (elongatoolithid); G, circular (elongatoolithid).

Fig. 10.9. Cross-section of a Megaloolithus *nest showing the approximate shape of the hole into which eggs were laid (A). The hole is reminiscent of a hole-nest made by turtles (B), whose young are precocial (C). (A, modified from Cousin and others, 1994.)*

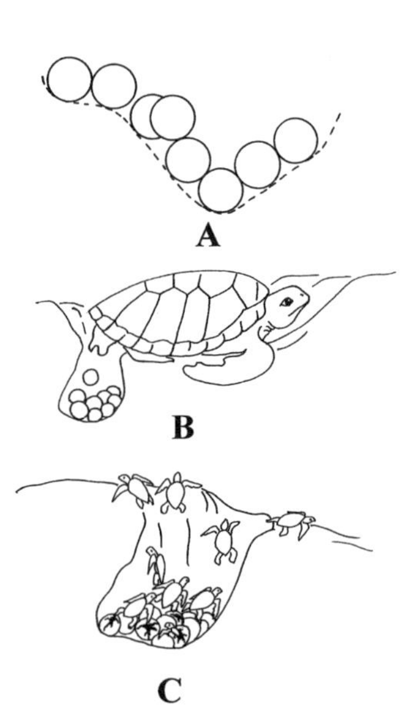

ostrich may push some eggs out of the nest toward the periphery for reasons that only the female ostrich knows (Fig. 10.11). Human suggestions for this behavior include "sacrificial" eggs for predators, getting rid of eggs in excess of what can be incubated, etc. (Bertram, 1992). Some megaloolithid eggs do seem to be peripheral to a clutch, but not all (Fig. 10.10). Yet another possible reason for these non-nested eggs is that they may have been dug up and scattered when another female excavated her own nest at a previously buried nest. These are just some of the many possibilities and I am sure you may have your own ideas. At present there are just too many possible causes for the scattering of single megaloolithid eggs to say anything definitive about them.

Among the largest clutches are the eggs of *Macroelongatoolithus*, which form a circle over 1.8 m (6') in diameter (Fig. 10.12). Embryos found with some of these eggs indicate the egg-layer was a theropod (see chapter 11), possibly a new, as yet unnamed giant oviraptorid. Remnants of a true nest may be known for another theropod, *Troodon* (Fig. 10.13). The nest rim was preserved as an ancient soil feature cemented together by calcium carbonate and is a little over a meter in diameter (Varricchio and others, 1999). The female scraped the dirt into a rim, then dug a smaller hole into

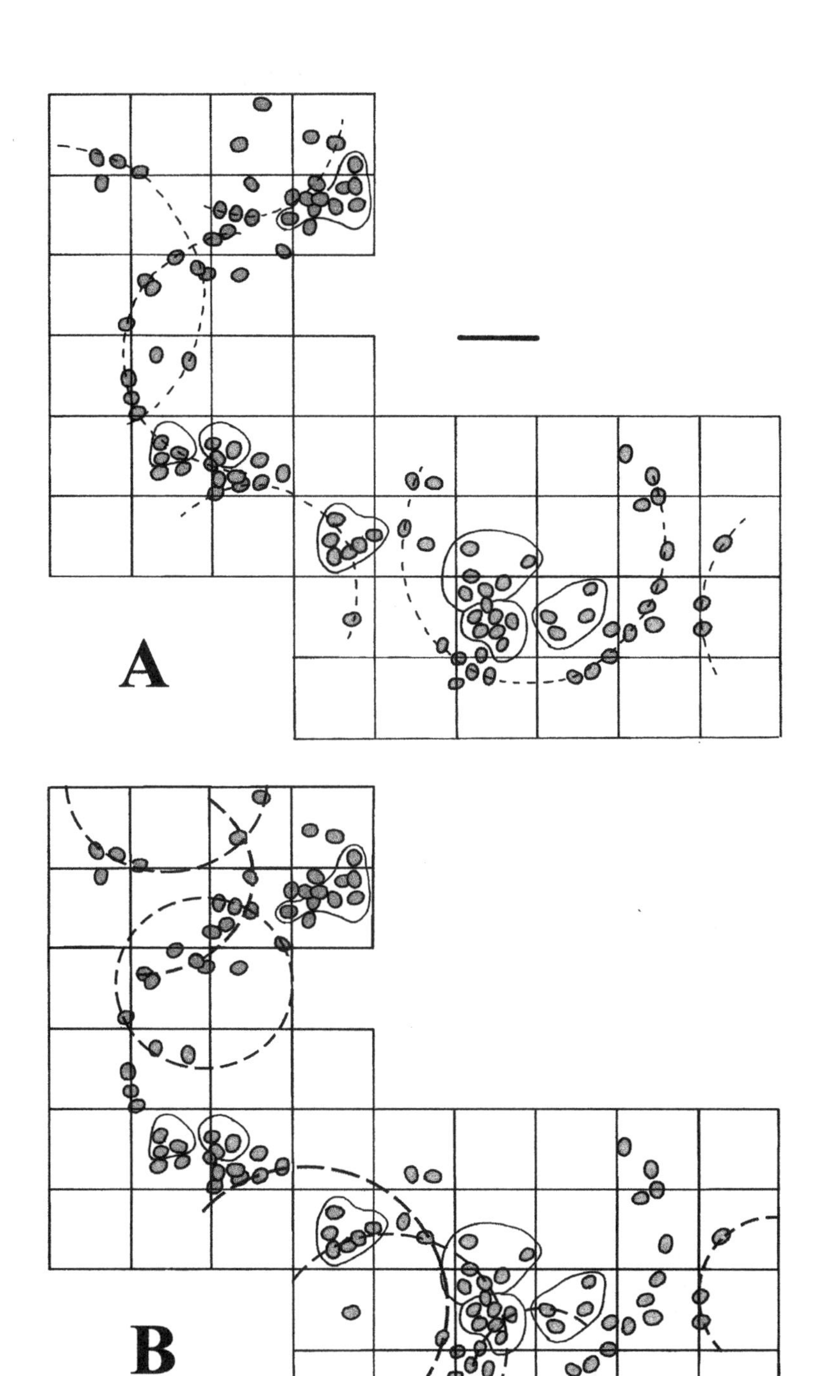

Fig. 10.10. A, map showing the distribution of megaloolithid eggs and nests. Dashed lines show arcs of eggs as determined by Cousin and his colleagues (1987). These arcs are suggested as being the rotational axis of a female laying eggs. B, alternative arcs are possible from the same data suggesting that these arcs are more apparent than real. See discussion in text. Heavy line = 1 m.

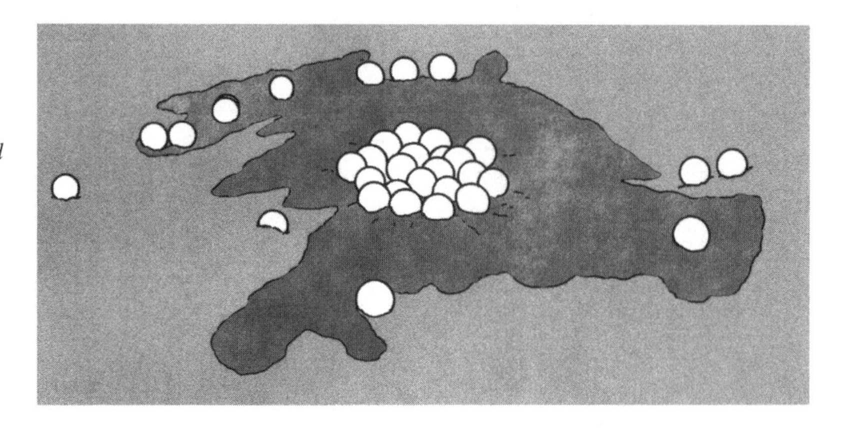

Fig. 10.11. Ostrich nest showing clutch being brooded and the scattering of rejected eggs around the periphery. The rejected eggs form arcs similar to those of some eggs in Fig. 10.10, opening the possibility that those eggs may also be rejects. (Adapted from a photograph in Bertram, 1992.)

which the eggs were laid upright (Fig. 10.8F). She must have partially buried the eggs in order to keep them in that position. Whether or not vegetation was piled on top of the eggs in the manner of the Bowerbird is unknown. The fact that the eggs are offset from the center of the nest suggests that the parent squatted within the nest and next to the eggs so they could be guarded. Perhaps this behavior was the evolutionary precursor to *Oviraptor* sitting among the eggs.

Jack Horner and Bob Makela (1979) reported that the eggs of *Maiasaura* occur in a bowl-shaped patch of green mudstone. This color is not due to copper as you might think, but to the iron atoms that are inevitably present in the ground. This iron is in what is called the reduced state, meaning that the outer electron shell is occupied. Typically, this indicates that the soil was originally high in moisture and blocked out oxygen, and/

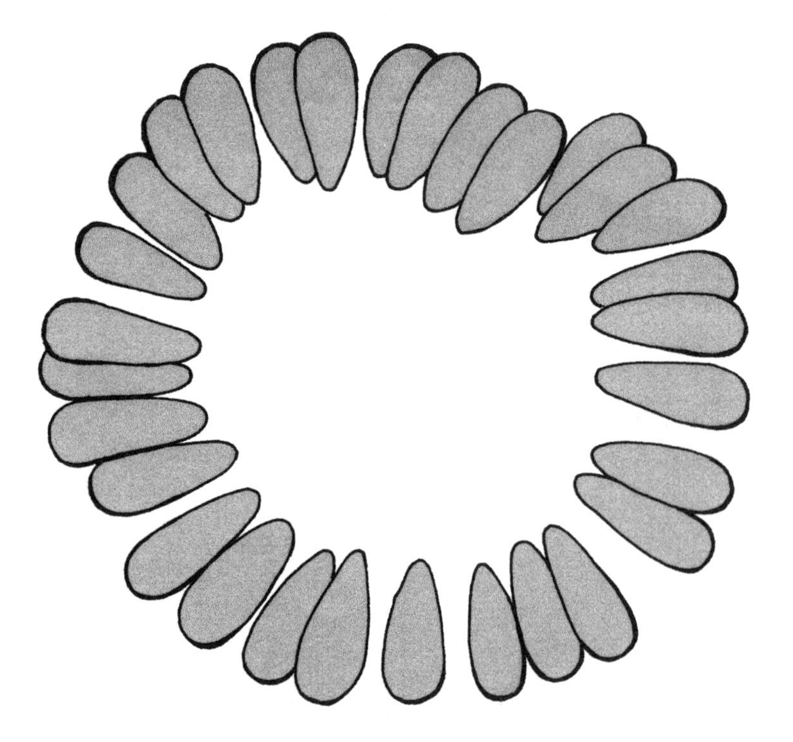

Fig. 10.12. Reconstructed ring of Macroelongatoolithus *eggs from the Lower Cretaceous of China. The ring is about 2 m in diameter.*

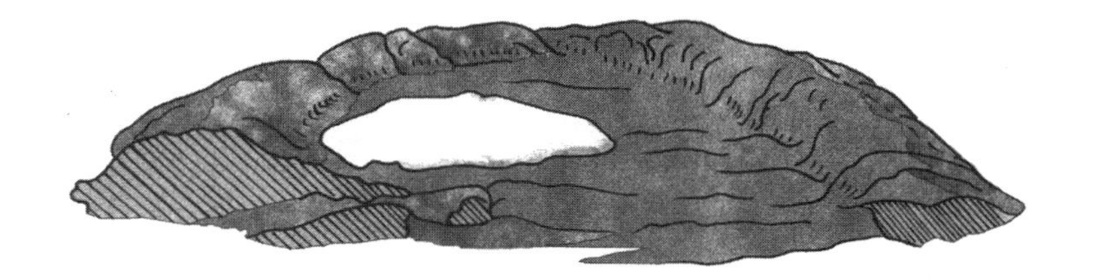

Fig. 10.13. "Fossilized" nest of the small theropod Troodon *consists of a slightly raised rim with the clutch located slightly off center (white mass is plaster of Paris and burlap covering the eggs). (From a photograph.)*

or that the decay of organic material (mostly plants) used up the soil's oxygen. If oxygen was present, it would attach to the iron atoms and "borrow" the electrons of the outer shell. When that has happened, the iron is said to be in an oxidized state, and the rock has a reddish rather than a greenish hue.

In a *Maiasaura* nest, the most likely cause for loss of oxygen in the soil is the decay of vegetation that might have been used to bury the eggs. As this vegetation decayed, it incubated the eggs much as it does for mound-building birds. Once the *Maiasaura* eggs hatched, however, the hatchlings trampled the inside of the nest, rapidly reducing the vegetation to a thin lining at the bottom of the nest. Perhaps this destruction of the vegetation is why I have not seen green mudstone associated with the *Maiasaura* nests (Jack has only reported this phenomenon at the original baby *Maiasaura* site). Nor do there appear to be green mudstones associated with any of the dozen or more nests of the hadrosaur *Hypacrosaurus* from the Two Medicine Formation at Devil's Coulee in southern Alberta—most curious. The other possibility is that vegetation was not used in dinosaur nests, and the presence of green mudstone associated with the first *Maiasaura* nest was due to some other cause, such as the decay of the babies.

The suggestion that some dinosaurs may have nested in vegetation or vegetation-mud mounds similar to those of megapode birds or alligators seems to be a popular idea (e.g., Sabath, 1991; Coombs, 1989; Fig. 10.14). But how can this be proven when all traces of vegetation have rotted away? Or how can we determine if vegetation was even used at all? An important clue may be the rate of gas exchange through the pores (called gas conductance), which we will explore in the next section.

Fig. 10.14. Albertosaurus *guarding her vegetation mound nest. The evidence for such nests is equivocal. See text. (Courtesy of G. Paul.)*

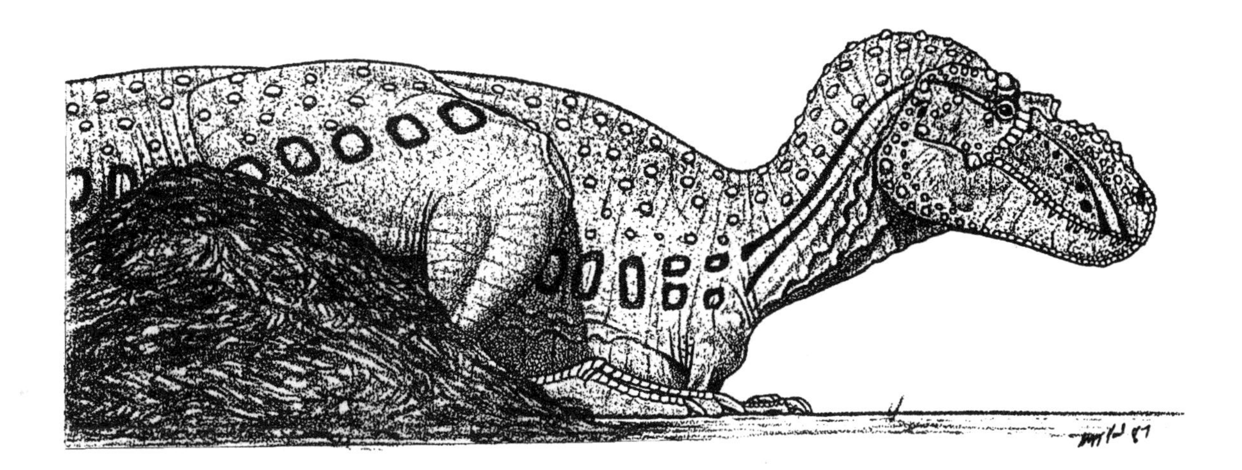

Nest Environment

Many birds lay eggs that are exposed to the air except when the eggs are being brooded. Gas exchange may occur freely between the egg and air (carbon dioxide out and oxygen in). But the female needs to stay with the eggs to incubate them, and if she leaves the eggs, they may get eaten by a predator. Mound-nesting birds, such as the Brush Turkey, lay their eggs in a vegetation mound. The adult occasionally stirs the nest to control mound temperature and gas levels, but is not required to be with the eggs constantly because the decay of vegetation provides the incubation heat (Fig. 10.15) and the eggs are hidden from predators. But being buried in a mound of debris, gas exchange is somewhat limited because the eggs are not exposed to the atmosphere. Although some oxygen is used by the embryo, most of it is used by decay of vegetation litter. Therefore, carbon dioxide levels are higher and oxygen levels are lower than those found around most bird eggs (Fig. 10.16). Naturally, the embryos have adapted to developing in the presence of higher carbon dioxide and lower oxygen.

Similar gas exchange problems are faced by some modern reptile eggs. Some eggs are laid in the leaf litter of the ground, in holes dug in the ground, or in mounds. Few reptiles guard their eggs, crocodiles being a notable exception, so eggs laid in the litter are prone to being eaten if discovered. Eggs laid in the open may have access to a lot of oxygen—but without brooding by the parent are subject to nighttime drops in air temperature. The python has gotten around this problem by using a primitive form of brooding which uses heat produced by the constriction of body muscles. For eggs laid in holes dug in the ground, the incubation temperature is identical to the ground temperature. Thus, egg laying is often timed to ensure that the ground has warmed to the optimal temperature for incubation. The ground is slow to warm or cool, so that eggs laid in holes are buffered from rapid temperature changes such as those associated with a passing rainstorm or nighttime cooling. If the hole in which the eggs were laid is open, then oxygen can readily get to the eggs. If, however, the hole is filled in, such as in the nest of a leatherback turtle, then the rate of gas

Fig. 10.15. Internal temperature of a Mallee Fowl mound approximately six weeks after the first egg was laid. (Adapted from Seymour and Ackerman, 1980.)

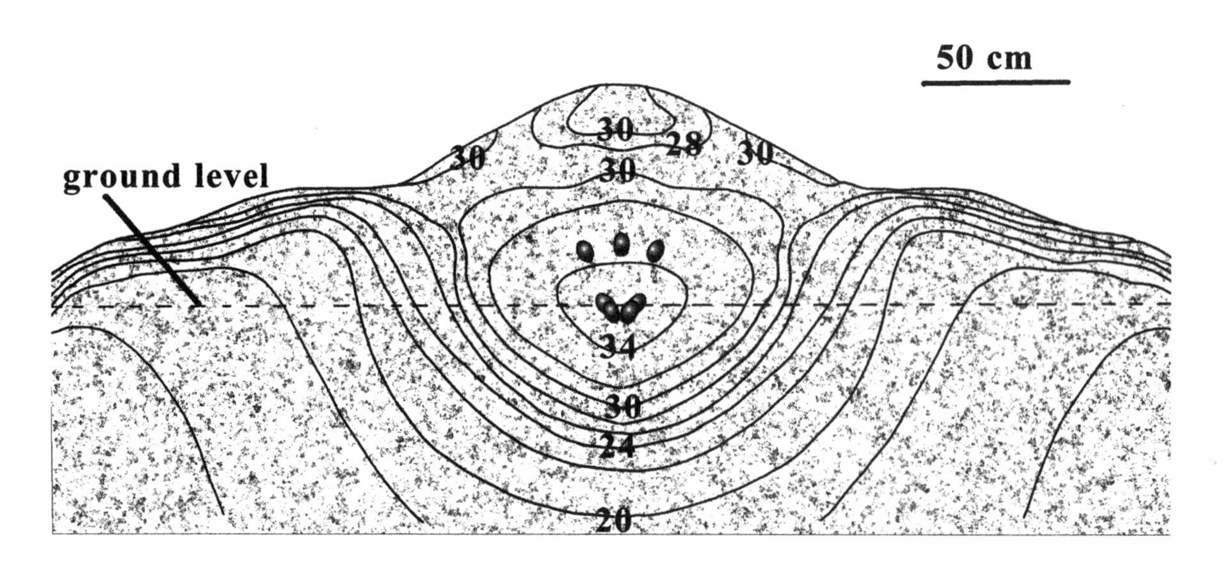

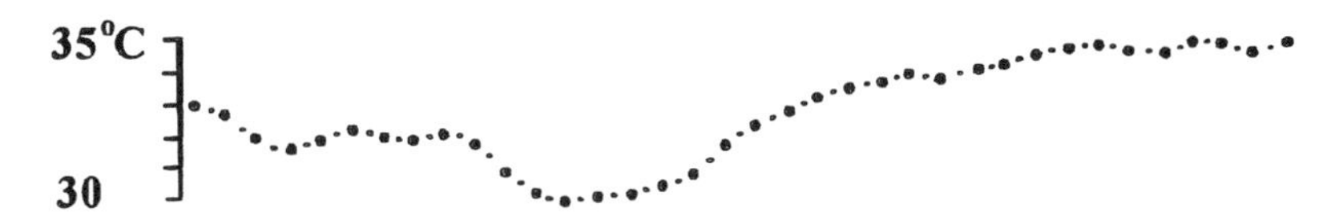

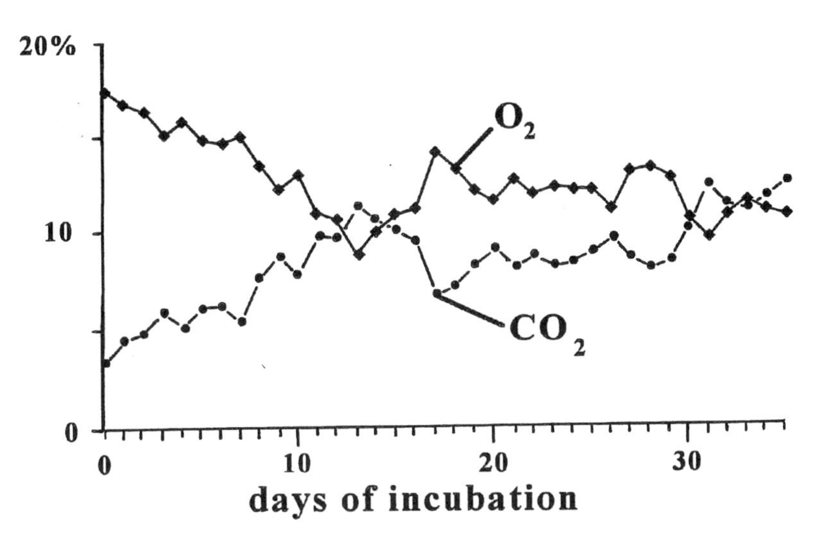

Fig. 10.16. Temperature and daily average of oxygen and carbon dioxide concentration within a Brush Turkey mound. (Adapted from Drent, 1975.)

exchange depends on the porosity of the soil. Reptile nests with the most stable temperature are those of crocodiles. As with the Brush Turkey, decay of vegetation ensures a steady supply of heat. The disadvantage, of course, is the low availability of oxygen.

Eggs laid in the ground or in a mound are subject to higher humidity. This isn't a problem with many reptile eggs, because they actually increase in weight through absorption of water after being laid. Bird eggs, on the other hand, tend to lose water during incubation, although water loss of mound-nesting birds tends to be slight (2–3%). The absorption of water by reptile eggs occurs through the pores of hard-shelled eggs, and across the shell membrane of soft-shelled eggs.

Let's look at dinosaur eggshell to see what it may tell us about the environment of dinosaur nests. Some of the earliest work on this subject was conducted by Roger Seymour (1979). It involves determining how rapidly oxygen and carbon dioxide can move back and forth through the shell. This speed goes by the fancy name of gas conductance rate. The rate actually reflects the porosity of the egg, with higher values indicating that there are more pores per cm^2 of surface area, or that the pores themselves are large. Knowing the conductance rate is important because the amount of water vapor (humidity) that gets into or out of the egg is also related to porosity. If the egg is laid in an arid environment, having a high conductance rate may mean the egg dehydrates.

The porosity of the shell is a compromise between the embryo's respiration and its water needs. For example, Seymour (1979) argues that the gas conductance rates are high for eggs of the Brush Turkey and crocodile because of the low oxygen levels within the very humid mounds. On the other hand, bird and reptile eggs in dry environments, where the risk of dehydration is high, have low gas conductance rates.

Sabath (1991) used Seymour's logic to estimate gas or water vapor conductance rates for dinosaur eggs to determine probable nest environments for clutches. This rate is determined with the formula (Ar and others, 1974):

$$G_{H_2 0} = 23.42 \frac{A_p}{L}$$

where the water vapor conductance $G_{H_2 0}$ is water in milligrams per torr (a pressure unit) per day, and 23.42 is a conversion constant based on the diffusion of water vapor in air. A_p refers to the pore area, which is the area in square millimeters of all the pores (estimated from the number of pores per $cm^2 \times$ the surface area of the egg), and L is the average length of the pore canals in millimeters (equal to shell thickness in the straight pore canals). If you didn't follow this, see Seymour's (1979) article where he goes into detail.

Now let's look at an example using *Megaloolithus,* which has a pore area of 9.47 cm^2 (Seymour, 1979) and an average pore canal length of 0.189 cm:

$$G_{H_2O} = 23.42 \frac{9.47cm^2}{0.189cm}$$

The result is 1173 mg/day torr (rounded off) of water vapor conductance.

How does this compare with a bird egg of the same size? For an egg the size of *Megaloolithus,* we need to determine the pore area and pore canal length. Pore area is related to egg weight in birds, which can be expressed as (Ar and others, 1974):

$$A_p = (9.2 \times 10^{-5}) \times W^{1.236}$$

where 9.2×10^{-5} (=0.000092) is a constant and W is the mass of the egg in grams. But first we must determine the mass for a *Megaloolithus*-sized bird egg. We can use 1900 cm^3 (748 cubic inches), the volume of a *Megaloolithus* egg, and 1.086 grams/cm^3, the average density of bird eggs (Ar and others, 1974). W then is 1900 $cm^3 \times 1.086$ g/cm^3, or 2,063 grams (roughly 4.5 pounds). The pore area is $(9.2 \times 10^{-5}) \times 2{,}063^{1.236}$, or a mere 1.15 cm^2. This is only 12 percent of the pore area for *Megaloolithus.*

Now we need to estimate the shell thickness. For this we can use the formula (Ar and others, 1974):

$$\text{shell thickness (in mm)} = (5.126 \times 10^{-2}) \times W^{0.456}$$

where 5.126×10^{-2} (=0.05126) is a constant, and W is the mass of the egg determined in the previous paragraph. The shell thickness, therefore, is 1.7 mm (=0.17 cm), the same as the pore canal length. So, water vapor conductance for a *Megaloolithus*-sized egg is:

$$23.42\ x\ \frac{1.15}{0.17} = 158\ mg\ per\ day\ torr$$

This result, which we can refer to as the predicted water vapor conductance value for a *Megaloolithus* egg, is only about 13 percent of the actual value (which is close to the 12 percent pore area that was determined). *Megaloolithus* eggs, then, have a very high conductance value implying that the nest humidity must be very high. As a result of this and other analyses, both Seymour and Sabath conclude that these eggs were not laid on the surface but in a mound or buried in the ground. We can now determine the conductance values for other dinosaur eggs (Table 10.2).

But how meaningful are these values in determining whether dinosaur eggs were laid in a vegetation mound? It is important to realize that many of the larger dinosaur eggs have branching or complex pore canals (Fig.

TABLE 10.2.

Water vapor conductance values for various dinosaur egg species compared to calculated values for a bird egg of the same size. Data from Williams and others (1984) and Sabath (1991).

Egg Species	Conductance Value (mg/day torr)	Predicted for Bird
Cairanoolithus dughii	1825	226
Megaloolithus mamillare	2893–4500	346–366
Megaloolithus siruguei	3144–4864	257–283
Dendroolithus sp.	540.3	89.5
Faveoloolithus	7580	104
Paraspheroolithus	734	47
Elongatoolithus	91	55
Protoceratopsidovum sp. A	32	33
Protoceratopsidovum sp. B	25	34
Protoceratopsidovum sp. C	29	174

8.20). Board and Scott (1980) have noted that in modern bird eggshell, branching of the pore canals only occurs in large, thick-shelled eggs, such as those of the ostrich and rhea, with the thinner rhea shell having less branching of the pore canals (Fig. 6.13). Branching pore canals produce more pore openings on the surface than do simple, straight pore canals.

Let's look again at pore area. For simplicity of analysis, we can combine the areas of the pore openings into a single pore area, which then can be thought of as the area of a single, giant pore. Once done, it is clear that there is a correlation between the pore area and the size of the egg. The formula for the relationship is:

$$A = 9.2 \times 10^{-3} \times M^{1.236}$$

where A is the pore area in mm^2 and M is the egg mass in g (Ar and others, 1974). What this means is that a 600 g rhea egg, which is ten times the size of a 60 g chicken egg, will actually have a pore area more than seventeen times the size:

$$0.0092 \times 600^{1.236}/0.0092 \times 60^{1.236} = 17.2$$

If this formula is applied to dinosaur eggs, we would most certainly expect to find larger dinosaur eggs having more pores, and indeed that is what we find.

But as mentioned above, having more pores also means greater water loss during incubation, as correctly noted by Sabath (1991). Again there is a formula that correlates egg mass to water vapor conductance, which is the rate of water loss as a gas through the pores:

$$G = 0.432 \times M^{0.78}$$

where G is the gas conductance for water vapor and M is the mass of the egg in grams. For a 60 g chicken egg, G = 10.5 mg/day torr.

We get very surprising results when we apply this formula to the eggs of two mound-nesting birds, the Brush Turkey and the Mallee Fowl; their eggs are 203 g and 170 g, respectively (Seymour and Ackerman, 1980). The predicted gas conductance values for these eggs are 27.2 mg/day torr for the Brush Turkey egg and 23.7 mg/day torr for the Mallee Fowl egg. However, the actual measured values are 47.3 mg/day torr and 21.4 mg/day torr (Seymour and Ackerman, 1980). Why this contradiction? It turns out that although the Brush Turkey egg is 3.4 times the size of a chicken egg, the shell thicknesses are about the same (0.34 mm vs. 0.35 mm for the chicken). Likewise, the Mallee Fowl egg is 2.8 times the size of the chicken egg, but the shell thickness is 0.27 mm (Seymour and Ackerman, 1980). Clearly, then, we need to be careful in making inferences about the nesting environment of dinosaur eggs based on the gas conductance values.

Rahn and his colleagues (1987) have also noted that larger eggs have more pores (i.e., greater pore area) because of the higher metabolic needs of the larger embryo. It is important to realize that an ostrich egg is not simply an enlarged chicken egg. Actually several things change with increased size. With every tenfold increase of mass, the number of pores per cm^2 increases 18 times, pore length increases 2.7 times, and gas conductance increases 6.5 times. This increase is due to the greater oxygen requirements of an embryo developing in a large egg, compared to one developing in a small egg.

These nonlinear changes in the egg are called scaling, a point not generally appreciated. Scaling can be better understood by considering that a three-foot child is half an adult male's six-foot height, but is not half of the adult's weight; rather, the weight is closer to a quarter or less. Not understanding scaling can have tragic consequences, as related by Schmidt-Nielsen (1972). In the 1960s, an elephant was given a dose of LSD based on the amount needed to put a cat into a rage. The dosage for the cat was 0.1 mg/1 kg, therefore the amount for the 2970 kg elephant was 297 mg. Simple, straightforward scaling, right? Wrong. This amount of LSD was injected into the elephant, who immediately started trumpeting and running around. About five minutes later, the elephant collapsed, went into convulsions, and died. Ah ha! said the scientists—elephants are peculiarly sensitive to LSD. But no, said Schmidt-Nielsen, you should have scaled on metabolism, which is much lower in the elephant—only 80 mg would have been needed. The point is that in using scaling to understand dinosaur eggs, we need to be careful that we are scaling the right features, and not scaling to support our pet hypotheses.

One final point of caution regarding pore area is that gas exchange is not possible through all pores in a bird egg (Board and Scott, 1980). Pores may be partially or completely blocked by organic or inorganic material and therefore the calculated results for gas conductance can be two or three times higher than actual values. Possible organic plugs have been reported from some *Megaloolithus* eggs, and their occurrence in other dinosaur eggs is expected.

So what do I conclude about the nesting environment of dinosaur nests? That it is difficult to prove that eggs were laid in vegetation mounds, buried in the dirt (or sand), or left partially exposed. The answer, if it is to be found, lies in very, very careful forensic study of the eggs as they lie in the ground. Perhaps then someone will be able to determine whether or not the rock covering the eggs is subtly different from the rock below the eggs. The differences might be used to denote the edges of the nest, reveal whether or not plants were used in the construction of the nest, etc.

Clutch Size

Although we don't know the environment of dinosaur nests, we do find eggs in clutches. If we assume that eggs were buried in a mound of either earth or vegetation for incubation and protection, the number of eggs in a clutch is probably limited by the gas conductance values. Too many eggs and the clutch may smother. The suggestion by Ted Case (1978) that sauropods may have laid one hundred eggs is unfounded and erroneously assumes that sauropods were like giant turtles in their egg laying. The greatest number of sauropod eggs, of *Megaloolithus,* recovered in a clutch is twenty eggs. This is more than the thirteen predicted by Seymour (1979), who assumed the eggs were laid by *Hypselosaurus.*

The least number of eggs in a clutch may be one, assuming that the solitary *Megaloolithus* eggs from France were laid by females in their first year of breeding. If these eggs are actually rejects from a larger clutch, then the smallest number of eggs in a clutch may be the paired eggs reported by Horner (1984). I suspect, however, that these eggs may have been part of a larger ring of eggs. Time will tell if these paired eggs do indeed represent the only eggs in the clutch. The largest number of eggs in a clutch is forty (eggs called *Protoceratopsidovum*); these eggs were found in a multi-tiered nest.

Among living reptiles, there is a strong correlation between body size and the number of eggs or young. The largest clutches in lizards (thirty to sixty eggs) occur in the largest lizards (e.g., *Iguana, Varanus*), and up to two hundred in large sea turtles. In smaller lizards, clutches range from one to four, with two being the most common (Porter, 1972). Even birds show a wide range of egg numbers in a clutch, from as few as one in the Peruvian booby, kiwi, and albatross to thirty-five in the Mallee Fowl of Australia (O'Connor, 1991a). Actually, the number of eggs laid and the clutch size among reptiles and birds can vary from individual to individual within a species, by the same individual from year to year, or from clutch to clutch if more than one clutch is laid per year. For example, the chameleon *Chamaeleo dilepis* can lay as few as twenty-three eggs or as many as fifty eggs, and in the ostrich as few as fifteen to as many as thirty-six (Porter, 1972; Bertram, 1992). Even dinosaur clutches show variation: three to six eggs of *Megaloolithus* in India, and six to twenty-four eggs in nests of *Prismatoolithus* described by Karl Hirsch and Betty Quinn (1990). With dinosaur clutches, however, there is often uncertainty about the number of eggs destroyed by the processes of erosion that exposed the eggs in the first place.

What, besides gas conductance, determines the size of a clutch? The most important factor is clutch size, which must be large enough to maintain the population of the species. Too small a clutch size, and the population size will gradually decrease through time because deaths exceed replacement. Too large, and the population may not be able to feed itself, resulting in starvation and mass deaths. Henry Fitch (1985) has examined the question of clutch size in reptiles, and Martin Cody (1966) in birds. Both agree that clutch size is dependent upon environmental stability. Thus in the tropics, on islands, or along the coast, clutch size tends to be smaller than in the less stable high latitudes where seasonal change is pronounced. The reasons for this variation in clutch size are not understood. Fitch suggested that reptiles in the tropics can breed several times a year, whereas those in more seasonal environments can only breed once. This change is noted among different species, as well as among individuals of a single, wide-ranging species. To illustrate the implications for this in dinosaurs, the hadrosaur *Edmontosaurus,* known from above the Arctic Circle, may have had larger clutches than the *Edmontosaurus* from Colorado. Perhaps someday we'll find clutches of eggs for this dinosaur in Alaska and Colorado, and prove or disprove this hypothesis.

Fitch also noted that clutch size correlates with the size of the female.

Because reptiles continue to grow throughout their lives, an older female lays more eggs than a younger, smaller female. Some dinosaurs also apparently did not stop growing (though growth may have slowed), so an older, more mature female undoubtedly laid more eggs. This may be the main reason why there is so much variation in the sizes of dinosaur clutches from the same general area (e.g., Djadokhta Formation of Mongolia). Among dinosaurs, a similar trend may be apparent. For example, the hadrosaur *Maiasaura* laid about eleven eggs (based on the number of hatchlings in a nest), whereas the smaller *Bactrosaurus* (assuming that it laid the eggs identified as *Spheroolithus irenensi*) laid about six or seven eggs. The size difference between these two dinosaurs can be seen by comparing adult femur length of *Bactrosaurus,* 80 cm (31½"), with that of *Maiasaura,* 110 cm (43¼").

Finally, the size of the hatchling determines the size of the egg, and this in turn determines the number of eggs the female can lay. Among precocial birds, which require no parental care, the chicks are larger than those of altricial species because they are better developed when they hatch (more in the next chapter). The eggs of precocial birds also tend to be larger to accommodate a larger food supply (yolk). As a result of the larger size, fewer eggs are laid in a clutch. Although all reptile hatchlings are precocial, the size of the hatchling also correlates with the size and number of eggs in a clutch. As for dinosaurs, we really can't say if the dinosaurs with larger hatchlings had fewer eggs because we have so few dinosaur embryos upon which to make a comparison. Stay tuned, however. . . .

The Clutch

The arrangement of the eggs within a nest is variable depending on the species of the egg-layer (Fig. 10.8, 10.17). The simplest arrangement of eggs is clutches of six to eight *Megaloolithus* eggs that were apparently laid randomly in a conical hole. These clutches have been reported from the Upper Cretaceous of France (Cousin and others, 1994). The hole was probably dug with the large thumb claw of a female sauropod (assuming that the eggs are indeed those of sauropods).

Most clutches show more deliberate placement of the eggs (Fig. 10.17). These eggs occur in one or two layers and were evidently laid in a shallow pit. The simplest arrangement occurs with some clutches of *Megaloolithus.* These eggs appear to have been dumped into a shallow hole, and are randomly piled on top of one another. A slightly more organized pattern occurs with the single-layer clutch of *Dendroolithus.* The nest consists of eleven or more eggs organized into alternating rows of three to five eggs. This arrangement is called close packing and maximizes the number of eggs that can be packed together in a confined space. We do not know if the female nosed the eggs into position, or whether they rolled into place in a shallow bowl-shaped pit. The eggs of *Spheroolithus maiasauroides* form a ring of about eight or nine eggs. *Spheroolithus irenensis* clutches are also arranged in a ring, but the inside of the ring is also filled with eggs. A variant of this pattern is seen in *Faveoloolithus* nests, in which the eggs are laid in two or three layers of about eleven eggs each.

The most complex nests are those containing elongatoolithid or *Protoceratopsidovum* eggs. These were also laid in a ring, but individual eggs occur in pairs. The eggs were apparently laid in vertical or near vertical position. *Protoceratopsidovum sincerum* clutches may contain twenty to thirty (perhaps forty) eggs in two layers. The eggs of *Prismatoolithus* are reported by Jack Horner (1984) to occur in a spiral. Actually, only one nest

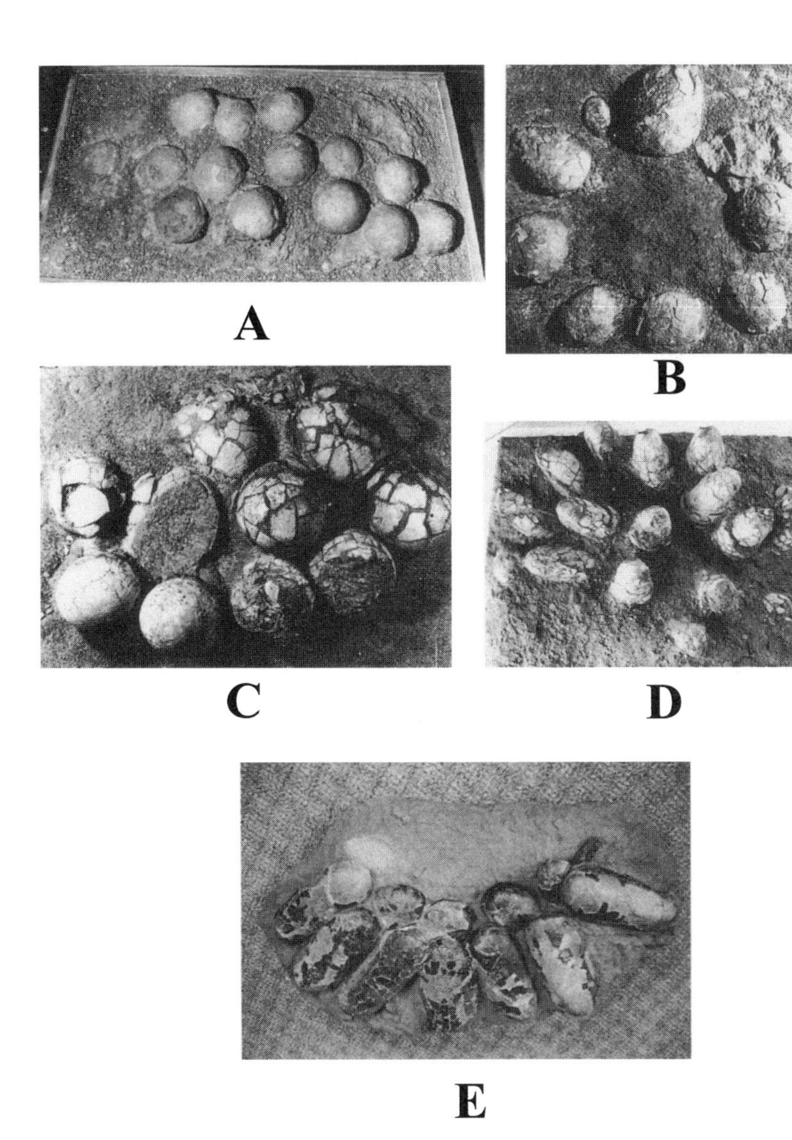

Fig. 10.17. Various types of clutches: A, close packing of Faveoloolithus; *B, ring of* Spheroolithus chianghiungtigensis; *C, filled ring of* Spheroolithus irenensis; *D, upright eggs of* Protoceratopsidovum; *E, half of a two-tiered ring of* Elongatoolithus. *(Courtesy of K. Sabath, K. Mikhailov.)*

shows this feature, which is not obvious to me. Several other nests show twelve to twenty-four eggs forming a ring filled with eggs similar to that of *Spheroolithus irenensis,* but with the elongate eggs standing nearly vertically.

Colonial Nesting

Clutches of dinosaur nests rarely occur singly. Instead, there are usually several clutches in the area. Some of the richest known sites include Nemegt, Bayn Dzak, and Toogreek in Mongolia, Henan, Nanxiong, and Laiyang Provinces in China, the Kheda District in India, Tash-Kumyr in Kyrgyzstan, Aix-en-Provence in southern France, and the recently discovered Auca Mahuevo, Argentina. Literally thousands of eggs and millions of eggshells are known from these sites. The richest of all may be Tash-Kumyr where there are over 629 m (2,095') of egg-bearing strata spanning most of the Cretaceous (Nessov and Kaznyshkin, 1986). Unfortunately, the area has only briefly been studied.

The most famous colonial nesting site is a few miles west of Choteau, Montana, in the vicinity of Egg Mountain and Camposaur. Here several hadrosaur nests in a relatively small area led Jack Horner (1982, 1984) to suggest that hadrosaurs may have nested in colonies. That would certainly be a sight: acres of hadrosaurs squatting near their nests, bellowing and grunting as they awoke at sunrise. Indeed, artist John Gurche reconstructed such a scene for *National Geographic* magazine (Gore, 1993). But what scientific evidence is there for such a scene? Very little, I'm afraid. The idea is based primarily upon Horner's claim that the nests of the hadrosaur *Maiasaura* are about 7 m (23') apart, about the length of an adult. Such a spacing might imply that the nests were occupied at the same time with room available for maneuvering through the colony. Having been to Choteau and seen the evidence, I am more convinced that the spacing was a fortuitous quirk of erosion, rather than deliberate planning on the part of the females.

My reason is that the nests are all exposed on very irregular ground with several small gullies or runnels separating them (Fig. 10.18). With so few nests and the nature of the rock record, it is difficult to prove that the nests were occupied contemporaneously. Only if some marker layer, such as volcanic ash, settled across the nesting ground could simultaneous occupation be proved, because then all the eggs buried by the same ash had to have been in their nests at the same time. Unfortunately, no such marker layer has been reported for any nesting site. An additional problem is that without a marker layer it would be difficult to separate nests of one year from those of another year. This, of course, assumes that the nesting grounds were reused year after year. I have no doubt that someone will eventually provide evidence for nests being occupied at the same time. For now, we only *infer* colonial nesting by the sheer numbers of eggs at a site.

Fig. 10.18. Irregular ground near Egg Mountain, Montana, due to erosion. Several Maiasaura *nests were discovered in the flat areas, but it is difficult to say that they were occupied at the same time.*

Even without good evidence yet for colonial nesting at Choteau, Horner's suggestion is worth examining further. Studies of crocodiles in the wild have shown that nests are not randomly scattered, but several of them are clumped together a few meters apart (Fig. 10.19; Cott, 1961; Woodward and others, 1984) in rather limited areas (Fig. 10.20). This clumping is apparently not due to any environmental factors (e.g., suitable sand for nests) but because the females want to be near one another. Among birds, thousands may congregate into rookeries. Why do nesting animals congregate like this? With so many bodies packed into a small area, disease and skin parasites can spread throughout the colony. In addition, colonial nesting would actually provide predators with a concentration of prey, both eggs and parents. While true at face value, the greater numbers of prey may be more than predators can deal with. The result is that egg loss or the number of parents killed is less than if the nests and parents were more scattered. Still, in a colony losses are higher for nests along the edges than nearer the center. But even for these nests, protection is provided by having a greater number of vigilant individuals alert for approaching predators. Once a predator is detected, the parents can mob it, driving it away, as seen in smaller birds harassing a single larger bird. It is hard to imagine a *Tyrannosaurus* driven away by a mob of hadrosaurs guarding their nests, but the idea may not be as far-fetched as it sounds.

Despite the advantages afforded by colonial nesting, there is no good evidence among crocodiles that a guard female vocally alerts others at the

Fig. 10.19. Distribution of Nile crocodile nests. Note the irregular clumping of the nests into a small area. (Adapted from Cott, 1961.)

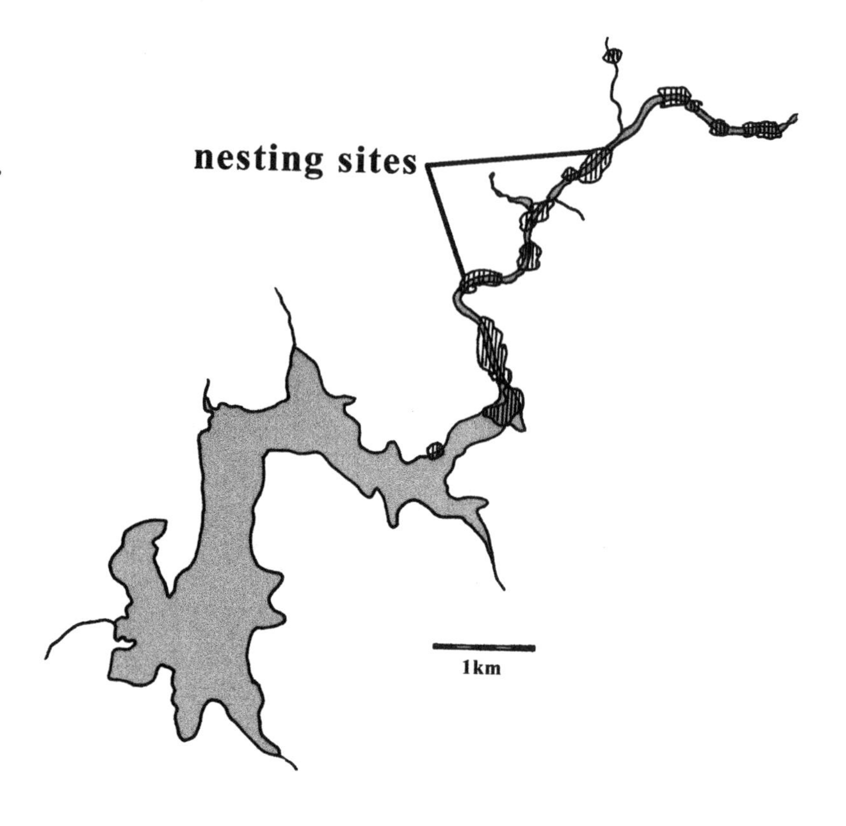

Fig. 10.20. Location of nesting sites for the Nile crocodile along the Ngezi River, near Lake Ngezi, Zambia. (Adapted from Hutton, 1989.)

approach of a predator toward a nest that is not hers. Furthermore, it is a myth that crocodiles always guard their nests. Diligence among individuals, as well as species, varies considerably. For example, the female Nile crocodile rarely leaves the nest unguarded even to eat (Cott, 1961), whereas 45–50 percent of alligator nests may be unguarded at any one time (Joanen, 1969). Naturally, egg loss is higher in nests that are not guarded.

So how diligent were dinosaurs in guarding their nests? We really don't know, but it is safe to assume that the degree of guarding varied considerably among species from none to a lot. For example, the eggs called *Megaloolithus* were apparently buried and left to be incubated by the sun. On the other hand, there are several instances in which the small Mongolian theropod *Oviraptor* was found associated with a nest of eggs. The original discovery of *Oviraptor* was made during the Central Asiatic Expedition (see chapter 1). At the Flaming Cliffs, a clutch of eggs was found that were thought to belong to *Protoceratops*. In excavating these eggs, a small theropod skeleton was found on top of them. Named *Oviraptor philoceratops,* this poor dinosaur has gotten a bum rap for over seventy years. In proposing the name, Henry Osborn wrote:

> The generic and specific names of this animal, *Oviraptor,* signifying the "egg-seizer," *philoceratops,* signifying "fondness for ceratopsian eggs." . . . The names are given because the type skull . . . was found lying directly over a nest of dinosaur eggs. . . . This immediately put the animal under suspicion of having been overtaken by a sandstorm in the very act of robbing the dinosaur egg nest. (Osborn, 1924)

We now know that the poor *Oviraptor* died protecting its own eggs from the ravages of a sandstorm, rather than attempting to steal them.

Egg Laying

When and how long it takes to lay eggs for a bird or reptile varies considerably. Some begin egg laying in the morning and take all day, and some take only part of a day, such as the afternoon. Even the time of year can be crucial for those living in the high latitudes. Increased light levels as winter gives way to spring trigger mating and egg laying. It is important that the eggs be laid early enough in the spring so that they will hatch by late spring. This gives the young the entire summer to grow and either prepare for winter or migrate toward lower latitudes with the parents. As A. J. Holman (1995) has noted, the northern geographic range for reptiles is determined by the number of days in the summer that are warm enough for incubation. Not enough days in a row and incubation will not occur.

In many birds, the first clutch is laid when the female is about one year old. Ostriches may continue to lay for thirty years, with an average of twenty-six eggs each year (Bertram, 1992)—that's a lot of eggs in a lifetime! Actually, the number of eggs in a clutch is variable and depends on the availability of food and the abundance of predators and parasites. In general, the fewer eggs in a clutch, the larger the hatchlings. Ostriches lay anywhere between fifteen and thirty-nine eggs. This range of variation suggests that the total number of dinosaur eggs in a clutch is an interesting feature—but not very meaningful for use in egg taxonomy.

Ratites, such as the ostrich and rhea, are unusual among birds in that several birds may lay eggs in the same nest. Among ostriches, the top female, called the major hen, lays her eggs first at two-day intervals (Bertram, 1992). This interval is longer than that of a chicken because it takes so much longer for the ostrich to form each egg. After the major hen is through, two to five other females, called minor hens, may add their eggs to the nest. The result is that it takes several weeks for all eggs in the clutch to be laid by all the females. Nevertheless, hatching of all the chicks occurs within a few hours of one another. Once about sixteen eggs are laid (about six weeks), incubation is started (Bertram, 1992). But before this, several days are spent arranging the eggs in the nest. The major hen pushes out eggs she is unable to incubate, usually those of the minor hens; the rejects become the "sacrificial" eggs for predators.

With fifteen to thirty-nine eggs in a nest, it is impossible for all the eggs to be incubated. Only about fourteen to twenty-five eggs (average nineteen eggs) near the center are incubated, while those on the edges die and rot. Eggs are incubated alternately by both the male and major hen. With up to a third of the eggs left to die, this would seem to be a wasteful use of energy on the part of the females. But it is these "sacrificial" eggs that get taken by predators first, thereby extending the time the incubated eggs have to develop. As mentioned earlier, some of the sauropod nests from France show a distribution pattern of eggs similar to that of an ostrich nest. Does this mean that sauropods laid more eggs than necessary in order to sacrifice a portion of them to predators? Possibly, but these scattered eggs do not show the higher-than-expected proportion of preburial breakage. Still, this possibility cannot be completely discounted without further study.

Among reptiles, only 60 percent or so of female crocodiles nest in a given year. The clutch size varies from two to fifty-eight eggs, with an average of thirty-nine eggs. In a colony, it takes about two weeks for all the females to lay their eggs because not all females begin laying at the same time. The time of year when the eggs are laid varies from year to year and depends on environmental conditions. The Komodo lizard (*Varanus komodoensis*) usually lays its eggs underground. About nineteen eggs may be laid, although as many as thirty have been reported (Auffenberg, 1981). The eggs are rarely laid all at once, but at one- to eight-day intervals with one to twenty eggs at a time. As a result, it may take up to two weeks for the entire clutch to be laid. The eggs take about eight to eight and a half months to incubate, being warmed by the sun. Because there is no parental care of the eggs, they may be eaten by other Komodo lizards or even by the female who laid them. Egg cannibalism among dinosaurs is unknown, but remains a distinct possibility among some of the predatory dinosaurs. Eggs are not laid every year by the Komodo lizard, although the reasons are not known. Furthermore, unlike some reptiles, female Komodo lizards actually lay fewer eggs at a time as they age.

Egg laying in dinosaurs probably began when the females reached maturity. But when this occurred undoubtedly varied among the species. For example, the small 34 kg (75 lb) *Protoceratops* probably matured faster than its bigger, 4,500 kg (5 ton) distant cousin, *Triceratops.* But how many months or years until maturity is unknown, partly because we still do not have a good idea of growth rates among the different dinosaur species. Horner claims growth rates of hadrosaurs approached that of birds, meaning that a 2,700 kg (3 ton) hadrosaur matured in about five years. Maybe he is right, but he hasn't presented the scientific data to support this claim yet. We'll explore growth in chapter 12.

Egg laying might have been a bit of a problem for those dinosaurs living in the higher latitudes. For example, the hadrosaur *Edmontosaurus* has been found along the Colville River, north of the Arctic Circle in Alaska. Because this part of Alaska was also north of the Arctic Circle during the Cretaceous, Nick Hotton (1980), a paleontologist at the Smithsonian Institution, suggested that the hadrosaurs migrated south with the approach of "winter" and its six months of darkness. Without sunlight, the plants would become dormant, leaving the large herbivores with little to eat. It makes sense, then, to expect the hadrosaurs to have migrated north-south seasonally in order to have sufficient food supplies. However, the presence of at least one baby hadrosaur and several juveniles among the fossils from the Colville River area indicates that the hadrosaurs spent at least a portion of their early life north of the Arctic Circle. Jack Horner suggests from his bone studies that growth in hadrosaurs was very rapid for the first few months. If true, the eggs would have to have hatched at a time when there was abundant young, nutritious plant growth. This would ensure that there was adequate food for the growing young. In addition, hatching must have occurred early enough for the young to grow large enough to join the adults in their southward migration. Because this migration was very long (estimated to have been over 3,000 kilometers, almost 1900 miles) and the time for growth was short (probably four months or less), there wasn't very much time for the eggs to be incubated. Most likely the embryos were well developed within the mother by the time the shell was produced and the eggs laid, much as occurs in some lizards.

1. Little Diablo Hill is situated in an arroyo cut into the prairie of southern Alberta, Canada. Several clutches of eggs were found containing the remains of hadrosaur embryos. Photo by author.

2. Excavating dinosaur eggs doesn't necessarily involve using delicate tools. A jack hammer was needed to remove the hard overburden of rock from eggs in the Morrison Formation in southern Colorado. Photo by author.

3. A group of young male Corythosaurus *sport bright colors they will later use to attract mates. Painting © by Douglas Henderson.*

4. Male Stygiomoloch *during the mating season engaged in flank butting. These pachycephalosaurs probably engaged in head display before resorting to butting. Photo by author.*

5. "Carnotaurus *sex!*" *Painting* © *1998 by Luis V. Rey.*

6. Unlaid egg mass from a wild alligator, just prior to the eggshell forming. Because crocodilians are related to dinosaurs, egg masses of primitive dinosaurs may have looked like this before being laid. However, best evidence suggests that dinosaurs, unlike crocodilians, produced eggs one at a time. Photo by J. O. Farlow.

7. A possible interpretation of hadrosaur behavior. Here two Kritosaurus *tend a nest formed by a mound of vegetation, a depiction based on the nests of alligators and megapode birds. Painting © 1995 by Gregory S. Paul.*

8. An artist's portrayal of events around a hadrosaur nesting ground. Near the center, an adult maintains the proper incubation temperature by adjusting the vegetation mound of the nest. On the right, a nest is being opened to release the newly hatched eggs. On the left, hatchlings are tended by parents in the nest, where they will stay until mature enough to leave. Such a scene is widely accepted, yet is little supported by fossil evidence. Painting © 1990 by Gregory S. Paul.

9. *A mother* Hypacrosaurus *releasing her precocial hatchlings from the confines of a vegetation mound nest. Painting © 1990 by Mark Hallett.*

10. Orodromeus *hatchling in what is now northern Montana. Painting © 1989 by D. Braginetz.*

11. *The little hypsilophodontid* Leaellynasaura *struggling out of an egg in southern Australia about 115 million years ago. At this time Australia was within the Antarctic Circle, so egg laying and hatching had to be timed so that the hatchling would have healthy growth. Painting by Peter Trusler © 1993 by Australia Post.*

12. Because hatchlings had such small body mass, they may have had a hard time retaining body heat. To prevent their core temperatures from dropping to lethal limits, they may have had a coating of "protofeathers." Alternatively, they may have been able to endure lower body temperatures much like some altricial birds. Painting © 1995 by Larry Felder.

13. To protect the young from predators, young sauropods may have joined the herd when about one-third grown. Diplodocus carnegii. *Painting © 1999 by Michael W. Skrepnick.*

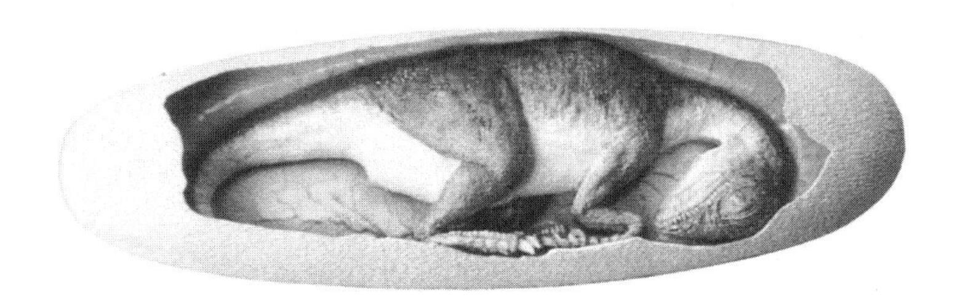

11 • The Embryo and Hatching

Growth of the embryo begins long before the egg is laid, and actually begins before the egg is even shelled. This growth begins with fertilization, which in some birds and many reptiles (although apparently not crocodiles; Ferguson, 1985) can sometimes be delayed for a week or several years, although a few weeks is more common. The sperm is stored and kept alive until used to fertilize the egg. The storage site may be a small pocket where the uterus and vagina connect or folds in the walls of the uterus. Here, the sperm is nourished and maintained until transferred to the top of the reproductive tract to fertilize the egg (see chapter 6). It is possible that some dinosaurs might also have had a similar means of storing sperm, in which case we would expect it among dinosaurs with low population levels living in marginal environmental conditions, such as in or around deserts. That is because with low population levels, it would be exceptionally hard for males and females to get together. One candidate would be the prosauropod *Ammosaurus,* which is known to have lived in a sand desert (it is found in the sand dune deposits of the Navajo Formation in Arizona). Of course, this is all speculation.

Embryo Development

Although we cannot study dinosaur embryo development firsthand, we can get important clues from embryonic studies of modern reptiles and birds, especially the lizard, *Lacerta* (Dufaure and Hubert, 1961); the *Alligator* (Ferguson, 1985, 1987); and chicken, *Gallus* (Patten, 1951; Hamburger and Hamilton, 1951; Freeman and Vince, 1974). These studies show that, despite their differences as species, their embryos undergo many of the same developmental processes. With this information, we can make inferences about dinosaur embryo development (Fig. 11.1, 11.2). The information for the lizard embryo is given in stages of development because the moment of fertilization is unknown. For crocodiles, development is given in hours or days since the egg was laid, and for birds in hours or days since start of incubation. The developing embryos of the various species are shown in Figures 11.2–11.6.

Early Tissue Formation

Atop each blob of yolk is a single cell in a region variously called the blastodisc or germinal disc from which the embryo will develop (Fig. 11.3,

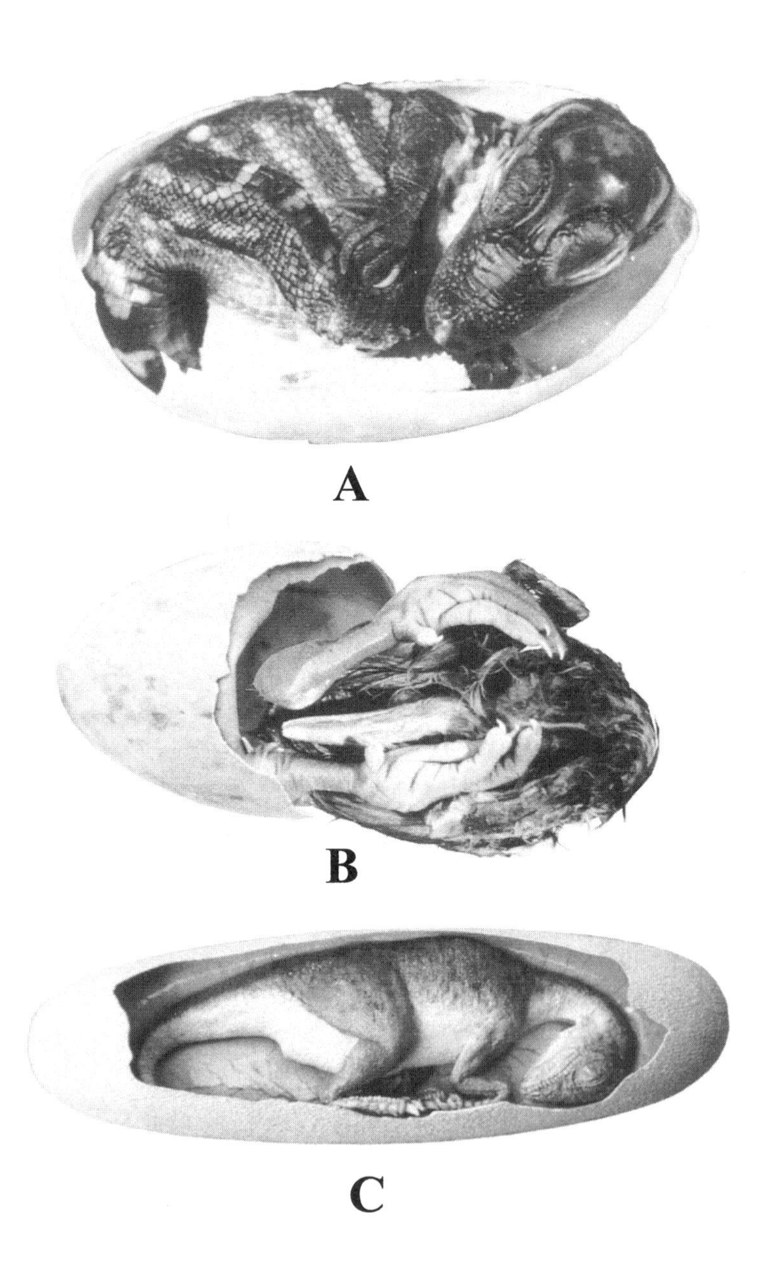

Fig. 11.1. Studies of the crocodilian embryo (A) and bird embryo (B) provide evidence for reconstructing a dinosaur embryo (C). (A, B courtesy of K. Hirsch; C, courtesy of the Stone Company.)

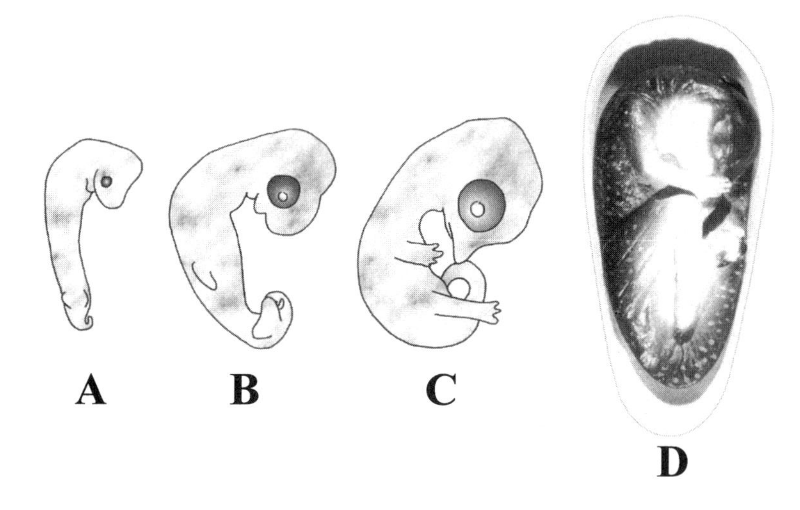

Fig. 11.2. Hypothetical dinosaur embryo made by 50 percent morphing of various stages of development of a crocodile and bird (see Figs. 11.4, 11.6): A, early embryo; B, middle stage embryo; C, late stage embryo; D, embryo just prior to hatching. (Model by Matt Smith.)

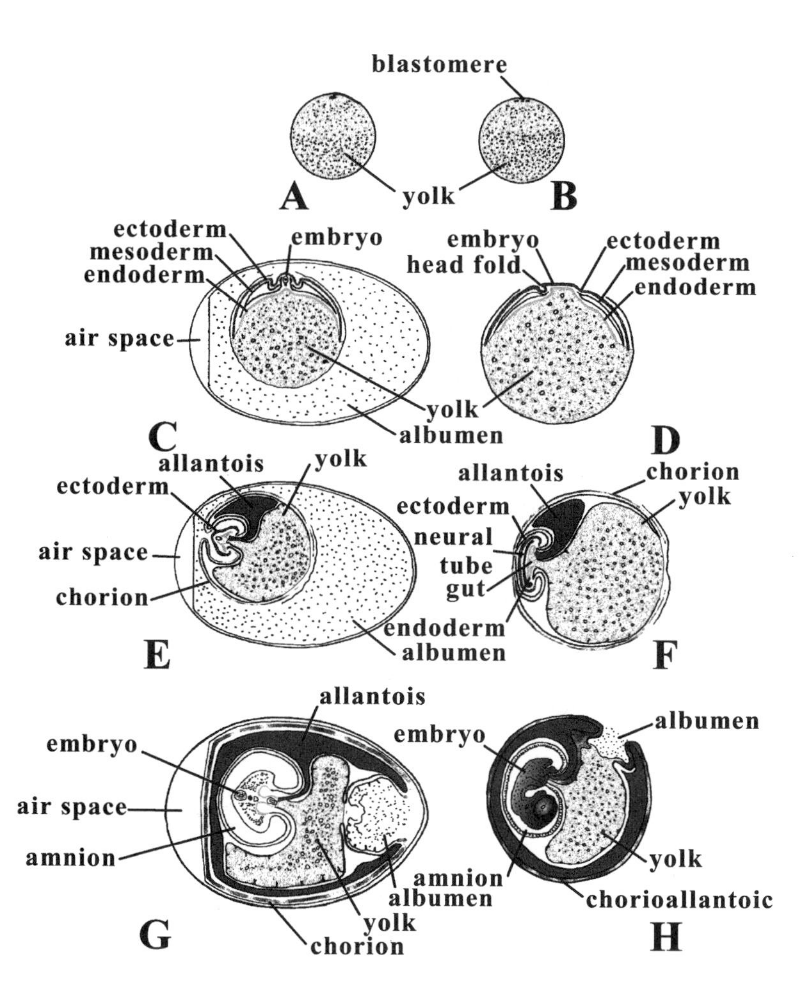

Fig. 11.3. Stages showing the development of the bird (chicken) embryo: (A) The blastodisc sits atop a ball of yolk. Once it has been fertilized, it begins to undergo cell division, first in half, then into quarters, then sixteenths, etc. (B) These early divisions form the blastomeres. Eventually, these cells begin to differentiate into the three major tissue types: ectoderm, mesoderm, and endoderm. These tissues begin to take on the shape of the embryo. (C) By making a cross-section of the embryo, we can see that it sits at right angles to the axis of the egg, (D) but if we make a cross-section of the egg, we can see the long axis of the embryo. Although the same tissues are present, their arrangement is slightly different. (E-H) Eventually, however, the embryo begins to rotate so that it has more space for growth and for the various developing membranes. (Adapted from Patten, 1951.)

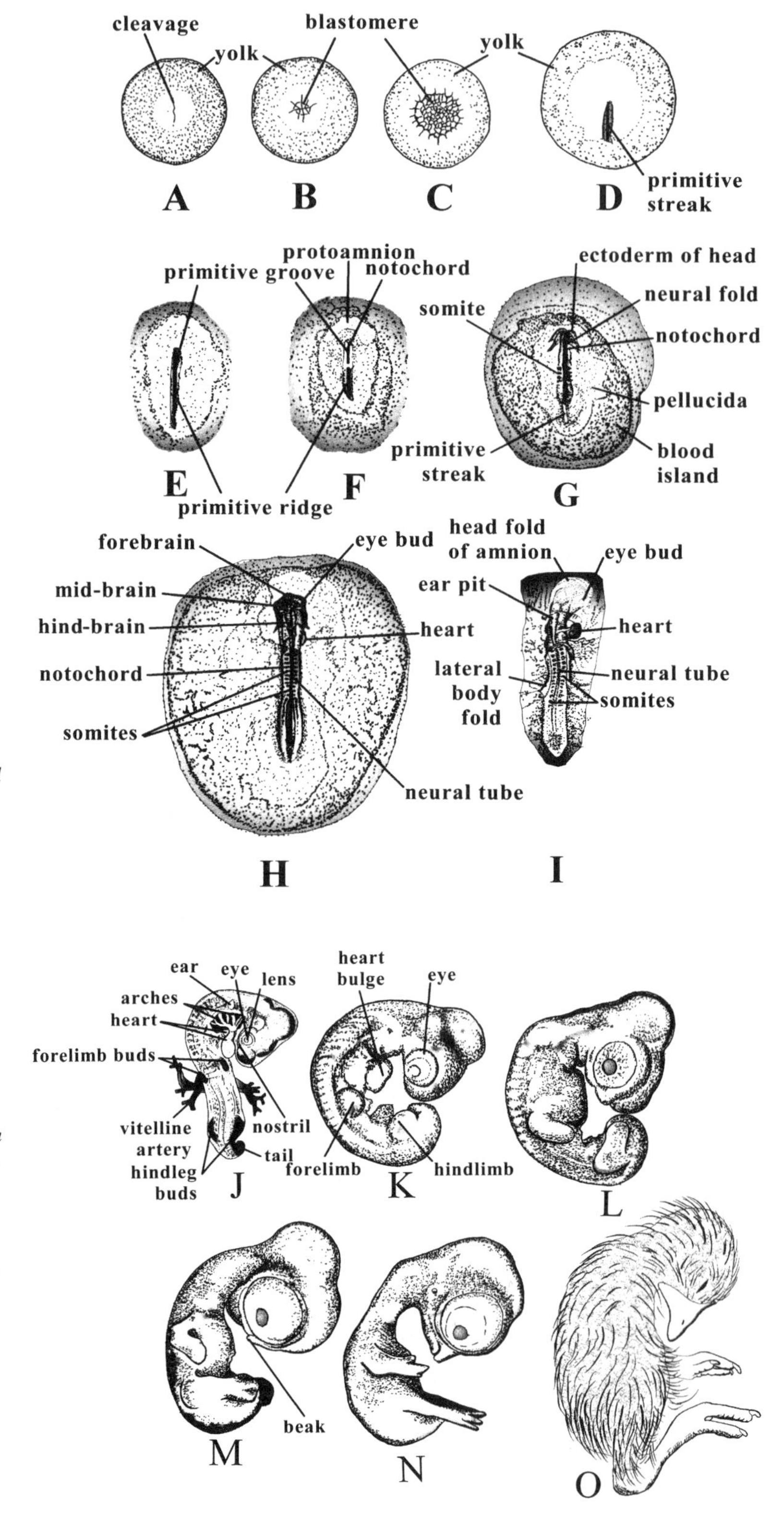

Fig. 11.4. More detailed look at the development of the chicken embryo. A-C, initial growth of the blastomere. D, stage 3, appearance of the primitive streak (individual cells of the blastomere not shown). E, stage 5, appearance of the primitive groove and ridges (only a small portion of the yolk is shown). F, stage 7, notochord appears, as does the precursor to the amnion over the head region. The primitive ridges begin to close over beginning at the head. G, stage 8, somites begin to appear, as do blood islets. The primitive streak is reduced in size as the neural ridges close over, forming the neural tube. H, stage 11, the primitive streak is gone, the neural tube almost completely formed, and somites continue to increase in numbers. The three regions of the brain appear, as do the eye buds. Part of the heart also appears. I, stage 14, the head of the embryo begins to rotate toward the right. The heart is expanding, as the amnion begins to form over the head. J, stage 18, the front half of the body has now rotated and the branchial arches begin to form. Limb buds appear. K, stage 21, the embryo has completed its rotation. Limb buds are well developed. The heart is still outside the chest. L, stage 25, joints appear in the limb buds. The eyes are well developed, and the heart has been pulled into the chest. Some of the branchial clefts have begun to close. M, stage 29, the eyeball is pigmented, the ear cleft is isolated from the branchial clefts, and most clefts have closed. N, stage 33, the limbs are almost completely formed. O, stage 39, feathers have appeared. Most subsequent changes are associated with growth of the embryo, rather than the formation of new structures. (Adapted from Patten, 1951; Hamburger and Hamilton, 1951.)

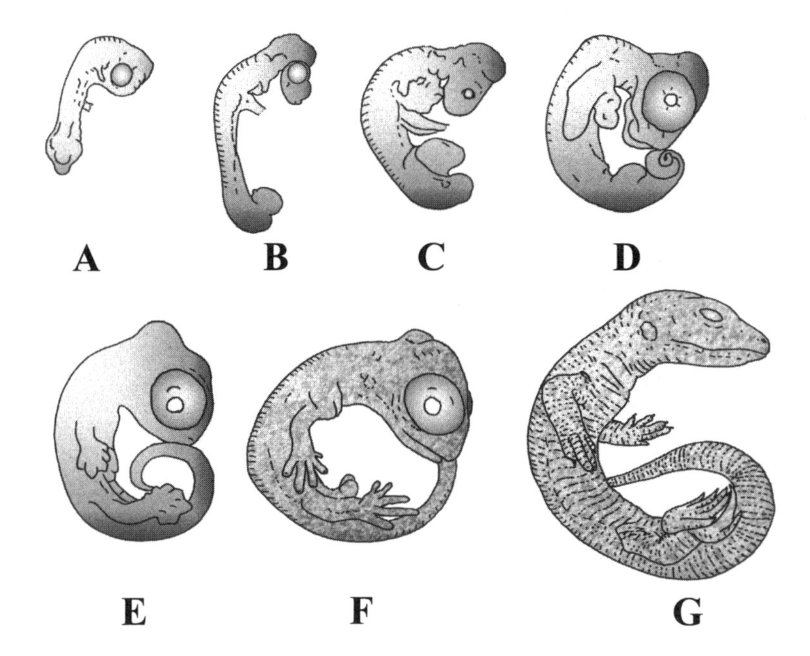

*Fig. 11.5. Stages of development for a lizard (*Lacerta*). Although the sequence of development is somewhat similar to the chicken, there is no one-to-one correspondence. A, stage 23; B, stage 26; C, stage 27; D, stage 31; E, stage 34; F, stage 36; G, stage 40. (Adapted from Porter, 1972.)*

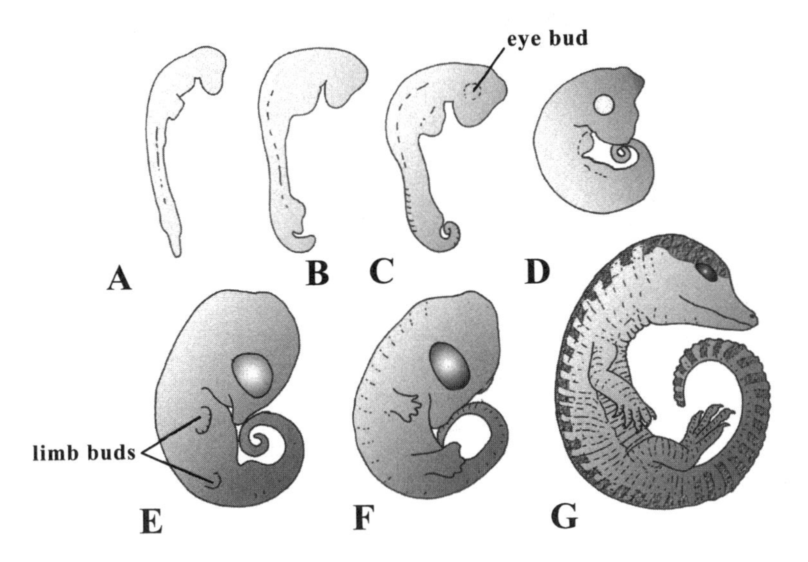

Fig. 11.6. Stages of development for the crocodile. As with the lizard and chicken, the stages of development are only roughly similar. A, stage 4 (day 4); B, stage 6 (day 6); C, stage 9 (day 9); D, stage 12 (day ~14); E, stage 17 (days 22–23); F, stage 19 (days ~27–28); G, stage 23 (days 41–45). (Adapted from Ferguson, 1987.)

11.4). Within a few hours the cell begins to divide into many cells, collectively called blastomeres, a process called cleavage or segmentation. By this point the egg begins its long descent down the uterus where it will get its membranes, be shelled (see chapter 6), and then laid.

The blastomeres of the expanding blastodisc meanwhile differentiate into various types of tissue, a process called gastrulation. The three major tissues formed are the ectoderm (outer surface of the embryo), endoderm (lower surface against the yolk, and mesoderm between them (Fig. 11.3; Table 11.1). Gastrulation differs between birds and lizards (there is inadequate data for crocodiles), so it is difficult to know which version occurred in dinosaurs; probably avianlike dinosaurs had bird gastrulation. In lizards, gastrulation begins at a hole through the blastodisc, called the blastopore, that develops at one end. This end will become the tail-end of the embryo (stage 5 of lizards). Gastrulation progresses below the surface of the blastodisc as a growing tube of mesodermal and endodermal tissue, the blastopore canal (stages 5 and 6) (Fig. 11.7). The canal reaches its maximum length at stage 7.

In birds, instead of a blastopore, an arc of thickened tissue appears, called Köller's sickle (also called an "embryonic shield"). Gastrulation progresses toward the head end of the blastodisc. The mesodermal layer migrates forward beneath the surface as a sheet, rather than along a tube as in lizards. As it migrates, a line of tissue called the primitive streak is produced (about six to seven hours after incubation) (Fig. 11.4). The mesoderm development reaches its maximum forward point in birds about

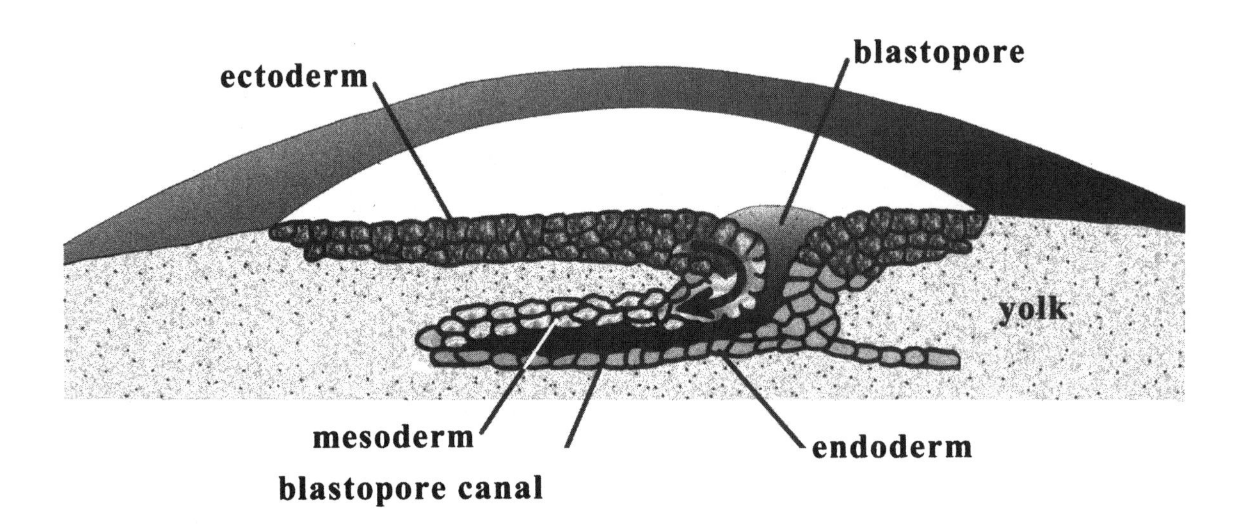

Fig. 11.7. Detail of the blastodisc of a lizard showing the infolding of the blastopore canal below the ectoderm.

TABLE 11.1.
Different organs formed from the three main tissue types

Ectoderm
skin, nail, teeth, feathers/scales, nerves, brain, eyes, nostrils, ears, skull bones

Mesoderm
muscle, blood, bone, kidneys, testes and ovaries, tissue lining the body cavities

Endoderm
tissue lining the gut, trachea, lungs, urinary system, liver, pancreas

eighteen to nineteen hours after incubation, then reverses direction (Bellairs, 1991). Interestingly, crocodiles have both a blastopore and a primitive streak, but how these tissues form is yet unknown (Bellairs, 1991). Once gastrulation has begun to differentiate the tissue types in both reptiles and birds, the various parts of the embryo begin to develop. We'll look at these next. They are discussed as anatomical sections and do not imply the order of appearance.

The Skeleton

The skeleton serves to protect the vital organs (brain, heart, etc.), to provide internal support for the body, and to provide attachment sites for muscles. What little we know about dinosaur embryos is based upon skeletons found within eggs. These few fossils indicate that the formation of the internal skeleton was very similar to that seen in living reptiles and birds. One of the first structures to appear is a rod of tissue, called the notochord, which is a precursor of the vertebral column. It develops from mesodermal tissue along the top of the neural canal in lizards (stage 11) and beneath the primitive streak in birds (19–22 hours after incubation); how it formed in dinosaurs is not known but pits in the center of embryonic vertebrae show that it was present (more below).

Once the notochord has formed, segmented blocks of mesodermal tissue, or somites, appear on each side (stage 16 in lizards; 23–26 hours in birds) (Fig. 11.4, 11.7). These somites appear every few hours for several days, by which time they extend to the end of the tail. In many reptiles, between seventy-five and one hundred or more somites are produced; fifty to fifty-two form in birds (Bellairs, 1991). The actual number of somites is very difficult to determine, because once the somites in the tail start forming, those in the neck begin to disappear as they become modified into vertebrae (about stage 31 in lizards, day 5 in crocodiles, and ~3½ days in birds).

The modification of the somites into vertebrae begins with each somite, called a schlerotome, dividing in half and fusing to the schleromote half behind it (Fig. 11.8). The combined schlerotomes fuse and grow around the notochord until they resemble a string of spools. The mesodermal tissue of the spools becomes cartilage, as do small segments of tissue above the spinal cord (more on the cord below). Interlocking projections develop from these segments, as do erect blades, called neural spines. These spines

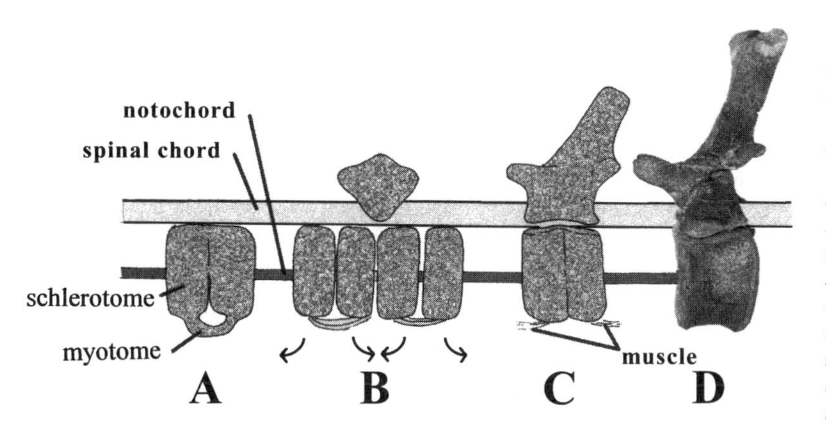

Fig. 11.8. Schematic showing the formation of a Stegosaurus *tail vertebra. A, a schlerotome appears around the notochord. B, the schlerotomes split and adjacent halves migrate toward each other. The myotomes still bind schlerotomes halves together and these will become muscles and other tissue that connect vertebrae together. Above the spinal cord, the neural crest begins its change into the neural arch. C, the co-joined schlerotome halves fuse, and the neural arch develops an upright spine and interlocking processes on the front and back. D, the cartilage precursor gets modified into bone. The bone is an actual baby* Stegosaurus *vertebra from the front part of the tail.*

provide attachment sites for the back muscles. In dinosaurs, remnants of the notochord are preserved as a dimple in the center of the vertebrae of embryos and hatchlings (Fig. 11.9A), called a notochordal pit. Although the pit is obliterated by the time the embryo emerges, traces can sometimes be seen even in adult dinosaurs.

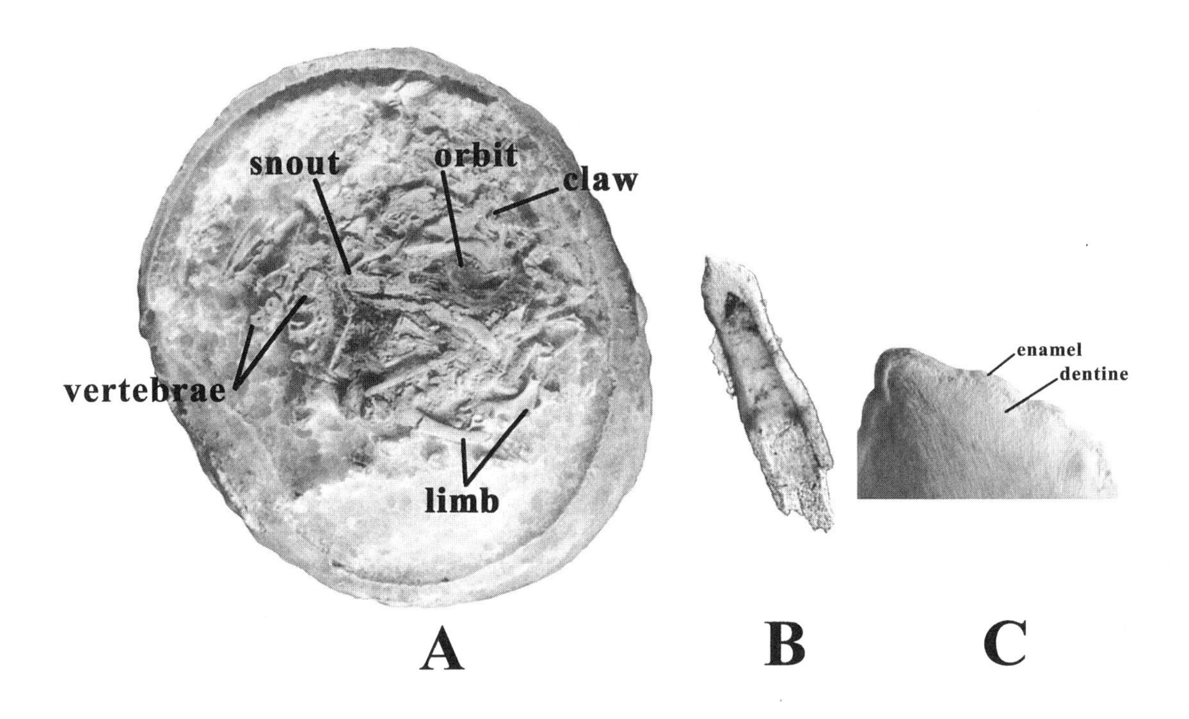

Fig. 11.9. A, a dinosaur embryo thought to be segnosaur. A dimple may be seen on the face of the vertebrae through which the notochord passed. B, a tooth from the segnosaur embryo. C, a detail of the tooth showing the hairlike tubules of dentine and the denser enamel forming on the edge. Eventually the enamel will cover the entire crown of the tooth. (Courtesy of Terry Manning.)

The formation of the tail in dinosaurs (Fig. 11.2) probably began as a little nub as it does in the lizard (stage 22), crocodile (day 3), and chicken (day 2½). As the tail lengthened, it probably coiled to take up less space in the egg. In reptiles, the time it takes to fully develop a tail is roughly proportional to the size of the species. For example, a lizard (hatchling 4 cm) takes about three to four days to form the tail, whereas the alligator (hatchling 22 cm long) takes about seven days. We may therefore safely assume that the embryo of a small dinosaur, like *Compsognathus,* took mere days to form a tail, whereas the giant *Apatosaurus* took over a week, possibly two.

The formation of the skull is slightly more complicated than for the rest of the skeleton because it involves both mesodermal and ectodermal tissue (Fig. 11.10). Most of the lower portions of the skull, such as where the nerves enter the braincase, are formed from the mesoderm in reptiles and birds, whereas the braincase is ectodermal. As the head grows and the sensory organs develop (more below), the face of the embryo begins to take shape. The muzzle begins to elongate in front of the eyes. Along the throat, a series of slits, called branchial clefts, appear. These are separated by the branchial arches (Fig. 11.4J; the first of these arches appears in stage 25 in the lizard, day 1 in the crocodile, and day 2 in the bird; the fifth and last arch appears by stage 30 in lizards, day 13–15 in the crocodile, and day 4 in birds). The first branchial arch becomes incorporated into the hinge between the lower jaw and skull, whereas the second arch is modified as a bone (called a hyoid) to support the tongue. Branchial cleft 1 becomes the ear (more below). Branchial arches 3, 4, and 5, and branchial clefts 3 and 4 are eventually absorbed and disappear (by stage 34 in lizards, day 16–19 in crocodiles, and day 4½–5 in birds). These arches and clefts do not all appear at once, nor are they equally developed. Their appearance and disappearance is timed to the formation of various structures in and around the head.

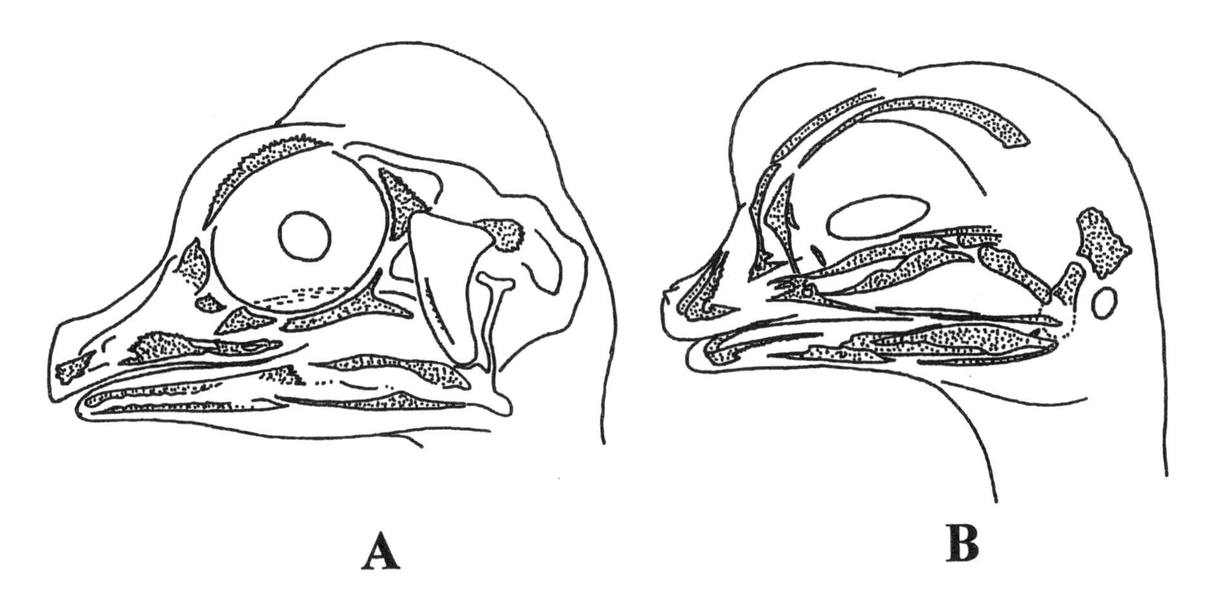

Fig. 11.10. Regions of bone formation (ossification) in the head of a crocodile (A) and a chicken (B). (Crocodile modified from Rieppel, 1993; chicken from Jollie, 1973.)

The limbs of the dinosaur embryo (Fig. 11.2) began development as small buds of mesodermal tissue (stage 5 in lizards, day 6 in crocodiles, and day 2½ in chickens). As the limbs elongated, the tissue diversified into masses of muscle (more below), nerves, and cartilage rods. The rods developed segments that separated the upper limb from the lower (stage 32 in lizards, day 11–12 in crocodile, and about day 3½–4 in birds). The hands and feet emerged first as paddlelike structures (stage 31 in lizards, day 16–19 in crocodiles, and about day 5 in birds), which then divided into fingers and toes (stage 35 in lizards, day 29–36 in crocodiles, day 7½ in birds). Interestingly, based on what happens in lizards and crocodiles, the toes probably became separated before the fingers.

The development of the internal skeleton in the dinosaur embryo was a slow process that began with the formation of cartilage templates within the embryo. These templates started as rods of mesodermal tissue that became cartilaginous versions of the various limb parts (Fig. 11.11). The mass of cartilage cells then underwent a change and arranged themselves into long columns. Deposits of calcium next surrounded the cartilage, killing it in the process. Blood vessels invaded this proto-bone and spread internally. As the vessels spread, they dissolved the dead cartilage. The blood also brought specialized cells, called osteoblasts, that laid down bone in place of the cartilage. This process of ossification began at the center and

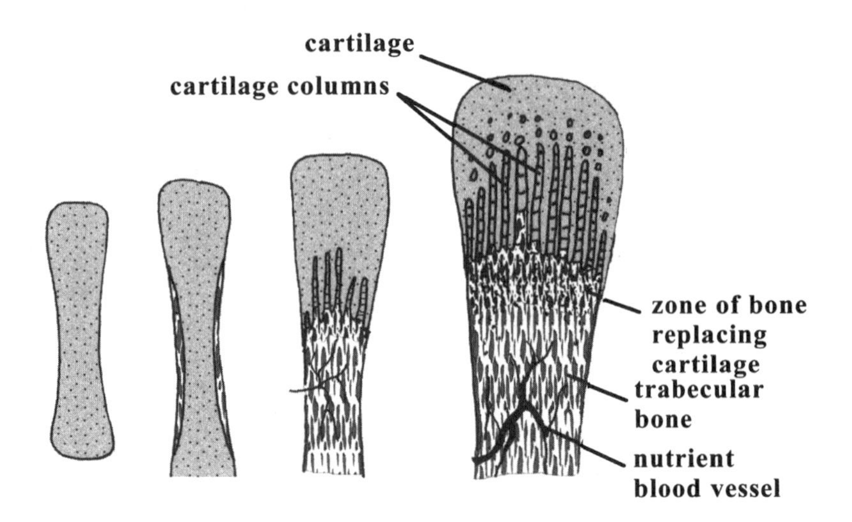

Fig. 11.11. Formation of limb bones from a cartilage template (see text for details). Once a blood vessel pierces the template, it introduces bone cells, which modify the cartilage.

proceeded toward the ends. The process of bone formation, or ossification, occurs late in embryo development (days 7½–8 in the chicken). At first, most of the calcium is obtained from the yolk, but later (day 12 in the chicken) is taken from the underside of the shell by the chorioallantois (more below).

Teeth in the dinosaur embryo formed from ectodermal tissue that turned outside-in along the margins of the jaws. Each tooth site folds inward creating a pocket within the mesoderm. The mesodermal tissue formed a bud at the bottom of the pocket and bulged upward. As it did so, it stimulated specialized cells lining the walls of the pocket to secrete enamel, the hard, shiny outer coating of teeth. In many reptiles the teeth are fully formed and erupted when the hatchling breaks out of the egg (but not in the alligator; Ferguson, 1987). Interestingly, the teeth of some embryonic dinosaurs (hadrosaurs, sauropods, *Psittacosaurus*) show wear on their tips indicating that the embryos were grinding their teeth (Horner and Currie, 1994; Chiappe and others, 1998). Perhaps the wear was incidental to exercising the jaw muscles—we can be sure that it wasn't because of nervous energy.

Because many of the embryo structures and organs began as small clumps of cells (e.g., the buds for the limbs), it is easy to see how evolution can make major changes in the adult. For example, increasing the number of cells that become vertebrae can increase the number of vertebrae in the tail from fifty-two in the short-tailed *Camarasaurus* to eighty-two or more in the whip-tailed *Diplodocus*. In a reverse process, reducing the number of times the cells for the fingers split can change the five fingers of the dinosaur ancestor to four seen in the primitive ceratosaurs, then later to three seen in allosaurs, two seen in tyrannosaurs, and one in *Mononykus*. However, for a structure, once lost, forever gone.

Brain and the Senses

As gastrulation proceeded in the dinosaur embryo, a furrow separating two ridges appeared (Fig. 11.12, 11.13; stage 12 in lizards; 18–19 hours in birds). The ridges folded toward each other over the groove, beginning at the head end and forming the head fold (stage 16 in lizards; 26–29 hours in birds). The ridges slowly came together like a zipper closing toward the tail end (completed by stage 21 in lizards; day 3 after laying in the crocodile; 2½ days in birds). The joining of the ridges above the groove formed a tube, the neural tube, which became the spinal cord. In both the shoulder and pelvic regions, the spinal cord expanded into a large lump of tissue called a plexus. In *Stegosaurus*, the pelvic plexus gave rise to the myth that this dinosaur had two brains. In reality, each plexus helps the brain control the limbs.

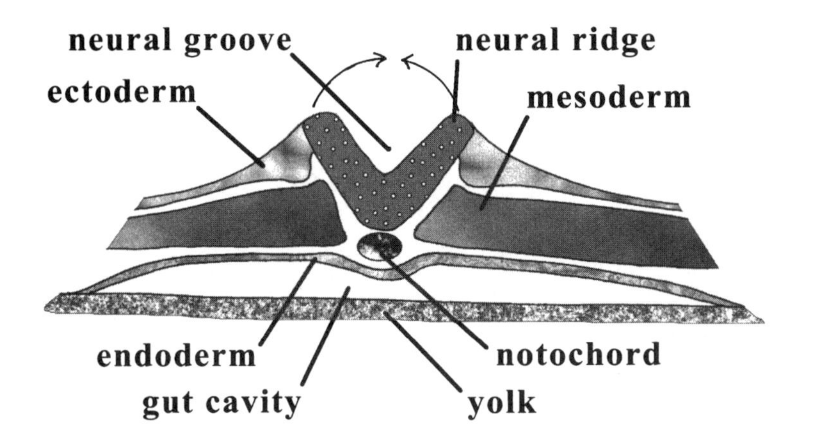

Fig. 11.12. The neural ridges on each side of the neural groove fold over and join forming a neural tube. This tube then becomes the spinal cord. Below the neural groove is the notochord. The mesodermal tissue closest to the notochord will become somites, precursors to the vertebrae and associated muscle.

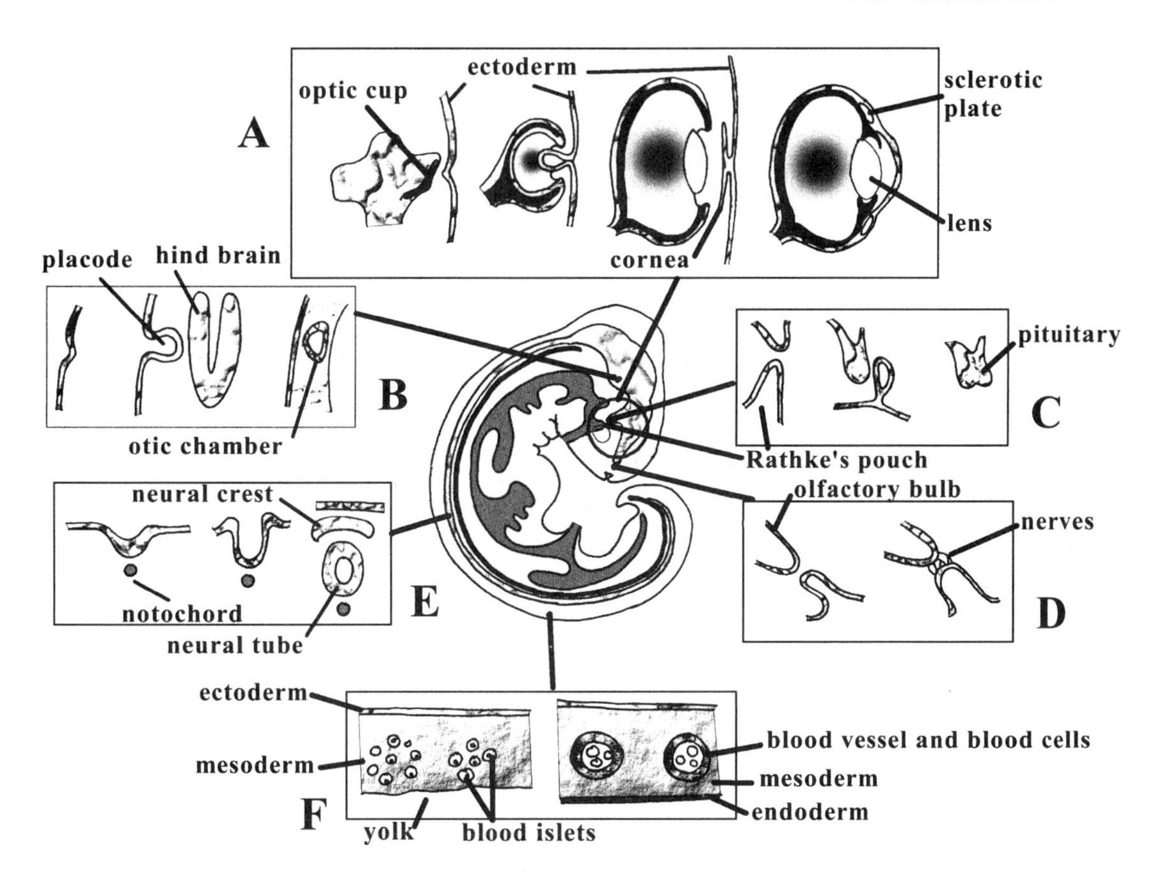

Fig. 11.13. The formation of various tissues in the embryo (see text for details): A, formation of the eye; B, formation of the inner ear; C, formation of the pituitary; D, connection of the olfactory or smell tissue with the brain; E, formation of the neural tube; F, appearance of blood islets and their modification into blood vessels and blood cells.

The front of the neural tube gradually expanded to form the brain and other sensory organs (Fig. 11.13). Brain development occurred early (stage 15 in lizards; already present in the crocodile at egg laying; after a day in birds) because a great deal of time was needed to develop this organ. The brain developed three major regions, the forebrain, midbrain, and hindbrain (stage 23 in lizards; day 3 in crocodiles; and day 2 in birds). The forebrain controlled, among other things, the sense of smell, the midbrain the region for sight, and the hindbrain processed hearing and controlled the jaws, viscera (heart, stomach, etc.), and muscles for locomotion. Considering how complicated the brain and the sensory system is, it is no wonder that hardwiring of the nervous system begins early in the embryo. The head region grows to accommodate the developing brain.

At about the same time that the brain began to form in the dinosaur embryo, or soon thereafter, the eyes started to develop. They first appeared as a pair of buds, called placodes, from each side of the developing brain (Fig. 11.13; stage 23 in lizards; already present when laid in crocodiles; and day 1 in the chicken). These buds grew outward on long stalks, the ends of which swelled into cuplike structures that form the rear of the eyeballs (stage 30, lizard; day 9 in crocodiles; and day 2 after incubation in the chicken). The eye cups (optic cups) fused to the ectoderm, the outer layer of tissue. Gradually, as development progressed, the ectoderm became transparent, but not before forming a pigmented ring, the iris (in lizards, this begins in stage 30; day 10–11 of crocodiles; and day 3 in the chicken). In chickens, the "wiring" between the eyes and the brain is completed four to five days after incubation begins. The eyelids in the dinosaur embryo did not form until after the eyeballs were almost completely formed. In lizards, this occurs in stage 33, day 27–33 in crocodile, and day 13 in birds.

Another sensory organ associated with the brain of the dinosaur embryo was the ear (Fig. 11.13). The outer tube portion grew from a dimple behind the eye. This dimple formed from the first of five branchial clefts that appeared in the throat region. The auditory tube connected to the inner ear region (otic chamber), which developed from a pocket off the pharynx. This pocket became sealed off from the pharynx by tissue and remained next to the developing brain. The pocket developed three small chambers, the semicircular canals, which helped the animal to maintain balance. The rest of the pocket was used to convert sound vibrations into electrical signals the brain could process. The ear begins developing during stage 30 in lizards, day 5 in crocodiles, and day 1 in birds. Nerves connect the brain with the inner ear during day 8–9 in birds, but hearing does not occur until day 13.

The sense of smell is associated with tissue that grows from the front of the brain. This tissue, the olfactory lobes, was created from the forebrain (Fig. 11.13D). Nerves eventually connected the nostrils to the olfactory lobe to transmit the electrical impulses of smell. Development of the olfactory lobes begins in stage 27 in lizards, day 6 in the crocodile, and day 2 in the chicken.

As the dinosaur embryo's brain began to develop, the front of the face turned outside-in along an indentation called Rakthe's Pouch (day 8 in crocodiles). This pouch deepened toward the brain and was eventually cut off (Fig. 11.13C). A small part lodged against the brain, where it developed into the pituitary gland. The pituitary and most glands are formed of ectodermal tissue. The pituitary begins secreting growth hormone on day 15 in birds.

The Gut

As the head fold of the dinosaur embryo was produced, another pocket, the foregut, developed beneath the head from endodermal tissue (Fig. 11.14A). This pocket opened to the rear into the midgut region, and together with the midgut formed the first part of the digestive system (lizards, stages 12–15; day 1 in birds). The hindgut later formed from a pocket beneath the tail as a mirror of the foregut (day 6 in crocodiles; and day 3 in birds). Eventually, the foregut became the throat, lungs, and stomach, the midgut became the small intestine, and the hindgut the large intestine (Fig. 11.14B). The midgut was initially in direct contact with the yolk, but eventually this contact was restricted to a short umbilical called the yolk stalk. The volume of the yolk naturally decreased as the dinosaur embryo slowly digested it (similar to what happens in crocodiles, Fig. 11.18). This of course left more room within the egg for the embryo to grow into. The development of the kidneys from mesodermal tissue in the hindgut region was coupled with formation of the genitalia (stage 31 in lizards, day 8 in crocodiles). The genitalia were slow to mature because they were not needed until the little dinosaur reached sexual maturity, probably many years after hatching.

The Heart

The circulatory system, including heart and blood vessels, developed in the dinosaur embryo very early from growing masses of mesodermal

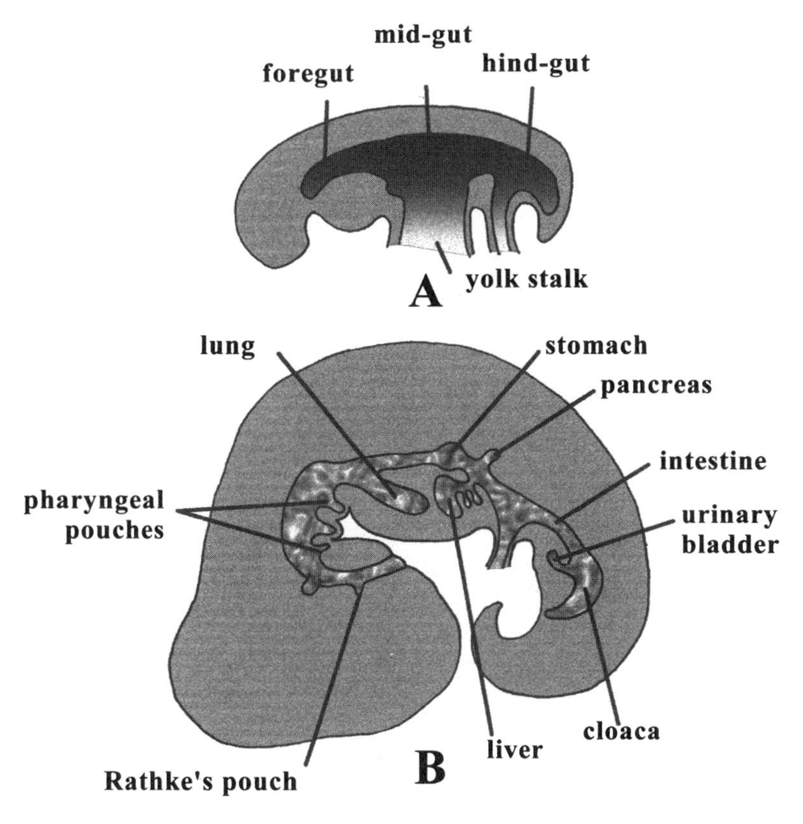

Fig. 11.14. Development of the gut: A, the simple three parts of the early-stage embryo; B, compartmentalization of the gut in a late-stage embryo.

tissue (Fig. 11.13F; after day 1 in birds). This tissue was initially scattered throughout the mesoderm like little islands, hence the name blood islets. Some of the islets came together to form the network of vitelline vessels, which grew out from the embryo over the surface of the yolk (Fig. 11.15). Other vessels penetrated the embryo and became extensive as a mesh of blood vessels (Fig. 11.16; after day 2 in birds). Near the center of the chest, the heart developed from a swelling (stage 23 in lizards; before laying in crocodiles; and after day 1 in birds). As it grew, it formed a series of loops and internal divisions that became the various chambers. Blood islets not used to form the blood vessels become blood cells (after 1½ days in birds). A few days or a week later, the heart was complete and was pumping blood through the system (day 5 in the chicken).

Muscles

There were three types of muscles in the dinosaur embryo's body: cardiac, smooth, and skeletal. Cardiac muscle was formed from mesodermal tissue and gave rise to the heart. Smooth muscles, such as around the stomach, were also produced from mesodermal tissue, but appeared as the various visceral organs developed. Skeletal muscles attached to the skeleton and made locomotion in the hatchling possible. The skeletal muscles (and the associated nerves to control them) developed from modified somite tissue, called myotome, and so were also mesodermal in origin (Fig.

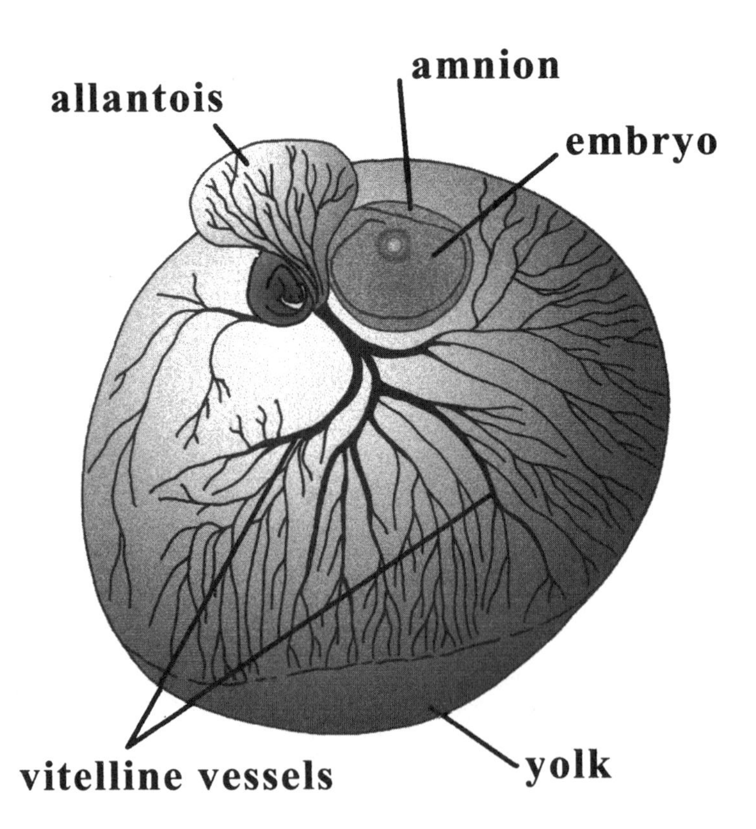

Fig. 11.15. The vast network of the vitelline vessels over the surface of the yolk. (Modified from Patten, 1951.)

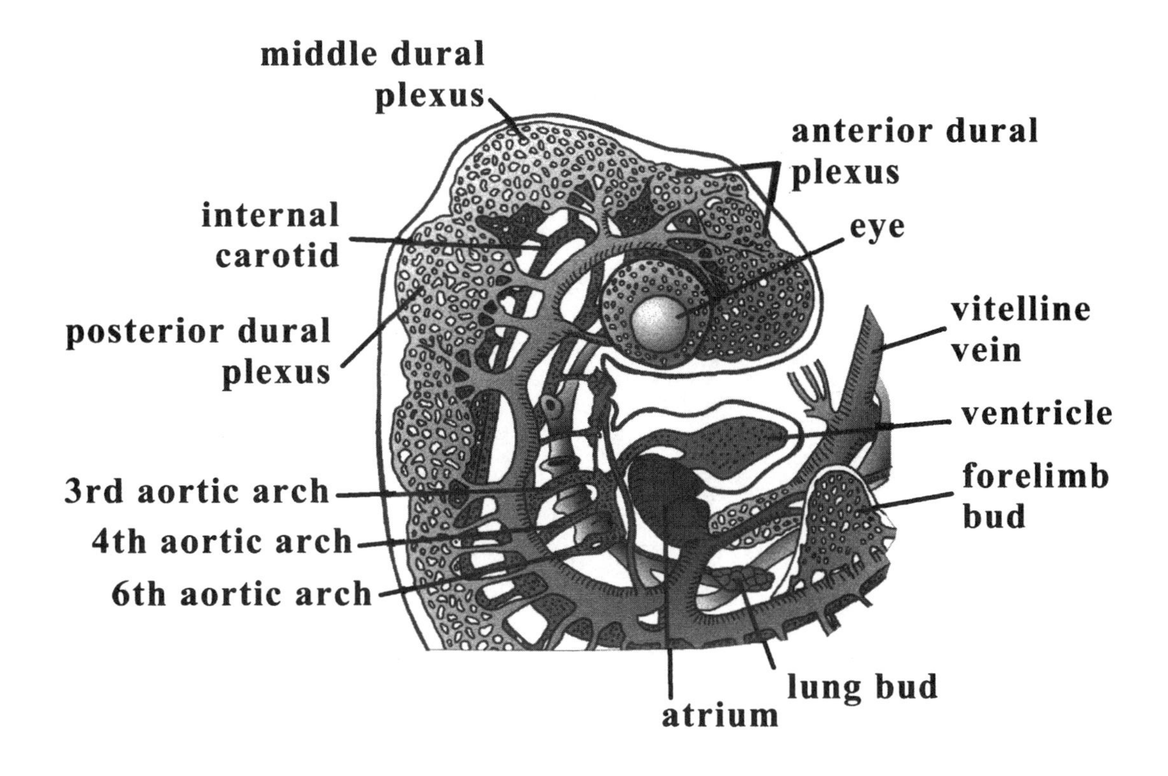

Fig. 11.16. Development of the circulatory system in the head of a bird embryo. (Adapted from Jollie, 1973.)

11.8). Initially, the muscles in the limbs formed into two groups, those located along the top and outside of the limbs, called the "dorsal mass," and those located along the insides of the limbs and closest to the body, called the "ventral mass" (Fig. 11.17). These masses of proto-muscles only vaguely resembled muscles (Romer, 1927; Sullivan, 1962). With time, the masses split into individual muscles, some to pull the limbs forward, some to pull the limbs back during the step, and some to keep the limbs from sprawling. In chickens, the limb muscles begin to appear around day 6 and are almost completely developed by day 12. Wing muscles also develop at about the same time. The pectoralis muscle, which is important in flight, is one of the largest muscle masses that forms. The first sign of skeletal muscle activity occurs in the chicken by day 7. A curious and temporary muscle in many birds appears on the back of the neck. Called the hatching muscle, it enables the head to push out of the egg. It appears just before hatching and disappears soon after hatching. Did dinosaurs have such a muscle? Possibly some of the smaller, more avianlike theropods did, but there is no evidence.

Skin

Skin serves to protect the underlying soft tissue from damage by drying or abrasion. During development in the dinosaur embryo, it was formed from ectodermal tissue. This tissue was also involved in the formation of teeth (see above), claws, scales (most dinosaurs), feathers (seen in some small theropods), and body armor (stegosaurs and ankylosaurs, among others).

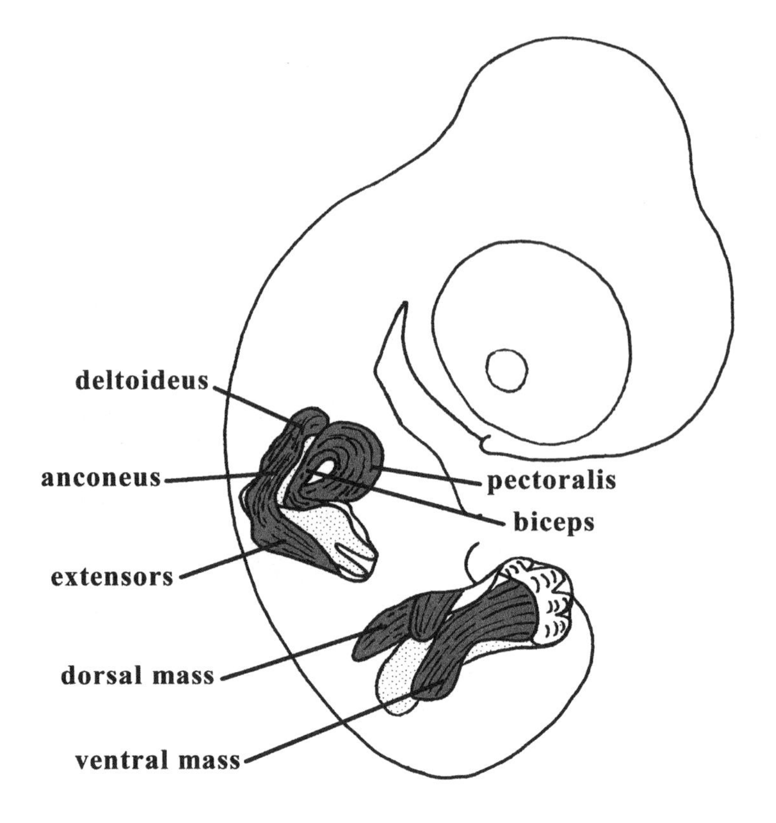

Fig. 11.17. Formation of fore- and rear-limb muscles in a chicken embryo at about six days after fertilization.

The claws are intimately linked to the formation of the fingers and toes, so were late in appearing in the dinosaur embryo (see above discussion on the skeleton). Scales and feathers are closely related, being formed from ectodermal (skin) tissue. Impressions of embryonic skin are known for a sauropod still preserved in its egg (Chiappe and others, 1998). This skin resembled that of various adult dinosaurs (*Carnotaurus, Diplodocus, Stegosaurus, Chasmosaurus, Edmontosaurus,* etc.) in that the scales were knobs and bumps of various sizes, not the flat, overlapping scales of snakes and lizards. In many ways, the scales resemble those of alligators.

The most primitive feathers are seen on some Early Cretaceous theropods from China (Chen and others, 1998). These primitive or protofeathers have hairlike bristles seen best along the back and tail. Embryological studies support the idea that feathers arose from the skin. In the chicken, the first indication of the feathers appears at 6½ days, long before the claws (day 10). In lizards, the claws and scales appear in stage 37. Color first appears in the scales during stage 39. In crocodiles, the scales first appear sometime between days 24–30 and color appears after day 36; claws begin to appear around day 30.

Bony armor, such as occurred on the backs of stegosaurs and ankylosaurs, probably did not form until after hatching, much like in the crocodile. This delay allowed the embryo to be tightly rolled up in the egg. The bone armor formed differently from the internal skeleton, in that it developed directly from ectodermal tissue without a cartilage template. Evidence for the delayed formation of armor can be seen in a group of baby *Pinacosaurus* found near the Chinese–Mongolian border. The individuals are about a meter long, so are definitely older than hatchlings. Yet the only armor on the bodies consists of two rings of bone on the neck. Based on comparisons with adult specimens, the body and tail should have been covered with armor plates, as well as a club of bone at the end of the tail. The absence of such armor suggests that it was still cartilage nubs embedded in the skin at the time of death. How much longer it took until all the armor developed is not yet known.

The Membranes

As the dinosaur embryo grew it made four membranes (Fig. 11.3). One of the first to appear enabled the embryo to absorb nutrients from the yolk. This membrane, the yolk sac, grew out and enveloped the yolk (Burley and Vadehra, 1989). As this sac grew, it was soon covered by a fine network of vitelline vessels that absorbed the yolk. These vessels were connected to the proto-circulatory system of the embryo by the yolk stalk.

Besides yolk, the embryo also needed oxygen to metabolically "burn" the yolk in its cells. Because at this stage of embryo development the lungs were still forming, the embryo could not breathe air. Instead, the embryo had an ingenious way of getting oxygen, using the lining of tissue beneath the surface of the shell. This membrane grew from the allantois, a saclike structure that developed from a bud off the midgut. This sack appears at stage 25 in lizards, day 4 in crocodiles, and day 2½ in birds. Eventually, the allantois grew down and out as a sack to store the waste produced from the newly formed kidneys. But it also grew outward until it met the chorion, another membrane, and the two fused into the chorioallantois (lizards, stage 30; crocodiles, day 9; chicken, day 8). The surface of the chorioallantois was covered by a fine network of capillaries and because it lined the underside of the eggshell, the capillaries had a very large surface area to maximize passive respiration. Why passive? Because the red blood cells in the capillaries had such an affinity for oxygen that they exerted a strong pull on oxygen molecules. Carbon dioxide molecules had the reverse effect and were easily "pushed" from the blood, out the pores. Basically then, the chorioallantois acted as the dinosaur embryo's lung because the lungs were too underdeveloped to function.

Oxygen consumption rose slowly at first as organs and other anatomical structures of the dinosaur embryo formed. But once formed, then most changes were in growth, not production of new organs, and oxygen consumption rose rapidly. Eventually, however, the embryo filled the egg and oxygen consumption slowed until hatching. Such a pattern in oxygen consumption may be seen in the chicken beginning around day 10 and continuing until around day 15, at which time consumption begins to level off (Freeman and Vince, 1974). The five days of rapid rise coincide with the most rapid growth of the chicken embryo. Similar changes can be seen in the crocodile (Fig. 11.18).

The chorioallantois had another function, being so close to the shell. It also absorbed calcium from the shell for the embryo's skeleton (see chapter 6). Based on what happens in crocodiles and birds, probably between 50 percent and 80 percent or more of the calcium needed by the dinosaur embryo was obtained from the shell, the rest from the yolk. In birds, an air space at one end of the egg prevents the chorioallantois from cratering the mammillae. Uncratered mammillae at one end of an elongated dinosaur egg have not yet been reported, so we do not know if a birdlike air space was present. We might expect such an air space in the eggs of *Oviraptor* or other birdlike dinosaurs. We'll explain the function of the air space below.

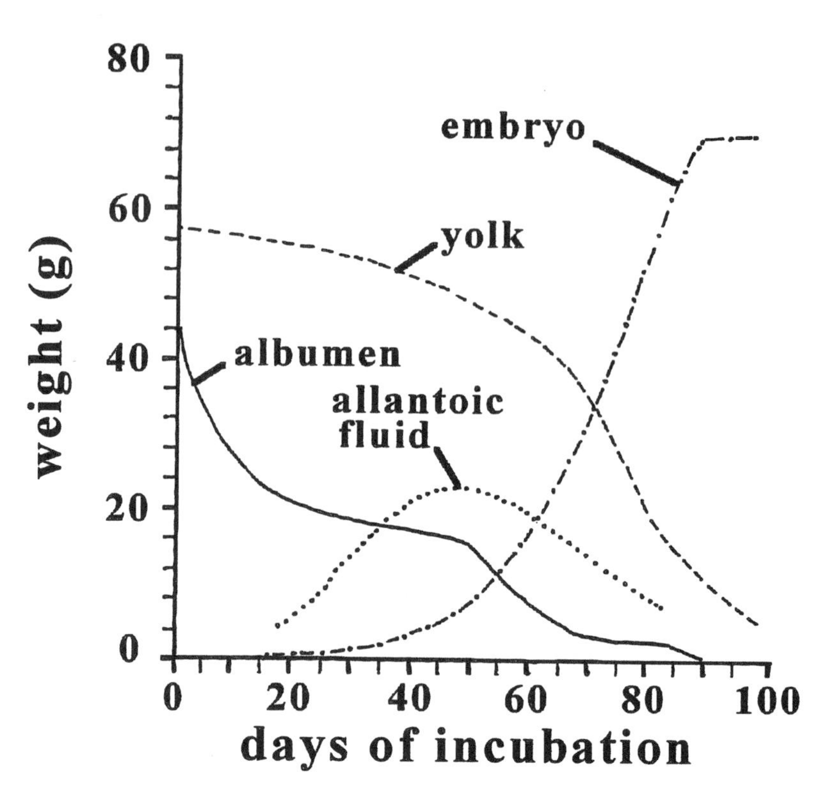

Fig. 11.18. Changes in the relative weights of various parts of a crocodile egg as it incubates. Note that as the weight (hence size) of the embryo increases, all other parts of the egg decrease to accommodate the embryo. This growth is initially slow as various tissues and organs form, but once formed growth of the embryo is very rapid. This rapid growth correlates with a rise in oxygen consumption. (Modified from Webb and others, 1987.)

Another membrane that was produced in the dinosaur embryo was the amnion, or amniotic sac. Its development takes from stage 17 to stage 25 in the lizard, over two weeks in the crocodile, and almost three weeks in the chicken. Fully developed, the sack acted as a fluid-filled cushion to keep the embryo moist and to protect it from sudden bumps. The amnion formed first just in front of the dinosaur embryo's head and extended backward as the embryo developed. In this manner, the amnion cushioned those parts of the embryo that had formed. In chickens, the embryo begins to ingest the amnion fluid around day 13 of development and this continues until hatching. The amnion is connected to the albumen, which also decreases as the embryo develops.

Changing Positions

During its growth, the dinosaur embryo probably changed its orientation several times. Initially, the blastodisc was oriented at right angles to the long axis of the egg as it is in modern lizard and chicken embryos (Bellairs, 1991). The modern embryo lies so that what will become the right side of the embryo is nearest the more pointed end of the egg, a recurring phenomenon in birds and reptiles called von Baer's Rule. The same rule probably applied to embryos in elongated dinosaur eggs, but not in spherical eggs—which have no orientation that we can determine.

For the bird embryo to develop properly, it must always stay on the upper side of the yolk. This situation is more of a problem for birds that rotate their eggs on occasion, when the embryo may suddenly find itself upside down. The yolk, however, slowly rotates, bringing the embryo back to the top. Why do birds subject the embryo to a constantly turning environment? Apparently it is to prevent the embryo from bonding or attaching itself to the shell membrane (Deeming, 1991). Experiments have shown that if eggs are not turned, the number of eggs that hatch is very low. In addition, the amount of unused albumen is high and the hatching weight of the embryo in unturned eggs is well below normal.

For some reason, the eggs of megapode birds and many reptiles develop quite well without turning—the eggs are left alone once they are laid. In fact, turning can be fatal because the various egg membranes can be ripped from each other. Even the embryo, which attaches itself to the shell membrane, can be torn apart (Webb and others, 1987). An additional reason why many reptiles do not turn their eggs is that a dry patch develops at the top of leathery eggshell. By drying the underlying shell membrane, oxygen flow to the embryo is less impeded. If the egg is turned, the dry spot is lost and embryo will not receive sufficient oxygen. An exception to this non-turning policy of most reptiles is the skink. The female guards her eggs and often licks and turns them to new positions when she has been away for a short time (Deeming, 1991). It seems doubtful, however, that most dinosaurs turned their eggs for the simple reason that, like reptiles and megapode birds, they most probably did not brood their eggs but buried them partially or entirely instead. On the other hand, small avianlike dinosaurs, such as *Mononykus*, may have started the egg-turning trend seen in modern birds.

As the dinosaur embryo took shape, it originally lay face down against the yolk (seen, for example, in birds, Fig. 11.4). Eventually, however, the embryo had to rotate onto its side because of the limited space available between yolk and shell (Fig. 11.4). This rotation began at the head and slowly extended down the rest of the body until the embryo had rotated (usually) onto its left side. This rotation begins in the lizard at stage 21, day 4 in the crocodile, and day 2 in the chicken. At the time that the head started to rotate, the body and tail had not yet completely formed. In birds, the embryo later rotates again so that it lies parallel to the length of the egg (around day 14). This new position gives the embryo the maximum available space for growth (Fig. 11.19). Dinosaur embryos in long eggs probably pivoted in a similar manner also.

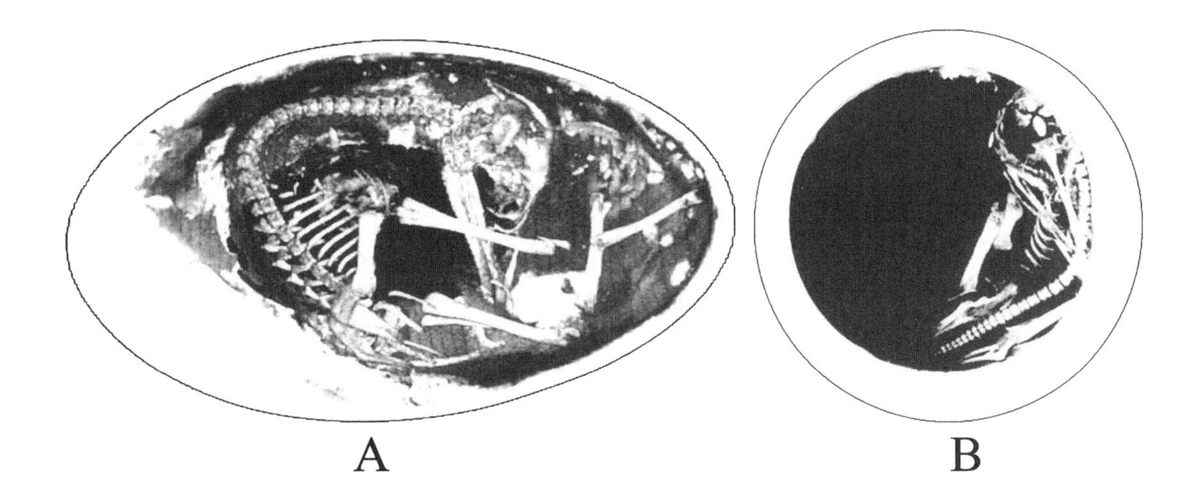

Fig. 11.19. CAT scan of an emu egg showing a late term embryo in side (A) and end view (B). Note that the embryo does not yet completely fill the egg at this stage (the absence of the forelimbs is probably due to decay; note that some of the toes are also displaced). (Courtesy of G. Grellet-Turner.)

Birth Defects

Sometimes developmental problems can arise. We know the horrors associated with the drug thalidomide, but the origins of other defects are less obvious, especially in the fossil record. At least two of them have been identified among baby dinosaur material. One of these is the lower jaw of *Troodon* (Fig. 11.20) (Carpenter, 1982). The defect is at the front of the jaw and would have caused the first tooth to point outward away from the snout. We don't know what other defects might have been present in this specimen because, except for one other bone, nothing else of the skeleton is known. We do not even know if this defect played a role in the early death of the hatchling.

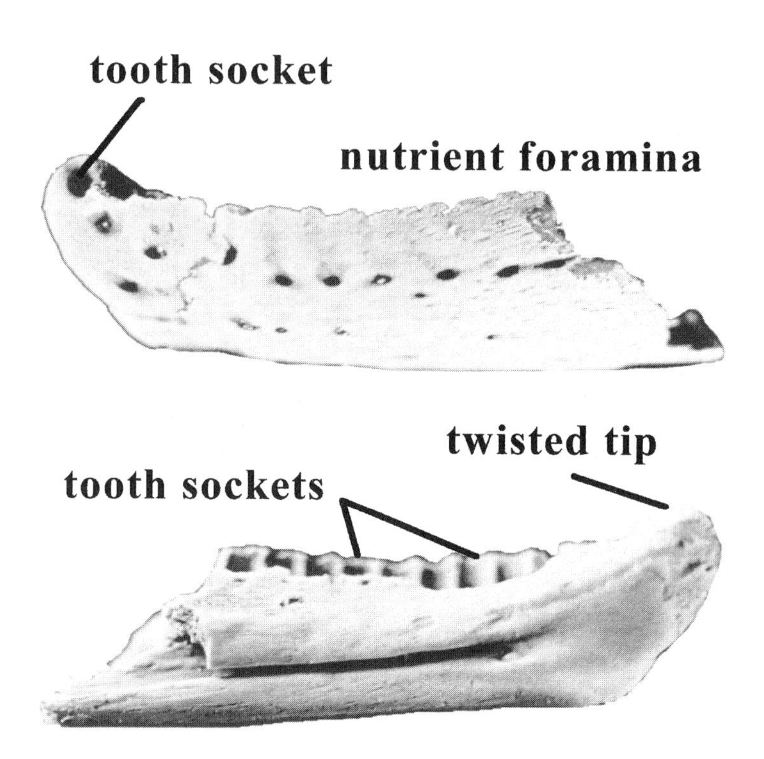

Fig. 11.20. Birth defect in the lower jaw of a hatchling Troodon. *The tip of the jaw is twisted outward as seen in the outside view (top). The tooth sockets seen on the inside view (bottom) should still be visible at the tip.*

A second example was found among the original material of baby *Maiasaura* that Jack Horner and Bob Makela collected. One of the foot bones has a kink in it (Fig. 11.21). Somewhat similar kinks are seen where one bone is pressed down over another by the weight of the overlying rock. But this appears not to have happened in this case because the defective bone does not show the usual marks due to crushing. Except for the kink, the bone is pristine.

We don't know the causes for the defects in these two examples, nor do we know how common defects were in the past. In Nature, birth defects tend to leave the individual susceptible to disease or predation, so are

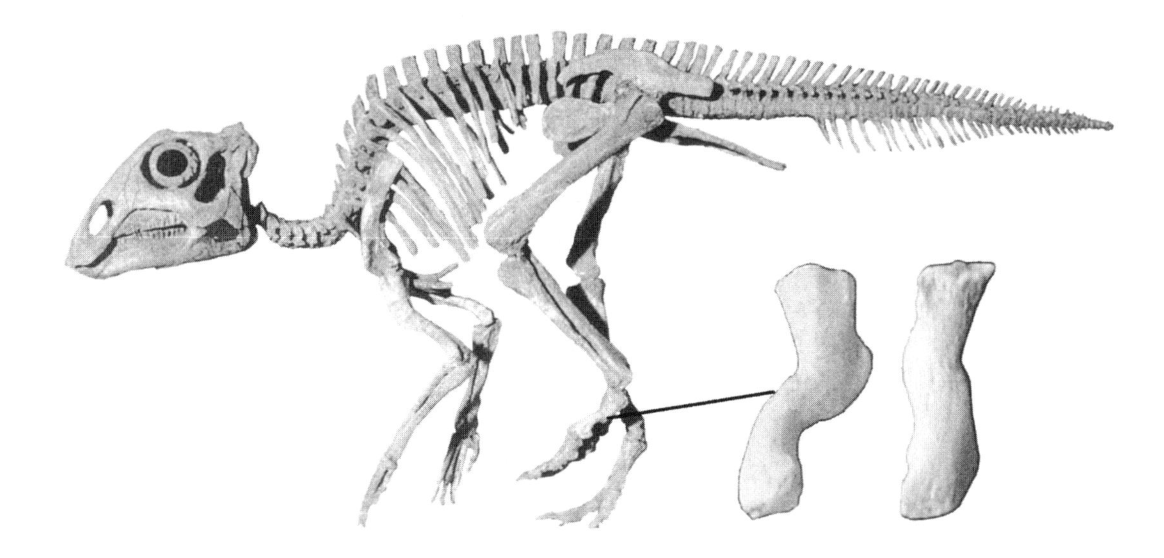

Fig. 11.21. "Crooked foot," named by Jack Horner for a baby Maiasaura *that has a pathological fourth metatarsal. Close-ups of this bone are seen in left lateral and front views.*

quickly weeded out. On this basis, it seems doubtful that we will ever find many examples in the fossil record.

Changes in the Egg

During the time that the dinosaur embryo developed, changes also occurred in the eggshell. As mentioned above, the chorioallantois was involved in removal of calcium from the underside of the eggshell. Removing calcium naturally leaves its mark on the eggshell. This is best seen in chicken eggshell where the mammillae are cratered (see chapter 6). A similar feature is seen in hatched dinosaur eggs, so embryonic dinosaurs must have also relied upon the eggshell for their calcium needs. With 50 percent or more of the calcium needs of the embryo taken from the eggshell, it is inevitable that the shell becomes thinner (Bond and others, 1988; Webb and others, 1987). This increases gas exchange at a time when the embryo needs it most.

As an aside, crocodile eggs develop a condition called banding. Banding begins as a small, white, chalky, opaque patch that appears immediately above the embryo within twenty-four hours of laying. Gradually, however, the band expands to first encircle the egg by day 7, and eventually encompasses the entire egg by day 52. Banding probably occurs to make the egg more porous to oxygen and water, and thus is analogous to the dry patch seen on the upper surface of leathery eggshell. Sometimes this banding is associated with cracking of the weakened shell. As yet, banding has not been reported for dinosaur eggs, possibly because the eggshell is proportionally thicker.

Incubation Time

The amount of time it takes for an embryo to mature to hatching, called the incubation time, varies considerably among different species. For example, the alligator hatches after 64–70 days of incubation, and the salt

Fig. 11.22. (facing page) A log-log plot shows the correlation between egg weight and incubation time in birds (dots) as a straight line (dashed lines denote +2 percent standard deviation on each side of the regression line and bracket 95 percent of all eggs). Plotting estimated mass of select dinosaur eggs (letters) along the regression line allows for a crude estimation of incubation time. Polygon in figure denotes region where most dinosaur eggs plot. A = Aepyornis; *D* = Dictyoolithus neixiangensis; *E* = Elongatoolithus elongatus; *L* = Laevisoolithus sochavi; *M* = Megaloolithus; *Mc* = Macroelongatoolithus xixia; O = "Oviraptor"; *S* = Spheroolithus irenensis. *The largest known egg is actually a bird,* Aepyornis. *(Plot for birds adapted from Rahn and Ar, 1974.)*

Fig. 11.23. (facing page) Assuming that the regression line in Figure 11.22 is valid for dinosaur eggs allows us to approximate the incubation time for various dinosaur eggs (diamonds). For example, a Spheroolithus *egg, with an estimated live weight (i.e., as it might have been 70 million years ago) of 152 g, would have an estimated incubation time (from time of egg laying until hatching) of thirty-five days. Because most bird eggs do not lie on the regression line (see Fig. 11.22), it is doubtful that all dinosaur eggs did as well. If 95 percent of all dinosaur eggs lay within +2 percent of the regression line, then the* Spheroolithus *egg might actually have taken anywhere from thirty to forty days to hatch, depending on whether the* Spheroolithus *embryo was altricial or precocial. Data for eggs taken from Appendix II.*

water crocodile after 92–97 days. In other reptiles, hatching can occur in as little as 28 days for the soft-shelled turtle *Trionyx,* or as long as 429 days in the tortoise *Geochelone pardalis.* The shorter incubation of *Trionyx* is due in part to its having a leathery-shelled egg, because such eggs tend to incubate more quickly (Ewert, 1991). In birds, incubation may take as little as 12 days in the cuckoos, to as long as 81 days in the royal albatross (Boersma, 1982; O'Connor, 1991b). Among ground birds, most megapode eggs hatch after 57 days. This vast difference is due in part to whether the chick is precocial or altricial (more in chapter 12), with the well-developed precocial birds taking the longest to form.

The incubation rate of dinosaurs can never be known with certainty, although some inferences can be made from bird models. For example, most birds show a strong correlation between incubation length and egg weight (Fig. 11.22; Rahn and Ar, 1974), although there are exceptions (Boersma, 1982). Not surprisingly, the larger the egg, the longer the incubation period. We can approximate the volume of an egg using the formula:

$$V = 0.51aw^2$$

where V = egg volume in cm^3, a = egg length in cm, and w = egg width in cm (Williams and others, 1984). Knowing the volume of an egg, we can guesstimate the weight by assuming that an egg weighs just a little more than water, or $1.09 g/cm^3$ (Williams and others, 1984). *Oviraptor,* an avianlike theropod, had an egg 18 cm × 6.5 cm (Norell and others, 1995) with a volume of $388 cm^3$ and a probable weight of about 423 g. Using Figure 11.23, we can deduce that the egg would take about forty-five days to incubate. On the other hand, the giant egg *Macroelongatoolithus,* also belonging to an avianlike theropod based on the shell structure, is about 45 cm × 15 cm, has a volume of $5{,}164 cm^3$, weighed about 5,628 g, and took about eighty days to incubate. In reality, these estimates for incubation are probably not very reliable. A 1,660 g ostrich egg should take fifty-nine days using Figure 11.22, but in actuality takes about forty-two days (Bertram, 1992). This short incubation period may be due to the ostrich sacrificing brain size for rapid body growth in order to minimize the time the embryo is vulnerable within the egg (Bertram, 1992). Maybe, but that is a difficult hypothesis to test. Another probable reason why the ostrich takes less time than expected to incubate its eggs is that the rate of incubation in both birds and reptiles depends a lot on temperature (Deeming and Ferguson, 1992). Considering the high temperature measured within ostrich eggs, up to 41°C (Bertram, 1992), this may be a more important factor than realized. Let's look more closely at temperature.

Effects of Incubation Temperature

The temperature at which embryos develop can have a profound influence upon incubation time and, for reptiles, even the sex of the hatchlings (Deeming and Ferguson, 1992). Because birds lay only one egg at a time, many days may pass between the first and last egg laid. One method to synchronize hatching is to keep all the eggs below incubation temperature until the last egg is laid (Bertram, 1992). Even after the last egg is laid, some birds such as the ostrich can continue to keep the eggs below incubation temperature if environmental conditions are not very good, e.g., a prolonged dry season. In this manner, the ostrich can delay hatching by two weeks (Bertram, 1992). The downside for the ostrich is that hatching rates decline if incubation is delayed more than fifteen days, and is very low after thirty days. An additional problem with delayed hatching is the

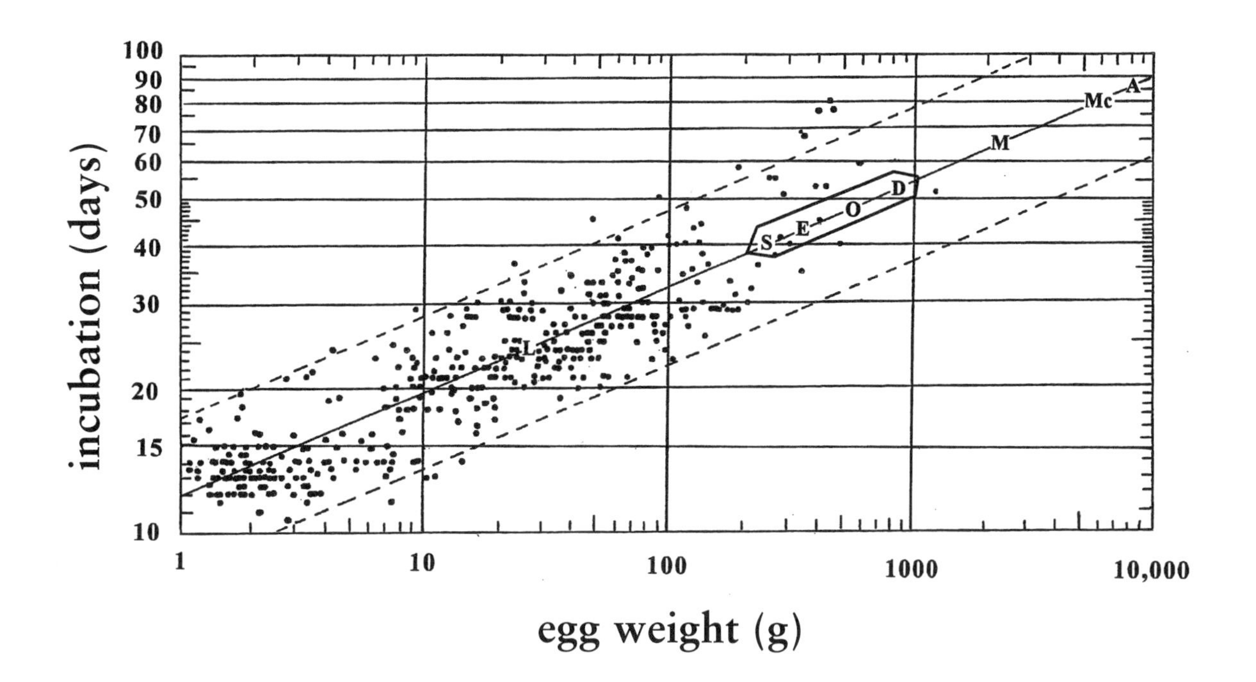

100
90
80
70
60
50
40
30
20
15
10
incubation (days)
1
10
100
1000
10,000
egg weight (g)
A
Mc
M
D
O
E
S
L

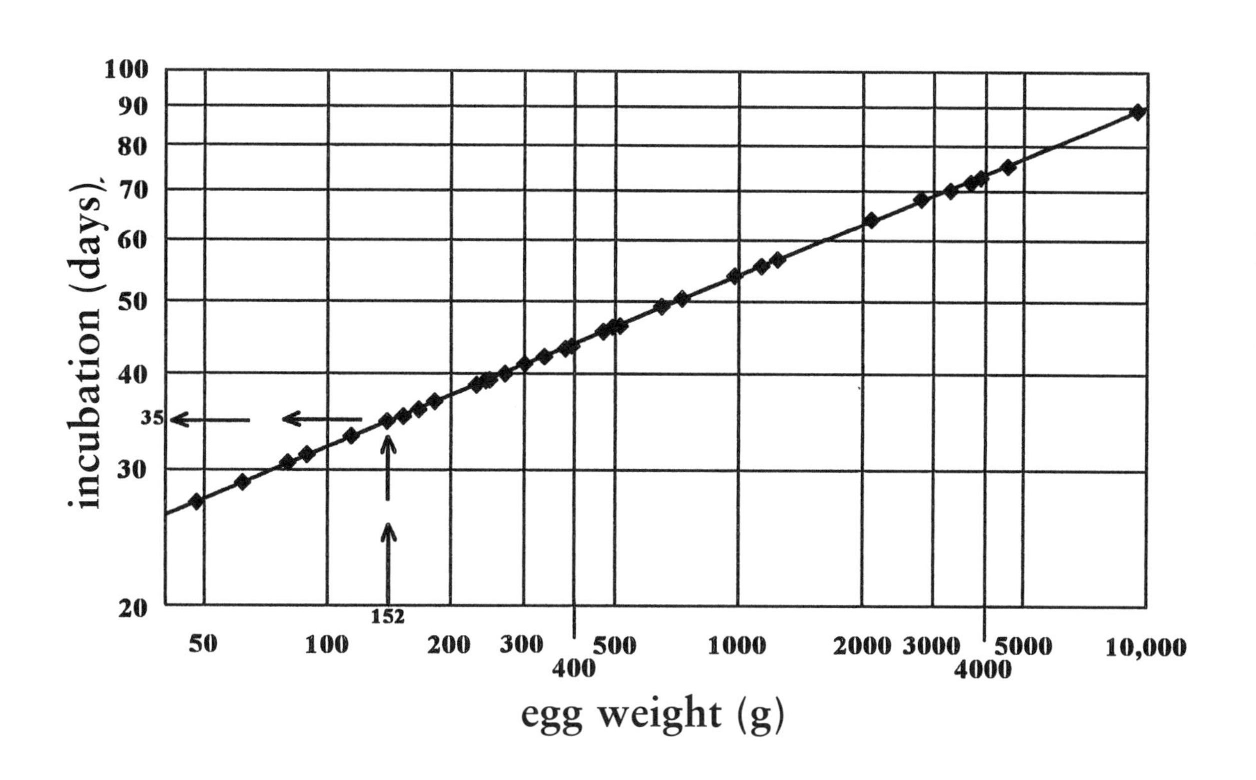

100
90
80
70
60
50
40
35
30
20
incubation (days)
152
50
100
200
300
400
500
1000
2000
3000
4000
5000
10,000
egg weight (g)

increased chance a predator will find the eggs. If keeping the eggs below incubation temperature can delay hatching, then elevating the incubation temperature can speed up embryo development by 10 percent or more. Where an ostrich embryo developing at 30°C (86°F) takes about sixty-two days to hatch, at 34°C (93°F) hatching occurs after fifty-five days (Bertram, 1992).

In reptiles, an increase in temperature can greatly change incubation time. For example, a 4°C (7°F) increase in temperature can cause a 33 percent *decrease* in incubation time in the desert iguana, *Dipsosaurus,* from eighty-three days to fifty-six days—that's a big difference! But even more profound than just prolonging or speeding up hatching is the effect temperature has on the sex of the embryo (Deeming and Ferguson, 1992). If alligator eggs are incubated at 29.4°C (85°F), the embryos are all females, and at a higher temperature, 32.7°C (91°F), all males (Ferguson and Joanen, 1983). With turtles, low incubation temperatures produce males, the reverse of the alligator. For example, sea turtles incubated below 28°C (82.5°F) produce all males, while incubation above 29.5°C (85°F) results in all females. Bird gender is not affected by temperature because sex is determined genetically. Another consequence of decreasing or increasing the optimal incubation temperature is a fourfold to fivefold increase in the number of smaller hatchlings, especially in crocodiles.

For dinosaurs, increasing incubation temperatures may have reduced incubation time much as it does in birds and reptiles. However, Paladino and others (1989) have considered a more sinister outcome, extinction, a possibility we'll examine in chapter 13.

Fossil Embryos

The fossil record for dinosaur embryos is much poorer than for eggs for three reasons. First, the skeleton is not well developed for a considerable portion of the embryo's growth, thus leaving little trace after death, scavenging, and decay. Second, an embryo may be present but hidden within the fossilized egg. Third, if eggshell is not present, it may be difficult to separate bones of a late term embryo from those of a hatchling. A list of some fossilized embryos is presented in Table 11.2.

How do we recognize these specimens as being embryos and how can we be certain as to their identification? First is association, either within an egg or within what is thought to be the mother. It is obvious that if bones are found *within* an egg (see Fig. 11.24), they are those of an embryo. But for animals that gave live birth, the evidence can be tricky. Association within an adult skeleton may indicate an embryo, as in the case of the marine reptiles of the Mesozoic, the ichthyosaurs. On the other hand, the association of small bones of *Coelophysis,* a small predatory dinosaur, within an adult is thought to be cannibalism because the bones are all jumbled up, not curled like a fetus.

A second clue to the bones being those of an embryo is the bones themselves. As discussed above, the embryonic skeleton undergoes considerable change as the cartilage skeleton is calcified. This immature bone looks profoundly different from adult bone. The outer surface is fibrous, lacking the dense outer cortical bone of more mature bone. Because the bones are still growing, sutures between adjacent bones are still very much open. For example, the upper portion of the vertebra that arches over the spinal cord, the neural arch, is not fused to the main body of the vertebra,

TABLE 11.2.
Mesozoic reptile and bird embryos and hatchlings

TAXA	FORMATION, LOCALITY	AGE	REFERENCE
		Chelonia	
turtle	Judith River Formation, Montana	Campanian	Hirsch, personal communication
turtle	Oldman Formation, Alberta	Campanian	Zelenitsky, personal communication
turtle	Nanchao Formation, China	Upper Cretaceous	Cohen and others, 1995
turtle	"red beds," Mongolia	Upper Cretaceous	Mikhailov and others, 1994
		Nothosauria	
Neusticosaurus	Monte San Giorgio Shales, Switzerland	Landinian	Sander, 1988
		Ichthyosauria	
Ichthyosaurus	Lias, England	Lower Jurassic	Deeming and others, 1993
Mixosaurus	Grenzbitumen, Switzerland	Middle Triassic	Bürgin and others, 1989
Shonisaurus	Luning Formation, Nevada	Upper Triassic	Massare and Callaway, 1988
Stenopterygius	Posidonia Shales, Germany	Lower Jurassic	Deeming and others, 1993
		Pterosauria	
Azhdarcho	Beleuta Formation, Uzbekistan	Upper Cretaceous	Nessov, 1991
Pterodactylus	Solnhofen Limestone, Germany	Kimmeridgian	Wellnhofer, 1991
		Dinosauria	
		Sauropodomorpha	
Titanosaurid	Rio Colorado Formation, Argentina	Upper Cretaceous	Chiappe and others, 1998
Camarasaurus	Morrison Formation, Colorado	Kimmeridgian	Britt and Naylor, 1994
Mussaurus	El Tanquilo Formation, Argentina	Upper Triassic	Bonaparte and Vince, 1979
?sauropod	Densuş-Ciula Formation, Romania	Maastrichtian	Grigorescu and others, 1994
		Theropoda	
Oviraptor	Djadokhta Formation, Mongolia	Upper Cretaceous	Norell and others, 1994
segnosaur	Nanchao Formation, China	Upper Cretaceous	Cohen and others, 1995
Scipionyx	Pietraroia Plattenkalk, Italy	Albian	Dal Sasso and Signore, 1998
Troodon	Two Medicine Formation, Montana	Upper Campanian	Horner and Weishampel, 1996
Troodon	Lance Formation, Wyoming	Maastrichtian	Carpenter, 1982
Velociraptor	Djadokhta Formation, Mongolia	Upper Cretaceous	Norell and others, 1994
		Ornithopoda	
Camptosaurus	Morrison Formation, Utah	Kimmeridgian	Chure and others, 1994
Dryosaurus	Morrison Formation, Colorado	Kimmeridgian	Scheetz, 1991
Edmontosaurus	Hell Creek Formation, Montana	Maastrichtian	unpublished
hadrosaur	Djadokhta Formation, Mongolia	Upper Cretaceous	Barsbold and Perle, 1983
Hypacrosaurus	Two Medicine Formation, Alberta	Campanian	Horner and Currie, 1994
Maiasaura	Two Medicine Formation, Montana	Campanian	Horner and Weishampel, 1988
?*Orodromeus*	Two Medicine Formation, Montana	Campanian	Horner and Weishampel, 1988
hypsilophodont	Twin Mountain Formation, Texas	Aptian	Winkler and Murry, 1989
unidentified	Amoreira-Porto Novo, Portugal	Kimmeridgian	Mateus and others, 1997a, b
		Ceratopsia	
Bagaceratops	Barun Goyot Formation, Mongolia	Lower Cretaceous	Maryanska and Osmolska, 1975
Protoceratops	Djadokhta Formation, Mongolia	Upper Cretaceous	Maryanska and Osmolska, 1975
Psittacosaurus	Oshih Formation, Mongolia	Lower Cretaceous	Coombs, 1982
		Avia	
Archaeornithoides	Djadokhta Formation, Mongolia	Upper Cretaceous	Elzanowski and Wellnhofer, 1992
Gobiopteryx	Djadokhta Formation, Mongolia	Upper Cretaceous	Elzanowski, 1981
unnamed	La Pedrera de Rúbies Formation, Spain	Lower Cretaceous	Sanz and others, 1997

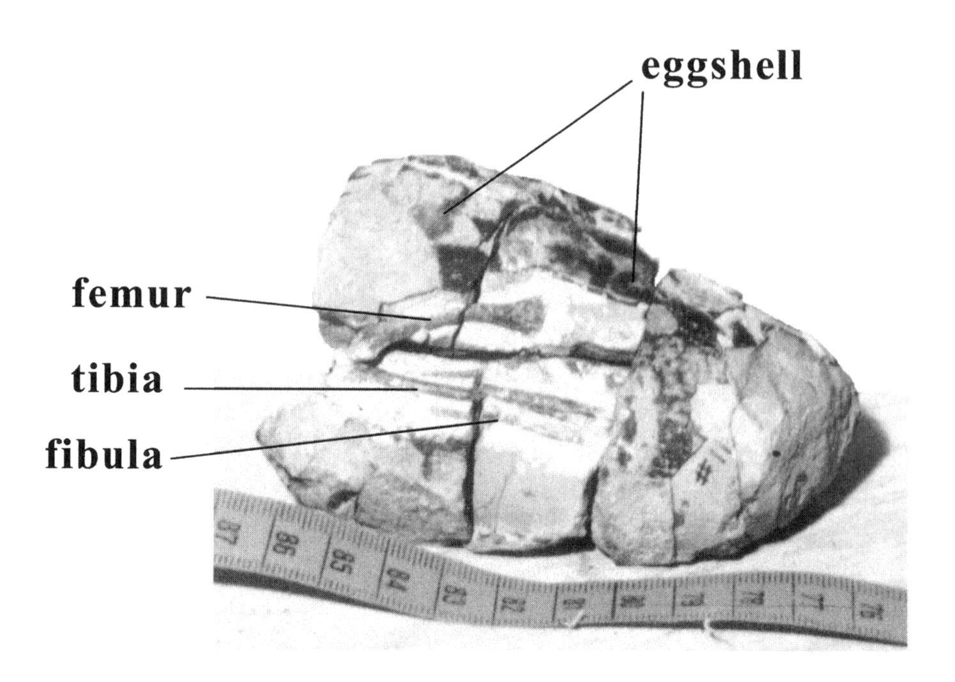

Fig. 11.24. The best proof that a dinosaur is an embryo is to find it within an egg. This is the skeleton of Troodon *from the Two Medicine Formation of Montana. The head was found tucked between the folded legs.*

the centrum (Fig. 11.8). The gap is very pronounced in embryos and becomes less so after hatching until it has completely fused when adult size is reached. The vertebra centrum also shows a pit where the bone grew around the notochord (Fig. 11.9). This pit usually disappears after hatching. The joint surfaces, where bone elongation occurs, are undeveloped as well. One final bone feature to consider is the teeth. These may not be erupted prior to hatching (e.g., alligator), but there are exceptions to this (e.g., crocodile).

A third feature of embryos is their very small size—after all, the embryo has to fit inside an egg. How small is small? To simply say "small" or "very small" is not enough because no point of reference is given (Fig. 11.25). We need to quantify the size, and different formulas have been developed (e.g., Currie and Carroll, 1984), but these formulas are often based on combining various reptile groups. As Andrews (1985) has shown, the relationship between average hatchling length and average adult female length differs markedly between crocodiles, lizards, snakes, and turtles (the relationship is closer between snakes and lizards, however) (Fig. 11.26). Not surprisingly, none of these correlations works for dinosaurs. I find, however, that dinosaur hatchling size correlates better with adult dinosaur size by the formula $12.5x^{0.38}$, where x is the average adult size in millimeters.

How could this correlation be used? Let us suppose we find a *Tyrannosaurus* skeleton 400 mm (16") long. Is it embryo or hatchling? Using the formula, the expected hatchling length for a 10,500 mm (10.5 m, 34½') *Tyrannosaurus* would be 422 mm; thus, it seems quite likely that the skeleton is indeed an embryo. The embryo could be easily accommodated in a *Macroolithus*-size egg. The guesstimated size for some dinosaur hatchlings is shown in Table 11.3.

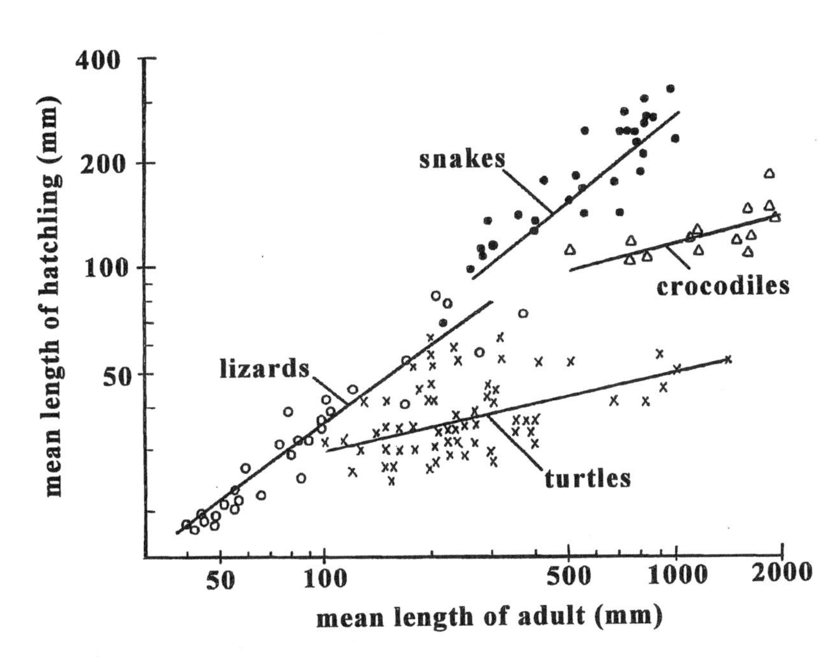

Fig. 11.25. Another feature of an embryo is its small size. However, size alone cannot be used to identify an embryo. This is a dinosaurling Protoceratops *compared with an adult.*

Fig. 11.26. Correlation between hatchling size and adult size in various reptiles. The straight lines lie along the average (regression lines) for each reptile group. The correlations are lizard: $1.16x^{0.74}$, snakes: $1.46x^{0.76}$, turtles: $12.40x^{0.20}$, and crocodiles: $24.13x^{0.20}$. Predictions for hatchling lengths can be made by replacing x *with the length (mm) of the adult female. Applied to dinosaurs, the crocodile correlation formula seems the most believable for estimating the length of a hatchling* Tyrannosaurus, *given that the female is 10.5 m (10,500 mm) long: 34.5 cm (the same results for the lizard formula is 60 cm, snake formula is 95 cm, and turtle formula is 1.2 cm). (Adapted from Andrews, 1985.)*

TABLE 11.3.

Estimated size for some hatchling dinosaurs using the formula $12.5x^{0.38}$, where x is the female adult length in millimeters. See text for discussion.

Adult	Adult Length (m)	Hatchling (cm) estimated	Hatchling (cm) actual	Reference
Allosaurus	9	40		
Brachiosaurus	21.5	55		
Camarasaurus	15.5	49		
Camptosaurus	5	32	24	Chure and others, 1994
Centrosaurus	5	32		
Coelophysis	2	22		
Deinonychus	3	26		
Diplodocus	24	57		
Dryosaurus	2	22		
Eoraptor	1	17		
Herrarasaurus	4	29		
Hypacrosaurus	9	40	52	Dong and Currie, 1993
Iguanodon	9	40		
Lesothosaurus	1	17		
Maiasaura	7	36	37	
Mussaurus	6	34	25	Bonaparte and Vince, 1979
Oviraptor	2.5	24	21	
Parasaurolophus	9	40		
Pinacosaurus	3.5	28		
Protoceratops	2.5	24	12–15	Barsbold and Perle, 1983
Psittacosaurus	2	22	26	Coombs, 1982
Sauropelta	5	32		
Stegoceras	2	22		
Stegosaurus	7.5	37		
Troodon	2	22	15	
Tyrannosaurus	10.5	42		

A fourth feature for recognizing embryos relates to the proportional differences of the skeleton as compared to the adult. For example, embryos have very large skulls in proportion to the rest of the skeleton (Fig. 11.27). That is due in part to the enormous eyes in the head, which in alligators reach half the skull length about midway through development and about 25 percent by hatching (Deeming and others, 1993). The skull above the eyes is usually domed to accommodate the large eyeballs. Also, the face of an embryo is short because the muzzle is stubby.

A fifth difference in reptiles and birds is reflected in different growth curves for embryos and hatchlings (e.g., Andrews, 1985). This difference occurs because growth is rapid as the embryo develops in a matter of weeks, whereas post-hatching growth is slower as the hatchling grows to adult size in months or years. The change in growth rate is due to the embryo's needing only to utilize yolk to grow, which requires minimal processing to be used (Fig. 11.28). Hatchlings, on the other hand, must

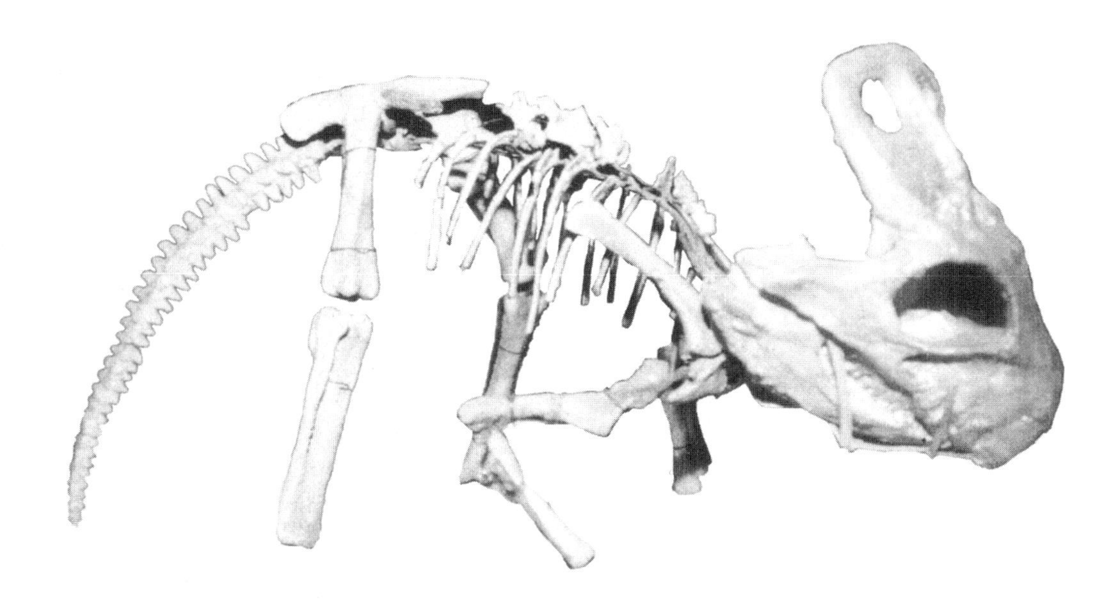

Fig. 11.27. *Baby and embryonic dinosaurs have very large heads and very large eye sockets compared to their body size. However, these features also occur in hatchlings as seen here with this* Protoceratops. *Skeleton about 50 cm long.*

use their own energy to obtain food, process the food—which has a large quantity of indigestible material (seen as feces)—and grow. As a result, growth is proportionately slower than for the embryo. We'll look at growth rates in the next chapter.

The use of bones to identify embryos is not without some major problems, not least of which is that the bones of hatchlings have many of the characters listed above. That is because the skeleton is still growing. It is best, therefore, to use a combination of characters, but still the most definitive answer is to find the bones inside an egg.

Identifying the Embryo

Having established that we have an embryo, the next problem is identifying who it is. You would think this would be easy, but this hasn't always been so. In 1988, Jack Horner and David Weishampel illustrated the embryo shown in Figure 11.24 as that of the hypsilophodontid *Orodromeus makelai.* This seemed a reasonable identification because the elongated egg with faint striations was fairly common at Egg Mountain where the embryo was found, as were isolated bones of juvenile and adult individuals of *Orodromeus*. But doubts arose when a nest of *Orodromeus*-like eggs was found farther north associated with a partial leg and tail vertebrae of an adult *Troodon* (Horner and Dobb, 1997). A former student of Horner, David Varricchio, and his colleagues described the association as evidence of *Troodon* brooding a nest of eggs just like *Oviraptor* on a clutch of eggs (Varricchio and others, 1997). Maybe, but because the setting is a floodplain rather than a sandy desert where sandstorms raged, I am skeptical. The association could easily be accidental, with the *Troodon* parts washed in by the flood that buried the eggs.

Nevertheless, the discovery led Horner to have the embryo in the egg

Fig. 11.28. *Growth measured as metabolic rate. Altricial birds have smaller yolks than precocial birds, hence the embryo grows continuously up till hatching. Precocial birds have larger yolks and remain in the egg longer. As a result, growth slows prior to hatching—otherwise the embryo would be too large for the egg. 100% incubation = hatching.*

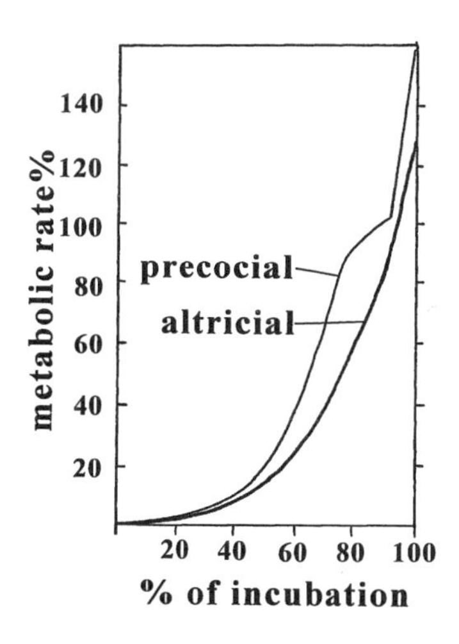

prepared further and this led to the discovery that the embryo was in fact that of the small theropod *Troodon* (Horner and Weishampel, 1996). The evidence included the presence of cavities in the vertebrae, called pleurocoels; long foot bones, called metatarsals, in which the central one is a tapering wedge that is braced on each side by the two outer ones; in the pelvis, a thick stump on the ilium for the pubis; and a large opening in front of the eye socket, called the antorbital fenestra. Admittedly, these features also occur in many advanced theropods such as *Tyrannosaurus*, but the outer of the three metatarsals is the largest, a unique *Troodon* feature. (Ah! you say, Varricchio was right after all, the *Troodon* was brooding her eggs. Not necessarily, because *Troodon* is a fairly common theropod in the Two Medicine Formation, so a chance association is still not improbable).

Other theropods have not been as difficult to identify. For example, a small partial skeleton was found plastered to the insides of an eroded *Elongatoolithus* egg in the Djadokhta Formation of Mongolia (Fig. 11.29). It has been identified as an embryo of an oviraptorid because of the lack of teeth in the jaws, and because the bones at the front of the skull, the premaxilla, are short and vertically oriented (Norell and others, 1994). The skull measures about 4 cm (1½") in length. The two halves of the lower jaw are not fused together as they are in adults. The vertebrae are all fully formed, although the upper parts, the neural arches, are not fused to the main body of the vertebrae. Interestingly, the furcula or wishbone is present indicating that it ossified before hatching. Two other fragmentary skulls found with the oviraptorid have been identified as *Velociraptor* (Fig. 11.30) because of the elongated premaxillary (Norell and others, 1994). The teeth are remarkable in that they are simple cones, rather than the serrated blades seen in the adult. The skulls were about the same size as the oviraptorid skull.

Fig. 11.29. Embryonic oviraptorid within an egg from Mongolia.

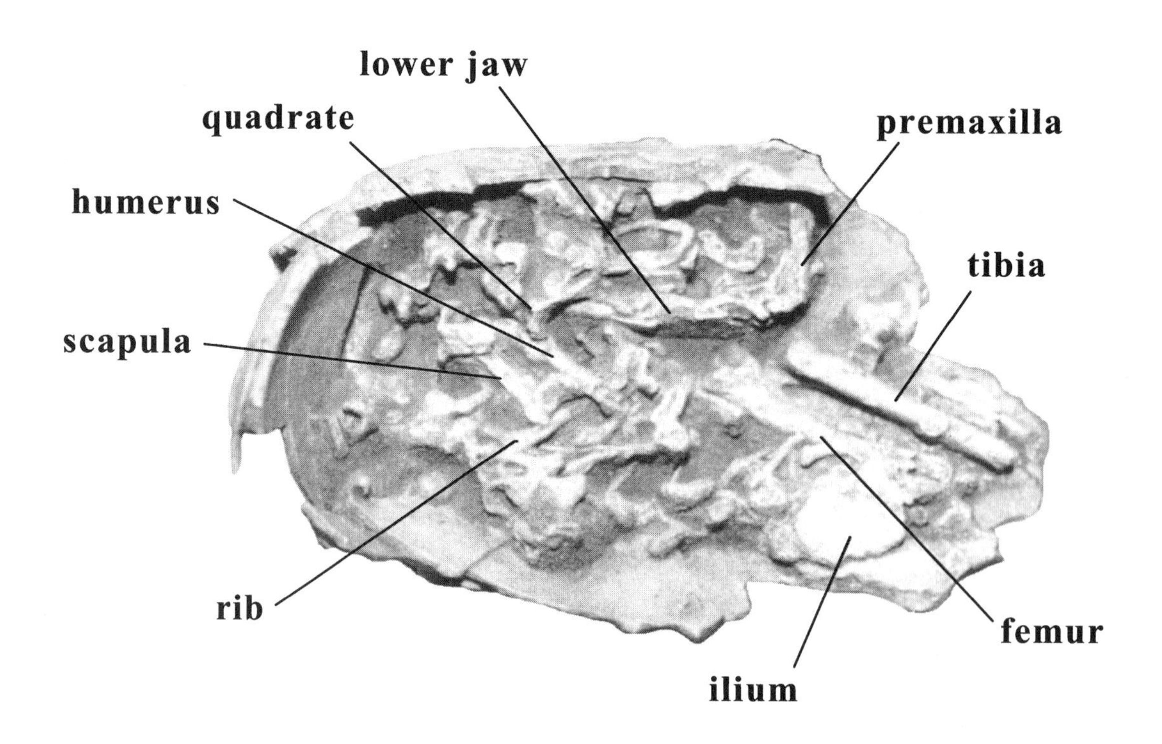

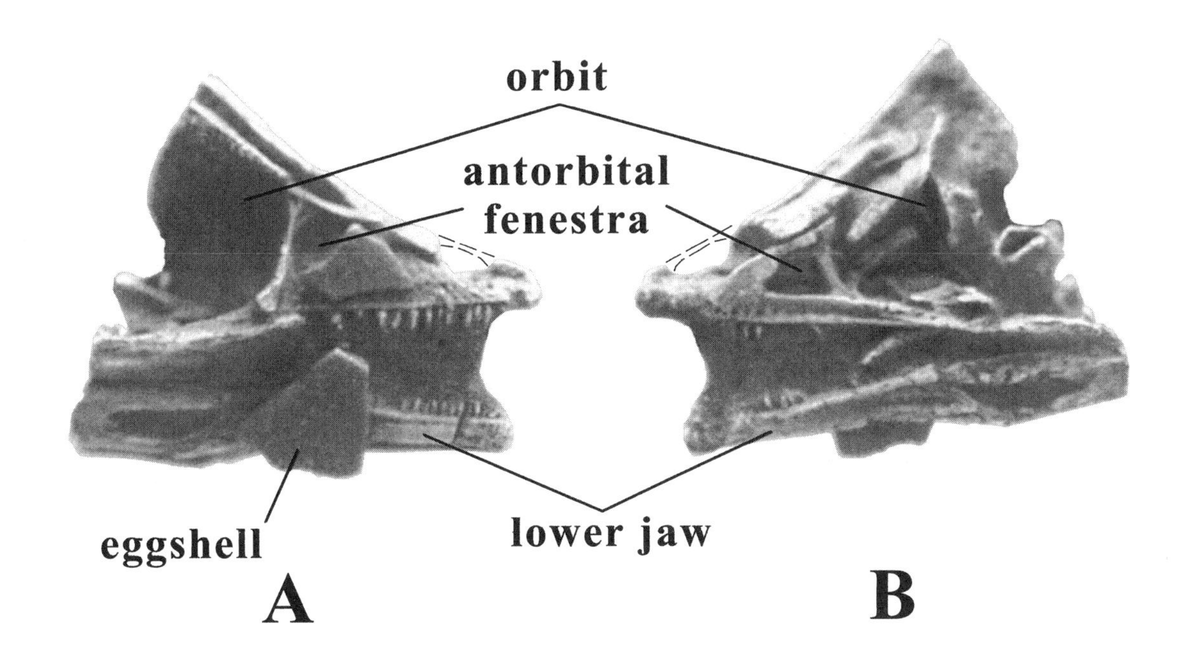

Fig. 11.30. Embryonic Velociraptor *skull from Mongolia. This and another skull were found associated with the oviraptorid embryo shown in Figure 11.29. Skull in right (A) and left (B) views.*

Fragmentary theropod embryonic bones have also been reported from the Upper Jurassic of Portugal (Mateus and others, 1997b). The bones occur in and around a nest of *Preprismatoolithus* eggs; at least one skeleton occurs within an egg. The bones (Fig. 2.8) have been identified as theropod based primarily upon the femur and skull fragments. Only two preliminary studies of the material have appeared and no name has been assigned to them (Mateus and others, 1997a, b).

The most remarkable of all embryonic theropods are of course the specimens that Terry Manning has revealed with acid (see chapter 8). The specimens have been identified as segnosaur, a very weird group of herbivorous theropods, mainly on the basis of the large claws. It remains to be seen whether or not the identification can be verified because there appear to be the tips of teeth in the premaxilla—segnosaurs aren't supposed to have any (Fig. 11.9, 11.31).

Sauropod embryos have recently been found in the Upper Cretaceous Rio Colorado Formation of Argentina (Chiappe and others, 1998). The embryos occur in about a dozen eggs and plastered against fragments of eggshell. The identification of these embryos from Argentina as sauropod is based upon the long, slender, chisel-like teeth, which are similar to those of titanosaurids found nearby. As yet, only a preliminary description of these remains has been presented (Chiappe and others, 1998). Also from Argentina is the discovery of prosauropod embryos from the Upper Triassic El Tanquilo Formation (Fig. 11.32). Seven, possibly nine, skeletons and partial skeletons were found. The skulls are about 3 cm (1") long and the entire skeletons about 20 cm (8"). They have been named *Mussaurus patagonicus* by Bonaparte and Vince (1979), which is unfortunate because of the changes that young undergo as they grow up (chapter 12). The large eye socket and short snout that were used to define this genus are characters

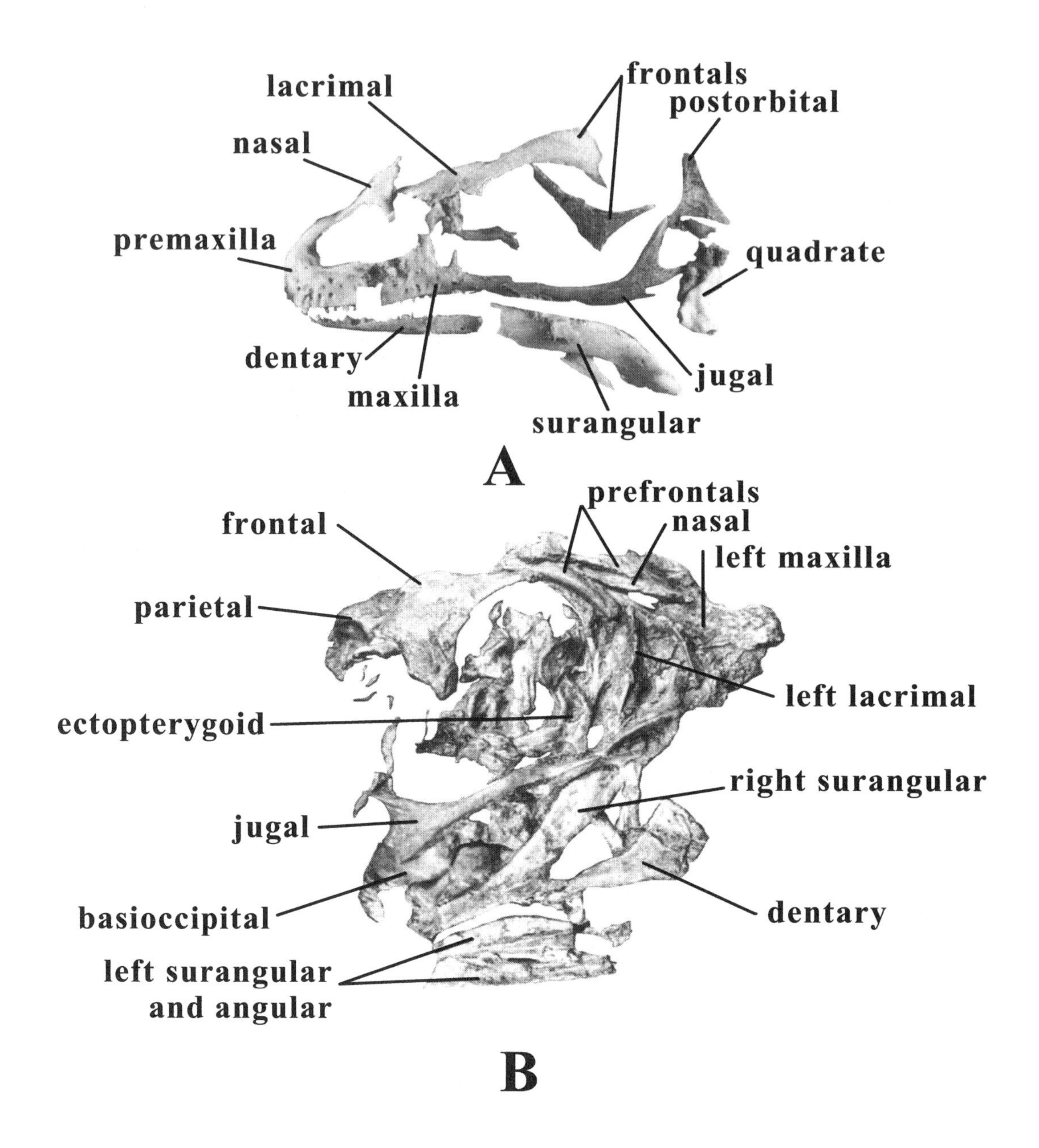

Fig. 11.31. Skulls of a possible embryonic segnosaur (A) and a hatchling of an unnamed giant oviraptorid (B).

not seen in adults. Nevertheless, the pelvis and foot have the characteristic form seen in adult prosauropod dinosaurs.

The most abundant embryonic dinosaurs are the hadrosaurs (Fig. 11.33, 11.34). They have been found associated with *Spheroolithus* eggs in Mongolia, Alberta, and Montana. Hadrosaur embryos are easily identifiable from the straight femur with prominent crest midway on the shaft (fourth trochanter), shape of the scapula, pelvic bones, and teeth with a ridge on the outer surface. A nearly complete growth series, from embryo to adult, is known for *Maiasaura* and *Hypacrosaurus;* we'll look at them in

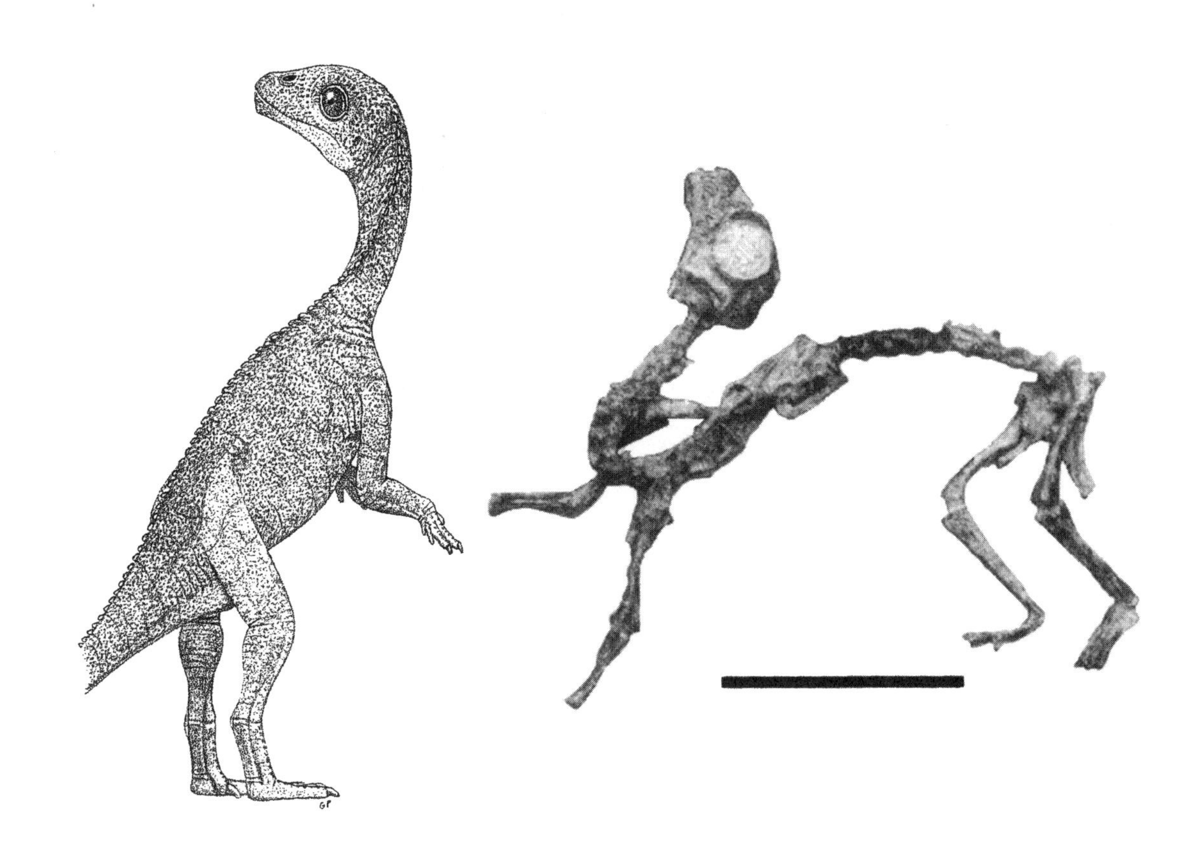

Fig. 11.32. One of several Mussaurus *embryos and a life restoration (by Greg Paul). Scale = 2 cm.*

greater detail in the next chapter. Remarkably, only the *Hypacrosaurus* embryos have been described in any detail (Horner and Currie, 1994). Nevertheless, these embryos are almost twice as big as those of *Maiasaura*. This difference is interesting because the adults are almost the same size, ~8 m (25'), yet just as birds of the same size do not always lay eggs the same size, there is no reason to assume that dinosaurs did either.

Other embryonic ornithischians include *Psittacosaurus* and *Protoceratops,* both from eastern Asia. The *Psittacosaurus* material includes at least a partial skeleton in fetal position (the shoulder region is tucked down between the legs), although there is no trace of eggshell. The skull associated with this specimen is 4.2 cm (1½") long, and it is larger than another skull, 2.8 cm (1") long. This second skull is less developed, suggesting that it belongs to an embryo of an earlier stage of development. Coombs (1982) described this material, but did not elaborate on why these bones are *Psittacosaurus.* I agree with his identification and note that the jugal of the larger skull has the characteristic spike (Fig. 11.35). Several *Protoceratops* skeletons have been found (Fig. 11.36) in the Djadokhta and equivalent formation in Mongolia. The skeletons are tiny, 12–15cm (4¾–6") long. The frill is very short, hardly the long structure of the adults. At one site near Tugriken-shire, a clutch of thirteen or so *Protoceratops* skeletons were found, but it is not known if they were hatchlings or embryos.

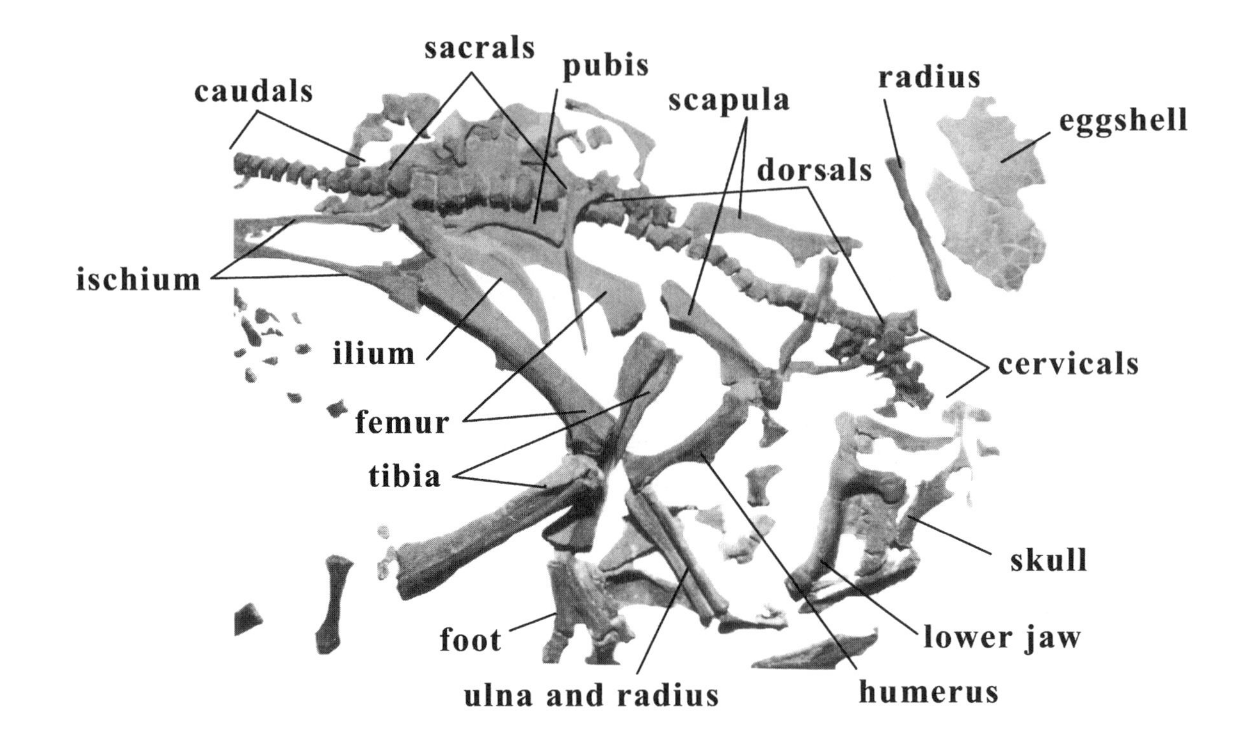
sacrals
pubis
caudals
radius
scapula
eggshell
dorsals
ischium
ilium
cervicals
femur
tibia
skull
foot
lower jaw
ulna and radius
humerus

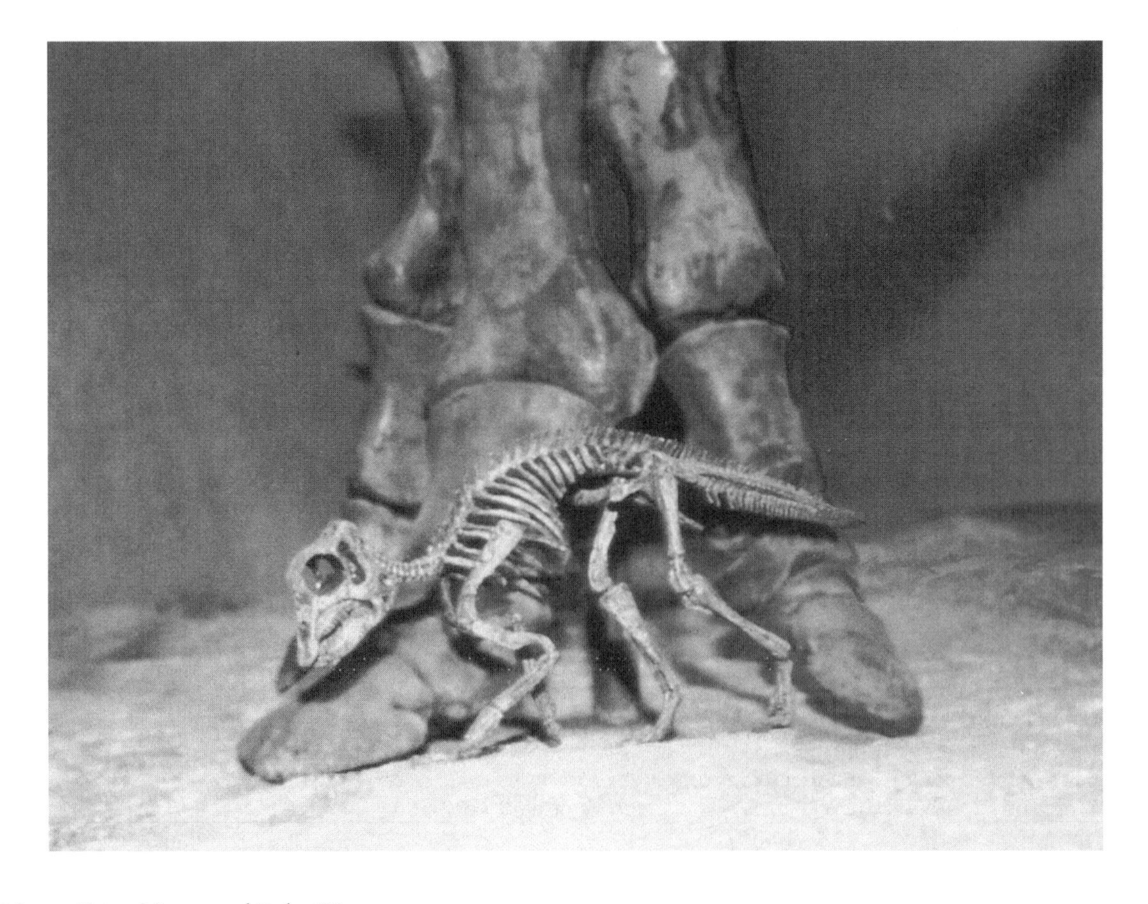

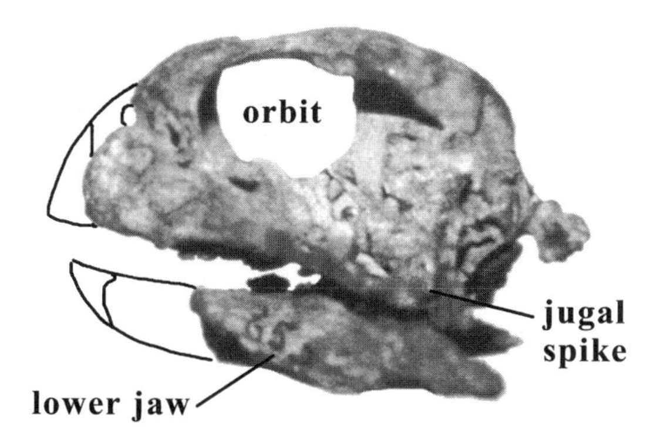

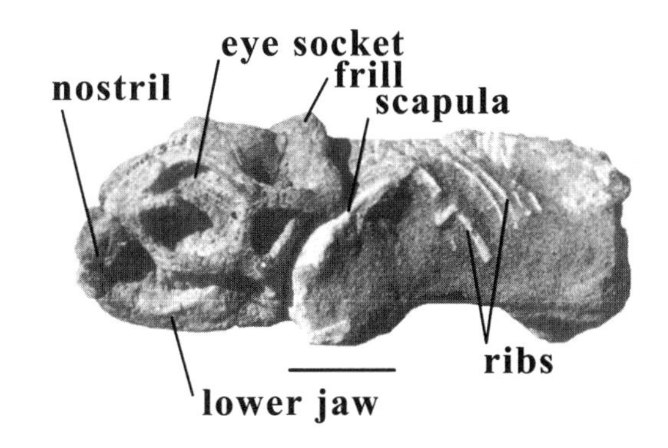

Fig. 11.33. (facing page) Composite skeleton of an embryonic Hypacrosaurus *from Devil's Coulee, Canada.*

Fig. 11.34. (facing page) Late term embryonic Maiasaura *skeleton mounted as if just hatched compared with an adult hadrosaur foot. The skeleton is 37 cm (14½") long.*

Fig. 11.35. (above left) Skull of an embryonic Psittacosaurus *from Mongolia.*

Fig. 11.36. (above right) An embryonic Protoceratops *from Mongolia. Note the short frill. Scale bar = 2 cm.*

Hatching

Inevitably, the dinosaur embryo developed to the point that it had to hatch. The act of hatching in birds cannot occur without some preparation for life outside the egg. The lungs of a bird may start working a day or two before hatching, when the embryo pokes its beak into the air space located at one end of the egg. This air space is present at the widest end of the egg, between the egg membrane and the shell membrane. As the lungs take an increasing part in respiration, the role of the chorioallantois is gradually reduced. The exchange is complete by the time of hatching (Whitehead, 1987). Crocodile eggs also develop an air pocket, but it apparently has no role in starting the lungs. Instead, the lungs begin working immediately upon hatching. As mentioned above, we do not yet know if any dinosaur eggs had an air space, although this is a testable hypothesis.

Hatching begins in many birds by tucking the head under the wings to get the bill against the shell. At this point, pipping or cracking of shell begins by pecking with the egg tooth. The egg tooth isn't really even a tooth, but a small spurlike growth made of keratin, the same substance as the bill. The tooth is located on top of the bill, so the head is moved backward using a powerful set of muscles on the neck called the hatching muscles. The crack is expanded by pushing the legs against the shell until the head or nape of the neck is forced out. After hatching, the egg tooth falls off or is absorbed by the bill. The time involved in hatching ranges from a few hours to a full day. A megapode chick lacks an egg tooth and uses the brute strength of its legs to push its shoulders through the shell. It is rather remarkable that although bird eggs may be laid over a span of days, hatching is fairly synchronous to within a few hours of each other.

Among reptiles, crocodile hatchlings make a croaking or chirping sound when they are ready to hatch. The mother, who often stays in the vicinity of the nest to guard it, will then uncover the eggs and carefully crack the shells with her teeth. She may then carry the hatchlings down to the water one by one where they will stay with her for up to a year or more. Turtle hatchlings, however, are left on their own to force their way out of the egg and to dig their way out of the nest. The Great Plains skink, *Eumeces obsoletus,* is an exception among most lizards who have adopted a lay-'em-'n'-leave-'em attitude toward their eggs. The skink remains with the eggs and stimulates the hatchlings to free themselves by touching them (Porter, 1972). Afterward, the hatchlings may remain with the mother for over a week.

How dinosaurs hatched is, of course, not known with certainty. However, from hatched eggs we can assume that the hatchling pushed its way out of the egg much like a megapode bird. We know these eggs are hatched because the upper portion of the egg is missing, the inside is filled with the same rock as surrounds the egg, and the mammillae are heavily cratered. The absence of eggshell within the egg shows that the eggshell exploded outward as the hatchling emerged; predation would have caused the eggshell to implode as the predator pushed its way into the egg. The opening of hatched dinosaur eggs is always on top, implying that the dinosaur hatchling (dinosaurling) pushed upward on its way out. The opening has been referred to as the "hatching window" by some French egg paleontologists (Cousin and others, 1994).

The timing of hatching in both birds and reptiles is crucial and must have been equally important for dinosaurs. It must occur when there is enough food for the many new mouths. For herbivores, the available vegetation must be of high enough quality to ensure proper growth. This means that hatching must have coincided with the season when food was near its maximum abundance, meaning spring or summer.

One problem faced by ground-nesting birds and reptiles is that spring floods can drown the embryos before they hatch. Nevertheless, among both reptiles and birds, of those eggs that survive predation and are incubated, hatching success can be as high as 80 percent to 100 percent. It seems reasonable to assume equally high hatching rates among dinosaurs, but not necessarily. A surprising number of whole dinosaur eggs are known, especially the big, spherical *Megaloolithus* eggs from France, India, and Argentina. Why these eggs failed to hatch is a subject of much speculation, most of it tied with the extinction of the dinosaurs, a topic we'll discuss in chapter 13.

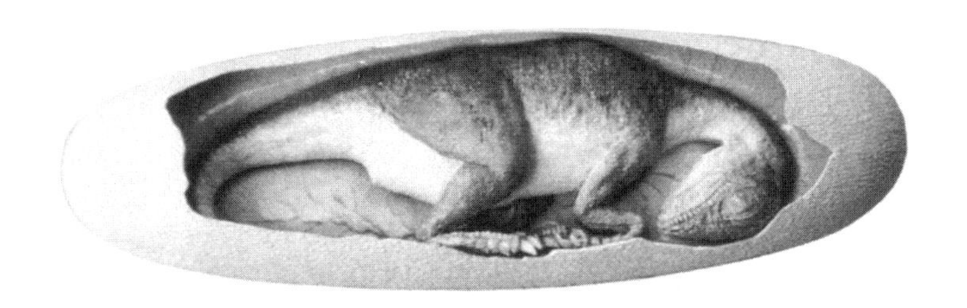

12 • Baby Grows Up

Once the dinosaurling emerged from the egg, it faced a whole new set of problems: getting food to grow, not getting eaten, not overheating or getting too cold, not getting stepped on, etc. Faced with many of the same problems, only 10–15 percent of bird and reptile hatchlings survive their first year. We can assume a similar survival rate for dinosaurlings. Even with parental care, survival rates among modern bird and reptile hatchlings are low. So why bother with parental care? Let's look at this in greater detail among living animals first, and then examine growth among the survivors.

Parental Care

Not all parental care is the same, and the degree of care depends on the state of the hatchling. For altricial birds such as the robin, the chick is not fully developed when it hatches. For example, the pelvis is still partially cartilaginous (Geist and Jones, 1996) and the associated muscles and nerves are not fully formed either. As a result, altricial chicks cannot move around on their own and require parental care to keep them warm and to feed them. They have little or no down covering their bodies and so consequently have a difficult time maintaining an even body temperature. This, however, is not necessarily bad because they can withstand low environmental temperatures. Initially their eyes are closed and they do not open for several more days. Complete development takes twelve to thirty-six days after hatching, by which time they may begin to fly. Until then, they are fully dependent upon the parents for food, warmth, and protection. Altricial birds typically nest in trees where the chicks are safer from predators than they would be if on the ground.

Precocial birds, which include the ostrich and megapode birds, nest on the ground. Consequently the hatchlings are like miniature adults and require minimal parental care. They hatch with their eyes open and have a down coat that helps them maintain a constant body temperature. Their pelvis and muscles are well developed, and they are able to leave the nest almost immediately or within a day or two (Ricklefs, 1983). Once they leave the nest, they are usually able to feed themselves, although how independent they are of their parents varies considerably among the species. Megapode chicks are completely ignored by their parents and are on their own from the moment they wriggle out of the nest. Other chicks, such as those of the ostrich, stay with the parents but feed themselves. Not all birds, however, are completely altricial or completely precocial. Many fall along the continuum between these two extremes (Nice, 1962).

Reptile hatchlings are all precocial, hatching fully developed and looking like miniature versions of the adult (Porter, 1972; Ewert, 1985; Ferguson, 1987). They may hatch with some help of the parent, as in crocodiles and some lizards, or without, as in turtles. Because reptiles lay their eggs all at once, hatching is usually synchronous. The advantage of synchroneity is that it improves the chances of at least some of the hatchlings' escaping predators. With so many hatchlings swarming at one time, predators are swamped with more prey than can be eaten before some of the hatchlings get to shelter. For example, gulls prey on hatchling sea turtles. If the turtles hatched at different times, the gulls would be able to eat them one by one as they hatched. But by synchronizing their hatching, at least some of the hundred or so baby sea turtles are able to reach the ocean before all are eaten.

Were dinosaurs precocial or altricial? That has been an ongoing debate since 1979 when Jack Horner and his childhood friend Bob Makela announced the discovery of a "nest" of hadrosaur babies, *Maiasaura* (Fig. 12.1). They concluded that some sort of parental care must have been involved because the teeth were worn, indicating that hadrosaurlings had been feeding (of course, we now know that teeth of embryos are often worn). Horner and Makela offered two possibilities for the type of parental care. One was that the hadrosaurlings were nest bound and the parents brought food; the other, that the hatchlings left the nest to feed, then returned under parental supervision. After other "nests" were found, Horner (1982) decided that the *Maiasaura* babies were altricial after all, despite the well-ossified skeleton. He reasoned that the young were safer in the nest from trampling by the much larger adults and from predators. Later, he also noted that eggshells associated with the babies were broken, implying that the eggs were broken up by the nest-bound hatchlings (Horner, 1982).

Anatomical evidence for altriciality in *Maiasaura* was presented by Horner and Weishampel (1988). They concluded that *Maiasaura* had underdeveloped knee joints because the surfaces had a thin calcified cartilage cap overlying very spongy internal bone. Such joints suggested to them that the hatchlings could not have ventured out of the nest in search of food, and therefore *Maiasaura* was altricial. As for a precocial dinosaur, Horner and Weishampel (1988) described the knee joint of *Troodon* (originally identified as *Orodromeus*), also found near the *Maiasaura* hatchlings, as having a smooth, well-formed, calcified cartilage surface. *Troodon* was first mentioned by Horner in 1982 as an unnamed hypsilophodont. The hatchlings were thought to have burst out of the eggs and immediately left the nest because the eggs did not show evidence of trampling, and because no hatchling skeletons were found in the ten nests found up to that time. Skeletons of different sizes were found in the vicinity, leading Horner to assume that some parental care was possible. Later, as more data were collected, he changed his mind and concluded that the absence of adults in the nesting sites suggested the hatchling *Troodon* were on their own (Horner, 1987).

The idea that *Maiasaura* was altricial and *Troodon* was precocial has

Fig. 12.1. Reconstruction of Maiasaura *babies in a nest based on the work of Jack Horner. These skeletons are based upon the actual bones collected by Jack and Bob Makela.*

been challenged recently by Nicholas Geist and Terry Jones (1996). They noted that all seven altricial bird hatchlings studied by them had an incompletely developed pelvis at the time of hatching. In contrast, the pelvis of the hatchling *Maiasaura* was as well developed as that of the nine precocial bird hatchlings they examined, as was the pelvis of the *Troodon* embryo. They also noted that the knee joint of the *Maiasaura* was not any more underdeveloped than that of hatchling precocial birds or alligators. All of this evidence they concluded demonstrated that *Maiasaura* was precocial as a hatchling.

I confess that I find their arguments compelling because I have long thought the knees were as developed as those of a young chicken (another point of argument between Jack Horner and me). At present, there is no convincing evidence that any dinosaur was altricial. However, a new clue to this problem may have been provided by the work of Bond, Board, and Scott (1988). They note that for altricial birds there is less change in the eggshell than in the shells of precocial birds. They suggest that the underdevelopment of altricial hatchling bones would cause less calcium depletion from the eggshell. Maybe we will be able to determine if a dinosaur was altricial or precocial by comparing the degree of calcium loss from the shell. Perhaps the hadrosaur *Maiasaura* (thought to be altricial by Horner) and *Troodon* (thought to be precocial) could provide a test of this hypothesis. Even if the eggshell evidence does not support *Maiasaura's* being altricial, this does not negate the evidence that suggests crèches of hatchling *Maiasaura* died together. Perhaps rather than the terms "altricial" and "precocial," which imply developmental attributes, the terms nidicolous, which means nest bound, and nidifugous, nest leaving, are more appropriate for dinosaurs.

Whether a parent *Maiasaura* maintained the crèche, as does the male ostrich, is not really known. Horner has long advocated this possibility based on the group of baby *Maiasaura*. He felt that a parent *must* have been present to maintain and guard the crèche (Horner, 1979, 1984, 1988;

Horner and Dobb, 1997). However, there remains the possibility that the parent was not active in overseeing the hatchlings. Hatchling iguanas congregate together without parental supervision, including individuals from other nests (Burghardt, 1977; Burghardt and others, 1977). By grouping together, survival is improved because a predator can only take one individual at a time. Furthermore, there is greater vigilance when more eyes are available to watch for predators. So simply finding a group of dinosaurling skeletons is not reason enough to assume parental care.

Is there evidence for nonsupervision of *Maiasaura* and other dinosaur young? Possibly, but it is based on negative evidence. Sites containing the skeletons of several individual hatchlings or juveniles less than half adult size don't contain adults. Even the skull of the adult *Maiasaura,* which gave "good mother lizard" its name, was found isolated and away from the little bone bed containing the dozen or so hatchlings. The inventory of sites lacking adults includes at least two bone beds of *Maiasaura* hatchlings, two bone beds of *Hypacrosaurus* (one of hatchlings, one of individuals one-third adult size, Fig. 12.2), one site of "*Brachyceratops*" juveniles less than half adult length, two bone beds of *Protoceratops* hatchlings, a bone bed of *Pinacosaurus* juveniles less than one-third adult size, and one bone bed of hatchlings of an unnamed hadrosaur (Horner and Makela, 1979; Gilmore, 1917). Although these occurrences are not proof that the young congregated independently of adult supervision, the frequency is high enough to cause some doubt. Nevertheless, we cannot extrapolate to say that no dinosaur young had adult guardianship, because several very young sauropods (length less than one-quarter adult size) have been found associated with adults: two different associations each of *Apatosaurus* (Carpenter and McIntosh, 1994) and *Camarasaurus* (see Fig. 10.3; Carpenter and McIn-

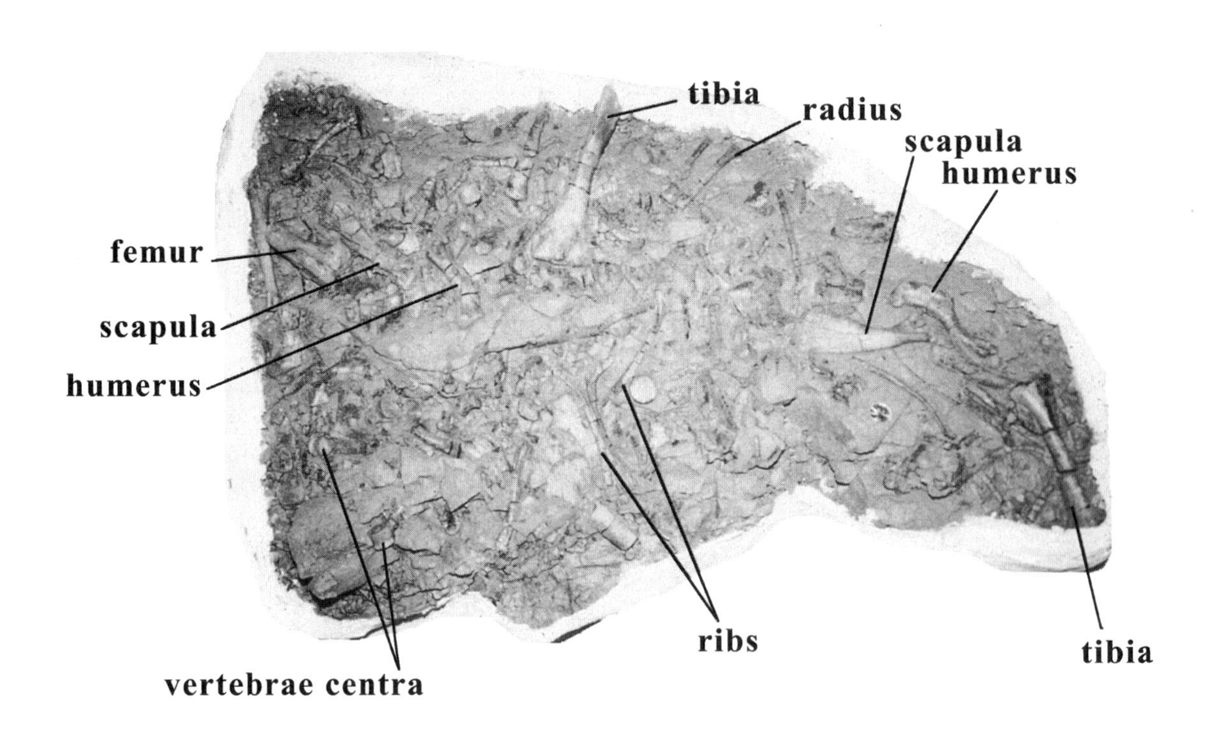

Fig. 12.2. A bone bed of juvenile Hypacrosaurus *from the Two Medicine Formation of Montana. The bones are from individuals over 1.5 m (5') long; an adult is about 8 m long. Although Horner and Currie (1994) think these bones are of nestlings, I think that they are older and were segregated juveniles.*

tosh, 1994; Britt and Naylor, 1994). In another case, four juvenile *Tenontosaurus* (less than one-quarter adult length) were found associated with an adult skeleton (Forster, 1990). We assume that this association of *Tenontosaurus* means that the adult was their parent, but we really don't know this. As far as we know, the association may have been coincidental (doubtful, but not impossible), the adult may have been an "aunt," hence only distantly related, or the dinosaurlings might have been from different nests overseen by a single parent (similar to the crèche of ostriches; Bertram, 1992). Without DNA testing, we will never know how closely related these young were to the adult with which they were found. What we can conclude from all of this is that no broad, sweeping statements can be made about parental care or the lack thereof in dinosaurs.

Let's assume that some dinosaurs did have parental care. What would compel such behavior? Horner and Dobb (1997) argue that the dinosaurling's "cuteness" factor triggered this care: the large eyes, the big, rounded head, the short snout (Fig. 12.3). It is true that these features in mammalian babies do trigger "Oooooh" and "How cute!"—but those are human responses (ever hear anyone, besides Jim Farlow, coo over a baby alligator?). There is no evidence for an "Ugly Duckling" response among nonhuman mammals, so to ascribe anthropomorphic response in *Maiasaura* is prob-

Fig. 12.3. Baby dinosaurs have big heads and big eyes. Although this might have made them "cute," as with the Troodon *on the left, that may not have been necessarily so, as with the* Hypacrosaurus *on the right. (Drawings by Greg Paul.)*

ably going too far. Most likely the "cuteness" factor is incidental to packing well-developed eyes and brain into a small skull. These organs grow little after birth (hatching) and the rest of the head and body "catch up" during growth (Fig. 12.4, 12.5).

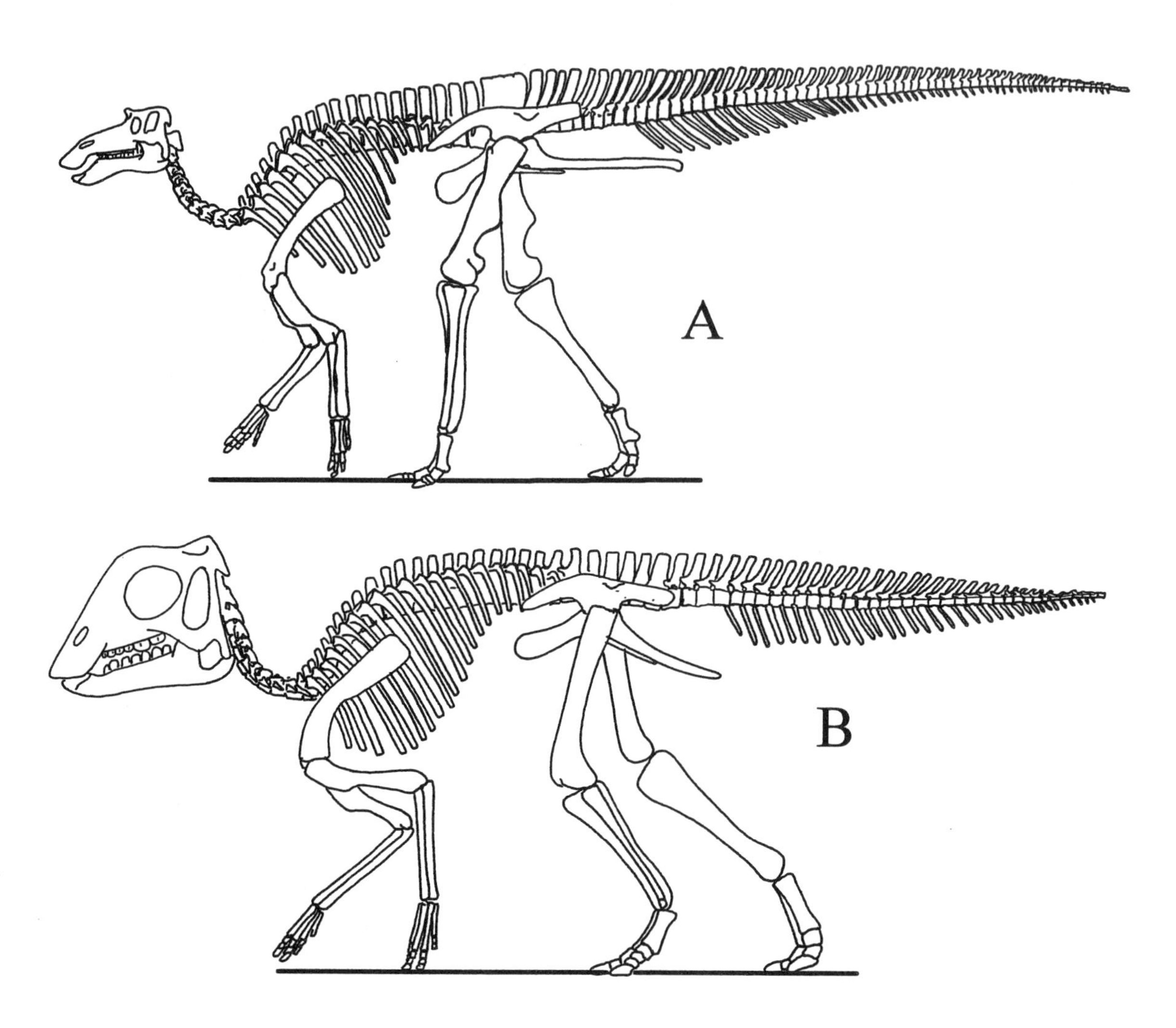

Fig. 12.4. To better appreciate proportional changes that accompany growth, we draw an adult Maiasaura *(A) and a hatchling (B) to the same size. It immediately becomes obvious that the hatchling has a huge head and short tail as compared to the adult. Basically then, growth is concentrated on catching up the body, but especially the tail, with the head.*

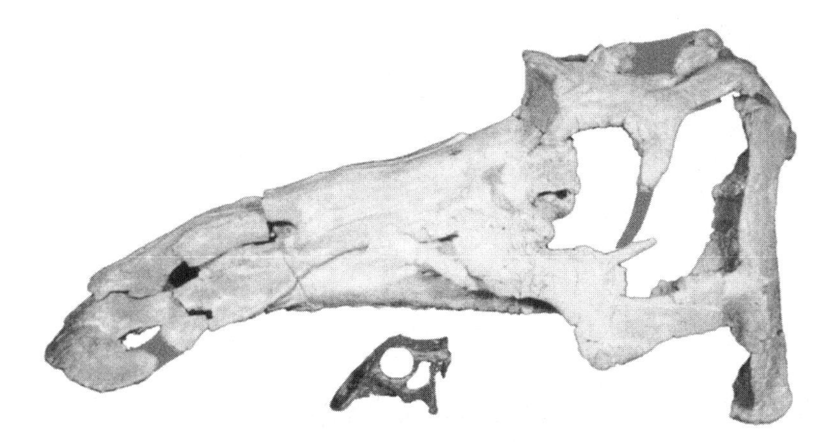

Fig. 12.5. If the skulls of an adult and baby Maiasaura *are shown to scale, the abnormally large size of the baby* Maiasaura *skull as compared with the body cannot be appreciated (see Fig. 12.4).*

A more likely reason for parental care is to protect the young from predators out for an easy meal (Fig. 12.6). After all, with hatchlings' being little more than a bite or two, protecting the young increases the chances that some of the young will attain breeding age. Among ostriches, hatchlings of different parents congregate into crèches guarded by the two senior adults (Bertram, 1992). But other birds such as ducks are not tolerant of hatchlings not of their own brood and may chase them away. The young know which parent to follow by imprinting. Imprinting, whereby a chick bonds with the adult, occurs in many birds. As a result, the chicks follow the parent to food, water, and shelter. The chicks also learn social behaviors that will enable them to interact with others of their kind. Although chicks from the same nest, called brood members, often bond, they can still be aggressively competitive toward one another over food. To exercise their

Fig. 12.6. Parental care in dinosaurs probably consisted mostly of protecting the young from predators. Here, a Tenontosaurus *protects the young from a* Deinonychus.

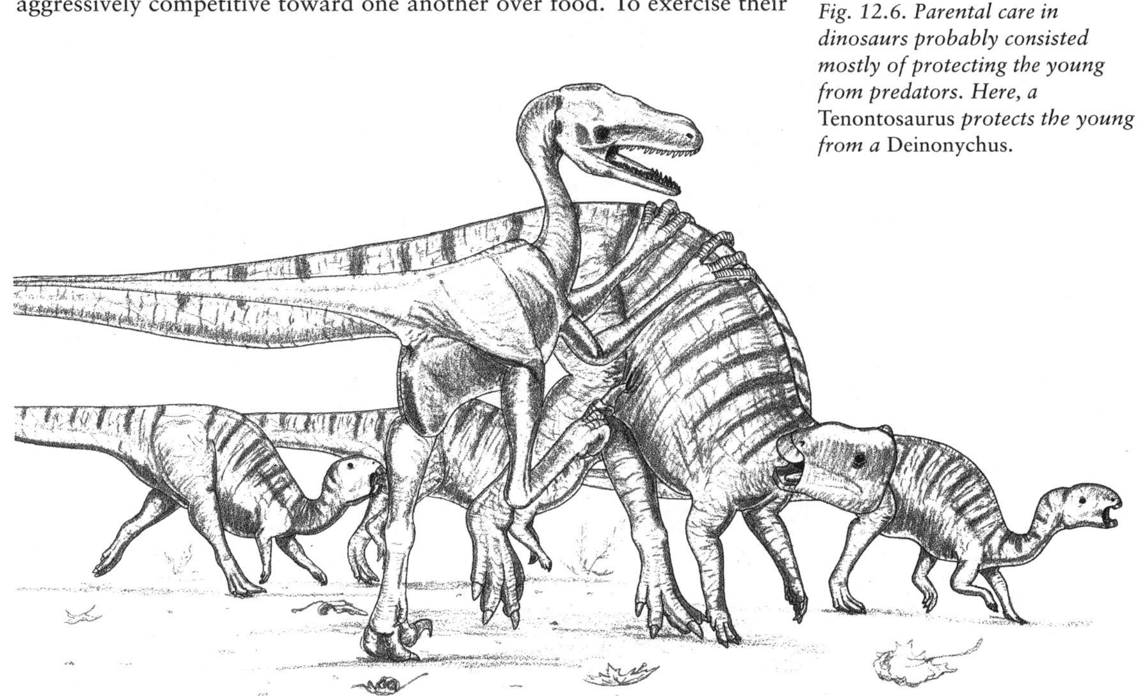

Fig. 12.7. Young animals will play, behavior that develops their muscles, coordination, and social skills. To convey that possibility in dinosaurs to the museum visitor, a pair of Coelophysis *skeletons are mounted like two dogs at play: the larger adult carries off the tail of an armored reptile,* Desmatosuchus, *while the smaller juvenile tries to pull the tail out of its mouth.*

muscles, hatchlings may engage in "play"—fleeing, running and flapping their wings, and making sharp turns. Perhaps dinosaurlings played to exercise as well (Fig. 12.7). As hatchlings and dinosaurlings grew, the general trend was presumably from greater to lesser dependence upon the parents, to complete independence.

In most reptiles, which lay 'em 'n' leave 'em, the parents are not around for the young to imprint on. Even if they were, imprinting does not appear to be a feature of most reptiles. A bonding of sorts develops in crocodiles

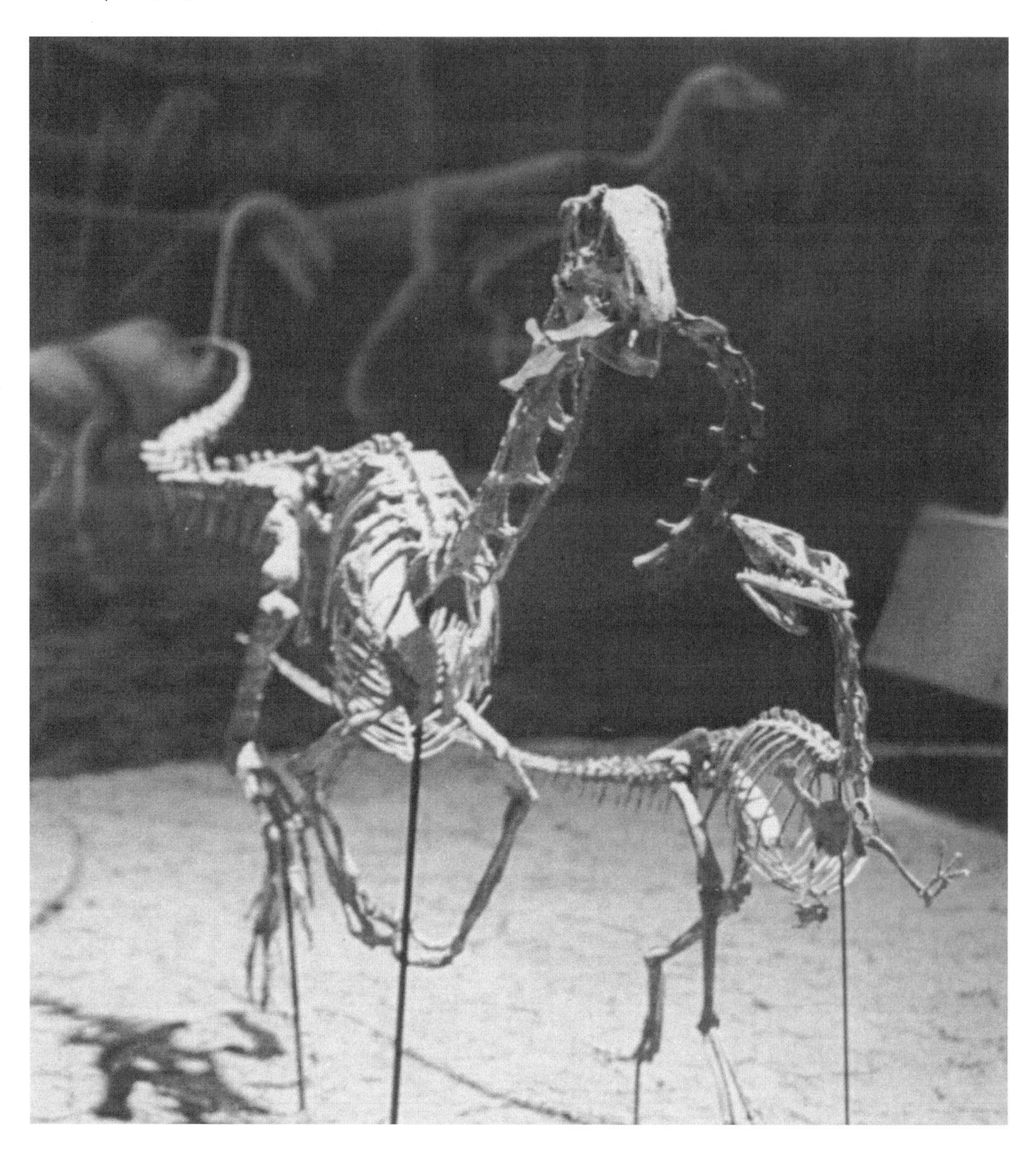

because the female helps the hatchlings escape from the eggs, carries them to the water, and guards them for a year or so. What about imprinting in dinosaurs? Not surprisingly, we just don't know. Perhaps in *Apatosaurus* and *Tenontosaurus* a crocodilian-style imprinting developed. This primitive imprinting might have been the precursor to a more advanced or avian-style of imprinting in some theropods, such as *Oviraptor,* and most probably developed in some Mesozoic birds.

Growth

Hatching is only the start for the dinosaurling, for now it must grow up. Growth outside the egg is very different from growth within the egg, because size is no longer limited to the capacity of the egg. Now energy is put into increasing the body size and not into developing new organs. Growth initially is very rapid and then slows as the young approaches adult size. Altricial hatchling birds grow two to three times more rapidly than precocial birds because they essentially hatch while still undergoing late stage embryological development. The rate of growth in birds and reptiles today is determined by a host of factors, including the quality and quantity of food, environmental temperature, brood size, egg size, and age experience of the parents (i.e., the number of previous clutches; Ricklefs, 1983; Porter, 1972).

Many of the factors affecting bird growth probably played a role in the rate of growth for the dinosaurling as well. For example, during a normal year, hatchling *Hypacrosaurus* would have lots of young plant shoots upon which to feed. But if the rainy season failed to bring a lush sprout of new vegetation, the hatchlings would have less food available, and what there was would be of very low quality (Carpenter, 1987). That is because during drought, plants remove their carbohydrates and protein from the leaves and store them in the roots. Thus, the poor little *Hypacrosaurus* might feed and feed, yet die with full stomachs because the vegetation had so little nutrition. Even if conditions did not get that severe, but rainfall was still below normal, the quality and quantity of vegetation would be below standard. Although the young *Hypacrosaurus* might eat, they might be getting less nutrition than their growing bodies required. The result, based on experimental studies of chicks, would be stunted growth and below normal body weight (Ricklefs, 1983). Up to a point, the young could recover and grow normally if the delayed rains came. But beyond a certain point, the effects could not be reversed. If the drought covered a great enough area, an entire generation could have been affected. These individuals would have been easy pickings for predators, such as *Daspletosaurus,* and the population levels would show a drop. Hypothetically, we should be able to find evidence of underweight juveniles, which we will discuss further below.

The rate of growth can also be affected by the type of food. Digestibility is highest for vertebrates and insects, and lowest for plants. That is because there are fewer nondigestible parts in vertebrates (mostly hair in mammals) and insects (chitinous shell), and more in plants (e.g., cellulose or woody tissue). You only have to compare the feces of a dog or cat with those of a horse or cow to note the difference. Because animal tissue is much higher in protein than plant tissue, many herbivorous birds will feed on insects during their early growth stage (Ricklefs, 1983). For those hatchlings that do not feed on insects, such as the Oilbird, growth is less than half the rate for insect eaters. Even some herbivorous lizards feed on insects when still hatchlings. From this, I conclude that insects or small vertebrates (lizards, frogs, mammals, etc.) could have been a significant, although variable, part of the diet for many herbivorous dinosaurs during their first year or two. They probably weaned onto plants gradually.

Another reason dinosaurlings may have fed on insects or small vertebrates their first year is that most animal tissue has more of the amino acids needed for hatchling growth (Ricklefs, 1983). Furthermore, feeding on small vertebrates or eggshell would also provide a source of calcium for the skeleton. The chicks of many birds will eat eggshell, bones, and teeth for the calcium supplement. Alternatively, if there was parental care in dinosaurs, the parent may have fed the young a protein-rich fluid called "crop milk." This secretion of the crop is almost 50 percent protein (Ricklefs, 1983). However, birds that feed their young crop milk are almost exclusively grain eaters. Because grains did not become part of the landscape until long after the dinosaurs, I doubt many if any dinosaurs fed their young crop milk.

Temperature can influence growth considerably (Ricklefs, 1983). If temperatures are above normal for extended periods, growth can actually accelerate somewhat. Not surprisingly, the reverse can also be true if temperatures are below normal for extended periods. These generalities, however, do not hold for all birds. Some, such as the Willow Ptarmigan, show no change, while below normal temperatures can actually stimulate growth in bantam chickens. Because of this unpredictability, we cannot state what dinosaurlings might have done. We could expect above normal temperatures during a drought and below normal temperatures during an extended rainy season. In addition, dinosaurlings at lower latitudes would be subjected to higher environmental temperatures than dinosaurlings at very high latitudes. We might be able to determine how temperatures affected growth in hadrosaurs because juveniles are known high in the Arctic Circle (Northern Alaska), as well as low in central Mexico. Might those from Alaska be statistically smaller?

The size of a brood can affect the size of some individuals in birds due to competition for food (Ricklefs, 1983). In any brood (or litter), not all will be the same size (Fig. 12.8). Some will be below normal size and these can find it difficult to get enough food during the growth period. For birds with parental care, the parents will sometimes selectively starve the smallest chicks, a process called "brood reduction" (Ricklefs, 1983). Regardless, competition is especially fierce the greater the number of individuals in a brood and we might expect such strife among dinosaurlings as well if they were nest bound and relied upon the parent to bring food. Another variable affecting growth is egg size. As mentioned in chapter 6, a clutch of eggs varies in size, with some being slightly but significantly larger than others. Larger eggs generally produce larger hatchlings in birds (Ricklefs, 1983) and the same must have been true of dinosaurs as well. Based on modern hatchlings, we would expect larger dinosaurlings to have a greater potential for survival, grow more rapidly, and attain a larger size more quickly.

Finally, the age of the parents and the level of their experience can influence the rate of growth in many birds (Ricklefs, 1983). The smallest eggs and/or smallest clutches are laid by females just reaching their breeding age, or by females past their peak (Romanoff and Romanoff, 1949). Hatchlings from such individuals will be smaller than normal. However, if food is abundant, growth can sometimes make up for the smaller hatching size. It seems probable that dinosaurlings also experienced similar influences of the age of their parents. Among birds, young, inexperienced parents may not adequately feed the hatchlings so that growth is slower

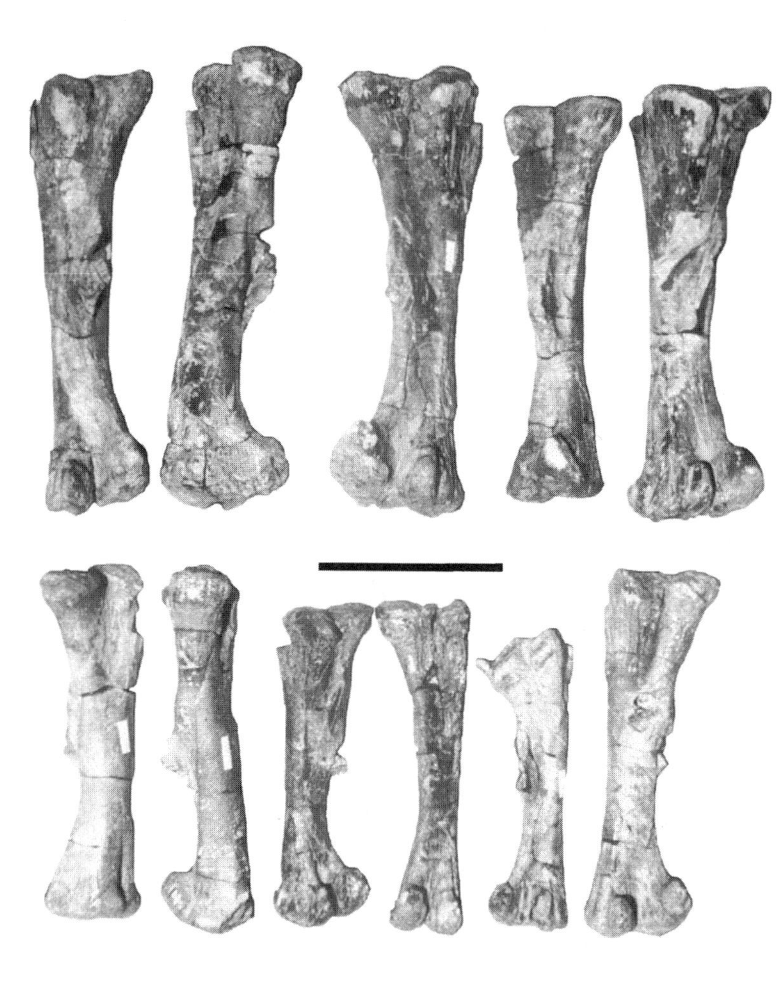

Fig. 12.8. An example of variation among dinosaurlings can be seen with these femora from a baby Hypacrosaurus *bone bed (see Fig. 12.2). The shortest femur, corresponding to the smallest individual, is about three-quarters of the largest. Scale = 10 cm.*

than for hatchlings of more experienced parents. Whether or not the same occurred in dinosaurlings depends on whether parents brought food to the young as discussed above.

Bone Growth

With dinosaurs, most of what we know comes from the skeleton, and that includes what we know about growth as well. To study growth in dinosaur bone requires that the bone be cut and ground thin enough for light to pass through. The technique is the same as that used to make thin sections of dinosaur eggshell (see chapter 8). Viewed through a microscope, a considerable amount of detail is visible (Fig. 12.9). Like dinosaur eggshell, dinosaur bones were thin sectioned soon after their discovery in the 1800s. Micro-photography had not yet been perfected, so the thin sections had to be meticulously drawn by hand. Curiosity about thin sections continued sporadically until the 1970s when it was thought that bones might reveal the physiology of the animals. Much of this work was done by French paleontologist Armand de Ricqlès at the University of Paris. Ricqlès built upon a major work on bone histology published in the 1950s by

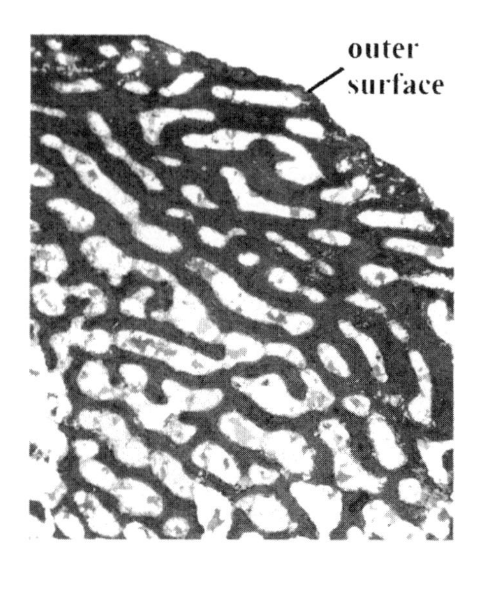

Fig. 12.9. Changes in dinosaur bone as it matures can be seen by these two microscopic thin sections. A, bone of a hatchling is composed of trabecular bone, made up of struts. As the hatchling grows, a dense outer surface will form. B, a thin section of this dense outer bone is full of minute holes, the Haversian canal system through which blood vessels pass.

D. H. Enlow and S. O. Brown (1956, 1957, 1958). Enlow and Brown made the remarkable observation that dinosaur bone more closely resembled that of mammals than it did that of reptiles. However, whereas they did not place any physiological interpretation upon this, Ricqlès did. He concluded that the bone did indeed most closely resemble that of rapidly growing mammals, and suggested that dinosaurs had a similar warm-blooded metabolism.

I should digress to mention that the 1970s were a heady time in dinosaur paleontology. The field was growing, mainly as a result of the baby boom generation reaching college age. Many of us grew up on books written by Edwin ("Ned") Colbert, such as *Dinosaurs: Their Discovery and Their World* (1961) and *Men and Dinosaurs* (1968). The traditional idea about dinosaurs was that they were nothing more than odd reptiles. But that concept began to crack in the late 1960s through the work of John Ostrom of Yale University. Ostrom was one of Colbert's students at Columbia University, and so was raised on the fine tradition of the "cold-blooded" reptile. But, as Ostrom tells it, it was his discovery of the fleet-footed *Deinonychus* that got him questioning the old ideas. The late 1960s and 1970s were a time of social revolution, with people wearing buttons proclaiming "Question Authority!" and "Don't Believe Anyone Over 30!" All the young bucks did precisely that and threw out the cold-blooded reptile dinosaur and replaced it with the hot-blooded "mammalian" dinosaur. The old guard did not take this lying down and a brawl erupted in the scientific journals. Much of this came to a head in 1980 with the publication of a symposium volume by the American Association for the Advancement of Science, *A Cold Look at the Warm-Blooded Dinosaurs*. Since that time things have quieted down somewhat, with most dinosaur paleontologists accepting that dinosaurs were dinosaurs and had their own unique physiology. Bone histology is no longer thought to have the final, definitive answer on dinosaur physiology (Farlow and others, 1995).

Dinosaur bone studies have in recent years concentrated on growth. Much of this work has been done by Anusuya Chinsamy from the South African Museum, but David Varricchio of the Old Trail Museum near Egg Mountain and Robin Reid of the Queen's University of Belfast have also made important contributions as well. Chinsamy, a slender woman of Indian heritage, has rather piercing eyes, all the more to pierce into the mysteries of dinosaur bone. Varricchio, on the other hand, is tall and gangling, with a carrot-top. The studies by Chinsamy and Varricchio are especially important because they involved comparing bones of different sizes, hence of different-aged individuals. As we shall see below, growth for some species was initially rapid until adult size was reached, but for others growth apparently never stopped.

The formation of bone in the embryo was briefly discussed in chapter 11. By the time the hatchling emerges, the skeleton either has been formed mostly in bone (precocial birds and reptiles) or still retains lots of cartilage (altricial birds). Hatchling bone has lots of blood vessels and is very fibrous because it is rapidly growing. This fibrous bone, called perichondrium, eventually matures and develops into a more dense outer bone called periosteal, cortical, or compact bone. The ends of limb bones, called the epiphyses, remain cartilaginous and it is here that most growth occurs (Fig. 12.10). New cartilage forms at the tip, just below the dense articular cartilage that forms the joint. Beneath the new cartilage, the old cartilage forms irregular pillars that eventually become calcified. The calcified cartilage is then replaced by bone, and in this way the bone elongates. The calcified cartilage can be fossilized, and is seen on the joint surface of dino-

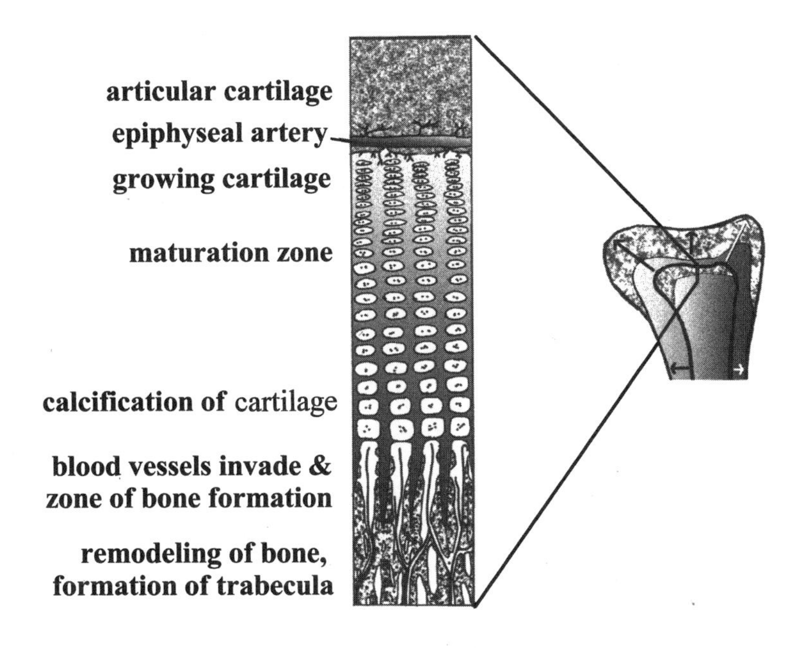

Fig. 12.10. Detail of the growth plate (left) at the end of a bone. It is here that the bone grows upward. The directions of growth are seen on the right (smaller version of the bone superimposed on a larger bone; arrows show direction and relative amounts of growth).

saur bones. This cartilage is what Horner and Weishampel (1988) reported as being very thin in the *Maiasaura* hatchlings. But as Geist and Jones (1996) have noted, the zone is very thin in young precocial birds and crocodiles as well. What we don't know is how thick the noncalcified cartilage was that used to cover the calcified cartilage. Considering how strong cartilage is (it covers the knee surfaces of an elephant), it would have been more than adequate to support any hatchling dinosaur.

Let's continue with bone growth. The main shaft of the bone, called the diaphysis, gradually thickens with the growth of periosteal bone. This growth in diameter occurs at a much slower rate than elongation so that the bone remains long and slender. At the center of the bone, endosteal bone forms and lines the marrow zone or cavity; it also extends to beneath the calcified cartilage at the ends of the bone. As the diameter of the bone grows, so too does the marrow zone (Fig. 12.11). As a result, all trace of the original fibrous bone is destroyed. In birds and theropods, in which the bone has a hollow core, the endosteal bone lines the marrow cavity. In sauropods and other dinosaurs, the marrow cavity is filled with a web of thin bone struts, called trabeculae or cancellous bone. Consequently, there is no internal cavity and the marrow tissue, which makes blood cells among other things, fills the voids throughout the cancellous bone.

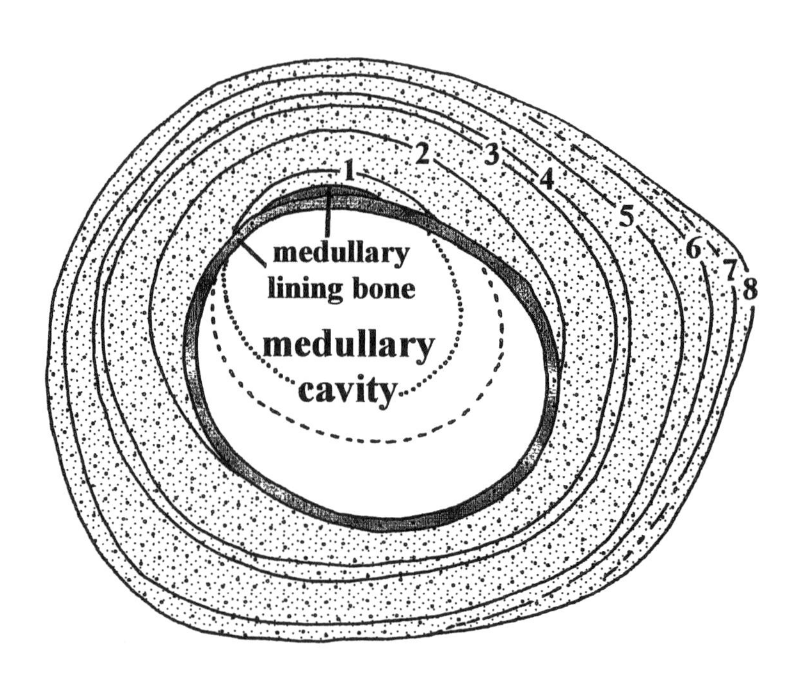

Fig. 12.11. Cross-section of a bone showing the growth rings and medullary (marrow) cavity. During the first year, the cavity was much smaller. Almost all traces of it and the first year's growth of bone have been lost through gradual expansion of the medullary cavity. (Adapted from Reid, 1993.)

Bone growth in dinosaurs is not simple and straightforward. Instead, several different variants or combinations have been identified microscopically, and these tell us something about growth of the animals (Fig. 12.12; Reid, 1997). The most common form shows no interruptions or textural variations in the periosteal bone, implying continuous growth at a constant rapid rate. This bone is called fibro-lamellar. The bone initially starts in the hatchling as a fibrous or very finely cancellous bone because the collagen fibers were laid down rapidly and haphazardly. It is around the collagen that the bone mineral is deposited. This early bone is then modified after

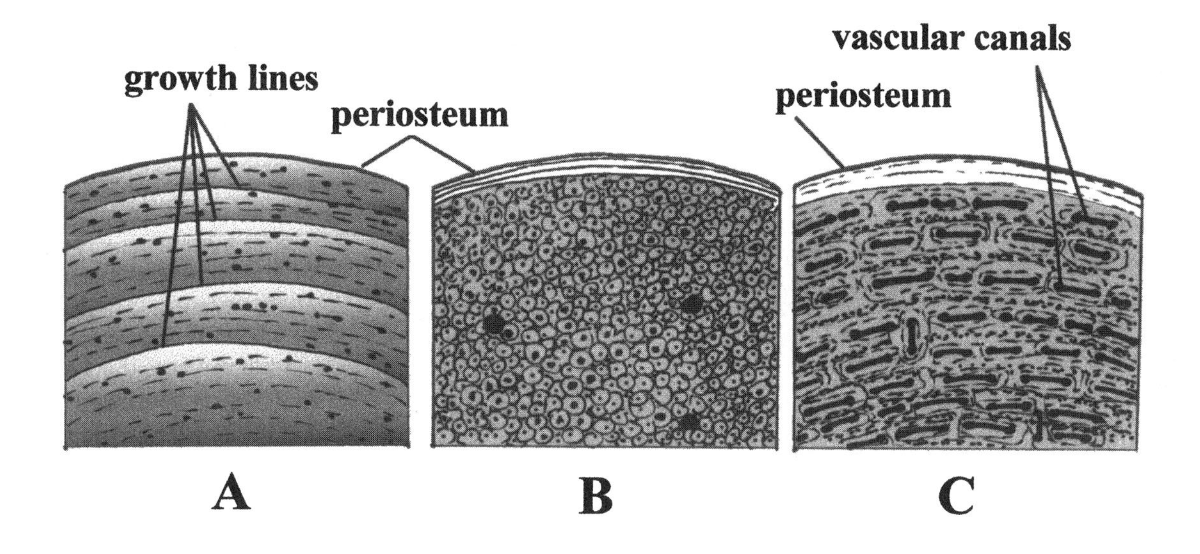

Fig. 12.12. Different types of bones: A, compact bone of a reptile with yearly growth rings. B, compact bone of a mammal, with rest lines near the surface. C, fibro-lamellar bone found in many dinosaurs. Some dinosaur bone is a cross between B and C.

hatching with bone being deposited in a more regular pattern. This new bone is deposited by bone cells called osteocytes living in minuscule cavities within the bone arranged in concentric layers (hence the lamellar part of the name) around tiny blood vessels. Minute tubes, called canaliculi, connect the central blood vessel with the osteocytes and in this way nutrients and oxygen are transferred to the cells. The bone that is deposited in each system of osteocytes and blood vessels is called the primary osteon.

Fibro-lamellar bone is characteristic of mammals and birds where growth is rapid. The presence of such bone in dinosaurs is why Enlow and Brown (1957) thought dinosaur bone resembled that of mammals and birds. Its presence has been used in the hot-blooded dinosaur debates, but such bone is sometimes also present in young alligators, so it is only useful in showing that the bone was rapidly growing, not that dinosaurs were warm-blooded like birds and mammals (Reid, 1990, 1993). Support for this hypothesis of rapid growth in dinosaurs is the observation by Barreto and others (1993) of an avianlike growth plate at the ends of the limb bones of *Maiasaura* juveniles. The growth plate is the cartilaginous cap where bone formation occurs (see above).

Chinsamy (1995) has described fibro-lamellar bone in a growth series of fourteen femora for the ornithopod *Dryosaurus*. The lengths of the bones were estimated to be 10–32 cm (4–12"), with weights of the living animals estimated to have been between 4 and 55 kg (9–121 lb) (Heinrich and others, 1993). The smallest animal is probably too large for a hatchling because it is about the same size as a juvenile *Dryosaurus* from Dinosaur National Monument (Carpenter, 1994). Chinsamy concluded that *Dryosaurus* grew at a continuous rate year round and so was unaffected by seasonal changes. A similar conclusion was reached for an unnamed hypsilophodontid (distantly related to *Dryosaurus*) from southern Australia (Chinsamy and others, 1998). What is remarkable is that 112 million years ago, this small dinosaur lived well within the Antarctic Circle, but was apparently not affected by the annual six months of darkness.

Another bone style is predominantly fibro-lamellar, but shows rings of growth much like wood. These rings, called annuli, are thin layers of slow growth bone that separate bands or zones of normal growth bone. Sometimes growth stopped completely and the resultant ring is called a line of arrested growth (LAG). LAGs can be preceded by lamellar bone indicating slowed growth just before growth stopped completely; if lamellar bone is not present, growth stopped suddenly. Bone with annuli or LAGs, but which is lamellar-zoned rather than fibro-lamellar, is commonly seen in reptiles, and the regular pattern of zones and annuli or LAGs represent seasonal cycles of normal and slowed or halted growth. As in trees, growth occurs during the spring, summer, and early fall when the animals are most active, and slows or halts during the late fall, winter, and early spring. The same seasonal cycle is thought to have formed rings in the fibro-lamellar bones of some dinosaurs (Chinsamy and Dodson, 1995). If true, then that means we can judge age in dinosaur bones much as we judge age in trees (the ages, however, are minimal ages, as we shall see). In some dinosaurs, the growth lines nearer the center of the bone are widely separated indicating early rapid growth. These lines may get closer either gradually or quickly toward the outer surface of the bone. This change is thought to reflect a slowing of growth once the animal reached sexual maturity (Ricqlès, 1980; Chinsamy and Dodson, 1995; Reid, 1997). Growth that stops at maturity is called determinate growth and typifies mammals and birds. Growth that continues with little or no decrease at maturity is called indeterminate growth and typifies reptiles. It is important to understand that determinate growth and indeterminate growth only refer to *growth,* not to the animals' physiology (warm-blooded vs. cold-blooded) (Ricqlès, 1980; Reid, 1990). Dinosaur teeth may also show growth rings (Fig. 12.13; Johnston, 1979). However, since dinosaurs periodically shed the teeth as they wear, the rings actually show the age of the tooth, not the age of the animal.

Fig. 12.13. "Growth" lines (numbered) in a ceratopsian tooth.

Age Profiles

How old did dinosaurs get? How quickly did they grow? Using growth rings as age indicators, Reid (1993) counted sixteen rings in a specimen of the small theropod *Saurornitholestes,* the last four of which are closely spaced. The zonal bone between them is mostly lamellar indicating slow growth. These growth rings means that the animal reached sexual maturity at twelve years and died at sixteen years, right? Well, not quite. The center of the femur is occupied by the cancellous or trabecular bone where the marrow once resided. The diameter of this cancellous bone is far greater than the pencil-diameter of the hatchling femur. Clearly, then, as the femur grew the cancellous bone also expanded—destroying all traces of the earlier bone. Chinsamy (1993), in her study of the prosauropod *Massospondylus,* attempted to determine the number of years lost by matching the first growth ring in an adult with the corresponding growth ring in the smallest, hence youngest individual. For example, assuming that the diameter of the first growth ring in a large femur was 10 mm, which corresponded to the diameter of the second growth ring of the juvenile, then at least one growth ring had been obliterated in the large bone. Is this system reliable? Actually no, because we know from bone beds that juveniles thought to be the same age can vary up to 25 percent in length (Fig. 12.8). This means that the runt of the litter will give the impression that more growth rings were missing than was actually true. Still, sometimes Chinsamy's method is all we have to go by.

A better way is to find traces of the younger bone, as Reid (1993) did with *Saurornitholestes.* He concluded that one or two years of early bone was lost. This means that the *Saurornitholestes* reached reproductive age at thirteen or fourteen years and died at seventeen or eighteen years of age. The age at death is not the maximum life span for this particular species of dinosaur, only the age of this particular specimen. After all, Chinsamy (1993) had individuals of the prosauropod *Massospondylus* as young as two years and as old as fifteen in her sample of nineteen specimens. Table 12.1 shows the probable age at sexual maturity and age at death for some dinosaurs based on growth rings. There are not many taxa on the list because this is still an area of current research.

The use of growth rings to judge age in dinosaurs must be employed with caution. First, the number of growth rings can vary in different parts of the skeleton because growth is not equal among the bones. Growth tends to be slower for bones nearest the surface of the body, such as the jaw (Ricqlès, 1980), than for bones deep within body tissue. These near-surface

TABLE 12.1.

Estimated ages in years at which various dinosaurs reached sexual maturity and died based on growth rings. Age at death does not necessarily imply maximum life span. Also presented is whether growth was determinate (Yes or No). See text for further discussion.

Taxa	Maturity	Death	Determinate?	Reference
Saurornitholestes	13–14	17–18	Y	Reid, 1993
Syntarsus	?	7–8	Y	Chinsamy, 1990
Troodon	3–5	~10	Y	Varricchio, 1993
Massospondylus	?	15	N	Chinsamy, 1993

bones will tend to show more "growth lines." To get around this problem, Chinsamy used the femur in her various studies. Second, superficial examination of bone can show what appear to be growth rings, such as reported by Jepsen (1964) for *Triceratops*. However, microscopic examination usually shows that these "growth lines" are in fact due to a regular arrangement of blood canals in lamellar or fibro-lamellar bone (Ricqlès, 1980).

Counting growth rings in dinosaur bone is perhaps the most reliable method, but it is also destructive because the bone must be cut for thin sections. Another—but nondestructive—method is to use relative size of certain bones, such as the femur or tibia. The technique requires a large sample of bones from a single-taxon bone bed so that there is little question that only a single species is represented. When the lengths are plotted, they are often found to form clusters that probably correspond to age classes (Fig. 12.14). These classes also reflect the population of the animals. Not

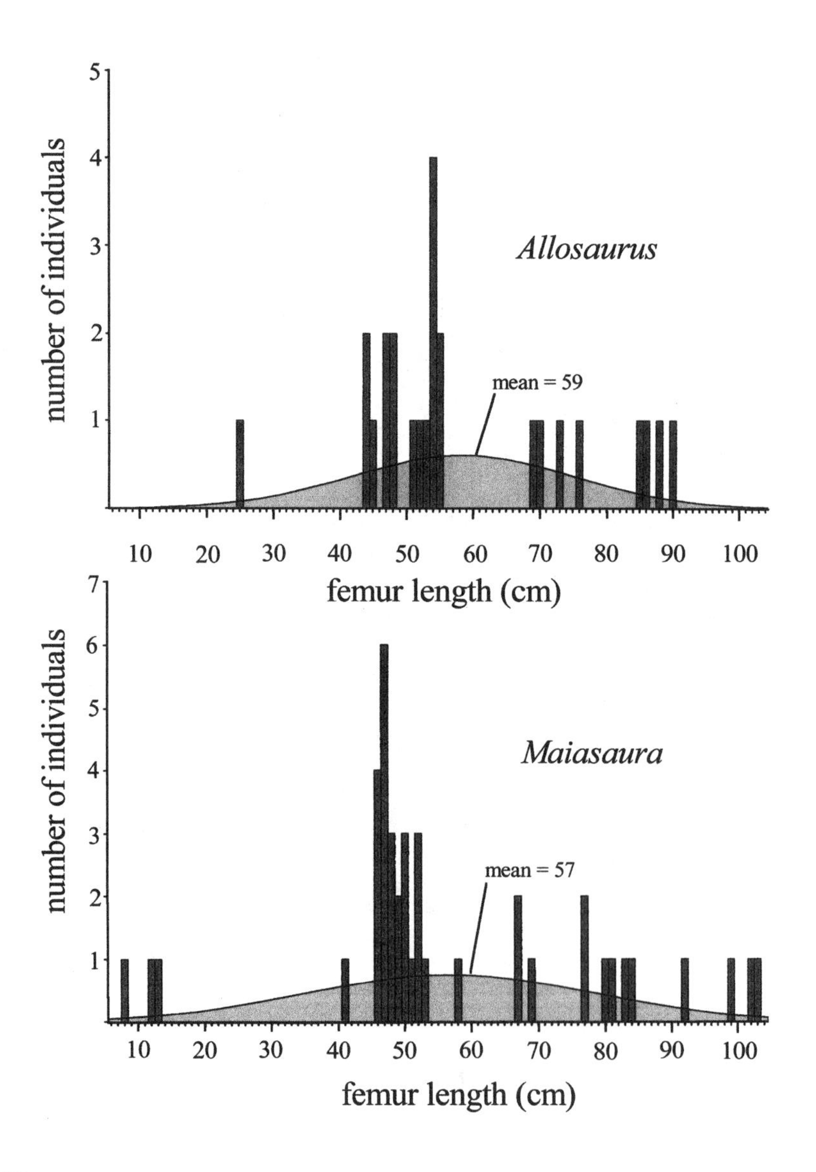

Fig. 12.14. Histograms of Allosaurus *and* Maiasaura *individuals from two bone beds (data in Table 12.2). Also shown are population curves based on this data. Note that very small and very large individuals make up only a small portion of the two populations and that most individuals are of intermediate size. The peak of the two curves is almost symmetrically placed on the curves indicating that the very young and very old animals tend to die off. The bone beds are assumed to be catastrophic assemblages, representing a "snapshot" of the population.*

surprisingly, the populations are dominated by young, intermediate-aged animals that were probably in their early reproductive years. Very young or very old individuals are less common because they are less fit. Most studies of growth in dinosaurs have used this technique because of its ease (e.g., Currie and Dodson, 1984; Colbert, 1989). Table 12.2 presents the raw data of measurements for the femur of *Maiasaura* and *Allosaurus;* this data was used to generate Figure 12.14. From this data, we can identify the size categories, which is the horizontal or X-axis of Figure 12.14. Plotting these categories on a log graph produces a growth curve for *Maiasaura* and *Allosaurus* (Fig. 12.15).

What happens if we combine the two techniques? The best sample that we can draw from is *Syntarsus.* Raath (1977) described this small dinosaur from a bone bed and gave measurements for the femora (Table 12.2).

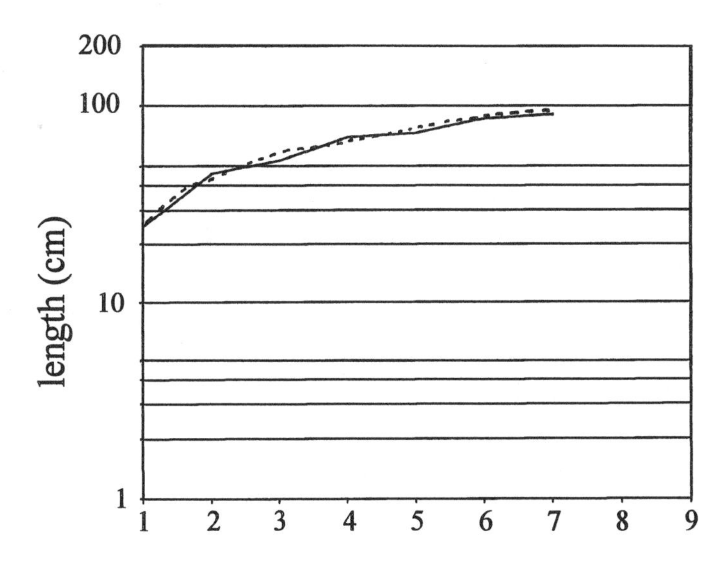

Fig. 12.15. Log plot of femur length for Allosaurus *(top) and* Maiasaura *(bottom). A solid line connecting the data points represents the growth curve based on the data from Table 12.2. The predicted growth curve (i.e., average curve) is shown as the dashed line. The solid and dashed curves for* Allosaurus *almost coincide, indicating that the sample represents all size (age) categories. For* Maiasaura, *the smaller size categories are either underrepresented (solid line below the dashed line) or overrepresented (above dashed line).*

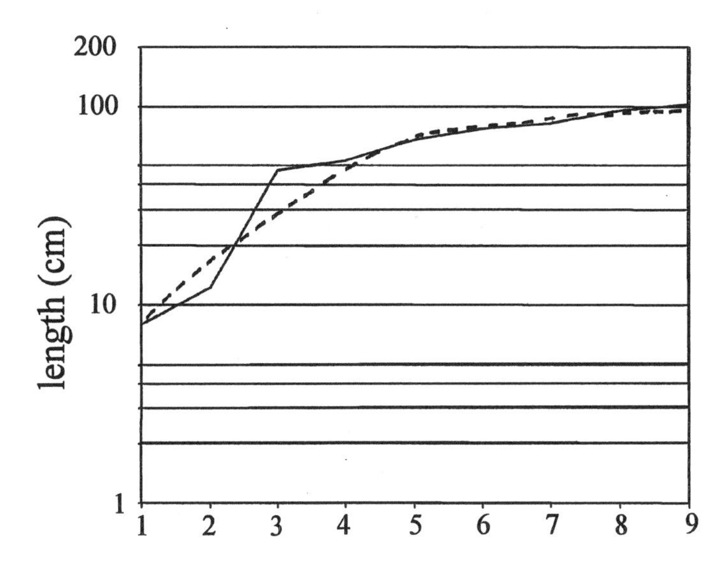

TABLE 12.2.
Femur lengths (cm) for populations of *Maiasaura, Allosaurus,* and *Syntarsus* from bone beds. *Alligator* measurements from museum collections. Data for *Syntarsus* from Raath (1977); *Allosaurus* from Madsen, 1976.

Maiasaura	*Allosaurus*	*Syntarsus*	*Alligator*
8	24.5	14.2	1.7
12	43.5	17.2	2.1
12.5	43.7	18.5	2.3
41	45	18.6	2.9
46	46.5	18.9	3.2
46.2	46.5	19.2	3.4
46.4	47.5	19.9	5.6
46.4	48		5.9
46.6	50.5		7.1
46.6	51.5		8.1
46.8	52.5		8.4
47	53.5		8.5
47	53.5		8.9
47.1	53.5		10
48	53.5		11
48	54.5		13.3
48.3	55.5		15.5
48.7	69.5		15.7
48.9	70		16.2
49.5	73		16.2
49.6	76.5		18
49.8	85		19.3
51.1	86.5		19.8
51.5	88		20.2
51.5	90.5		20.3
51.6			21.3
53			
58			
67			
67			
69			
77			
77			
80			
81			
83			
84.5			
92.6			
99			
102			
103.3			

Chinsamy (1990) used these specimens in her study and concluded that the series of thin sections showed that this small theropod had determinate growth. Plotting the femur lengths on a log graph (Fig. 12.16) does show that growth was initially fast and then slowed. However, whereas Chinsamy concludes that once a certain size was reached growth stopped, the graph of femur lengths shows that some growth continued. Why this apparent contradiction? There may be three reasons for this problem. First is the small sample size of measurements available. It is possible that more measurements might change the profile of the curve. Second, if the sample is adequate, the amount of growth is so slight that it might amount to determinate growth. Third, growth might have continued on the ends as the bone got longer but slowed on the sides so that the diameter did not increase much. Personally, I doubt this.

What we don't know is the time represented by the growth curves. Jack Horner (1992) believes that some hadrosaurs reached eight meters in only five years after an initial rapid growth the first year (Fig. 12.17,

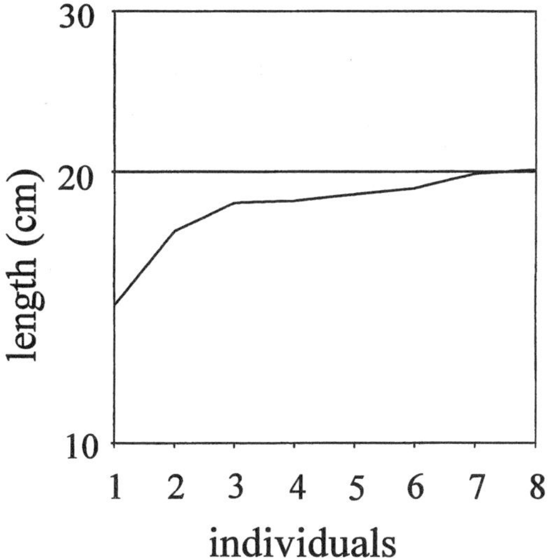

Fig. 12.16. (left) Growth curve for Syntarsus *based on femur length. Note that growth is initially rapid, then slows but never really stops. This is in marked contrast to what bone histology shows (see text). Data from Table 12.2.*

Fig. 12.17. (below) Growth curves for Hypacrosaurus *and* Maiasaura *based on data given by Jack Horner (1992).*

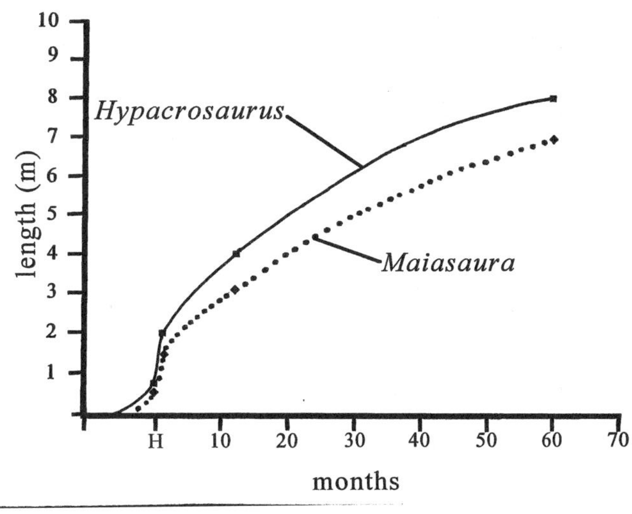

12.18). How feasible is this? If we compare the growth curves of the dinosaurs with that of the *Alligator* (Fig. 12.19), we note that the *Alligator* growth curve is almost a straight line. That is because growth is indeterminate and fairly constant, although it does slow in older individuals. In dinosaurs, there is an initial burst of rapid growth followed by slowed growth. Whether the initial burst represents months or a few years depends on how we interpret the size clusters in Table 12.2. If the mean of each size

Fig. 12.18. Hatchling-sized Maiasaura *skeleton (actually, near-term embryo) and a "nestling." Horner believes that the hatchling (38 cm) grew to nestling size (65 cm) in only one month.*

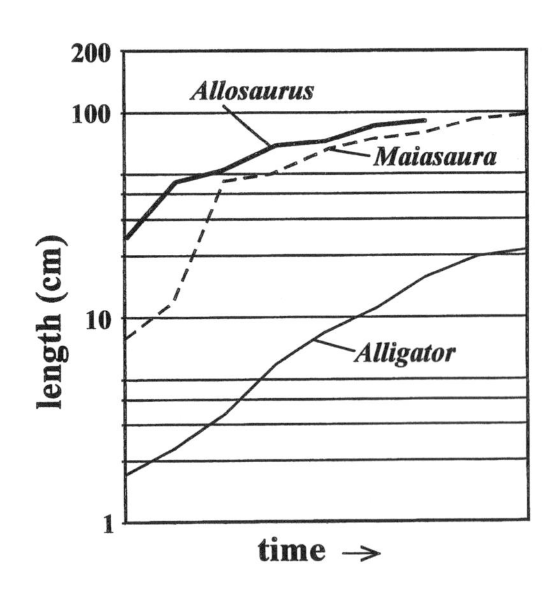

Fig. 12.19. Growth curves for two large dinosaurs, Allosaurus *and* Maiasaura, *and a reptile,* Alligator. *Note that for the dinosaurs, growth begins rapidly, then slows but does not completely stop. In contrast, growth in the* Alligator *remains constant with only a slight slowing at old age. Data based on femur lengths given in Table 12.2.*

group, which represents individuals from a single breeding season, is calculated, then the difference between each mean is the amount of growth in a year. However, the measurements don't really cluster very well, so it is difficult to know where to set the boundaries of each size/age group.

Changes with Growth

We all know that a baby is recognizable as human despite the fact that it has different proportions: the head and eyes are larger relative to body size, and the limbs proportionally shorter as well. Changes also occurred in young dinosaurs as they grew, a process called ontogeny. Surprisingly, this obvious point was not fully appreciated until rather recently. As a result, small dinosaurs were often given different names from the adults. For example, Peterson and Gilmore named a new genus and species of sauropod dinosaur, *Elosaurus parvus,* in 1902. It was very small, with a humerus 22 cm (8 3/4") long. Restudy of the material (a few vertebrae and some limb material) proved that the specimen was a baby *Apatosaurus* (Carpenter and McIntosh, 1994). This specimen was one of those occurrences of a juvenile sauropod found with an adult, in this case *Apatosaurus excelsus.* The adult skeleton, now mounted at the University of Wyoming, has a humerus 110 cm (43") long, making it five times the size of the baby's; i.e., the baby was about 20 percent of the adult size.

Recognition that many small dinosaurs are simply immature has allowed us to appreciate the changes that many dinosaurs underwent as they matured. Much of the early work was conducted by Peter Dodson of the University of Pennsylvania. Another of Ostrom's students, Dodson examined skull ontogeny in the crested hadrosaurs (lambeosaurs) and *Protoceratops* as part of his Ph.D. in the early 1970s at Yale University. He also studied the ontogeny of the modern alligator and two species of lizard for a standard of comparison. The method used in these studies involved a form of number crunching called morphometrics. This method involves measuring lots of variables (e.g., length of bones, length of the eye socket, and depth of the skull), then analyzing the data with what today would be a rather primitive and slow computer (if you must know, the analysis was a form of statistics called principal components analysis. Rather than put you and me to sleep, see Dodson's papers for the techniques; Dodson 1975a, b, c, 1976).

Dodson's work demonstrated that indeed some of the lambeosaurs were very likely juveniles of other forms. For example, the low-crested *Procheneosaurus* was a juvenile *Corythosaurus.* Actually, the work did more than just clean up the taxonomy of lambeosaurs; it also demonstrated that crest or frill size was dependent on age, with younger lambeosaurs and *Protoceratops* having small crests or frills and adults large crests (Fig. 12.20).

A variant of morphometrics relies upon digitally comparing the changing position of landmarks as an animal ages. The technique, with the long-winded name Resistant Fit Theta Rho Analysis, RFTRA for short, has been advocated by Ralph Chapman at the computer lab at the Smithsonian Institution. A large, always cheerful man, Ralph raves about the simplicity and potential of the technique. The method uses a computer to plot changes in the position of landmarks. By landmarks, I mean features that are easily and consistently identifiable. For example, in the skull, the tip of the snout might be one landmark, while another might be the frontmost edge of the eye socket. By comparing the position of these landmarks in

Fig. 12.20. Changes in the skull of Hypacrosaurus *is demonstrated by these three skulls. The juvenile skull (18 cm) at the bottom shows a slight doming of the forehead. This doming increases in size in an older juvenile. In the adult (77 cm), the dome has expanded into a crest. Lines 1–4 connect the same little-changing reference points (diamonds) on all three skulls. Lines A, B connect the same reference points (squares), but show a shift as the crest develops.*

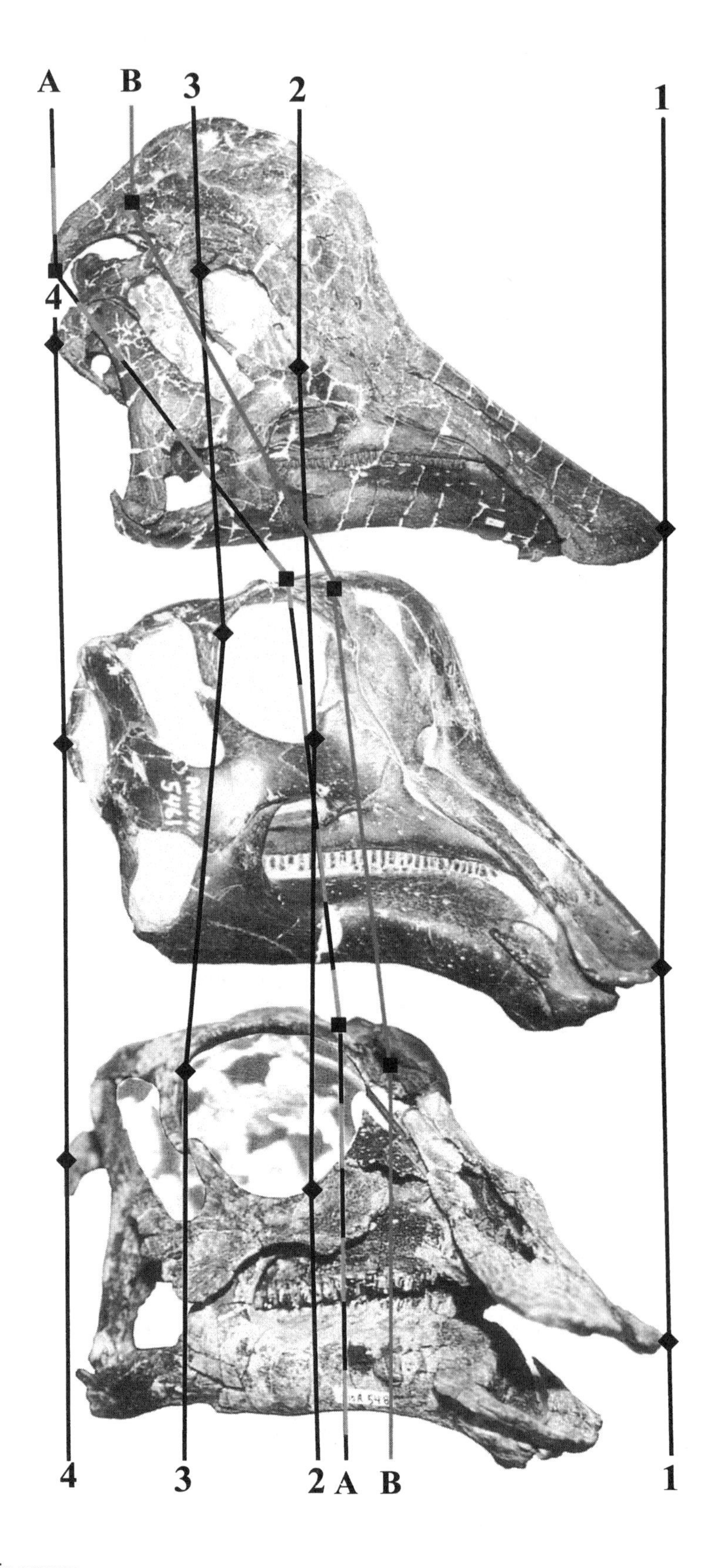

baby and adult specimens, it is possible to see what has changed, how much relative change has occurred, and the direction of change that accompanied growth (Fig. 12.21, 12.22). I'm not certain why I need a computer to tell me all of this, however. I've tried to discuss this with Ralph, but he gets a pained expression. I confess I spoofed the technique in a scientific paper

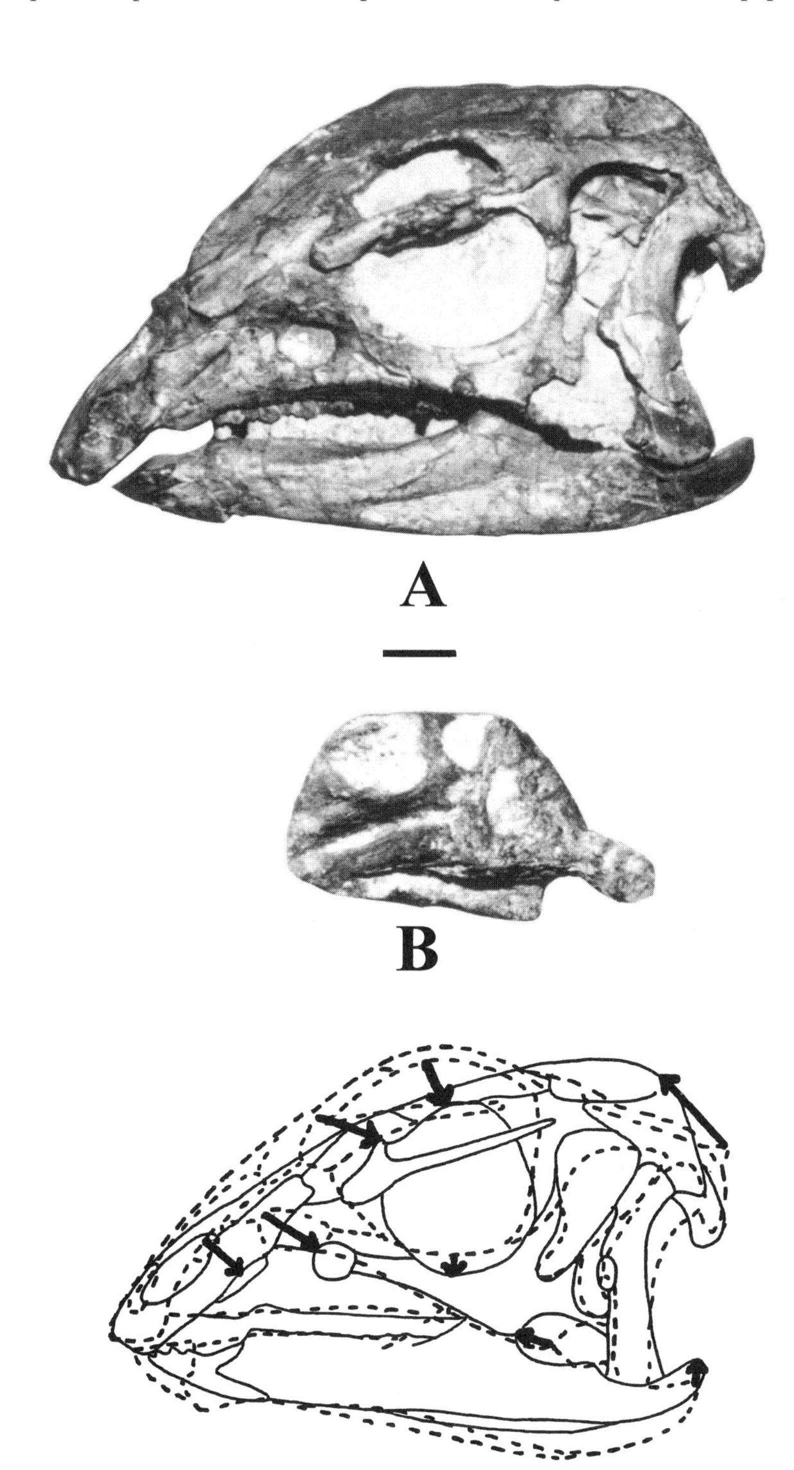

Fig. 12.21. Comparison of the skulls of an adult Dryosaurus *(A) and baby (B). Scale bar = 2 cm.*

Fig. 12.22. Outline overlay of the baby Dryosaurus *skull (dashed line) and adult (solid line). Arrows show the* relative *direction of movement as growth occurs. The most prominent are the reduction in size of the eye socket and rise of the back of the skull.*

(Carpenter, 1994; Ralph is probably grimacing now). Still, the technique of shape analysis is useful because it really does show at a glance the changes that a baby undergoes to become an adult.

Lastly, the teeth of some dinosaurs are known to change with growth. For example, the teeth of an adult *Troodon* are very coarse steak knives, whereas those of the embryo are leaflike and resemble the teeth of some primitive ornithischians (Horner and Weishampel, 1988). In another example, the teeth of an adult *Velociraptor* are thin, pointed, serrated blades, but the embryo has simple, pointed crowns (Norell and others, 1994). At what point the change occurred in *Troodon* and *Velociraptor* from the embryonic teeth to the serrated adult teeth is not known because teeth for hatchling specimens are not yet known (the hatchling *Troodon* mentioned earlier has no teeth). Why the teeth differ so much from the adults is a puzzle, but is not totally unique. For example, several species of lizards have small pointed teeth when young, but big fat teeth as adults (Estes and Williams, 1984). This change corresponds to a change in diet, from insect-eaters as young to omnivores or snail-eaters as adults. Cott (1961) has also noted shifting diets in crocodiles as they mature (Fig. 12.23). It is not unreasonable, therefore, that the teeth in embryonic *Troodon* and *Velociraptor* reflect adaptations for different diets in the hatchlings than in the adults.

The peglike teeth of the embryonic *Velociraptor,* however, resemble the teeth of the adult *Archaeopteryx* (Norell and others, 1994). *Archaeopteryx* and *Velociraptor* are believed to share a common ancestor, and the peglike teeth in *Archaeopteryx* may be the retention of an embryonic character, a process called pedomorphosis. An example is that the skulls of many sauropods look very similar to the skull of the hatchling prosauropod *Mussaurus*. These examples raise the possibility that new species in dinosaurs arise by changing the timing of changes that accompany growth, a process called heterochrony (Weishampel and Horner, 1994; Long and McNamara, 1997). The example of the embryonic-like teeth in *Archaeo-*

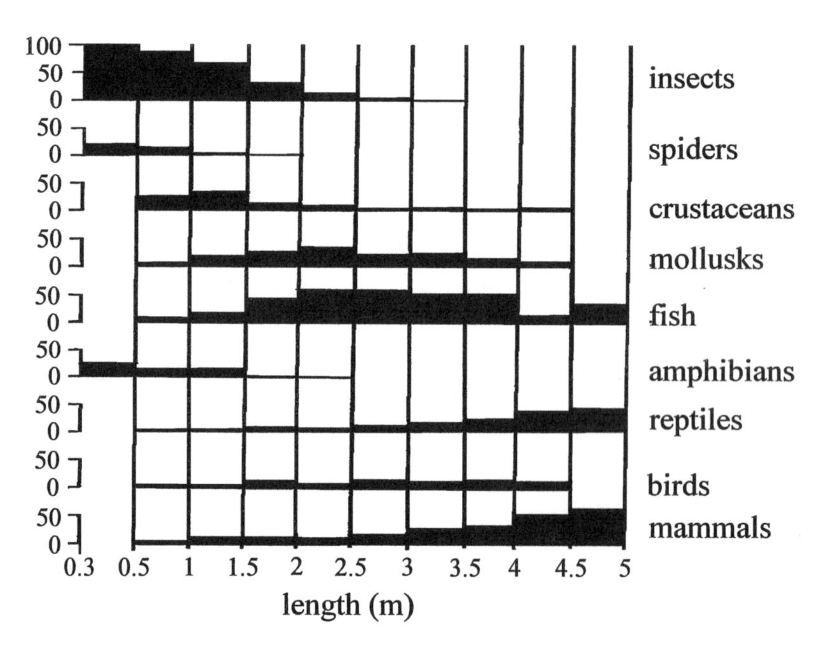

Fig. 12.23. Changing diet with growth in the Nile crocodile. The youngest (i.e., smallest) individuals feed predominantly on insects, subadults feed mostly on fish, and adults feed mostly on mammals. The shift partially reflects dietary needs and partially reflects ease of catching prey. (Adapted from Cott, 1961.)

pteryx might be due to an early cessation of development, meaning that serrated blades like those in the adult *Velociraptor* never formed. But what might happen if the development of some body part was accelerated or begun sooner? Such a rapid development is called peramorphosis. One of the most obvious examples is the general trend in increased body size in theropods: *Coelophysis* ⟶ *Allosaurus* ⟶ *Tyrannosaurus*. Heterochrony is still a relatively new field of research in dinosaur paleontology, but may provide an important clue to understanding dinosaur evolution (Weishampel and Horner, 1994).

Life in the Nest

Once the hatchling climbs out of the egg, life gets very difficult. There are predators to contend with, and if the young are restricted to the nest as Horner thinks, there are also the harsh extremes of the weather to contend with. James Kirkland, a portly, talkative individual with unkempt hair, has examined the predators that eggs and dinosaurlings had to contend with (Kirkland, 1994). Most mammals during the Mesozoic were very small, mousy things that kept a low profile. But *Gobicodon* from the Lower Cretaceous and *Didelphodon* from the Late Cretaceous were large, about the size of a well-fed house cat (Fig. 12.24). With their crushing teeth and powerful jaws, they could have easily tackled small eggs and most dinosaurlings. And predatory dinosaurs were definitely a threat. Small theropods could dispatch a hatchling in a few bites, while larger theropods could have swallowed the hatchling whole. The discovery of hatchling *Velociraptor* skulls in an *Oviraptor* nest raised the question whether they might represent the prey of the adult *Oviraptor* (Norell and others, 1994). Another example is the juvenile *Coelophysis* bones within the ribcage of two adults, which Colbert (1989) concluded represented examples of cannibalism.

Lizards are another potential predator of the nest. The Nile monitor shows little fear of crocodiles and even works in pairs. One will distract the

Fig. 12.24. The cat-sized mammal Gobicodon, *from the Lower Cretaceous of Montana, probably ate small dinosaur eggs and hatchling dinosaurs, and scavenged dead juvenile dinosaurs.*

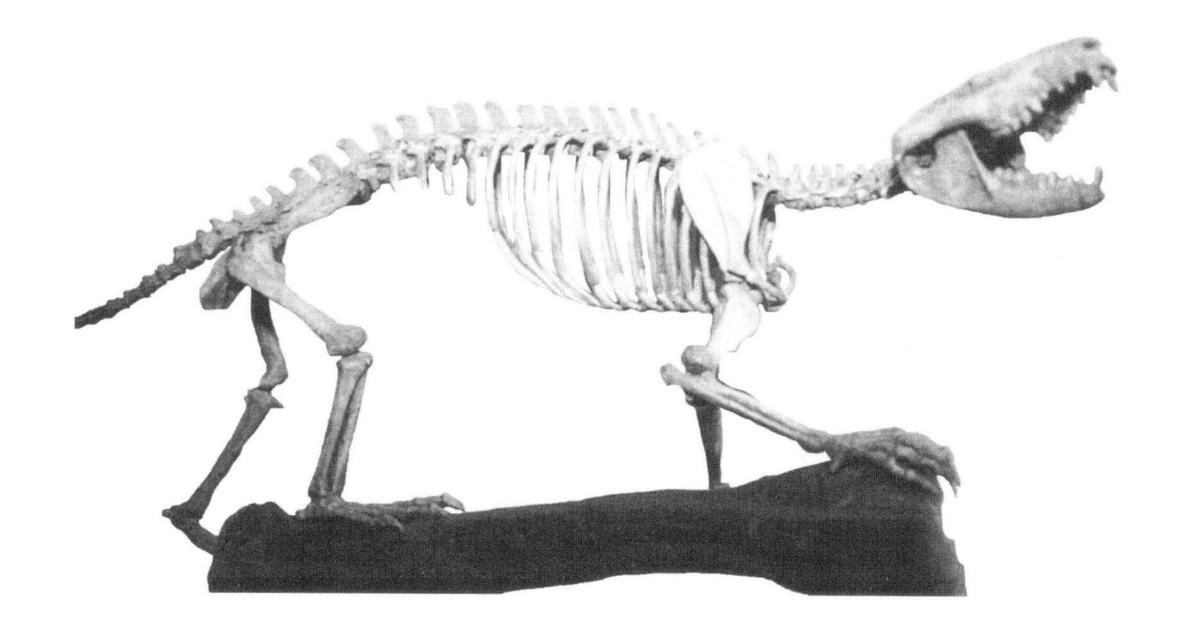

mother crocodile, while the other rushes in to dig up the nest (Cott, 1961). Magnusson (1982) reported that where the range of the Indian monitor overlaps that of the saltwater crocodile, almost all of the nests of the crocodile are raided. Large monitor lizards are known from the Cretaceous of Asia and North America and they were probably one of the top predators of dinosaur eggs and hatchlings. Crocodiles are important predators around water, and they would most certainly have taken any young dinosaur that didn't exercise caution when drinking (Fig. 12.25). But crocodiles of the Mesozoic also included several small, long-limbed terrestrial forms

Fig. 12.25. Crocodiles probably preyed upon young, unsuspecting baby dinosaurs.

Fig. 12.26. Large snakes such as this Dinilysia *may have preyed on baby dinosaurs.*

and these probably stalked nesting grounds. Finally, snakes are known to have been present in the Cretaceous, and one of these, *Dinilysia* from Argentina, was the size of a large boa. It could easily have eaten an unwary dinosaurling (Fig. 12.26).

For those dinosaurlings that remained in the nest, Greg Paul (1994b) argues that conditions were rather unpleasant. In his usual feisty, contentious, and somewhat contradictory manner, Greg takes exception to the pastoral scene of the dinosaur nest. On one hand, he talks about the environmental extremes of the nest—baking under the hot sun by day, freezing under the open sky at night—and on the other, he shows an illustration of a butterfly flitting above the head of an inquisitive *Hypacrosaurus* hatchling. The nest probably was a miserable place to be during a drought: hot, exposed to the sun, and without water. It was probably also miserable if the rainy season persisted too long. Assuming that the eggs did not drown, the hatchlings would have been soaked by cold rain. For these reasons, Greg argues that dinosaurlings were covered with downy feather insulation. Greg views the recent discovery of feathers or featherlike structures on small theropods from China (Qiang and others, 1998; Chen and others, 1998) as a vindication. Greg presents some interesting ideas, although their validity cannot be tested. Furthermore, the ideas are based on the supposition that many dinosaurlings were altricial. If dinosaurlings were precocial, then all discussions about the hazards faced in the nest are moot. At present we cannot say which case is true, but it is possible that there may have been altricial or semialtricial dinosaurlings. Thus discussions such as those by Greg Paul (1994a, b) are not without merit.

Staying Together

Regardless of whether dinosaurlings were altricial or precocial, we do know that they eventually left the nest. There is some evidence that once they did so, some remained together and separated from the adults (Fig. 12.27). In fact, some juvenile dinosaurs apparently did not join adults until they were about half grown. For example, although tracks of hatchling hadrosaurs are known (Fig. 12.28; Carpenter, 1992), these do not occur with adult prints. Instead, the smallest prints with adults are about half adult size (Lockley and others, 1983). The same pattern is seen with hadrosaur tracks from Canada (Currie, 1983) and sauropod tracks from Texas (Bird, 1944; Farlow and others, 1989). The segregation of the young from adults is also seen in small ornithischian tracks from Australia (Thulborn and Wade, 1984; Thulborn, 1990). These tracks suggest a pod (pack? gaggle? herd? mob? flock?) comprising a mixed-age group of fifty-seven individuals. It is debatable whether these individuals represent small adults or juveniles. The incredibly small size of some individuals suggests juveniles to me, although Thulborn and Wade (1984) preferred small adults because it was the simplest hypothesis (Thulborn, 1990). Because so much evidence indicates that the very young were often segregated from the adults, it seems more likely that the very small size of the individuals represents young (sorry, Tony).

Other evidence for segregation of young comes from bone beds. The most notable examples, of course, are the *Maiasaura* babies found together by Horner and Makela (1979; Fig. 12.29), and the *Hypacrosaurus* baby bone beds (Fig. 12.2). Eventually, however, the young grew large enough to join the adults. Evidence from other bone beds shows that the smallest of these is about half the size of the largest adult. Using femur length as the

Fig. 12.27. A segregated pack of baby Dryosaurus *skeletons.*

Fig. 12.28. Cast of baby hadrosaur footprints from a coal mine in the Blackhawk Formation, Utah. The footprints are about 5 cm across and match a foot of a "nestling"-sized individual (about the size of the skeleton seen in Figure 12.18, left side). These footprints indicate that hadrosaurs this size (70 cm) were mobile, not nest bound.

standard, because the bone is usually well preserved and easy to identify and measure, we find the following smallest and largest from bone beds: *Camptosaurus* (Quarry 13, Como Bluffs, Wyoming) 25.8 cm vs. 59.2 cm (Gilmore, 1909); *Coelophysis* (Ghost Ranch, New Mexico) 11.8 cm vs. 20.9 cm (Colbert, 1989); *Syntarsus* (Zimbabwe) 14.2 cm vs. 20.8 cm (Raath, 1977); *Allosaurus* (Cleveland-Lloyd Quarry, Utah) 43.5 vs. 91 cm (Madsen 1976; the smallest femur in his sample is actually that of *Coelu-*

rus). As you can see, the smallest individuals are almost half the largest adult size.

Why this apparent segregation of the very young from the adults? Possibly it is because the adults were far-ranging and the small legs of the young would have prevented them from keeping up with adults. For example, a 67 cm (26") hadrosaur, 32 cm (12½") at the hips, took about a 34 cm (13") step. In contrast, an adult, 7 m (23') long, 2.3 m (7½') at the hips, would take about a 2.4 m (8') step. The poor baby would have to take seven steps for each step of the adult. So, even if the adult was walking leisurely, the baby would be running all the time. Don't believe me? Watch a parent (human) walking while holding their child's hand (say, at a mall)—the poor kid is jogging. However, at half the adult size, young dinosaurs would have an easier time of keeping up (Fig. 12.30).

Fig. 12.29. A sample of ilia from a baby Maiasaura *bone bed. The babies were probably segregated from the adults. Scale = 2 cm.*

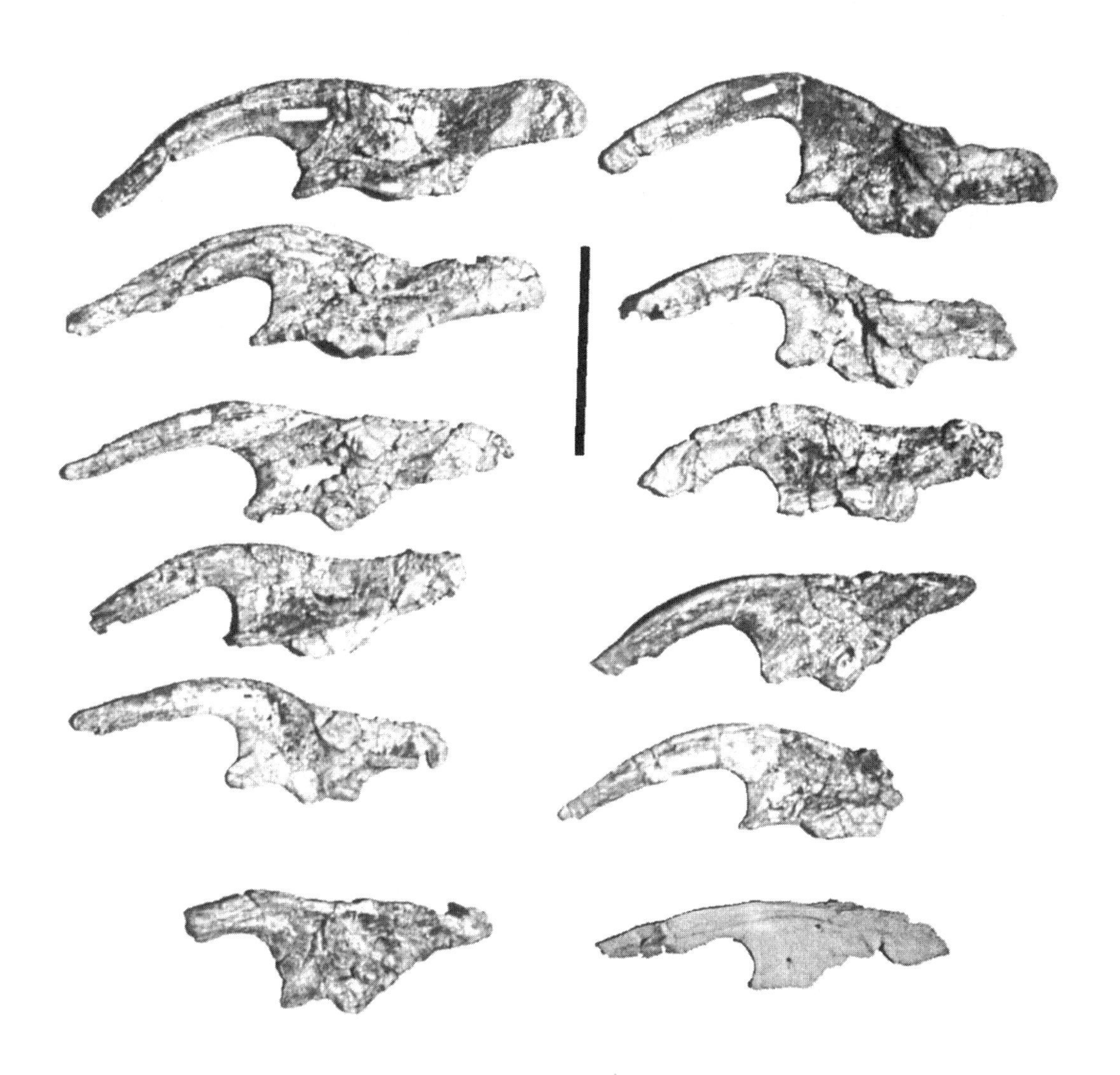

Fig. 12.30. When the baby hadrosaurs grew to about half the adult size, they joined the herd. At this size, it was easier for them to stay with the adults. (Courtesy of Gregory Paul.)

Fig. 12.31. (facing page, top) Not all baby dinosaurs were segregated from the adults. The occurrence of sauropod babies with adults suggests that at least some *sauropod babies stayed with adults. (Courtesy of Gregory Paul.)*

*Fig. 12.32. (facing page, bottom) When juveniles join adults, they learn the social skills needed for them to get along with the rest of the herd. (*Orodromeus, *courtesy of Gregory Paul.)*

These examples from the footprint record, as well as bone beds, do not constitute proof that all dinosaurs segregated their young. As mentioned earlier, the young of *Apatosaurus* and possibly other sauropods apparently stayed in the vicinity of the adults (although this does not automatically mean parental care; Fig 12.31). Still, those dinosaurs that were segregated joined their "clan" to finish growing and probably develop their social skills, i.e., learning their place in the herd, learning mating behavior, and in males, developing their strength for dominance (Fig. 12.32).

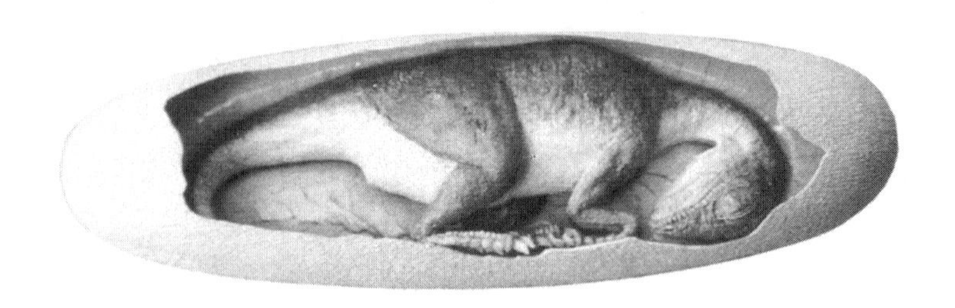

13 • Eggs and the Death of the Dinosaurs

All good things must come to an end, and that was certainly true of the Age of Dinosaurs. What was the cause, an asteroid impact? major volcanic eruption? a sinister plot of mammals to take over the world? Ever since it was realized that dinosaurs are extinct, there have been all sorts of reasons suggested (Table 13.1). We won't discuss most of these, only those that pertain directly to dinosaur eggs and dinosaur babies.

Reason 1. Too Many Males—Too Many Females

Imagine a world in which two-thirds of the babies born are male and that this trend continued for several generations (Fig. 13.1). The problem is that every year the number of adults decreases because of death (predation, disease, old age, etc.) and fewer babies are born to replace them. Eventually the population level would be so low that extinction is a very real possibility. Such a scenario has been presented as a possible cause of dinosaur extinction by Paladino and others (1989). The hypothesis is based upon work done on crocodiles that show that sex of the hatchlings is controlled by incubation temperature (see chapter 11). Paladino and his colleagues suggest that a rapid temperature increase or decrease by a few degrees centigrade near the end of the Cretaceous could have resulted in more of one sex than another. Once these dinosaurlings matured, they might have found it difficult to find a mate. This change in sex ratios might not have been too bad initially if there were more females than males because the males could breed with more than one female. But a point might be reached when there were just too many females who were unable to find a mate and fewer eggs would be laid. Too few eggs would also result if there were too few female dinosaurlings every year. Eventually, there would be a population crash with the result being the possible extinction of that species.

TABLE 13.1.
Various causes suggested for the extinction of the dinosaurs (modified from Molnar and O'Reagan, 1989)

Biological Causes

eggshells too thick
eggshells too thin
eggs developing into only one sex
loss of eggs to egg-eating mammals
loss of eggs to egg-eating dinosaurs
inability to mate (pre-Viagra)
sexual frustration ("not tonight, I have a headache")
competition with mammals (Mickey the Conqueror)
sterilization resulting from excessive heat
runaway effects of group selection acting to check excessive reproductive rates
collapse of intervertebral disks (because too big)
malfunction of endocrine systems (got too big)
too large to hibernate when it got cold
small brain and consequent stupidity (walk into tree)
absence of redundant DNA
chromosomal upsets
overaggression because of overcrowding (the New York City effect)
psychotic suicidal factors
competition with leaf-eating caterpillars
poisonous substances in plants (in cycads or angiosperms)
absence of natural laxatives ("full of s—")
loss of food plants
deleterious effects of the Mesozoic climate on endothermy (assumes warm-blooded dinosaurs)
collapse of the trophic (food) web
predator overkill (too many predatory species)
racial "senility"
overspecialization (too many frills)
increased size and resultant clumsiness
not sufficiently progressive
parasites
disease

Terrestrial (Nonbiological) Causes

climatic change (getting warmer, colder, drier, or wetter, or less equable seasonally or latitudinally)
changes in atmospheric pressure
changes in atmospheric composition (excessive oxygen production by plants; drop in oxygen levels)
floods
drainage of swamps and lakes (early land developers)
poisonous water
absence of necessary trace minerals (e.g., calcium and selenium)
presence of poisonous minerals
too high a level of natural radiation
earthquakes
poisonous gases
volcanic dust
mountain building (the Rockies!)
fall of sea level
rise of sea level
spillage of freshwater from Arctic Basin into warm southern seas
breakup of supercontinents by sea-floor spreading

Extraterrestrial Causes

blindness from solar radiation (walked into trees, walked off cliffs)
reversal of terrestrial magnetic field allowing flood of cosmic radiation
shift of rotational poles
origin of the Moon from the Pacific Basin
meteorites or comets
sunspots
supernovae
fall to earth of a (hypothetical) previous moon

Miscellaneous Causes

gravity greater or lesser
entropy
hunting by little green men from flying saucers
Noah's Flood
God's will
unfitted to the Marxist dialectic
ennui (been there, done that)

Fig. 13.1. A world with an unbalanced sex ratio could spell doom for the dinosaurs. Here a nubile Protoceratops *female is fleeing from a mob of males.*

How plausible is this hypothesis? First, it makes the assumption that all dinosaurs laid temperature-dependent eggs. Maybe, but there is no real proof of this. Besides, the very avianlike ornithoid eggs suggest that many of the theropods, especially the more avianlike *Velociraptor, Oviraptor,* and *Sinosauropteryx,* might have been more like birds in their development, i.e., non-temperature-dependent embryos. Second, because we know crocodiles are temperature-dependent and we know they were present in the Cretaceous, why didn't they go extinct? This latter point raises a major problem with many extinction models. They are often elaborately built to explain extinction, not to explain survival, and it is at the survivors that we should look. Why did crocodiles survive the extinction at the end of the Cretaceous? Why did many non-dinosaurian birds survive, but not the birdlike dinosaurs (*Oviraptor,* etc.)? Why were frogs and all other amphibians unaffected if they are so sensitive to minor environmental changes today?

Reason 2. Chemical Changes in the Environment

Because eggshell is produced within the female's body from material she takes in (a "you are what you eat" sort of thing), then perhaps eggshells might contain some clues about the environment, right? That was the assumption made by Chinese egg paleontologist Zhao Zi-Kui and his colleagues (1991, 1993; Zhao, 1994) in their studies of eggshells near the Cretaceous–Tertiary boundary. Of twelve egg species, only four, *Macroolithus yaotunensis, M. rugustus, Elongatoolithus andrewsi,* and *E. elongatus,* were abundant enough (n = 15,435 eggshells) to provide statistically meaningful data. The pathologies they reported included abnormal thickening or thinning (more than the normal range of variation), multilayered eggshell, irregular spaces within the eggshell, and disorderly arrangement of the crystalline layer (mostly in *Elongatoolithus andrewsi*). Zhao and others (1991) noted a high incidence of pathological eggshells in the Pingling Formation (Fig. 13.2), with a spike of almost 75 percent for *Macroolithus yaotunensis* about 50 m (165") below the Cretaceous–Tertiary (a.k.a. K-T) boundary.

To determine why these pathologies occurred, Zhao and his colleagues analyzed the chemical content of the eggshell. They concluded that eggshell from the upper horizons contained higher amounts of certain trace elements (elements that normally occur in minute amounts), including strontium, manganese, zinc, lead, copper, cobalt, nickel, and vanadium; these elements have a strong peak at 90 m (300') below the K-T boundary. They suspected an environmental change as the culprit for the higher amounts of trace elements entering into the Cretaceous food chain. To verify this hypothesis, they next analyzed the oxygen and carbon isotopes in the shell. Oxygen naturally comes in two forms, ^{16}O and ^{18}O, with ^{18}O having two more neutrons in its nucleus than ^{16}O. Water molecules (H_2O) formed with this heavier isotope are kept by gravity from evaporating and eventually a lake will become enriched in this "heavier" water. Dinosaurs drinking this water would then incorporate ^{18}O into the calcium carbonate ($CaCO_3$) of the eggshell. When might this trend toward enrichment of the water occur? Under normal conditions, lighter oxygen isotope rainwater would replenish any light isotope water that evaporated from the lake. But during a drought or if climatic conditions became more arid, then heavier isotope water would be expected. Zhao and his colleagues noted a sharp increase in heavier isotope oxygen in eggshells from 80 m (265') below the K-T boundary. Another study by Zhao and his colleagues (1993) examined the amino acids of *Macroolithus yaotunensis* from the same study area. They again noted that amino acids were abnormal in the pathological eggshells as compared to the non-pathological eggshells.

Taking all of this data together, Zhao and his colleagues (1991) concluded that there was a drastic climatic change beginning about 200,000 to 300,000 years before the end of the Cretaceous. This change was a sudden shift toward more aridity, which enriched ^{18}O in water and concentrated trace minerals in the soil. These trace elements were then taken up by the plants, which were in turn eaten by herbivorous dinosaurs. In the body, these trace elements caused eggshells to be produced that were thicker than normal or thinner than normal. Embryos reaching the bone-producing

Fig. 13.2. Diagram showing changes in shell thickness and variations in eggshell oxygen isotope and trace elements at different stratigraphic levels in Nanxiong Basin, China. Starting point for data was designated as 0 m, with samples collected all the way to the K-T boundary at 160 m above the starting point. Zhao and his colleagues argue that the spikes in various trace elements (jagged spike toward the right) indicate that the local environment became more arid during the latter part of the Late Cretaceous. See text for details. (Modified from Zhao and others, 1991.)

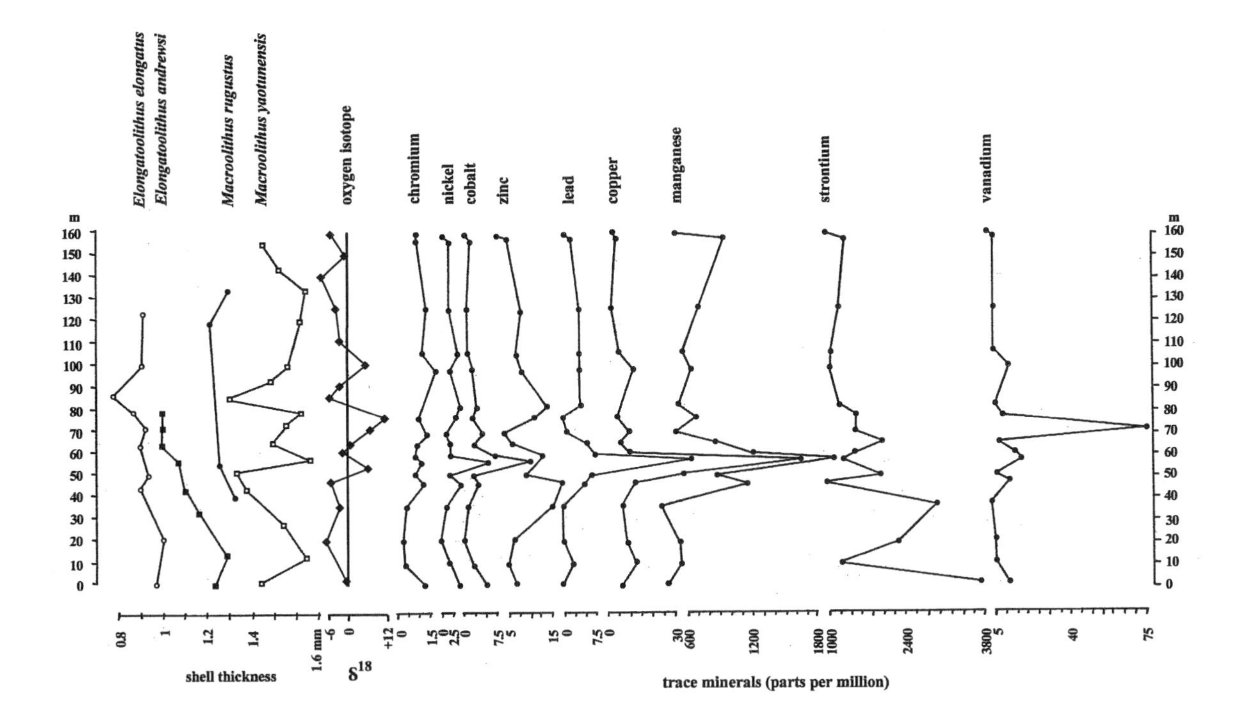

stage would contaminate themselves with trace elements incorporated into the eggshell and this would cause developmental defects. With each generation, fewer healthy dinosaurs would hatch and eventually they went extinct. The first of these that went extinct is the dinosaur that laid *Elongatoolithus andrewsi* eggs (Fig. 13.2). The last eggshells are found just above the ^{18}O and trace element spike. *E. elongatus* last appears ~30 m (100') below the K-T boundary, and *Macroolithus rugustus* 15 m (50') below. *M. yaotunensis* last appears at the K-T boundary. Zhao and his colleagues interpreted the data to mean that the extinction of the dinosaurs was not sudden, but took at least 200,000 to 300,000 years.

This hypothesis for a climatic change leading to the extinction of the dinosaurs does seem to have supporting evidence for a gradual extinction of the dinosaurs, at least in southern China. In North America, however, the K-T boundary is marked by coal beds indicative of swampy, wet conditions (Parrish and others, 1982). Such an interpretation is also supported by the fossil leaf record (Wolfe and Upchurch, 1987). So, whereas aridity may have developed in part of China, there is no evidence the effect was worldwide.

Let's look again at the Chinese data (Fig. 13.2). How reliable is the chemical data used by Zhao and his colleagues? Folinsbee and his group from the University of Alberta (1970) demonstrated that chickens laid eggs with isotopically heavy shells (i.e., high in ^{18}O) if fed water enriched with ^{18}O, or isotopically light if fed water high in ^{16}O. Thus, enrichment of ^{18}O in drinking water during the Late Cretaceous could have occurred as suggested by the Chinese. Furthermore, this water would have resulted in the female dinosaurs laying eggs with shells high in ^{18}O. A similar conclusion was reached by work done on dinosaur eggs from the Upper Cretaceous of India (Sarkar and others, 1991; Tandon and others, 1995). But if we look carefully at Figure 13.2, we note that the oxygen isotope ratio for ^{18}O was lower than "normal" for a considerable amount of time *before* the end of the Cretaceous. The evidence for terminal Cretaceous aridity just isn't there.

What about the trace elements? French geochemist Yannicke Dauphin (1990a, b, 1992) has come to a different conclusion from that of Zhao and his associates. First, his study of various bird eggshells led him to conclude that "what you eat, is what you lay" doesn't always hold, i.e., there isn't always a correlation between diet and chemical composition of the eggshell. Instead, these elements were concentrated in the dinosaur eggshell by ground water seeping through after burial. If Dauphin is correct, then the spike of the various trace elements probably did not result from the diet of the egg layers. Is this for certain? No, because how these trace elements could get concentrated from ground water is not yet known. At present the ideas presented by Zhao and his colleagues are interesting, but the correlation isn't as strong as they suggest.

Reason 3. Eggshell Too Thin, Eggshell Too Thick

For a long time, farmers have known that chickens occasionally lay eggs with abnormal shell thickness. For eggs that are abnormally thick, the chicks fail to develop and the egg rots (the classic rotten egg smell). The inquisitive farmer breaking open such eggs would find the shell to be formed of two or more layers, instead of the usual one (see chapter 6). When the additional layers were formed, the chances of getting the pores lined up was infinitesimally small (Fig. 13.3). Unable to breathe (see chapter 11), the embryo dies before it could ever really form. On the other hand, if the shell is abnormally thin, the egg is crushed by the weight of the brooding hen. During the late 1960s and early 1970s, various pesticides, such as DDT, caused an abnormal thinning of eggshell in many birds (McFarland and others, 1971; Cooke, 1973, 1975; Kiff and others, 1979). With fewer young hatching each year, population levels for these birds dropped drastically. This drop eventually led to the ban of DDT and other pesticides.

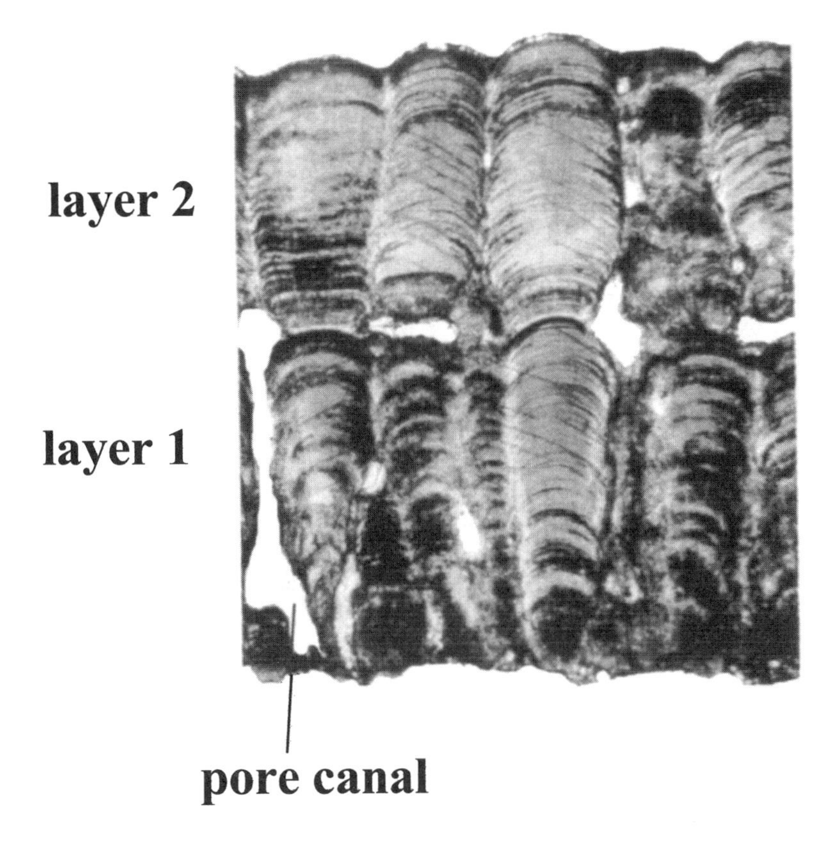

Fig. 13.3. Pathological Megaloolithus *eggshell showing two layers. Note that the second layer blocks the pore canal, thus smothering the embryo.*

The whole issue of abnormal eggshell leading to the near extinction of some bird species also led to an interest in abnormal eggshell as a factor in the extinction of the dinosaurs. Some of the earliest work along these lines was by Heinrich Erben, a German egg specialist, and his colleagues working in the Aix Basin of southern France. They sampled eggshell, thought to be of the sauropod *Hypselosaurus,* throughout the Upper Cretaceous deposits and claimed that an unusual percentage of eggshells in the upper beds (hence closer to the Cretaceous–Tertiary boundary) were abnormal in thickness (Fig.13.4). Some eggshells were thicker than normal because of multiple layers, and these smothered the embryo, whereas other eggshells were abnormally thin and these resulted in the dehydration of the egg under the hot ancestral Mediterranean sun.

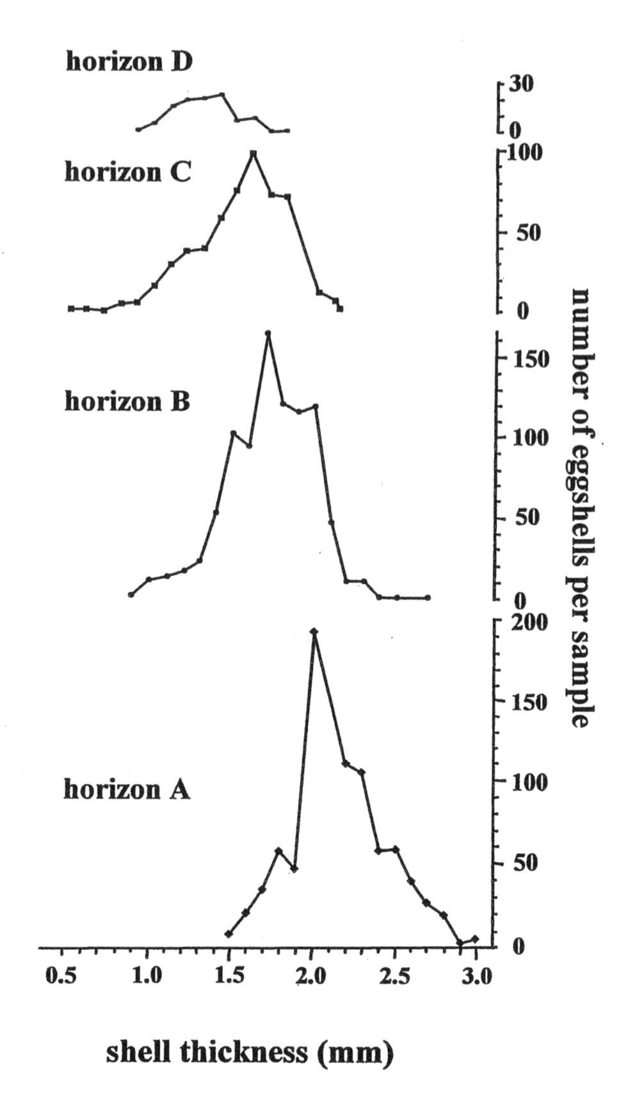

Fig. 13.4. Variations in shell thickness at different stratigraphic levels from the Upper Cretaceous of southern France. Erben and his colleagues argued that the shift of eggshell toward the left indicated the presence of something that caused thinner eggshell. See text for discussion. (Modified from Erben and others, 1979.)

Erben and others (1979, p. 396) argued that the multilayered eggshells were the result of a genetic mutation that "could have overcome [natural] selection and could have propagated throughout the population." But this is a silly statement because a genetic mutation is not something that can be passed around like the flu. If a female dinosaur lays multilayered eggs that smother the embryo, there are no young to grow up and pass the mutation to another generation. End of story. Erben and others do backpedal a little and admit that stress can cause multilayered shells in captive turtles, so maybe stress caused the multilayered shell. What that stress was and why it wasn't present before the last few thousand years of the Cretaceous is not said (I guess they don't know).

As for abnormally thin eggshells, which they claim is the most dominant form of egg abnormality, Erben and his colleagues suggested stresses brought about by improved climatic conditions. Yep, things got so good that vegetation was more lush, *Hypselosaurus* mated like rabbits and before you knew it, the countryside was awash in sex-crazed hypselosaurs.

There were so many of them that it got to be rather stressful (sort of like living in New York City).

> Individual territories diminished in proportion to the rising number of animals. This, in turn, would have led to an automatic increase in the number of territorial interactions of aggressions with which each individual was confronted and the resulting nervous tension and irritability would have represented a permanent emotional stress. Every act of aggression, however, be it actual physical violence or merely menacing, inevitably evokes in vertebrates a pituitary-adrenocortical activity which in the case of excess is apt to upset the estrogen level of females. Such a mechanism could, of course, easily explain the formation of pathologically thin-shelled eggs. (Erben and others, 1979, p. 410)

This is quite a "just-so" story. Unfortunately, there is little evidence to support it. The great herds envisioned should have left tons of bones (literally) in the fossil record, yet *Hypselosaurus* remains a very poorly known dinosaur. Furthermore, detailed studies of eggshell by French paleontologists have demonstrated that these supposed thin eggshells are of different morphotypes, hence represent different species of egg types (see Appendix II). A statistical study by French egg paleontologist Philippe Kérourio (1981) showed that multiple-layered eggshells averaged only 0.5–2.5 percent throughout the deposits containing dinosaur eggshells; they were certainly not more abundant near the K-T boundary.

Reason 4. Carbon Dioxide/Oxygen Imbalance

The development of the embryo was discussed in chapter 11. One important need of the embryo is respiration, that is, oxygen must be "inhaled" by the embryo and carbon dioxide must be "exhaled." The lungs of bird and reptile embryos don't start working until hatching or just before hatching; until then, they rely upon the chorioallantoic membrane to act on behalf of their lungs. Gas exchange through the shell occurs at the tiny pores that are scattered over the shell. For bird and reptile eggs that are buried, oxygen requirements of the embryo are less than for eggs laid above ground. Nevertheless, a slight decrease in oxygen can be more devastating for the embryo in a buried egg than for one above ground (Ackerman, 1980). Imagine, then, a scenario in which atmospheric oxygen decreases and carbon dioxide increases. This would probably have a drastic impact upon dinosaur eggs buried in nests. Such is the idea put forth by B. Oelofsen, from South Africa. Oelofsen (1978) argued that volcanoes, especially kimberlite volcanoes (which produce diamonds), were especially common during the Late Cretaceous. These CO_2-rich volcanoes spewed millions of tons of carbon dioxide into the atmosphere.

The buildup of atmospheric CO_2 coincided with a worldwide drop in sea level at the end of the Cretaceous. Normally, phytoplankton would scrub the atmosphere of CO_2, but with a drop in sea level (and a reduction in the sea surface area), there was also a decrease in phytoplankton populations. The gradual buildup of CO_2 made it progressively harder for embryonic dinosaurs to obtain sufficient oxygen for development. Eventually the number of dinosaurs hatching was less than the number of dinosaurs dying every year. With fewer adults being replaced, the global dinosaur population finally crashed. Crocodiles and other reptiles survived, Oelofsen (1978) argues, because being ectotherms ("cold-blooded"), the oxygen requirements of their embryos were less than for the endothermic ("hot-blooded") dinosaurs.

Is there any supporting evidence for this hypothesis? Actually, yes. There are massive lava flows in central India called the Deccan Traps that date from the K-T boundary. Covering tens of thousands of square kilometers, they indicate one of the most extensive volcanic episodes in the history of the earth. There was also a major drop in global sea level at the end of the Cretaceous that exposed more land. In addition, there was also a major extinction of the phytoplankton. The problem, however, is that the Deccan Trap lavas are not kimberlitic, so the atmospheric input of CO_2 was not anywhere as high as Oelofsen hoped. Second, the model assumes and requires that all dinosaurs were endothermic, which has yet to be proven. Although an interesting and innovative suggestion, the evidence supporting Oelofsen's hypothesis is not as compelling as it might at first appear to be.

Reason 5. Killer Rats

Ever since fossil mammals were found in the same deposits as dinosaurs, the suggestion has been made that perhaps the ravenous mammals dined on dinosaur eggs. Indeed, one of the largest mammals during the Late Cretaceous was a marsupial, *Didelphodon,* the size of a house cat (Fig. 13.5). It has been likened to the opossum, which—being an omnivore—will eat everything, including eggs. But this prehistoric opossum is a giant among Cretaceous mammals; most were the size of a mouse or shrew. Considering the small size of their mouths, it seems doubtful that any of these could have cracked open a dinosaur egg. Furthermore, many of these mammals have tall, pointed-cusped teeth that indicate a diet of insects. It seems doubtful, therefore, that mammals caused the extinction of the dinosaurs.

*Fig. 13.5. Life-size models of an early, house cat–sized mammal (*Didelphodon*) and a hatchling hadrosaur. The largest mammals during the Late Cretaceous were these opossum-like marsupials. They could easily have preyed upon baby dinosaurs when given the chance.*

Reason 6. Killer Dinosaurs

One idea, best forgotten, has it that carnivorous dinosaurs got big because they ate eggs! As G. Wieland (1925, p. 559) put it, "*Tyrannosaurus,* amongst all carnivorous dinosaurs, most suggests a growth and habitat due to some increased abundance of prey. What more likely than that the immediate ancestors of this dinosaur got their first impulse towards gigantism on a diet of sauropod eggs, and that . . . the theropods were the great egg-eaters of all time?" Leave it to a fossil plant guy to say (p. 560), "the view is presented is that the tooth-set gape of *Tyrannosaurus* is not inconsistent with nurture on sauropod eggs." Of course, never mind that sauropods don't co-occur with *Tyrannosaurus.* And so, because *Tyrannosaurus* was so big, it ate a lot of eggs (and babies), thus wiping out the dinosaurs (honest!).

These six extinction models are not the only ones that have been suggested, but these are the only ones that specifically involve dinosaur eggs. The most voguish model today is that an asteroid impact had a direct or indirect effect, but even this hypothesis is not universally accepted. For a discussion of the pros and cons of the impact hypothesis, see Russell and Dodson (1997). Regardless of which extinction model you personally believe, we can all agree that the extinction of the dinosaurs is due to there not being any young to replace the dead adults. So directly or indirectly, dinosaur reproduction did play a role in dinosaur extinction. Well almost—there is still the robin sitting outside your window. . . .

Appendix I. Where Dinosaur Eggs, Nests, and Baby Dinosaurs Have Been Found

Localities by time where dinosaur eggs, clutches, and bones of babies and young dinosaurs have been found. See maps, Figures 2.1–2.3. Names ending in *-oolithus* are egg names (see chapter 9).

Triassic

1. Ghost Ranch, New Mexico; Chinle Formation, Norian: *Coelophysis bauri* young (Colbert, 1989; Carpenter, unpublished notes).
2. Laguna La Colorada, Argentina; El Tranquilo Formation, Norian: *Mussaurus patagonicus* skeletons (Bonaparte and Vince, 1979).

Jurassic

3. Sheep Creek, Wyoming; Morrison Formation, Kimmeridgian: *Apatosaurus excelsus* baby bones (Carpenter and McIntosh, 1994).
4. Como Bluffs, Wyoming; Morrison Formation, Kimmeridgian: *Camarasaurus grandis* baby bones (Carpenter and McIntosh, 1994).
5. Dinosaur National Monument, Utah; Brushy Basin Member, Morrison Formation, Kimmeridgian: *Dryosaurus altus* baby skeleton, *Camptosaurus* sp. embryo, *Stegosaurus* sp. baby skeleton (Carpenter, 1994; Chure and others, 1994; Galton, 1982).
6. Cleveland-Lloyd Quarry, Utah; Brushy Basin Member, Morrison Formation, Kimmeridgian: *Preprismatoolithus coloradensis* egg (Hirsch and others, 1989; Hirsch, 1994).
7. Moab, Utah; Brushy Basin Member, Morrison Formation, Kimmeridgian: spheroolithid eggshells (Zelenitsky, personal communication).
8. Fruita, Colorado; Morrison Formation, Kimmeridgian: *Preprismatoolithus coloradensis* eggshells (Hirsch, 1994).
9. Young Locality, Colorado; Salt Wash Member, Morrison Formation, Kimmeridgian: *Preprismatoolithus coloradensis* egg and eggshells (Hirsch, 1994).
10. Dry Mesa, Colorado; Morrison Formation, Kimmeridgian: *Camarasaurus* sp. hatchling (Britt and Naylor, 1994).
11. Uravan Locality, Colorado; Brushy Basin Member, Morrison Formation, Kimmeridgian: *Preprismatoolithus coloradensis* eggshells and *Dryosaurus altus* baby bones (Hirsch, 1994; Scheetz, 1991).

12. Garden Park, Colorado; Morrison Formation, Kimmeridgian: *Preprismatoolithus coloradensis* clutch of eggs and eggshell, embryonic bones; ornithopod hatchling (Hirsch, 1994; Carpenter, unpublished notes).

13. Arch Mesa, New Mexico; Morrison Formation, Kimmeridgian: eggshells (Bray and Lucas, 1997).

14. Kenton, Oklahoma; Morrison Formation, Kimmeridgian: *Apatosaurus* sp. baby bones and *Camarasaurus* sp. baby bones (Carpenter and McIntosh, 1994).

15. Charmouth, England; Lias Formation, Sinemurian: *Scelidosaurus* juvenile (Carpenter, unpublished notes).

16. Petersborough, England; Oxford Clay, Oxfordian: camptosaurid(?) egg (van Straelen, 1928).

17. Guimarota, Portugal; unnamed formation, Kimmeridgian: eggshells (Hirsch, personal communication).

18. Forte Pai Mogo, Portugal; Sobral Formation, Kimmeridgian-Tithonian: *Preprismatoolithus* eggs with theropod embryos (Mateus and others, 1997a, b).

19. Peralta, Portugal; unnamed formation, Kimmeridgian-Tithonian: Megaloolithidae and *Preprismatoolithus* sp. eggshells (Dantas and others, 1992).

20. Paimogo, Portugal; unnamed formation, Kimmeridgian-Tithonian: Megaloolithidae eggshells (Dantas and others, 1992).

21. Valemitao, Portugal; unnamed formation, Kimmeridgian-Tithonian: Megaloolithidae eggshells (Dantas and others, 1992).

22. Tendaguru, Tanzania; Tendaguru Formation, Kimmeridgian: egg and eggshells (Swinton, 1950).

Early Cretaceous

23. Bridger, Montana; Cloverly Formation, Aptian-Albian: eggshells? and baby bones (Maxwell and Horner, 1994).

24. Cashen Ranch, Montana; Cloverly Formation, Aptian-Albian: *Tenontosaurus tilleti* baby skeletons (Forster, 1990).

25. Middle Dome, Montana; Cloverly Formation, Aptian-Albian: eggshells (Maxwell and Horner, 1994).

26. Wayan, Idaho; Wayan Formation, Albian: eggshells (Dorr, 1985).

27. Thomas Fork Creek, Wyoming; Thomas Fork Formation, Albian: Ornithopod(?) eggshells (Dorr, 1985).

28. Coalville, Utah; Kelvin Formation, Berriasian-Albian: eggshells (Jensen, 1970).

29. Castle Dale, Utah; Cedar Mountain and Dakota formations, Aptian-Albian and Cenomanian: *Macroelongatoolithus carlylei* eggshells.

30. Moore Road, Utah; Mussentuchit Member, Cedar Mountain Formation, Cenomanian: eggshells (Cifelli, personal communication). *Macroelongatoolithus carlylei.*

31. Mussentuchit Wash, Utah; Mussentuchit Member, Cedar Mountain Formation, Cenomanian: eggshells (Cifelli, personal communication). *Macroelongatoolithus carlylei.*

32. Ken's Locality, Utah; Yellow Cat Member, Cedar Mountain Formation, Neocomian: eggshells (Carpenter, unpublished notes).

33. Poison Strip Locality, Utah; Yellow Cat Member, Cedar Mountain Formation, Neocomian: eggshells (Carpenter, unpublished notes).

34. Proctor Lake, Texas; Twin Mountains Formation, Aptian: Hypsilophodontidae baby bones (Winkler and Murry, 1989).

35. Fort Worth, Texas; Paw Paw Formation, Albian: Nodosauridae baby bones (Jacobs and others, 1994).

36. Sunnydown Farm, Dorset, England; Cherty Freshwater Member, Purbeck Formation, Berriasian: four different types of eggshells (Anonymous, 1997).

37. Galve, Spain; Las Zabachoras Beds, Barremian: eggshells (Kohring, 1990a).

38. El Montes, Spain; La Pedrera de Rúbies Limestone, Berresian-Barremian: nestling bird (Sanz and others, 1997).

39. Naryn River, Kyrgyzstan; Khodzhaosmansk Formation, Aptian-Albian(?): clutches (Nessov and Kaznyshkin, 1986).

40. Kuren Tsav, Mongolia; unnamed formation, Neocomian: eggshells (Nessov and Kaznyshkin, 1986).

41. Oshi Nur, Mongolia; Ondai Sayr, Lower Cretaceous: *Psittacosaurus mongoliensis* clutch with baby bones (Fig. 11.35; Coombs, 1982; Nessov and Kaznyshkin, 1986).

42. Dushih Ula, Mongolia; Dushihin Formation, Lower Cretaceous: clutches (Mikhailov and others, 1994).

43. Builyasutuin Khuduk, Mongolia; Dushihin Formation, Aptian: *Trachoolithus faticanus* eggshells (Mikhailov, 1994a).

44. Ehr Chia Wu Tung (= Jianguanz), China; unnamed formation, Lower(?) Cretaceous: *Peishansaurus philemys* (= *Psittacosaurus*?) baby bones (Bohlin, 1953).

45. Ningxia Province, China; unnamed formation, Lower(?) Cretaceous: eggs (Young, 1979).

46. Xixia Basin, China; Kaoguo or Siguo Formation, Lower Cretaceous: *Macroelongatoolithus xixianensis, Faveoloolithus* sp., *Spheroolithus* cf. *S. irenensis,* and *Phaceloolithus taohensis* clutches (Zhao, 1979a; Li and others, 1995).

47. Liaoning, China; Yixian Formation, Lower Cretaceous: eggs of *Sinosauropteryx prima* (Chen and others, 1998).

48. Kamioka, Japan; Tedori Formation, Lower Cretaceous: group of dinosaur(?) eggs (Anonymous, 1995).

49. Phu Wiang, Thailand; Sao Khua Formation, Lower Cretaceous: *Phuwiangosaurus sirindhornae* baby bones (Martin, 1994).

50. Dinosaur Cove, Australia; Otway Group, Aptian-Albian: *Atlascopcosaurus loadsi* and *Leaellynasaura amicagraphica* baby bones (Rich and Rich, 1989).

Late Cretaceous

51. Ocean Point, Alaska; Prince Creek Formation, Maastrichtian: *Edmontosaurus* sp. bones (Nelms, personal communication).

52. Dinosaur Provincial Park, Alberta; Judith River Formation, Campanian: eggshells and hadrosaur baby bones (Sternberg, 1955; Brinkman, 1986).

53. Crowsnest Pass, Alberta; Belly River Formation, Campanian: hadrosaur baby bones (Currie, personal communication).

54. Cardston, Alberta; Willow Creek Formation, Maastrichtian: eggshells (Currie, personal communication).

55. Lethbridge, Alberta; Saint Mary River Formation, Maastrichtian: eggshells (Currie, personal communication).

56. Milk River, Alberta; Oldman Formation, Campanian: eggshells (Currie, personal communication).

57. Devil's Coulee, Alberta; Oldman Formation, Campanian: *Porituberoolithus warnerensis, Continuoolithus canadensis, Tristraguloolithus cracioides, Dispersituberoolithus exilis, Prismatoolithus levis, Spheroolithus albertensis, Hypacrosaurus stebingeri* clutch and embryo/hatchling skeletons (Zelenitsky and others, 1996; Zelenitsky and Hills, 1996).

58. Knight's Ranch, Alberta; Oldman Formation, Campanian: *Porituberoolithus warnerensis, Continuoolithus canadensis, Tristraguloolithus cracioides, Spheroolithus albertensis* (Zelenitsky and others, 1996; Zelenitsky and Hills, 1996).

59. Orion, Alberta; Judith River Formation, Campanian: hadrosaur baby bones and clutch (Currie, personal communication).

60. Killdeer Badlands, Saskatchewan; Frenchman Formation, Maastrichtian: Ceratopsidae baby bone (Tokaryk, 1990).

61. Landslide Butte, Montana; Two Medicine Formation, Campanian: hadrosaur clutches and baby bones (Carpenter, unpublished notes).

62. Cut Bank, Montana; Saint Mary River Formation, Maastrichtian: clutches (Weishampel, personal communication).

63. Browning, Montana; Saint Mary River Formation, Maastrichtian: hadrosaur baby bones and clutches (Weishampel, personal communication).

64. Two Medicine River, Montana; Two Medicine Formation, Campanian: eggshells, hadrosaur clutch, and baby bones (Carpenter, unpublished notes).

65. Blacktail Creek, Montana; Two Medicine Formation, Campanian: *Hypacrosaurus stebingeri* clutch and baby bones, hadrosaur clutches and baby bones (Horner and Currie, 1994).

66. Dupuyer Creek, Montana; Two Medicine Formation, Campanian: Lambeosauridae clutch and baby bones; Hypsilophodontidae baby bones (Carpenter, unpublished notes).

67. Salmon Ranch, Montana; Horsethief Formation, Campanian: eggshells (Bibler and Schmitt, 1986).

68. Egg Mountain area, Montana; Two Medicine Formation, Campanian: *Maiasaura peeblesorum* clutches and baby skeletons, *Troodon formosus* clutches and baby skeletons, and ornithoid and *?Orodromeus* sp. eggs (Hirsch and Quinn, 1990).

69. Red Rock Locality, Montana; Two Medicine Formation, Campanian: *Troodon* sp., *Maiasaura peeblesorum,* and *Orodromeus makelai* eggshells (Hirsch and Quinn, 1990).

70. Augusta, Montana; Saint Mary River Formation, Maastrichtian: clutches (Weishampel, personal communication).

71. Fresno, Montana; Judith River Formation, Campanian: hadrosaur clutch and baby bones (Carpenter, unpublished notes).

72. Fort Peck Reservoir, Montana; Hell Creek Formation, Maastrichtian: baby Dromaeosauridae teeth, baby *Aublysodon mirandus* teeth, baby *Tyrannosaurus* sp. tooth, baby Theropoda teeth, hadrosaur baby bones, baby ?Ankylosauridae teeth, and Ceratopsidae baby bone (Brown and Schlaikjer, 1940; Carpenter, 1982 and unpublished notes).

73. Corson County, South Dakota; Hell Creek Formation, Maastrichtian: hadrosaur bones (Carpenter, unpublished notes).

74. Judith River, Montana; Judith River Formation, Campanian: eggshells (Sahni, 1972).

75. Shawmut, Montana; Judith River Formation, Campanian: *Stegoceras* baby bones (Goodwin, personal communication).

76. Careless Creek, Montana; Judith River Formation, Campanian: Lambeosauridae baby bones (Fiorillo, 1987).

77. Red Lodge, Montana; "Lance" Formation, Maastrichtian: eggshells (Jepsen, 1931).

78. Polecat Bench, Wyoming; "Lance" Formation, Maastrichtian: eggshells (Carpenter, 1982).

79. Lance Creek, Wyoming; Lance Formation, Maastrichtian: baby *Troodon* sp. jaw and teeth, baby Dromaeosauridae teeth, baby Theropoda teeth, baby *Aublysodon mirandus* teeth, baby *Thescelosaurus* sp. teeth, baby hadrosaur teeth, baby Ceratopsidae teeth, baby Ankylosauridae teeth (Carpenter, 1982).

80. Wasatch Plateau, Utah; North Horn Formation, Maastrichtian: eggshells (Jensen, 1966).

81. Price, Utah; Blackhawk Formation, Campanian: hadrosaur baby footprints (Carpenter, 1992).

82. Kaiparowitz Plateau, Utah; Wahweap and Kaiparowitz formations, Campanian: eggshells (Carpenter, unpublished notes).

83. San Juan Basin, New Mexico; Fruitland Formation, Campanian: eggshells (Wolberg and Bellis, 1989).

84. Harell Station, Alabama; Mooreville Chalk, Campanian: Dinosaur(?) egg (Dobie, 1978).

85. El Rosario, Mexico; El Gallo Formation, Campanian: eggshells (Hernandez, personal communication).

86. Uberaba, Brazil; Bauru Formation, Campanian-Maastrichtian: eggshells (Price, 1951).

87. Soriano, Uruguay; Mones-Ascencio Formation, Senonian: *Sphaerouvum erbeni* and "*Tacuarembouvum oblongum*" eggshells and clutch (Faccio, 1994).

88. Algorta, Uruguay; Mones-Ascencio Formation, Senonian: eggshells (Faccio, 1994).

89. Pongo de Renterna, Peru; Bagua Formation, Campanian-Maastrichtian: eggshells (Mourier and others, 1988).

90. El Pintor, Peru; Bagua Formation, Campanian-Maastrichtian: eggshells (Mourier and others, 1988).

91. Fundo el Triunfo, Peru; Bagua Formation, Campanian-Maastrichtian: eggshells (Mourier and others, 1988).

92. Morerillo, Peru; Bagua Formation, Campanian-Maastrichtian: eggshells (Mourier and others, 1988).

93. Laguna Umayo, Peru; Vilquechio Formation, Maastrichtian: eggshells (Kérourio and Sigé, 1984).

94. Auca Mahuevo, Argentina; Anacleto Member, Rio Colorado Formation, Upper Cretaceous: titanosaurid eggs and embryonic bones (Chiappe and others, 1998).

95. Moreno and Ojo de Agua, Argentina; Allen Formation, Maastrichtian: Titanosauridae clutch (Powell, in press).

96. General Roca, Argentina; Allen Formation, Maastrichtian: clutch(?) (Powell, in press).

97. Casa de Piedra, Argentina; Angostura Colorada Formation, Senonian: eggshells (Powell, 1987).

98. Fox-Amphoux, France; unnamed formation, Maastrichtian: eggs (Lapparent, 1967).

99. Montmeyan, France; unnamed formation, Maastrichtian: eggs (Lapparent, 1967).

100. Sillans-Salernes, France; unnamed formation, Maastrichtian: eggs (Lapparent, 1967).

101. Rognette, France; unnamed formation, Maastrichtian: eggs. (Lapparent, 1967).

102. Rians, France; unnamed formation, Maastrichtian: eggs (Lapparent, 1967).

103. Pourcieux, France; Rognac Formation, Maastrichtian: clutches (Williams, Seymour, and Kérourio, 1984).

104. Rousset, France; unnamed formation, Maastrichtian: eggs (Lapparent, 1967).

105. La Cairanne, France; unnamed formation, Maastrichtian: eggs (Penner, 1985).

106. Rogues-Hautes, France; unnamed formation, Maastrichtian: eggs (Lapparent, 1967).

107. Rognac, France; Rognac Formation, Maastrichtian: clutch (Lapparent, 1967).

108. Eygalieres, France; unnamed formation, Maastrichtian: eggs (Lapparent, 1967).

109. Maussane, France; unnamed formation, Maastrichtian: eggs (Lapparent, 1967).

110. Argelliers, France; unnamed formation, Maastrichtian: eggs (Lapparent, 1967).

111. Montpellier, France; unnamed formation, Maastrichtian: eggs (Lapparent, 1967).

112. Villeveyrac, France; unnamed formation, Maastrichtian: eggs (Lapparent, 1967).

113. St. Chinian, France; unnamed formation, Maastrichtian: eggs (Lapparent, 1967).

114. Castigno, France; unnamed formation, Maastrichtian: eggs (Lapparent, 1967).

115. Fontfroide, France; Rognac Formation, Maastrichtian: eggshells (Penner, 1985).

116. St. André-de-Roquelongue, France; unnamed formation, Maastrichtian: eggs (Lapparent, 1967).

117. Vitrolles (= Velaux), France; Rognac Formation, Maastrichtian: clutches (Cousin, Breton, and Gomez, 1987).

118. Albas, France; unnamed formation, Maastrichtian: eggs (Lapparent, 1967).

119. Quintanilla del Coco, Spain; Calizas de Lychnus Formation, Maastrichtian: eggshells (Moratalla, 1992).

120. Suterranya, Spain; Tremp Formation, Maastrichtian: *Megaloolithus pseudomamillare* eggshells and clutches (Vianey-Liaud and Lopez-Martinez, 1997).

121. Moro, Spain; Arenisca de Arén Formation, Maastrichtian: *Megaloolithus petralta* eggshells (Vianey-Liaud and Lopez-Martinez, 1997).

122. Bastus, Spain; Arenisca de Arén Formation, Maastrichtian: *Ageroolithus fontllongensis* eggshell, *Megaloolithus mamillare* clutches (Vianey-Liaud and Lopez-Martinez, 1997).

123. Abella, Spain; Tremp Formation, Maastrichtian: *Megaloolithus mamillare* eggshells (Vianey-Liaud and Lopez-Martinez, 1997).

124. Biscarri, Spain; Tremp Formation, Maastrichtian: *Megaloolithus sirguei* eggshells and clutches (Vianey-Liaud and Lopez-Martinez, 1997).

125. Fontllonga 6, Spain; Arenisca de Arén Formation, Maastrichtian: *Megaloolithus petralta, Megaloolithus* cf. *M. aureliensis, Prismatoolithus tenuis, Prismatoolithus* aff. *P. matellensis, Pseudogeckoolithus nodosus, Ageroolithus fontllongensis* eggshells (Vianey-Liaud and Lopez-Martinez, 1997).

126. Pietraroia, Italy; Pietraroia Plattenkalk, Albian: *Scipionyx samniticus* baby (Sasso and Signore, 1998).

127. Tustea, Romania; Densuş-Ciula Formation, Maastrichtian: sauropod clutch (Grigorescu and others, 1994).

128. Daugyztai, Uzbekistan; unnamed formation, Upper Cretaceous: clutch(?) (Nessov and Kaznyshkin, 1986).

129. Central Kazakhstan; unnamed formation, Upper Cretaceous(?): eggs (Nessov and Kaznyshkin, 1986).

130. Shakhaftan, Kyrgyzstan; Nichkesaisk Formation, Santonian-Campanian: eggshells (Nessov and Kaznyshkin, 1986).

131. Arslanbob, Kyrgyzstan; Ialovatch Formation, Santonian: *Ovaloolithus chinkangkouensis* eggshells (Nessov and Kaznyshkin, 1986).

132. Charvak, Kyrgyzstan; Ialovachsk Formation, Turonian-Santonian: eggshells (Nessov and Kaznyshkin, 1986).

133. Tash Kumyr, Kyrgyzstan; Ialovachsk and Nichkesaisk series, Turonian-Campanian: Spheroolithidae, Ovaloolithidae, Protoceratopsidovum, and Elongatoolithidae eggs and eggshells (Nessov and Kaznyshkin, 1986).

134. Chimket, Kyrgyzstan; unnamed formation, Upper Cretaceous(?): eggshells (Nessov and Kaznyshkin, 1986).

135. Tayzhuzgen River, Kazakhstan; Manrakskaja Formation, Upper Cretaceous: *Macroolithus rugustus* eggshells and *Elongatoolithus* sp. eggs (Bazhanov, 1961; Sochava, 1971; Mikhailov, 1994a).

136. Mirakheri, State of Gujarat, India; Lameta Formation, Maastrichtian: *Megaloolithus* eggs (Mohabey and Mathur, 1989).

137. Dholidhanti, State of Gujarat, India; Lameta Formation, Maastrichtian: *Megaloolithus phensaniensis, Megaloolithus matleyi,* and *Megaloolithus megadermus* clutches, eggs (Mohabey, 1998).

138. Paori, State of Gujarat, India; Lameta Formation, Maastrichtian: *Megaloolithus phensaniensis,* and *Megaloolithus megadermus* clutches (Mohabey, 1998).

139. Waniawao, State of Gujarat, India; Lameta Formation, Maastrichtian: *Megaloolithus phensaniensis, Megaloolithus matleyi, Megaloolithus mohabeyi* eggshells and clutches (Khosla and Sahni, 1995; Mohabey, 1998).

140. Bagh, State of Gujarat, India; Lameta Formation, Maastrichtian: *Megaloolithus jabalapurensis, Megaloolithus dhoridungriensis* and *Megaloolithus baghensis* eggshells and clutches (Khosla and Sahni, 1995; Mohabey, 1998).

141. Anjar, State of Gujarat, India; Lameta Formation, Maastrichtian: *Megaloolithus baghensis* and *Subtiliolithus kachchhensis* eggshells and clutches (Khosla and Sahni, 1995).

142. Dhoridungri, State of Gujarat, India; Lameta Formation, Maastrichtian: *Megaloolithus dhoridungriensis* and *Megaloolithus matleyi* eggshells (Mohabey, 1998).

143. Khempur, State of Gujarat, India; Lameta Formation, Maastrichtian: *Megaloolithus rahioliensis, Megaloolithus khempurensis,* and *Megaloolithus cylindricus* eggshells and clutches (Khosla and Sahni, 1995; Mohabey, 1998).

144. Daulatpoira, State of Gujarat, India; Lameta Formation, Maastrichtian: *Megaloolithus megadermus* clutches (Mohabey, 1998).

145. Kevadiya, State of Gujarat, India; Lameta Formation, Maastrichtian: *Megaloolithus rahioliensis* clutches (Mohabey, 1998).

146. Rahioli, State of Gujarat, India; Lameta Formation, Maastrichtian: *Megaloolithus rahioliensis* clutches, and *Ellipsoolithus khedaensis* egg (Mohabey, 1998).

147. Phensani, State of Gujarat, India; Lameta Formation, Maastrich-

tian: *Megaloolithus phensaniensis* and *Megaloolithus mohabeyi* eggshells and clutches (Khosla and Sahni, 1995; Mohabey, 1998).

148. Dhuvedia, State of Gujarat, India; Lameta Formation, Maastrichtian: *Megaloolithus* clutches (Srivastava and others, 1986).

149. Werasa, State of Gujarat, India; Lameta Formation, Maastrichtian: *Megaloolithus khempurensis* clutches (Mohabey, 1998).

150. Balasinor, State of Gujarat, India; Lameta Formation, Maastrichtian: *Megaloolithus phensaniensis, Megaloolithus balasinorensis, Megaloolithus mohabeyi,* and *Megaloolithus baghensis* eggshells and clutches (Khosla and Sahni, 1995; Mohabey, 1998).

151. Sonipur, State of Gujarat, India; Lameta Formation, Maastrichtian: *Megaloolithus phensaniensis* clutches (Mohabey, 1998).

152. Jabalapur, State of Madhya Pradesh, India; Lameta Formation, Maastrichtian: *Megaloolithus cylindricus, Megaloolithus rahioliensis, Megaloolithus matleyi,* and *Megaloolithus jabalapurensis* eggshells and clutches (Khosla and Sahni, 1995; Mohabey, 1998).

153. Lametaghat, State of Madhya Pradesh, India; Lameta Formation, Maastrichtian: *Megaloolithus baghensis* clutches (Sahni and others, 1994).

154. Walpur, State of Madhya Pradesh, India; Lameta Formation, Maastrichtian: *Megaloolithus walpurensis* and *Megaloolithus cylindricus* shell (Khosla and Sahni, 1995).

155. Dholiya, State of Madhya Pradesh, India; Lameta Formation, Maastrichtian: *Megaloolithus cylindricus, Megaloolithus dholiyaensis,* and *Megaloolithus mohabeyi* shell (Khosla and Sahni, 1995).

156. Padiyal, State of Madhya Pradesh, India; Lameta Formation, Maastrichtian: *Megaloolithus jabalapurensis* and *Megaloolithus padiyalensis* eggshells (Khosla and Sahni, 1995).

157. Takli, State of Maharashtra, India; Lameta Formation, Maastrichtian: *Megaloolithus* eggshells (Sahni and others, 1994).

158. Piraya, State of Maharashtra, India; Lameta Formation, Maastrichtian: *Megaloolithus* eggshells (Sahni and others, 1994).

159. Tidakepar, State of Maharashtra, India; Lameta Formation, Maastrichtian: *Spheroolithus* sp. eggshells (Mohabey, 1996).

160. Pisdura, State of Maharashtra, India; Lameta Formation, Maastrichtian: *Megaloolithus baghensis* and *Spheroolithus* sp. eggshells (Mohabey, 1996).

161. Pavna, State of Maharashtra, India; Lameta Formation, Maastrichtian: *Megaloolithus matleyi* and *Spheroolithus* sp. eggshells and clutches (Mohabey, 1996, 1998).

162. Asifabad, Andhra Pradesh, India; Lameta Formation, Maastrichtian: *Megaloolithus* eggshells (Sahni and others, 1994).

163. Dongargaon, State of Maharashtra, India; Lameta Formation, Maastrichtian: *Spheroolithus* sp. eggshells (Mohabey, 1996).

164. Polgaon, State of Maharashtra, India; Lameta Formation, Maastrichtian: *Spheroolithus* sp. eggshells (Mohabey, 1996).

165. Nand, State of Maharashtra, India; Lameta Formation, Maastrichtian: *Spheroolithus* sp. eggshells (Mohabey, 1996).

166. Ariyalur, State of Tamil Nadu, India; Kallankurichi Formation, Maastrichtian: *Megaloolithus* sp. (Kohring and others, 1996).

167. Shiregin Gashun, Mongolia; Bayn Shiren Formation, Cenomanian-Campanian: *Spheroolithus irenensis* clutch (Mikhailov 1994b, as Nemegt Formation).

168. Khermeen Tsav I and II, Mongolia; Barun Goyot and Nemegt Formations, Santonian-Campanian, and Maastrichtian: *Ingenia yanshini* baby bone; *Bagaceratops rozhedestvenskyi* baby bones; *Protoceratopsi-*

dovum sincerum, Protoceratopsidovum fluxuosum clutches, *Faveoloolithus ningxiaensis* eggshell, *Dendroolithus microporosus* eggshell, *Dendroolithus verrucarius, Macroolithus rugustus* eggshells and clutches, Elongatoolithidae eggshells (Maryanska and Osmolska, 1975; Barsbold and Perle, 1983; Mikhailov 1994a, b; Mikhailov and others, 1994).

169. Bambu Khudu, Mongolia; Barun Goyot Formation, Campanian: eggshells (Nessov and Kaznyshkin, 1986).

170. Undurshil Ula, Mongolia; Barun Goyot Formation, Santonian-Campanian: *Elongatoolithus* sp. eggs (Mikhailov and others, 1994).

171. Udan Sayr, Mongolia; Djadokhta Formation, Campanian: *Elongatoolithus subtitectorius* and *Ovaloolithus dinornithoides* eggshells (Mikhailov, 1994b).

172. Toogreek, Mongolia; Djadokhta Formation, Campanian: *Protoceratopsidovum sincerum, Protoceratopsidovum minimum, Elongatoolithus frustrabilis* clutches, protoceratopsid and hadrosaur baby bones (Barsbold and Perle, 1983; Mikhailov, 1994a).

173. Ologoy Ulan Tsav, Mongolia; Barun Goyot Formation, Campanian: *Faveoloolithus ningxiaensis* clutch (Mikhailov, 1994b).

174. Gurlin Tsav, Mongolia; Nemegt Formation, Maastrichtian: *Spheroolithus irenensis* eggshell, *Elongatoolithus sigillarius, Macroolithus rugustus* eggshells and clutch (Mikhailov, 1994b).

175. Khaichin Ula I, Mongolia; Nemegt Formation, Maastrichtian: *Macroolithus rugustus* and *Subtiliolithus* sp. eggshells and clutches (Mikhailov, 1994a).

176. Tsagan Khushu, Mongolia; Nemegt Formation, Maastrichtian: *Elongatoolithus excellens, Ovaloolithus dinornithoides* eggshells (Mikhailov, 1994b).

177. Altan Ula, Mongolia; Nemegt Formation, Maastrichtian: Elongatoolithidae clutches and Protoceratopsidae eggshells; *Ovaloolithus dinornithoides* (Sabath, 1991; Mikhailov, 1994a).

178. Naran Bulak, Mongolia; Nemegt Formation, Maastrichtian: eggshells (Nessov and Kaznyshkin, 1986).

179. Nemegt, Mongolia; Nemegt Formation, Maastrichtian: *Laevisoolithus* sp. eggs (Mikhailov and others, 1994).

180. Dzamyu Khond, Mongolia; Djadokhta Formation, Campanian: *Protoceratops andrewsi* baby skeletons (Barsbold and Perle, 1983).

181. Ukhaa Tolgod, Mongolia; Djadokhta Formation, Campanian: *Oviraptor* with *Elongatoolithus* eggs and embryos (Norell and others, 1994, 1995).

182. Bayn Dzak, Mongolia; Djadokhta Formation, Campanian: *Elongatoolithus frustrabilis, Spheroolithus maiasauroides, Protoceratopsidovum sincerum* clutches; *Oviraptor philoceratops* on eggs, *Protoceratops andrewsi* baby skeletons, and *Archaeornithoides deinosauriscus* baby bones (Osborn, 1924; Maryanska and Osmolska, 1975; Elzanowski and Wellnhofer, 1992; Mikhailov 1994a, b).

183. Gilbent, Mongolia; Barun Goyot Formation, Campanian: eggs (Mikhailov and others, 1994).

184. Khulsan, Mongolia; Barun Goyot, Campanian: *Dendroolithus microporosus* eggshell, *Dendroolithus verrucarius* eggshells and clutch, *Protoceratopsidovum fluxuosum* clutches, *Breviceratops kozlowskii* baby bones (Maryanska and Osmolska, 1975; Mikhailov, 1994a, b).

185. Bugin Tsav, Mongolia; Nemegt Formation, Maastrichtian: *Laevisoolithus* sp. eggs (Mikhailov and others, 1994).

186. Baynshin Tsav, Mongolia; Bayn Shiren Formation, Cenomanian-Campanian: eggshells (Nessov and Kaznyshkin, 1986).

187. Shiljust Ula, Mongolia; Barun Goyot Formation, Santonian-Campanian: *Dendroolithus microporosus* clutch; *Spheroolithus tenuicorticus* eggshells (Mikhailov, 1994b).

188. Ikh Shunkht, Mongolia; Barun Goyot Formation, Campanian: *Protoceratopsidovum sincerum, Protoceratopsidovum minimum, Macroolithus mutabilis, Faveoloolithus ningxiaensis, Dendroolithus verucarius,* and *Spheroolithus* sp. eggshells and clutches (Mikhailov, 1994a, b).

189. Bagamod Khuduk, Mongolia; Barun Goyot Formation, Campanian: *Protoceratopsidovum minimum* clutches (Mikhailov, 1994b).

190. Khara Khutul, Mongolia; unnamed formation, Cenomanian-Turonian: eggshells (Nessov and Kaznyshkin, 1986).

191. Tel Ulan Ula, Mongolia; Bayn Shiren Formation, Cenomanian-Santonian: eggshells (Nessov and Kaznyshkin, 1986).

192. Darigana, Mongolia; Bayn Shiren Formation, Cenomanian-Campanian: *Ovaloolithus chinkangkouensis* eggs (Mikhailov and others, 1994).

193. Ekhin Tukhum (= Ikh Eren?), Mongolia; Bayn Shiren Formation, Cenomanian-Santonian: eggshells (Nessov and Kaznyshkin, 1986).

194. Moyogn Ulagiyn Khaets (= Mogoin Bulak?), Mongolia; Bayn Shiren Formation, Cenomanian-Campanian: *Ovaloolithus* cf. *O. chinkangkouensis* clutches with hadrosaur baby bones (Mikhailov, 1994b).

195. Baga Tarjach, Mongolia; Djadokhta Formation, Santonian-Campanian: *Protoceratopsidovum minimum* and *Spheroolithus maiasauroides* clutches (Mikhailov, 1994a, b).

196. Ulungar, China; Donggou(?) Formation, Maastrichtian(?): eggs (Russell, unpublished notes).

197. Shenjinkou, China; Subash Formation, Maastrichtian: *Paraspheroolithus* sp. and *Elongatoolithus* sp. eggs (Hao and Guan, 1984).

198. Qitai, China; Subashi Formation, Upper Cretaceous: *Elongatoolithus andrewsi* eggs (Young, 1965).

199. Tsondolein-Khuduk (= Ondor Mod?), China; Minke Formation, Upper Cretaceous: *Microceratops gobiensis* baby bones (Bohlin, 1953).

200. Alxa, China; unnamed formation, Upper Cretaceous: eggshells (Zhao and Ding, 1976).

201. Bayan Mandahu, China; Djadokhta Formation, Campanian: *Prismatoolithus gebiensis* clutch; *Oviraptor* on eggs; *Protoceratops andrewsi* baby bones and *Pinacosaurus grangeri* baby skeletons (Dong and others, 1989; Zhao and Li, 1993; Dong and Currie, 1996).

202. Jiangjunmiao, China; Hongshaguan Formation, Maastrichtian: "*Protoceratops*" and *Paraspheroolithus* cf. *P. irenensis* clutches (Young, 1965).

203. Erenhot, China; Iren Dabasu Formation, Campanian(?): *Spheroolithus* cf. *S. irenensis* clutch (Dong and others, 1989).

204. Changchun, China; Quantou Formation, Cretaceous: Spheroolithidae eggs (Zhao, personal communication).

205. Quantou, China; Quantou Formation, Cretaceous: Spheroolithidae eggs (Zhao, personal communication).

206. Chiangchungting, China; Jiangjunding Formation, Campanian(?): *Spheroolithus chiangchiungtingensis* clutches (Chao and Chiang, 1974).

207. Laiyang, China; Jiangjunding Formation, Campanian(?): *Spheroolithus chiangchiungtingensis, ?S. megadermus, Ovaloolithus laminadermus, O. chinkangkouensis,* and *Paraspheroolithus irenensis* eggs (Chao and Chiang, 1974).

208. Chaochun, China; Jingangkou Formation, Upper Cretaceous: *Elongatoolithus elongatus, Spheroolithus* sp., and *Ovaloolithus* sp. clutches (Chow, 1951).

209. Chinkangkou, China; Jingangkou Formation, Upper Cretaceous: eggs (Young, 1965).

210. Jiaozhou, China; Jingangkou Series, Cretaceous: Spheroolithidae, Ovaloolithidae and Elongatoolithidae eggs (Zhao, personal communication).

211. Zhucheng, China; Jingangkou Formation, Upper Cretaceous: Spheroolithidae, Ovaloolithidae and Elongatoolithidae eggs (Zhao, personal communication).

212. Xuanzhou, China; Xuannan Formation, Upper Cretaceous: Elongatoolithidae eggs (Zhao, personal communication).

213. Tiantai, China; Laija B Formation, Upper Cretaceous: Faveoloolithidae and Spheroolithidae eggs (Mateer, 1989; Zhao, personal communication).

214. Gaotangshi, China; Quxian Formation, Upper Cretaceous: clutches (Mateer, 1989).

215. Quzhou, China; Qujiang Group, Upper Cretaceous: Faveoloolithidae and Spheroolithidae eggs (Zhao, personal communication).

216. Anwen, China; Xiaoyan Formation, Upper Cretaceous: eggs (Dong, 1980).

217. Yixing, China; unnamed formation, Cretaceous: Elongatoolithidae eggs (Zhao, personal communication).

218. Anqing, China; unnamed formation, Cretaceous: Spheroolithidae eggs (Zhao, personal communication).

219. Anlu, China; Gong An Zhai Formation, Upper Cretaceous: *Dendroolithus wangdianensis* eggs (Zhao and Li, 1988).

220. Shanyang Basin, China; Shanyang Formation, Maastrichtian: Spheroolithidae and Elongatoolithidae eggs (Zhao, personal communication).

221. Lingbao, China; Nanzhao Formation, Upper Cretaceous: Elongatoolithidae eggs (Zhao, personal communication).

222. Liguanqiao Basin, China; Hugang Formation, Upper Cretaceous: Elongatoolithidae eggs (Zhao, personal communication).

223. Xinyang, China; unnamed formation, Upper Cretaceous: Elongatoolithidae eggs (Zhao, personal communication).

224. Xiaguan Basin, China; unnamed formation, Cretaceous: *Youngoolithus xiaguanensis* clutch (Zhao, 1979b).

225. Yunxian, China; unnamed formation, Cretaceous: Dendroolithidae eggs (Zhao, personal communication).

226. Taohe Basin, China; Majiachun Formation, Cretaceous: Dendroolithidae and Dictyoolithidae eggs (Zhao, personal communication).

227. Nanxiong, China; Yuanpu and Pingling formations, Campanian-Maastrichtian: *Macroolithus yaotunensis, M. rugustus, M.* sp. nov., *Elongatoolithus andrewsi, E. elongatus, E.* sp. nov., *Nanshiungoolithus chuetienensis, Ovaloolithus* cf. *O. chinkangkouensis, Ovaloolithus* cf. *O. laminadermus, O.* sp. nov., *Shixingoolithus erbeni,* and *Stromatoolithus pinglingensis* clutches and eggs (Zhao and others, 1991).

228. Taoyuan and Dongting Basin, China; Fenshui'ao Formation, Maastrichtian: *Phaceloolithus hunanensis, Elongatoolithus magnus,* and *E.* sp. clutches (Zeng and Zhang, 1979; Zhao, personal communication).

229. Chaling, China; Daijiaping Formation, Maastrichtian: *Elongatoolithus andrewsi* eggs; and Elongatoolithidae eggs (Young, 1965; Zhao, personal communication).

230. Taihe-Ganzhou area, China; Yuanpu Formation, Maastrichtian: *Macroolithus* sp. and *Nanhsiungoolithus chuetienensis* eggs (Young, 1965).

231. Gao'an, China; Qingfengqiao Formation, Upper Cretaceous: Elongatoolithidae eggs (Zhao, personal communication).

232. Xinyu, China; Qingfengqiao Formation, Upper Cretaceous: Elongatoolithidae eggs (Zhao, personal communication).

233. Heyuan, China; Nanxiong Group, Upper Cretaceous: Elongatoolithidae eggs (Zhao, personal communication).

234. Huizhou, China; Nanxiong Group, Upper Cretaceous: Spheroolithidae eggs (Zhao, personal communication).

235. Shiguguan, China; Hugang Formation, Upper Cretaceous: Elongatoolithidae eggs (Zhao, 1979b; personal communication).

236. Tsatzeyuanhsu, China; unnamed formation, Upper Cretaceous: eggshells (Young, 1965).

237. Guangzhou, China; Sanshui Formation, Cretaceous: Elongatoolithidae and Spheroolithidae eggs (Zhao, personal communication).

Appendix II. Dinosaur Egg Types

This is great stuff to know for a cocktail party because it drives people away from you so you can hog all those cute little hors d'oeuvres. Anyway, this guide should help in the identification of dinosaur eggs and eggshell (I also threw in Cretaceous bird eggshell). I have tried to translate the really technical descriptions into more readable sentences, but there still remain all those awful technical terms discussed in chapters 8 and 9. A list of dinosaur and Cretaceous bird egg taxa is presented in Table AII.1.

I have not illustrated every egg species because the most noticeable differences occur at the genus level. Many species of a genus are separated on subtle differences or by thickness alone. I have, however, tried to illustrate most egg genera, especially in thin section (the most useful view). Some of these are photographs of specimens, but others are drawings. These drawings are schematic, even when traced from photographs, and are meant to convey a sense of the shell in thin section. Most horizontal lines in these drawings represent growth lines, and not every growth line was drawn in so as not to clutter the figure. (So don't count them!) Vertical lines usually represent the shell units; these may be accurately drawn in (usually irregular). Structures seen within the shell units or mammillae may be stylized (usually as straight lines, sometimes in a slightly radiating pattern).

Keep in mind that it is not always possible to identify every egg or eggshell fragment. Quality of preservation (alteration, weathering, etc., as discussed in chapter 8) can greatly affect the eggshell, making it difficult or impossible to separate between two different genera. The most difficult to identify are those taxa that may have faint shell units but distinct growth rings, such as in ovaloolithid, spheroolithid, and prismatic eggshell. These eggshells can look ornithoid (i.e., avian with a continuous layer). Staining or difference in mineralization can make these eggshells appear to have a separate mammillary layer (which is really not present in ovaloolithid, spheroolithid, or prismatic shell). However, a polarizing microscope can help distinguish a true mammillary layer from a wanna-be mammillary layer. Finally, avoid identifying an egg by shape alone, or by one character. As you will see below, the eggs of *Faveoloolithus, Dictyoolithus,* and a small *Megaloolithus* may be spherical (or nearly so) and have a bumpy surface. Separating them requires a look at the edge of the eggshell with at least a good hand lens, but preferably with a microscopic thin section.

TABLE AII.1.
Alphabetized parataxonomic list of dinosaur eggshells ("incertae sedis" means "uncertain" or "unknown")

I. Dinosauroid-Spherulitic Basic Shell Type
- A. DENDROOLITHIDAE Zhao and Li, 1988
  - *Dendroolithus* Zhao and Li, 1988
    - *Dendroolithus wangdianensis* Zhao and Li, 1988; ?Lower and Upper Cretaceous, China
    - *Dendroolithus verrucarius* Mikhailov, 1994b; Upper Cretaceous, Mongolia
    - *Dendroolithus microporosus* Mikhailov, 1994b; Upper Cretaceous, Mongolia
- B. DICTYOOLITHIDAE Zhao, 1994
  - *Dictyoolithus* Zhao, 1994
    - *Dictyoolithus hongpoensis* Zhao, 1994; Upper Cretaceous, China
    - *Dictyoolithus neixiangensis* Zhao, 1994; Upper Cretaceous, China
  - *Stromatoolithus* Zhao, Ye, Li, Zhao, and Yan, 1991
    - *Stromatoolithus pinglingensis* Zhao, Ye, Li, Zhao, and Yan, 1991; Upper Cretaceous, China
- C. FAVEOLOOLITHIDAE Zhao and Ding, 1976
  - *Faveoloolithus* Zhao and Ding, 1976
    - *Faveoloolithus ningxiaensis* Zhao and Ding, 1976; ?Lower and Upper Cretaceous, China and Mongolia
  - *Youngoolithus* Zhao, 1979b
    - *Youngoolithus xiaguanensis* Zhao, 1979b; Upper Cretaceous, China
- D. MEGALOOLITHIDAE Zhao, 1979a
  - *Cairanoolithus* Vianey-Liaud, Mallan, Buscail, and Montgelard, 1994
    - *Cairanoolithus dughii* Vianey-Liaud, Mallan, Buscail, and Montgelard, 1994; Upper Cretaceous, France
  - *Dughioolithus* Vianey-Liaud, Mallan, Buscail, and Montgelard, 1994
    - *Dughioolithus roussentensis* Vianey-Liaud, Mallan, Buscail, and Montgelard, 1994; Upper Cretaceous, France
  - *Megaloolithus* Vianey-Liaud, Mallan, Buscail, and Montgelard, 1994
    - *Megaloolithus aureliensis* Vianey-Liaud, Mallan, Buscail, and Montgelard, 1994; Upper Cretaceous, France
    - *Megaloolithus baghensis* Khosla and Sahni, 1995; Upper Cretaceous, India
    - *Megaloolithus balasinorensis* Mohabey, 1998; Upper Cretaceous, India
    - *Megaloolithus cylindricus* Khosla and Sahni, 1995; Upper Cretaceous, India
    - *Megaloolithus dholiyaensis* Khosla and Sahni, 1995; Upper Cretaceous, India
    - *Megaloolithus dhoridungriensis* Mohabey, 1998; Upper Cretaceous, India
    - *Megaloolithus jabalpurensis* Khosla and Sahni, 1995; Upper Cretaceous, India
    - *Megaloolithus khempurensis* Mohabey, 1998; Upper Cretaceous, India
    - *Megaloolithus mammilare* Vianey-Liaud, Mallan, Buscail, and Montgelard, 1994; Upper Cretaceous, France
    - *Megaloolithus matleyi* Mohabey, 1996; Upper Cretaceous, India
    - *Megaloolithus megadermus* Mohabey, 1998; Upper Cretaceous, India
    - *Megaloolithus mohabeyi* Khosla and Sahni, 1995; Upper Cretaceous, India
    - *Megaloolithus padiyalensis* Khosla and Sahni, 1995; Upper Cretaceous, India
    - *Megaloolithus petralta* Vianey-Liaud, Mallan, Buscail, and Montgelard, 1994; Upper Cretaceous, France
    - *Megaloolithus phensaniensis* Mohabey, 1998; Upper Cretaceous, India
    - *Megaloolithus problematica* Mohabey, 1998; Upper Cretaceous, India
    - *Megaloolithus rahioliensis* Mohabey, 1998; Upper Cretaceous, India
    - *Megaloolithus siruguei* Vianey-Liaud, Mallan, Buscail, and Montgelard, 1994; Upper Cretaceous, France
    - *Megaloolithus trempii* Vianey-Liaud, Mallan, Buscail, and Montgelard, 1994; Upper Cretaceous, France
    - *Megaloolithus walpurensis* Khosla and Sahni, 1995; Upper Cretaceous, India
  - *Sphaerovum* Mones, 1980
    - *Sphaerovum erbeni* Mones, 1980; Upper Cretaceous, Uruguay
- E. OVALOOLITHIDAE Mikhailov, 1991
  - *Ovaloolithus* Zhao, 1979a (= *"Oolithes spheroides"* Young, 1954)
    - *Ovaloolithus chinkangkouensis* (Zhao and Jiang, 1974); Upper Cretaceous, China and Mongolia
    - *Ovaloolithus dinornithoides* Mikhailov, 1994b; Upper Cretaceous, Mongolia

Ovaloolithus laminadermus (Zhao and Jiang, 1974); Upper Cretaceous, China
Ovaloolithus mixtistriatus Zhao, 1979a; Upper Cretaceous, China
Ovaloolithus monostriatus Zhao, 1979a; Upper Cretaceous, China
Ovaloolithus tristriatus Zhao, 1979a; Upper Cretaceous, China

F. SPHEROOLITHIDAE Zhao, 1979a

Shixingoolithus Zhao, Ye, Li, Zhao, and Yan, 1991

Shixingoolithus erbeni Zhao, Ye, Li, Zhao, and Yan, 1991; Upper Cretaceous, China

Spheroolithus Zhao, 1979a (= "*Oolithes spheroides*" Young, 1954)

Spheroolithus albertensis Zelenitsky and Hills, 1997; Upper Cretaceous, Canada and Montana
Spheroolithus chiangchiungtingensis (Zhao and Jiang, 1974); Upper Cretaceous, China
Spheroolithus (= Paraspheroolithus) irenensis (Zhao and Jiang, 1974); Upper Cretaceous, China and Mongolia
Spheroolithus maiasauroides Mikhailov, 1994b; Upper Cretaceous, Mongolia
Spheroolithus megadermus (Young, 1959); Upper Cretaceous, China
Spheroolithus tenuicorticus Mikhailov, 1994b; Upper Cretaceous, Mongolia

II. Prismatic Basic Shell Type

A. PRISMATOOLITHIDAE Hirsch, 1994

Preprismatoolithus Zelenitsky and Hills, 1996

Preprismatoolithus coloradensis (Hirsch, 1994); Upper Jurassic, Colorado

Prismatoolithus Zhao and Li, 1993

Prismatoolithus gebiensis Zhao and Li, 1993; Upper Cretaceous, China
Prismatoolithus levis Zelenitsky and Hills, 1996; Upper Cretaceous, Canada and Montana, USA
Prismatoolithus matellensis Vianey-Liaud and Crochet, 1993; Upper Cretaceous, France
Prismatoolithus tenuius Vianey-Liaud and Crochet, 1993; Upper Cretaceous, France

Protoceratopsidovum Mikhailov, 1994a

Protoceratopsidovum fluxuosum Mikhailov, 1994a; Upper Cretaceous, Mongolia
Protoceratopsidovum minimum Mikhailov, 1994a; Upper Cretaceous, Mongolia
Protoceratopsidovum sincerum Mikhailov, 1994a; Upper Cretaceous, Mongolia

B. OOFAMILY INCERTAE SEDIS

Pseudogeckoolithus Vianey-Liaud and Lopez-Martinez, 1997

Pseudogeckoolithus nodosus Vianey-Liaud and Lopez-Martinez, 1997; Upper Cretaceous, Spain

III. Ornithoid Basic Shell Type

A. ELONGATOOLITHIDAE Zhao, 1975

Ellipsoolithus Mohabey, 1998

Ellipsoolithus khedaensis Mohabey, 1998; Upper Cretaceous, India

Elongatoolithus Zhao, 1975

Elongatoolithus andrewsi Zhao, 1975; Upper Cretaceous, China
Elongatoolithus elongatus (Young, 1954); Upper Cretaceous, China
Elongatoolithus excellens Mikhailov, 1994a; Upper Cretaceous, Mongolia
Elongatoolithus frustrabilis Mikhailov, 1994a; Upper Cretaceous, Mongolia
Elongatoolithus magnus Zeng and Zhang, 1979; Upper Cretaceous, China
Elongatoolithus sigillarius Mikhailov, 1994a; Upper Cretaceous, Mongolia
Elongatoolithus subtitectorius Mikhailov, 1994a; Upper Cretaceous, Mongolia

Macroelongatoolithus Li, Yin, and Liu, 1995

Macroelongatoolithus xixia Li, Yin, and Liu, 1995; Lower Cretaceous, China
Macroelongatoolithus carlylei (Jensen, 1970); Lower and Middle Cretaceous, Utah

Macroolithus Zhao, 1975

Macroolithus mutabilis Mikhailov, 1994a; Upper Cretaceous, Mongolia
Macroolithus rugustus (Young, 1965); Upper Cretaceous, China
Macroolithus yaotunensis Zhao, 1975; Upper Cretaceous, China

Nanhsiungoolithus Zhao, 1975

Nanhsiungoolithus chuetienensis Zhao, 1975; Upper Cretaceous, China

Trachoolithus Mikhailov, 1994a

Trachoolithus faticanus Mikhailov, 1994a; Upper Cretaceous, Mongolia

B. OBLONGOOLITHIDAE Mikhailov, 1996

Oblongoolithus Mikhailov, 1996

Oblongoolithus glaber Mikhailov, 1996; Upper Cretaceous, Mongolia

C. PHACELOOLITHIDAE Zeng and Zhang, 1979?

Phaceloolithus Zeng and Zhang, 1979?

Phaceloolithus hunanensis Zeng and Zhang, 1979?; Upper Cretaceous, China

D. LAEVISOOLITHIDAE Mikhailov, 1991

Laevisoolithus Mikhailov, 1991

Laevisoolithus sochavai Mikhailov, 1991; Upper Cretaceous, Mongolia

Subtiliolithus Mikhailov, 1991

Subtiliolithus kachchhensis Khosla and Sahni, 1995; Upper Cretaceous, India

Subtiliolithus microtuberculatus Mikhailov, 1991; Upper Cretaceous, Mongolia

E. OOFAMILY INCERTAE SEDIS

Ageroolithus Vianey-Liaud and Lopez-Martinez, 1997

Ageroolithus fontllongensis Vianey-Liaud and Lopez-Martinez, 1997; Upper Cretaceous, Spain

Continuoolithus Zelenitsky, Hills, and Currie, 1996

Continuoolithus canadensis Zelenitsky, Hills, and Currie, 1996; Upper Cretaceous, Canada

Dispersituberoolithus Zelenitsky, Hills, and Currie, 1996

Dispersituberoolithus exilis Zelenitsky, Hills, and Currie, 1996; Upper Cretaceous, Canada

Porituberoolithus Zelenitsky, Hills, and Currie, 1996

Porituberoolithus warnerensis Zelenitsky, Hills, and Currie, 1996; Upper Cretaceous, Canada

Tristraguloolithus Zelenitsky, Hills, and Currie, 1996

Tristraguloolithus craciodes Zelenitsky, Hills, and Currie, 1996; Upper Cretaceous, Canada

If you can identify your specimen to the genus you are doing really well, and if to species, congratulations! You are an expert!

I. Dinosauroid-Spherulitic Basic Shell Type

Eggshells having distinct shell units with a radial-tabular ultrastructure.

DENDROOLITHIDAE Zhao and Li, 1988

These small to medium-size (6–20 cm, 2¼–8") eggs are spherical to slightly oval, and in a nest are arranged irregularly in a single layer (Fig. AII.1). The eggshell has a dendrospherulitic morphotype, meaning that the shell units in thin section are irregularly shaped, branching fans. Three-dimensionally, the shell units are irregular, branching columns like a tree (which is what gave the eggshell its name). The shell units are interlocked near the surface where they may fuse together. The pores are widely scattered over the surface. The pore canals are prolatocanaliculate and form a network between and within the shell units. Smaller canals form a network among the shell units in the upper part of the shell. The shell units actually form the rough and irregular ornamentation on the surface. These eggs contain the spectacular embryos thought to be of segnosaurs.

Dendroolithus *Zhao and Li, 1988*

The characters that define this egg genus are the same as those that define the family. The reason is that there is only one genus in the family. Three species have been named based on egg shape, egg size, surface texture, and shell thickness.

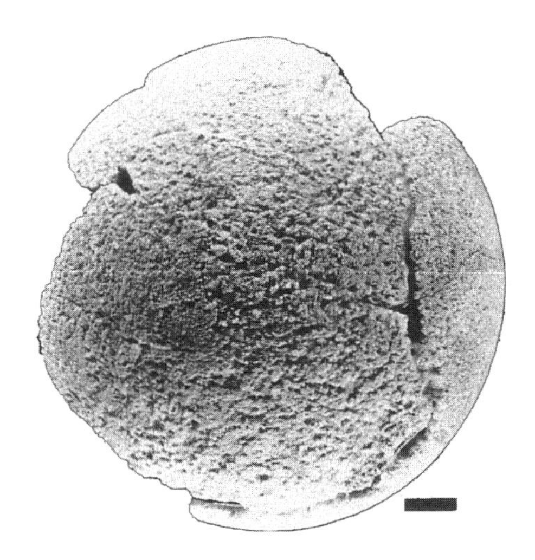

Fig. AII.1. Dendroolithus microporosus *egg (left) and thin section (right—polarized light). Scale = 1 cm. (Courtesy of K. Mikhailov.)*

Dendroolithus wangdianensis Zhao and Li, 1988. The eggs are oval, not spherical, and slightly pointed at one end. They are 14.5–16.2 cm (5¾–6½") long and 11–13 cm (4¼–5") in diameter. The shell is thin, 1.7–2.1 mm thick, of which about a third is the mammillary layer. Clutch consists of at least eleven eggs. From the Upper Cretaceous Gong An Zhai Formation of northern China.

Dendroolithus verrucarius Mikhailov, 1994b. Known only from shell fragments. The surface ornamentation is nodose. The shell is thicker than in *D. wangdianensis* and may range from 1.8–3.8 mm in the same egg. There is a second layer of spherulitic shell units that occupy $^1/_5$–$^1/_7$ the shell thickness. Collected from the Upper Cretaceous Barun Goyot Formation of Mongolia.

Dendroolithus microporosus Mikhailov, 1994b. The eggs are 7 cm (2 ¾") long, and 6 cm (2¼") in diameter, and with a shell 2–2.7 mm thick (Fig. AII.1). Unlike the other two species, the surface is smooth. From the Upper Cretaceous Barun Goyot Formation of Mongolia.

DICTYOOLITHIDAE Zhao, 1994

Dictyoolithid eggs are oval to spheroidal, and with a surface that looks grainy due to very small nodes or bumps formed by the tops of the shell units. The units are irregular and branching, much like *Dendroolithus,* although with more branching prolatocanaliculate pore canals that are intertwined between the shell units.

Dictyoolithus *Zhao, 1994*

These eggs are characterized by branching shell units that give the impression of superimposed layers of shell units in thin section (Fig. AII.2). The eggs are easily confused with those of *Dendroolithus.* Mikhailov (1997) suspects that *Dictyoolithus* is so similar to *Dendroolithus* in size, shape, surface texture, and range of thickness that it might be better to place this genus in the family Dendroolithidae, rather than in its own distinct family. Two species have been named from the Upper Cretaceous of China.

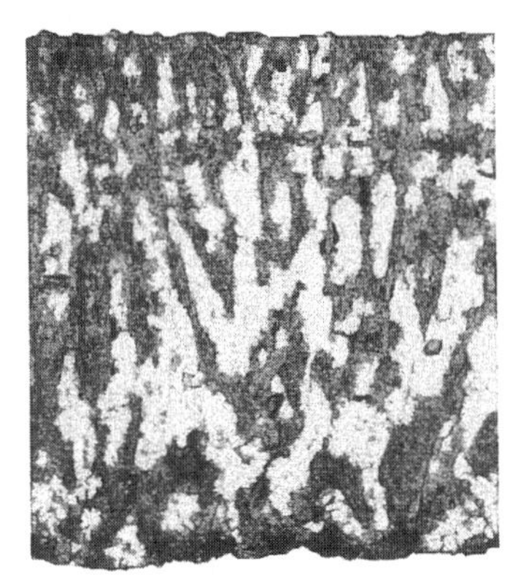

Dictyoolithus hongpoensis Zhao, 1994. The eggs are oval, with shell about 2.5–2.8 mm thick. The shell column is composed of five or more "superimposed shell units" (Fig. AII.2). From the Sigou Formation, Henan Province, China.

Dictyoolithus neixiangensis Zhao, 1994. Slightly older than *Dictyoolithus hongpoensis,* this species is characterized by spherical eggs 12 cm (4¾") in diameter. The shell is 1.5–1.7 mm thick, with columns composed of two or three "stacked shell units." From the Gaogou Formation, Henan Province, China.

Stromatoolithus *Zhao, Ye, Li, Zhao, and Yan, 1991*

The original description is very brief and leaves much to be desired. The surface ornamentation consists of "wormlike" ridges, suggesting that it is sagenotuberculate. The eggshell is dendrospherulitic and is most similar to *Dictyoolithus.* The shell units are branched and give the impression of two or three stacked prisms or columns. Pores are well developed, although nothing is said about the pore canals. This genus is not well differentiated from *Dictyoolithus* and may yet prove to be the same.

Stromatoolithus pinglingensis Zhao, Ye, Li, Zhao, and Yan, 1991. The characters that define this shell species are the same as for the genus. From the Maastrichtian Pingling Formation, Guangdong Province, China.

Fig. AII.2. Dictyoolithus hongpoensis *in thin section (shell is the darker portion).*

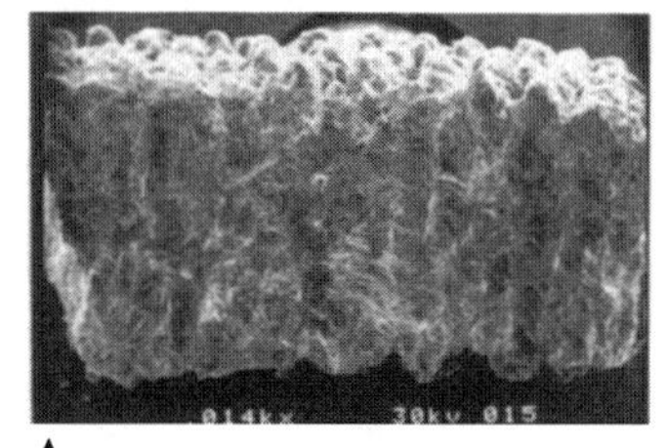

A

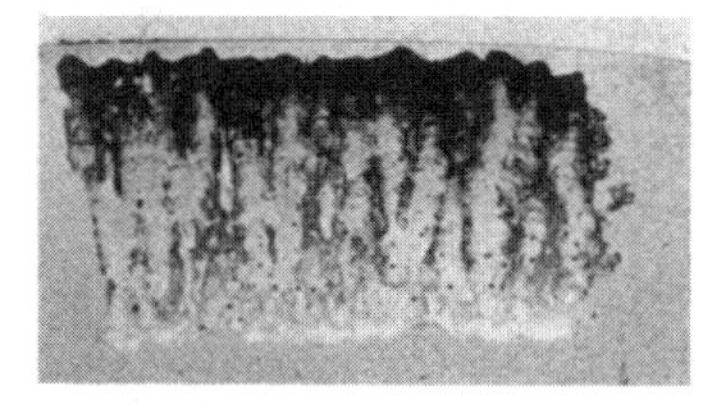

B

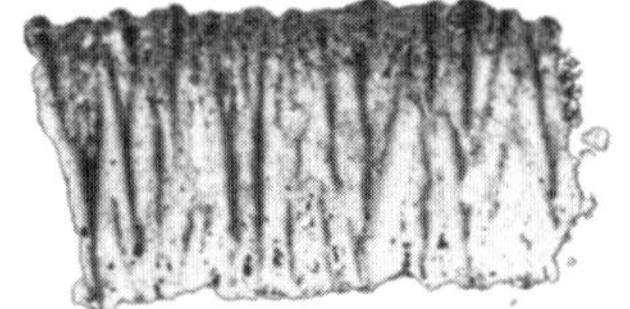

C

Fig. AII.3. Faveoloolithus ningxiaensis: *A, SEM of edge of shell (note also the lumpy surface formed by the tops of the shell units); B, thin section (normal light); C, thin section (polarized light). (Courtesy of K. Mikhailov and K. Hirsch.)*

FAVEOLOOLITHIDAE Zhao and Ding, 1976

Large (12–20 cm, 5–8") spherical to ellipsoidal eggs with a smooth or slightly rough surface. Eggshell of the filispherulitic morphotype and with multicanaliculate pore canals. Eggs are arranged in a disorderly fashion and stacked into one to three layers. Two genera are known from the Upper Cretaceous of Mongolia and China.

Faveoloolithus *Zhao and Ding, 1976*

These eggs are almost spherical and are about 12.6 to 14.2 cm (5–5½") in diameter. The shell is 1.8–2.6 mm thick, with a smooth outer surface. Pores are very abundant, in fact more so per square cm than any other known egg (Fig. AII.3). A clutch consists of about fifteen eggs in each of two to three layers, for a total of thirty to forty-five. The eggs in each layer alternate so that the eggs are tightly arranged. Only a single species is known.

Faveoloolithus ningxiaensis Zhao and Ding, 1976. The characters describing the species are the same as for the genus. From the Upper Cretaceous of Ningxia Province, China.

Youngoolithus *Zhao, 1979b*

The eggs are oval, about 16–17.5 cm (6 ¼–7") long and 9–10.2 cm (3½–4") in diameter. Shell thickness is about 1.38–1.75 mm. The genus has small, irregular shaped units of different lengths and widths, two or more of which are superimposed upon one another. The shell surface is smooth with numerous pores scattered throughout. The pore canals are straight, branching, extremely numerous, and closely spaced, forming a honeycomb (Fig. AII.4). Only a single species is known.

Youngoolithus xiaguanensis Zhao, 1979b. The characters describing the species are the same as for the genus. Clutch contains sixteen eggs arranged in alternating rows. From the Upper Cretaceous of Henan Province, China.

MEGALOOLITHIDAE Zhao, 1979a

These are large, spherical to subspherical eggs (diameters 10–20 cm) with a surface that is usually covered with compactituberculate ornamentation. These closely spaced nodes are actually the tops of the shell units. The nodes show so much variation in size and shape as to be worthless as a character to separate different species. The eggshell is a discretispherulitic (= tubospherulitic) morphotype with shell units composed of distinct, irregular columns. The units have erroneously been called spheroliths by some Indian paleontologists because of the superficial resemblance of the shell units to those of turtle eggs. The pore system is tubocanaliculate or multicanaliculate. The shell thickness ranges from 1.0–3 mm. Most occurrences of the eggs are in disorganized clutches, and the rings or arcs of eggs reported from France (Cousin and others, 1994) are probably coincidental (see chapter 10). Mikhailov (1997) expresses doubt that all the egg species are valid because the ranges of thickness of many overlap. I suspect that he may be partially correct because the diversity of large, closely related dinosaurs (e.g., all sauropods, or all hadrosaurs) in the respective countries does not even remotely approach the diversity seen in the eggshells.

The large size of these eggs is the major reason why they are thought to belong to sauropods. However, megaloolithid eggshell has not been found in the Upper Jurassic Morrison Formation of the western United States, where sauropods are numerically and taxonomically the most abundant dinosaur. Possibly sauropods from the Morrison laid their eggs in areas unfavorable to fossilization.

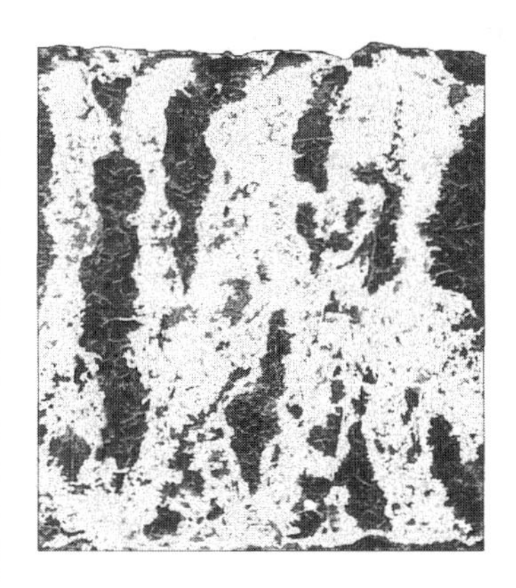

Fig. AII.4. Youngoolithus xiaguanensis, *thin section. Note the large, branching canals (light areas).*

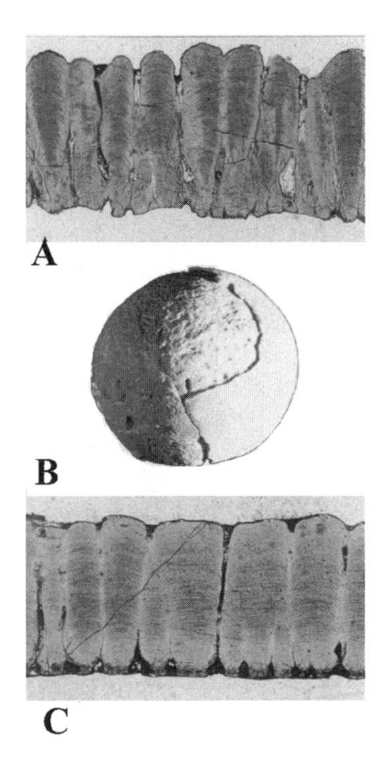

Fig. AII.5. A, Cairanoolithus dughii *in thin section (normal light); B,* Megaloolithus mammilare *(typical megaloolithid-type egg); C, thin section of* Megaloolithus mammilare. *(Courtesy of M. Vianey-Liaud.)*

Cairanoolithus *Vianey-Liaud, Mallan, Buscail, and Montgelard, 1994*

The shell has a smooth surface or is slightly tubercular (i.e., small bumps or nodes), but I cannot rule out the possibility that the surface is eroded. The shell units are occasionally fused together (Fig. AII.5A, B). Accretion lines parallel the surface of the shell units. The pore canals are rather small, being about 25 microns in diameter. No complete eggs are known yet, but the fragments are moderately thick, about 1.5–2.4 mm. Only a single species is known. It is questionable whether the specimens originally used to create this genus warrant being separated from *Megaloolithus,* considering the wide range of variation in that genus (see below). All the features of *Cairanoolithus* fall within the range of variability for *Megaloolithus.*

Cairanoolithus dughii Vianey-Liaud, Mallan, Buscail, and Montgelard, 1994. The characters that define this species are the same as that for the genus. Known only from eggshells from the lower Maastrichtian of southern France (Fig. AII.6A).

Dughioolithus *Vianey-Liaud, Mallan, Buscail, and Montgelard, 1994*

This eggshell is similar to *Cairanoolithus* and it is questionable whether a distinct genus is warranted for the specimens (assuming that *Cairanoolithus* is even valid). *Dughioolithus* is supposed to differ in the slightly tuberculate surface. The accretionary lines undulate and parallel the shell surface (Fig. AII.6B). The shell is also thinner than *Cairanoolithus,* being 1.4–1.7 mm thick, but still overlapping in part the range of *Cairanoolithus.* The pore canals are larger, at 50–100 microns in diameter, with many of the pores clustering into arcs around the base of the nodes. Perhaps the pores and size of the pore canals warrant a distinct species, or perhaps further study will show that the pores and canals vary too much.

Dughioolithus roussentensis Vianey-Liaud, Mallan, Buscail, and Montgelard, 1994. The characters that define this species are the same that define the genus. Shell fragments are known from the Lower Maastrichtian of southern France.

Megaloolithus *Vianey-Liaud, Mallan, Buscail, and Montgelard, 1994*

The most common eggs in this family belong to the genus *Megaloolithus*. Named by French paleontologists for spherical eggs in the Upper Cretaceous of southern France, similar eggs are also known from India, Asia, and South America. The genus is characterized by fan-shaped shell units that project as nodes on the surface (tuberculate ornamentation). Growth lines are usually present within each shell unit and these arc upward (Fig. AII.5C). The various species are mostly separated on the basis of egg size, shell thickness, and pore canal diameter.

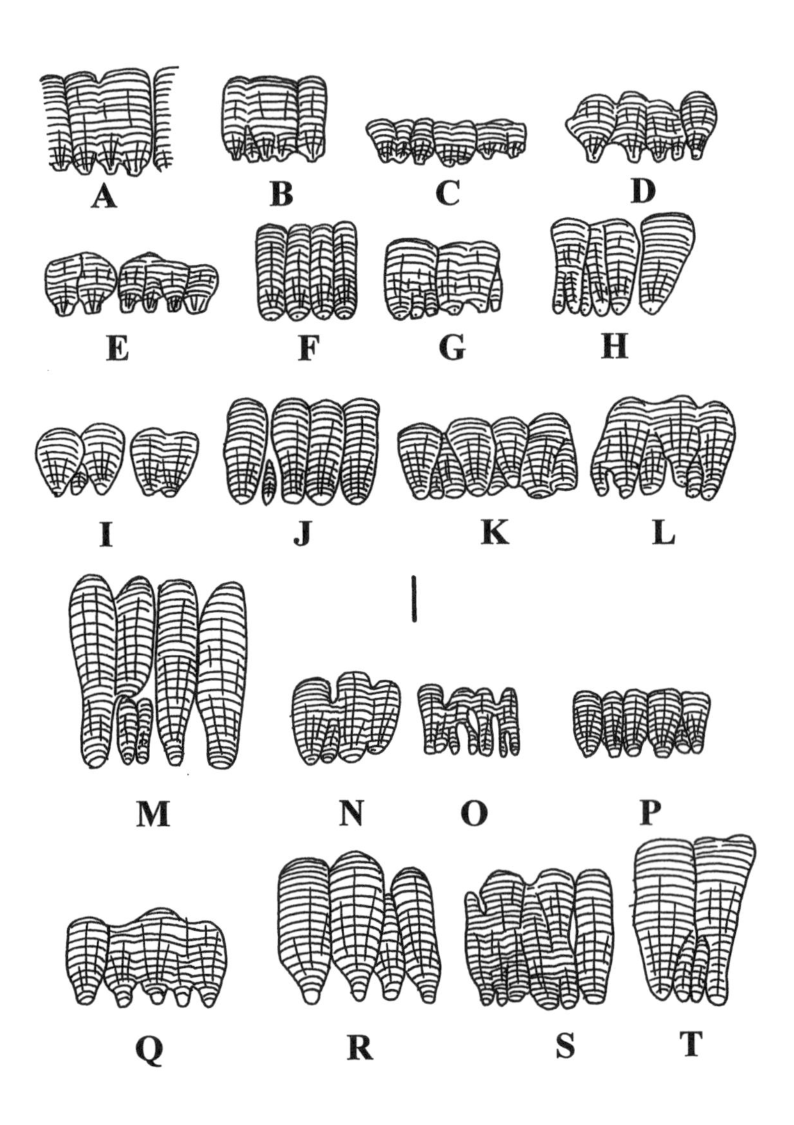

Fig. AII.6. Schematic thin sections of megaloolithid eggshells: A, Cairanoolithus dughii; *B,* Dughioolithus roussentensis; *C,* Megaloolithus aureliensis; *D,* Megaloolithus baghensis; *E,* Megaloolithus balasinorensis; *F,* Megaloolithus cylindricus; *G,* Megaloolithus dholiyaensis; *H,* Megaloolithus dhoridungriensis; *I,* Megaloolithus jabalpurensis; *J,* Megaloolithus khempurensis; *K,* Megaloolithus mammilare; *L,* Megaloolithus matleyi; *M,* Megaloolithus megadermus; *N,* Megaloolithus mohabeyi; *O,* Megaloolithus padiyalensis; *P,* Megaloolithus petralta; *Q,* Megaloolithus phensaniensis; *R,* Megaloolithus rahioliensis; *S,* Megaloolithus siruguei; *T,* Megaloolithus walpurensis. *All to scale. Scale = 1 mm.*

Megaloolithus aureliensis Vianey-Liaud, Mallan, Buscail, and Montgelard, 1994. The eggs are slightly elliptical, measuring 18.5 cm (7¼") long and 14.5 cm (5¾") in diameter. The shell is thin, about 0.8–1.5 mm, and shell unit height 1.75 times width (Fig. AII.6C). Despite its thinness, the shell has rather large pore canals 100–120 microns in diameter. From the Campanian of southern France.

Megaloolithus baghensis Khosla and Sahni, 1995. These eggs are 14–20 cm (5½–8") in diameter, with shell 1–1.7 mm thick. The shell units are irregularly rotund and sometimes partially fused together; height about 2.33 times width (Fig. AII.6D). On the inner side of the shell, the mammillae are very irregularly spaced. From the Maastrichtian Lameta Formation of India.

Megaloolithus balasinorensis Mohabey, 1998. Eggs 14–18 cm (5½"–7") in diameter, with a shell thickness 1.45–1.65 mm. The shell units are short and wide, with shell unit height equal to shell unit width (Fig. AII.6E). These units may be partially fused along their tops. About twelve eggs in a single-layered clutch. From the Maastrichtian Lameta Formation of India.

Megaloolithus cylindricus Khosla and Sahni, 1995. Spherical eggs 12–20 cm (4¾–7¾") in diameter, shell 1.7–3.5 mm in thickness (Fig. AII.6F). Shell units tall and narrow with height about four times the width. From the Maastrichtian Lameta Formation of India. I am not convinced that this species is distinct from *M. siruguei* from the Maastrichtian of France.

Megaloolithus dholiyaensis Khosla and Sahni, 1995. Known only from eggshells, which range in thickness from 1.5–1.75 mm. Shell units are fused together, but appear to be three times taller than wide (Fig. AII.6G). From the Maastrichtian Lameta Formation of India. This species is so similar to *Cairanoolithus* and *Dughioolithus* that it adds further doubt about the validity of those two genera.

Megaloolithus dhoridungriensis Mohabey, 1998. Spherical eggs, 14–18 cm (5½–7") in diameter. Shell thickness 2.26–2.36 mm, with distinct shell unit height about 2.75 times width (Fig. AII.6H). A supposed juvenile sauropod skeleton associated with one nest (Mohabey, 1987) is that of a snake (Jain, 1989). From the Maastrichtian Lameta Formation of India.

Megaloolithus jabalpurensis Khosla and Sahni, 1995. Spherical eggs 14–16 cm (5½–6") in diameter. Shell thickness 1–2.3 mm. Shell units with height 2.5 times width (Fig. AII.6I). Probably the same as *M. mammilare* from the Maastrichtian of France. From the Lameta Formation of India.

Megaloolithus khempurensis Mohabey, 1998. Clutch containing eggs in one or two layers. Eggs are spherical, 17–20 cm (6½–8") in diameter. The shell thickness is 2.36–2.6 mm. The units are columnar, with a height about 3.5 times the width (Fig. AII.6J). From the Maastrichtian Lameta Formation of India. I am not convinced that this species is distinct from *M. siruguei* from the Maastrichtian of France.

Megaloolithus mammilare Vianey-Liaud, Mallan, Buscail, and Montgelard, 1994. Known only from shell fragments, the shell measures 1.2–2.1 mm thick; unit height about twice unit width (Fig. AII.6K). The pore canals are about 75–120 microns in diameter. On the surface, the nodes are about 1 mm in diameter. From the Upper Maastrichtian of southern France.

Megaloolithus matleyi Mohabey, 1996. Spherical eggs 16–18 cm (6¼–7") in diameter and shell thickness 1.5–1.8 mm. Height of shell units about three times width (Fig. AII.6L). Clutch consisting of a single layer (number of eggs unknown). Probably the same as *M. mammilare* from the

Maastrichtian of France. *M. matleyi* is from the Maastrichtian Lameta Formation of India.

Megaloolithus megadermus Mohabey, 1998. Single layer of eggs in a clutch. Spherical eggs 13–18 cm (5–7") in diameter (Fig. AII.6M). Shell thickness 4–5 mm, and composed of units that are tall and narrow, with a height about 9.5 times the width. From the Maastrichtian Lameta Formation of India.

Megaloolithus mohabeyi Khosla and Sahni, 1995. Spherical eggs that are 16–19 cm (6¼–7½") in diameter and with shell thickness 1.8–1.9 mm. The shell units are tall and often fused to their neighbors. The shell unit height is three times width (Fig. AII.6N). From the Maastrichtian Lameta Formation of India.

Megaloolithus padiyalensis Khosla and Sahni, 1995. Known only from eggshell that is 1.12–1.68 mm thick. The shell units are irregular, of different sizes and many are fused together. The units are about four times taller than wide (Fig. AII.6O). The pore canals are irregular, variable in diameter, with an occasional branch. From the Maastrichtian Lameta Formation of India.

Megaloolithus petralta Vianey-Liaud, Mallan, Buscail, and Montgelard, 1994. Known only from eggshell that ranges from 1.1–2 mm thick. Shell unit height about 2.75 times unit width (Fig. AII.6P). From the Upper Maastrichtian of southern France. I am skeptical that this species is valid because the specimens that were used to create *M. petralta* were found with *M. aureliensis*.

Megaloolithus phensaniensis Mohabey, 1998. Spherical eggs 16–19 cm (6¼–7½") in diameter. Shell thickness 1.65–1.9 mm, with tall, narrow, distinct shell units whose height is about three times width (Fig. AII.6Q). Pore canals moderately wide and of variable width. From the Maastrichtian Lameta Formation of India.

Megaloolithus rahioliensis Mohabey, 1998. Spherical eggs 12.5–16 cm (5–6") in diameter. Shell thickness 2.8–3.5 mm, with distinct shell units that are four times taller than wide (Fig. AII.6R). The pore canals are narrow. The number of eggs in a clutch is unknown, but they appear to have been laid in a single layer. From the Maastrichtian of India.

Megaloolithus siruguei Vianey-Liaud, Mallan, Buscail, and Montgelard, 1994. Known only from shell fragments that range from 2.3–2.6 mm in thickness. Shell unit height is about three times unit width (Fig. AII.6S). The pores measure 50–80 microns in diameter. From the Maastrichtian of southern France.

Megaloolithus walpurensis Khosla and Sahni, 1995. Known only from eggshells that are thick, about 3.5–3.6 mm. The shell units are of different sizes and may fuse together near the outer surface (Fig. AII.6T). Shell height is three times shell width. Pore canals are slender with slitlike openings. From the Maastrichtian Lameta Formation of India.

Sphaerovum erbeni Mones, 1980. Eggs 17–20 cm (6¾–8") in diameter, with a shell thickness of 2.5–5 mm. From the Upper Cretaceous Mercedes Formation, Uruguay. Some new eggs are poorly preserved, with much of the calcium replaced by chalcedony, a form of silica (Faccio, 1994). As a result, much of the ultrastructure has been destroyed. What remains suggests similarities with *Megaloolithus siruguei,* in that the shell units are tall and slender, with accretionary lines curving down along the margins. Because *Sphaerovum* (1980) was named before *Megaloolithus* (1994), technically this is the proper name for large, spherical megaloolithid eggs. I doubt, however, that it will replace *Megaloolithus* because this name is more ingrained in the scientific literature.

OVALOOLITHIDAE Mikhailov, 1991

Eggs are oval to spherical depending on the genus. The shell is of the angustispherulitic morphotype. The shell units are tall and slender, and fused together forming a continuous layer, except in the lower portion. The outer surface is smooth or sagenotuberculate. The pore canals are varied, mainly rimocanaliculate, prolatocanaliculate and angusticanaliculate. Mikhailov (1997) notes that in thin section, poorly preserved ovaloolithid eggshell can be difficult to separate from prismatoolithid or elongatoolithid eggshell, probably because the shell units are fused together so as to resemble the continuous layer of the other two. However, the elongated shape of the prismatoolithid and elongatoolithid eggs easily separates them from ovaloolithid eggs.

Ovaloolithid eggs are the most common type found in the Upper Cretaceous of China and Mongolia, and Kyrgyzstan as well (Mikhailov, 1997). Mikhailov (1997) also speculates that these eggs belong to an ornithopod dinosaur just because they co-occur in the same deposits as lambeosaurine hadrosaurs. Although possible, this does not constitute proof. As we shall see below, hadrosaur eggshell is spherulitic.

Ovaloolithus *Zhao, 1979a*

The eggs are oval (ellipsoidal) with length 1.2–1.4 times the egg diameter. The mammillae occupy about half of the shell thickness. The mammillae are closely packed together, about thirty-five per square millimeter, making it difficult to separate them in tangential view. Growth lines are distinct in radial, or cross-section. Pore canals in the equatorial region are of the rimocanaliculate type, and of angusticanaliculate and prolatocanaliculate at the poles. The genus *Ovaloolithus* was named by Zhao Zi-Kui in 1979a for some eggs previously named *Oolithes spheroides* by Young in 1954.

Ovaloolithus chinkangkouensis (Zhao Zi-Kui and Jiang Yuan-Kai, 1974). The eggs range from 7.1–9.4 cm (2¾–3¾") long and 6.2–6.8 cm (2½") in diameter. Egg length is 1.1–1.4 times diameter. The shell, 1.7–2.6 mm in thickness, has numerous nodes separated by irregular depressions or grooves (Fig. AII.7C). From the Upper Cretaceous Yuanpu and Pingling Formations, Guangdong Province; Upper Wangshih Group, Shandong Province, China.

Ovaloolithus dinornithoides Konstantin Mikhailov, 1994b. Oval eggs 10.5 cm (4") long and 7.4 cm (3") in diameter, with length 1.4 times width. Shell surface with fine sagenotuberculate ornamentation (Fig. AII.7A, B). Shell thickness 1.1–1.8 mm. From the Upper Cretaceous Nemegt and Djadokhta Formations of Mongolia.

Ovaloolithus laminadermus (Zhao and Jiang, 1974). Known only from shell fragments that are thin, about 0.6–1.2 mm. From the Upper Cretaceous Pingling Formation, Guangdong Province; Upper Wangshih Group, Shandong Province, China. Originally called *Oolithes laminadermus* by Zhao and Jiang (1974).

Ovaloolithus mixtistriatus Zhao Zi-Kui, 1979a. Shell fragments from the Upper Cretaceous Upper Wangshih Group, Shandong Province, China. The shell is 2.1–3.3 mm in thickness.

Ovaloolithus monostriatus Zhao Zi-Kui, 1979a. Known only from some shell fragments found in a nest of elongated eggs. It is doubtful that this species is valid and mostly likely represents poorly preserved fragments of *Elongatoolithus* (see comments above). From the Upper Cretaceous Upper Wangshih Group, Shandong Province, China.

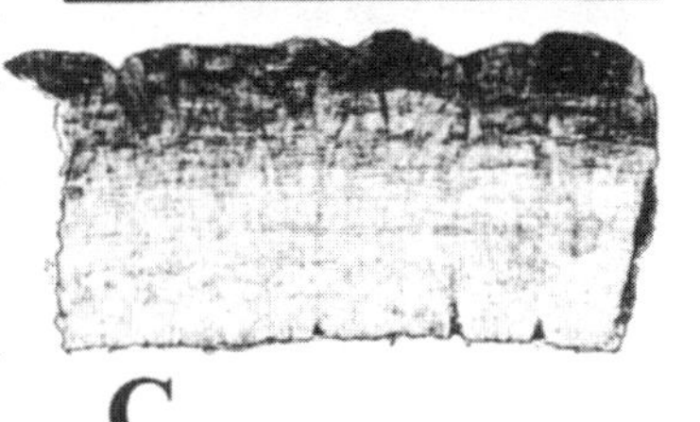

Fig. AII.7. A, B, Ovaloolithus dinornithoides *showing variation in shell surface; C,* Ovaloolithus chinkangkouensis *thin section. (Courtesy of K. Mikhailov.)*

Ovaloolithus tristriatus Zhao Zi-Kui, 1979a. The eggs are oval, measuring 9 cm (3¾") long and 7.3 cm (3") in diameter, with length about 1.25 times width. Shell is 1.1–3 mm thick. The surface is covered by nodes and valleys. This species is probably the same as *O. chinkangkouensis*. From the Upper Cretaceous Upper Wangshih Group, Shandong Province, China.

SPHEROOLITHIDAE Zhao, 1979a

Spherical or oval eggs that are rather small, about 7–10 cm (2¾–4") in length and 6–8 cm (2¼–3¼") in diameter (length/width ratio = 1.2–1.25). The shell is the prolatospherulitic morphotype, with prolatocanaliculate pore canals. Outer surface smooth, rough, or sculptured (sagenotuberculate ornamentation).

Shixingoolithus *Zhao, Ye, Li, Zhao, and Yan, 1991*

Fig. AII.8. Shixingoolithus erbeni *thin section.*

Nearly spherical eggs 10.5–12.5 cm (4–5") long, 9.9–12.3 cm (3¾–4¾") in diameter. The thickness of eggshell is 2.3–2.6. The shell units are composed of very tall, very narrow prisms. The mammillae are about 0.6 mm thick, which is about one-fourth the thickness of the shell. The pore canals are narrow and irregular in shape. A faint herringbone or cross-net pattern is visible, although this may be due to fossilization processes.

Shixingoolithus erbeni Zhao Zi-Kui, Ye Jie, Li Huamei, Zhao Zhenhua, and Yan Zheng, 1991. Named for a nest of thirty-four eggs, the characters that define this species are the same that define the genus. From the Upper Cretaceous Pingling Formation, Guangdong Province, China.

Spheroolithus *Zhao, 1979a*

Slightly oval eggs with either smooth, sagenotuberculate, or ramotuberculate ornamentation, 7–12 cm (3–5") long. Shell thickness ranges from 1–3 mm. The prolatocanaliculate pore canals are extremely variable in diameter along their lengths from slits to expanded chambers. The variants include foveocanaliculate, which has parts of the canals with wide pore openings that taper downward, and lagenocanaliculate, which are the expanded chambers. The result of the canal variability is to give the shell a superficial resemblance to dendroolithid and faveoloolithid eggshell. However, the absence of cavities within the shell units distinguishes this genus. Whole eggs can be confused with *Ovaloolithus,* but the two can be separated by thin sections. Known from the Upper Cretaceous of China, Mongolia, Kyrgyzstan, and North America (Mikhailov, 1994b). This genus was named by Zhao Zi-Kui (1979a) for some of the eggs called *Oolithes spheroides* by Young (1954).

Spheroolithus albertensis was named by Darla Zelenitsky and Len Hills (1997). The eggs are 10–12 cm (4–4¾") long, 7–9 cm (2¾–3½") in diameter, with a length/diameter ratio of 1.33–1.46 (Fig. AII.9). The eggshell is 0.96–1.46 mm thick, although the majority are .96–1.22 mm. The surface ornamentation is sagenotuberculate and ramotuberculate. The eggshell may be the same as *Spheroolithus maiasauroides* from Mongolia, but seems to differ mostly in shell thickness and finer ornamentation. From the Campanian Oldman Formation of Alberta and the Two Medicine Formation of Montana. These eggshells were described previously by Hirsch and Quinn (1990) from Montana specimens, but not named. The fragments were collected from sites containing baby *Maiasaura* bones, so are thought to have been laid by that dinosaur. If correct, the implication is that the various *Spheroolithus* eggs were laid by hadrosaurs.

Fig. AII.9. Clutch of Spheroolithus albertensis. *Scale in cm. (Courtesy of D. Zelenitsky.)*

Fig. AII.10. Clutches: A, Spheroolithus irenensis; *B,* Spheroolithus chiangchiungtingensis. *(Courtesy of K. Mikhailov.)*

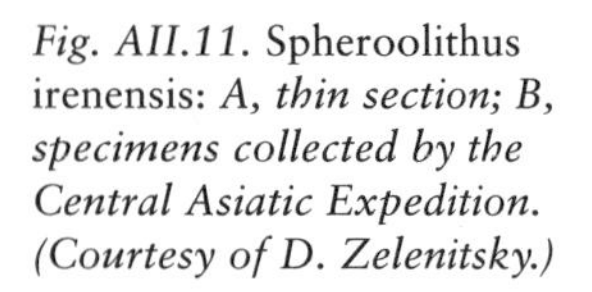

Fig. AII.11. Spheroolithus irenensis: *A, thin section; B, specimens collected by the Central Asiatic Expedition. (Courtesy of D. Zelenitsky.)*

Fig. AII.12. Spheroolithus maiasauroides: *A, thin section in polarized light; B, normal light; C, SEM of an etched thin section.*

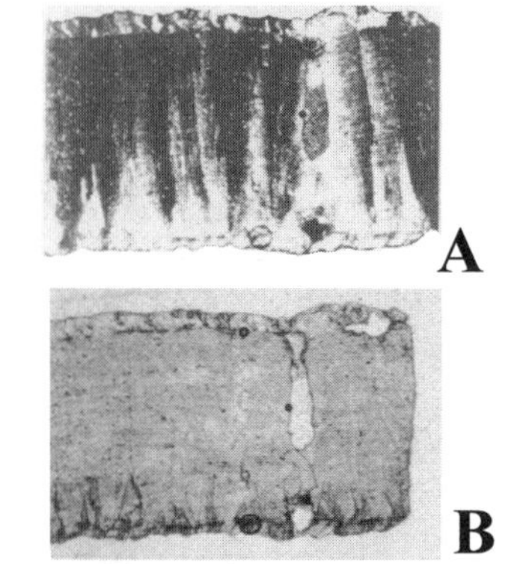

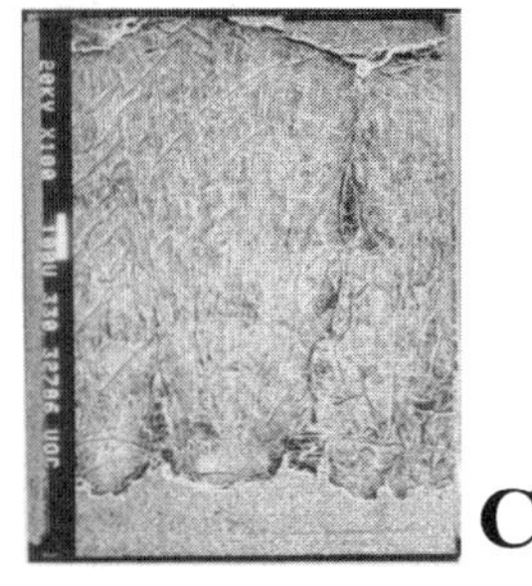

Spheroolithus chiangchiungtingensis (Zhao and Jiang, 1974). Oval eggs 6.8–8.8 cm (2 3/4–3 1/2") long, 5.5–7.1 cm (2–2 3/4") in diameter (length/width = 1.2). Shell thickness 1.1–3 mm, and considerably variable within the same egg, 2.1–3 mm (Young, 1954). The mammillary layer is 0.8–0.1 mm, or about a third of the shell thickness. In addition, there are about seven to fourteen mammillae per square millimeter. The clutch consists of at least seven eggs in two layers, but probably consisted of more (Fig. AII.10B). Known from the Upper Cretaceous Lower Wangshih Group, Shandong Province China. Originally called *Oolithes chiangchiungtingensis* by Zhao and Jiang (1974).

Spheroolithus irenensis (Zhao and Jiang, 1974). Smooth, oval eggs, 8.0–9.1 cm (3–3½") long and 6.6–6.9 cm (2½–2¾") in diameter; egg length is 1.2–1.3 times egg diameter. Shell thickness 1.5–2.2 mm, of which 0.6–0.8 mm consists of mammillae. Mammillary layer a little over one-third shell height; there are about twelve to twenty-one mammillae per square millimeter. At least eight eggs in a clutch (Fig. AII.10A, AII.11). Known from the Upper Cretaceous Iren Dabasu Formation of Inner Mongolia and Wangshih Group, Shandong Province, China; Nemegt Formation, Mongolia.

These specimens were originally called *Oolithes irenensis* by Zhao and Jiang (1974). Zhao (1979a) created the genus *Paraspheroolithus* for this species, but Mikhailov (1994b) concluded that the differences did not warrant a different genus name. *Spheroolithus irenensis* eggs were first described but not named by van Straelen from eggs collected at Iren Dabasu, China, by the American Museum of Natural History. It was left to Zhao Zi-Kui to name the eggs in 1979. He also reported the eggs from northeastern China, at the famous localities at Laiyang. Zhao has recently noted that the eggs are similar to those of the hadrosaur *Maiasaura,* and suggests that they might have been laid by *Bactrosaurus* (known from the same deposits at Iren Dabasu), although they could belong to *Gilmoreosaurus,* which is also present.

Spheroolithus maiasauroides was named by Konstantin Mikhailov (1994b). Eggs estimated to be oval, 9 cm ($3\frac{1}{2}$") long, 7 cm ($2\frac{3}{4}$") in diameter, with a length/diameter ratio of 1.3. The shell thickness is 1–1.6 mm, although most are 1.2–1.5 mm (Fig. AII.12). The equatorial surface is covered with small squiggly ridges and nodes of sagenotuberculate ornamentation, whereas the poles are smooth or nearly so. The pores are not as evenly distributed as in *S. irenensis,* are smaller in diameter, and occur in clusters. Known from the Upper Cretaceous Djadokhta Formation of Mongolia.

Spheroolithus megadermus (Young, 1959). Known only from very thick eggshells—5.5–5.8 mm. The mammillary layer is about 1.7 mm thick, or about 30 percent of the shell thickness. Known from the Upper Cretaceous Middle Wangshih Series, Shandong Province, China. Originally called *Oolithes megadermus* by Young (1959).

Spheroolithus tenuicorticus was named by Konstantin Mikhailov (1994b) for shell fragments that strongly resemble *S. irenensis* and differ only in the thinner shell, 0.8–1.8 mm. Known from the Upper Cretaceous Barun Goyot? Formation of Mongolia. This eggshell species may prove to be the same as *S. irenensis.*

II. Prismatic Basic Shell Type

Eggshell with two parts, a lower radial-tabular ultrastructure (mostly the mammillae), and an upper tabular ultrastructure forming the prismatic units.

PRISMATOOLITHIDAE Hirsch, 1994

Elongated or oval eggs, with a prismatic (angustiprismatic or obliquiprismatic) morphotype of shell, and an angusticanaliculate or obliquicanaliculate pore canal. The surface is smooth or with fine linearituberculate ornamentation in the mid or equatorial section. Eggshell thickness 0.8–1.0 mm (medium thick).

Preprismatoolithus *Zelenitsky and Hills, 1996*

Oval eggs with one end slightly pointed. The shell is 0.7–1.0 mm in thickness and with obliquicanaliculate pore canals. The outer surface is smooth. The eggs are estimated to be 11 cm ($4\frac{1}{4}$") long and 6–8 cm ($2\frac{1}{2}$–3") in diameter. The clutch has at least a dozen eggs in a ring. Known from the Kimmeridgian Morrison Formation of the western United States and possibly Portugal.

Preprismatoolithus coloradensis (Hirsch, 1994). One of the oldest known egg species from North America. It is known from several eggs and thousands of shells from the Upper Jurassic Morrison Formation of Colorado and Utah.

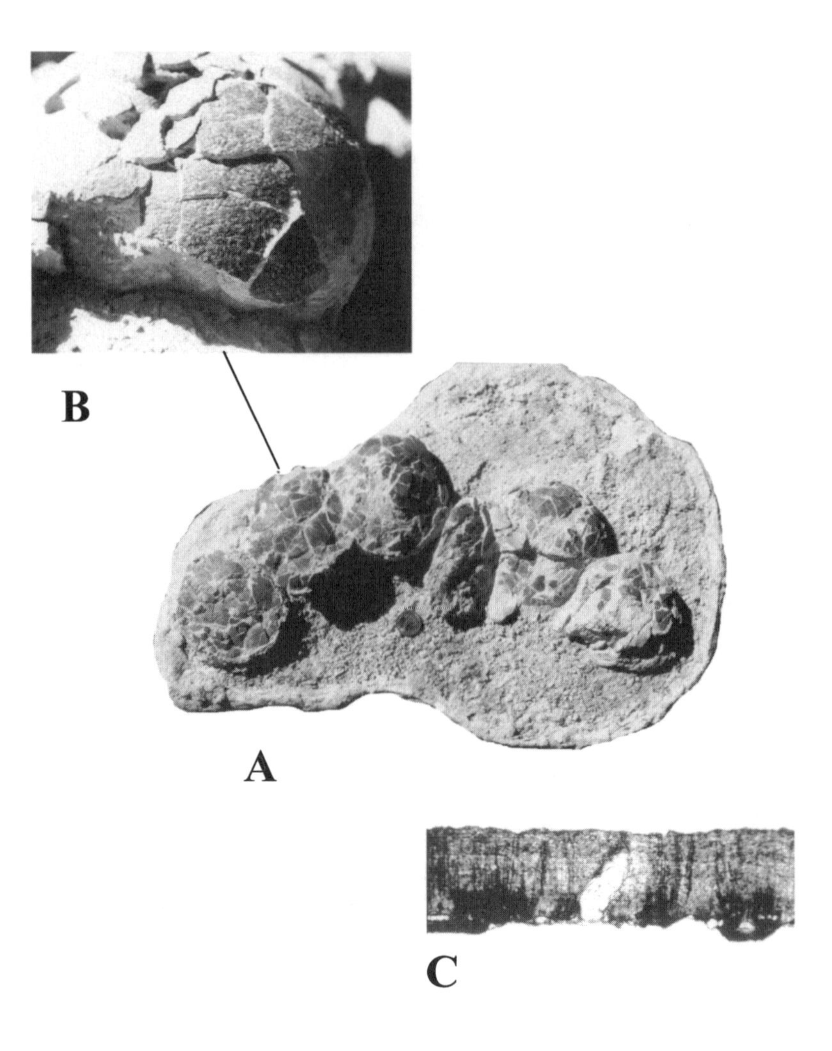

Fig. AII.13. Preprismatoolithus coloradensis: *A, partial clutch; B, close-up of shell surface; C, thin section. Note the partial angled pore canal in the lower center of* C.

The same or a very similar species is also known from the Upper Jurassic of Portugal where it is associated with embryonic or dinosaurling theropods (Mateus and others, 1997a, b). The clutch consists of thirty-four eggs ranging in size from 10–13.7 cm long (4–5 1/4") long and 8–10 cm (3–4") in diameter. The eggshell has both angusticanaliculate and obliquicanaliculate pore canals.

Prismatoolithus *Zhao and Li, 1993*

Elongate oval eggs, 10.5–15 cm (4–6") long, 5–7 cm (2–2 3/4") in diameter, with length 1.8–2.4 times diameter. Eggshell with angustiprismatic morphotype with shell units consisting of slender, interlocking prisms (hence the name), and angusticanaliculate pore canals. The outer surface is smooth or finely ornamented. Shell thickness 0.3–1.2 mm.

Prismatoolithus gebiensis. Named by Zhao Zi-Kui and Li Zuocong

(1993) for a clutch of seven eggs. The eggs are elongated, smooth, and have one end more pointed than the other. The eggs are about 12 cm (4¾") long, and about 5 cm (2") in diameter. The shell is thin, about 0.7–0.9 mm. The mammillae occupy only 1/7 the shell thickness. Pores are oval to circular. Pores are absent on the pointed end, and very abundant on the blunt end. The eggs stood in the nest slightly inclined, with their pointed ends down. From the Upper Cretaceous Djadokhta Formation, Inner Mongolia, China.

Prismatoolithus levis Zelenitsky and Hills, 1996. Elongated eggs 10–12 cm (4–4¾") long and 5–6 cm (2–2¼") in diameter. Shell thickness 0.7–1 mm, with the mammillary layer about 1/6–1/8 total shell thickness. The pores occur in pairs. There are about twenty-four eggs in a clutch arranged in rough concentric rings with the eggs standing almost vertically. These eggs were once thought to belong to the hypsilophodont *Orodromeus* by Hirsch and Quinn (1990), but reidentification of the embryo within the egg indicates that these are the eggs of the small theropod *Troodon* (see chapter 11). From the Campanian Oldman Formation of Alberta and the Two Medicine Formation of Montana.

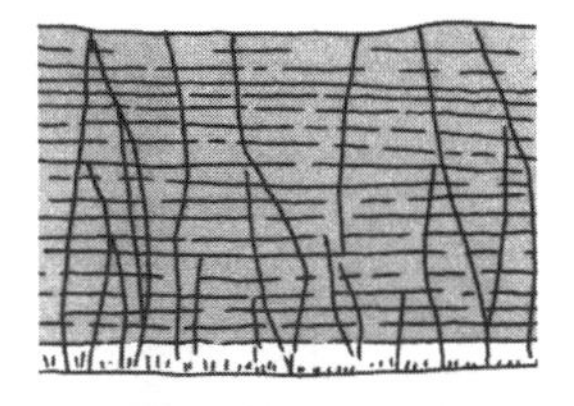

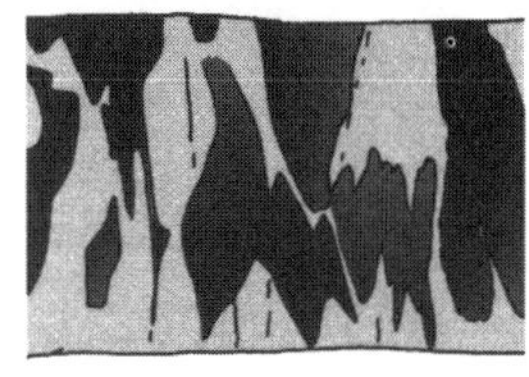

Fig. AII.14. Schematic drawings of Prismatoolithus tenuius *in normal light (top) and polarized light (bottom). Scale = 1 mm.*

Prismatoolithus matellensis Vianey-Liaud and Crochet, 1993. Eggshell 1.06–1.22 mm in thickness, with a smooth outer surface. The mammillary layer is about 1/10 the shell thickness. The pore canals vary from about 25–70 microns in diameter. Known only from shell fragments from the Maastrichtian of France and possibly the Tremp Formation of northern Spain.

Prismatoolithus tenuius Vianey-Liaud and Crochet, 1993. Very thin shell, 0.24–0.6 mm thick, with mammillary layer about 1/6 shell thickness (Fig. AII.14). Surface mostly smooth, but with some scattered small irregular, flat-bottomed pits. Known only from shell fragments from the Maastrichtian of France and the Maastrichtian Tremp Formation of northeastern Spain. The extreme thinness of the shell makes it doubtful that these specimens belong to the genus *Prismatoolithus*. A restudy is in order and I would not be surprised if these specimens are reallocated to a new family close to the Laevisoolithidae.

Protoceratopsidovum *Mikhailov, 1994a*

Eggs elongated, with one end more pointed than the other. The length is two to three times the diameter. Shell of the angustiprismatic morphotype, with angusticanaliculate pore canals. Surface of shell smooth or with lineartuberculate ornamentation around equator only, not at the poles. The shell is thickest at the lower pole and thinnest at the upper.

The name was proposed by Konstantin Mikhailov, implying that it was laid by a protoceratopsian dinosaur. However, embryonic bone is not known in any of the eggs and there is no certainty that the eggs actually belong to these dinosaurs. Considering that the closely related *Prismatoolithus* eggs are now known to be theropod (see above), *Protoceratopsidovum* must either be theropod as well, belong in a different family, or be part of *Prismatoolithus* and not a distinct genus. It is also unfortunate that Mikhailov departed from the accepted practice of using *-oolithus* as an ending or suffix for the genus, using *-ovum* instead.

Protoceratopsidovum fluxuosum was named by Konstantin Mikhailov (1994a). The eggs are 13–15 cm (5–6") in length, and 5–5.7 cm (2–2¼") in diameter, for a length/diameter ratio of 2.6; shell thickness is 0.6–0.7 mm near the middle, 0.3 mm on the pointed end, and 1.4 mm on the blunt end. The shell is characterized by fine lineartuberculate ornamentation consisting of ridges that are 30–40 mm long (Fig. AII.15). Isolated

Fig. AII.15. Protoceratopsidovum fluxuosum: *A, B, clutches; C, single egg; D, close-up of shell; E, F, thin sections in normal light. Approximate margins of shell units noted between white lines in E and F. Scale in B = 10 cm; others in mm. (Courtesy of K. Mikhailov and K. Sabath.)*

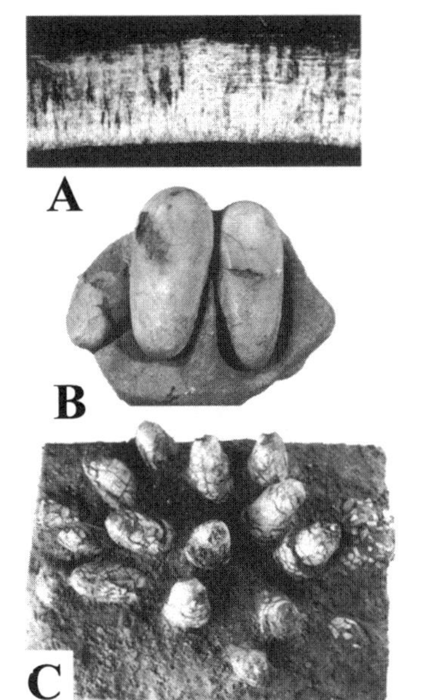

Fig. AII.16. Protoceratopsidovum sincerum: *A, thin section in polarized light; B, two eggs; C,* Protoceratopsidovum minimum *clutch. (Courtesy of K. Mikhailov.)*

nodes or bumps are not present as in some other species. There are twenty eggs in a clutch, and these are inclined outward in two concentric rings. This species is known from the Upper Cretaceous Barun Goyot Formation of southern Mongolia.

Protoceratopsidovum minimum was named by Konstantin Mikhailov (1994a). Eggs are 10–11 cm (4–4¼") long, and 4–5 cm (1½–2") in diameter, for a length/diameter ratio of 2.2–2.5. The shell is 0.3–0.7 mm thick and has a smooth surface, although a very faint striation is sometimes visible around the middle. Eggs in a clutch stand vertically, although the number of eggs in a clutch is unknown (Fig. AII.16C). These eggs are known from the Upper Cretaceous Djadokhta and Barun Goyot Formations of Mongolia.

Protoceratopsidovum sincerum, named by Mikhailov (1994a) for eggs from Djadokhta Formation of Mongolia. The eggs are 11–12 cm (4¼–4¾") long, and 4–5 cm (1½–2") in diameter, with a length/diameter ratio of 2.4–2.75. They also have a smooth shell, although a faint striation is sometimes visible around the middle (Fig. AII.16A, B). The shell thickness is 0.6–0.7 mm in the middle, 0.3 mm on the pointed end, and 1.2 mm on the blunt end. There are about twenty to thirty eggs in a clutch, and these stand vertically in two concentric circles. *P. sincerum* eggs are known from the Upper Cretaceous Djadokhta and Barun Goyot Formations of Mongolia. It is possible that this species is the same as *Prismatoolithus gebiensis* because the two are nearly identical.

OOFAMILY INCERTAE SEDIS

Pseudogeckoolithus *Vianey-Liaud and Lopez-Martinez, 1997*

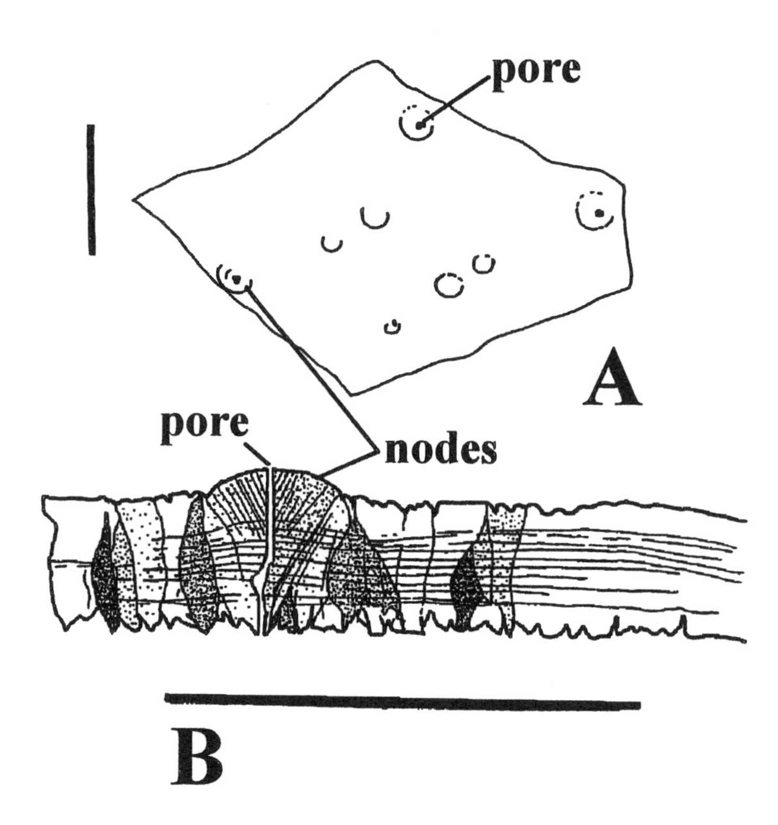

Fig. AII.17. Schematic sketches of Pseudogeckoolithus nodosus: *A, showing the surface nodes and location of pores on the nodes (scale = 1 mm); B, thin section showing the pore extending through the pore (scale = 1 mm). (Modified from Vianey-Liaud and Lopez-Martinez, 1997.)*

Very thin shell 0.3 mm thick, with a short "mammillary" layer about $^1/_8$–$^1/_{10}$ shell thickness. The outer surface is covered with scattered nodes. Pores often located on top of nodes. I am not totally convinced that this genus was laid by a dinosaur. The extreme thinness matches that for the lizard *Gekko,* although the shell units are better defined than in gecko eggs (see Packard and Hirsch, 1989).

Vianey-Liaud and Lopez-Martinez (1997) thought that the eggshell looked prismatic. However, I suspect that it is actually ornithoid (the two morphotypes can be confused sometimes) and that this genus is the same as *Porituberoolithus* Zelenitsky, Hills, and Currie, 1996 (see below). Both have the peculiar scattered nodes (dispersituberculate ornamentation) with a pore located on them. The two differ enough in thickness and relative thickness of the mammillary layer to the total shell thickness that a distinct species is indicated.

Pseudogeckoolithus nodosus. Named by Vianey-Liaud and Lopez-Martinez (1997) for two very small pieces of eggshell from the Upper Cretaceous Tremp Formation of Spain.

III. Ornithoid Basic Shell Type

Eggs with a birdlike shell structure containing a distinct mammillary layer and continuous layer; distinct shell units are not defined. The pore canals are usually angusticanaliculate.

ELONGATOOLITHIDAE Zhao, 1975

As their name implies, these eggs are all elongated. One end is usually rounded, while the other is somewhat pointed. The shell has the ratite morphotype, meaning that it is very similar to that seen in the eggs of ratite birds (e.g., ostrich). Because birds and theropod dinosaurs are closely related, Mikhailov believes elongatoolithid eggs were laid by theropod dinosaurs. Indeed, at least one of the theropods is *Oviraptor* based on a partial embryonic skeleton found adhering to the inside of an elongatoolithid shell fragment (see chapter 11 for more).

The mammillary layer is about $1/3$–$1/5$ of the shell thickness. The pore canals are angusticanaliculate and in tangential section appear as ovals or circles. The surface of the shell is covered by linearituberculate ornamentation around the equator and grading through ramotuberculate to dispersituberculate at the poles. The accretionary lines are parallel to the surface of the shell, thus are wavy as seen in radial section. The clutches are formed by concentric circles of outwardly inclined eggs with their larger ends pointing inward.

Ellipsoolithus *Mohabey, 1998*

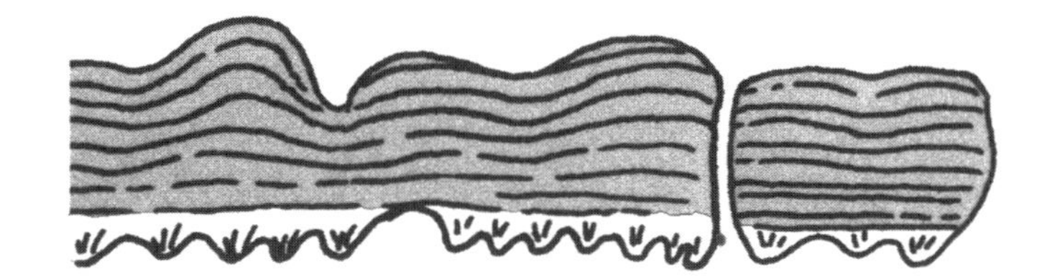

Fig. AII.18. Ellipsoolithus khedaensis, *schematic drawing of thin section. Scale = 1 mm.*

Oval eggs 9.8–11 cm ($3\frac{3}{4}$–$4\frac{1}{4}$") long, 6.5–8 cm in diameter, with a length/diameter ratio of 1.4–1.5. The shell is 1.2–1.6 mm thick, with a mammillary layer possibly about $1/4$ shell thickness.

Ellipsoolithus khedaensis was named by Indian paleontologist D. M. Mohabey in 1998. The characters that define the species are the same as for the genus. The eggshell is highly altered due to diagenesis, making it almost impossible to determine the internal structure. There are about thirteen eggs in a clutch. From the Maastrichtian Lameta Formation of India.

Elongatoolithus *Zhao, 1975*

The eggs are up to 17 cm ($6\frac{1}{2}$") long and have a length/diameter ratio of 2–2.2. The surface is covered with a fine linearituberculate ornamentation. There is a sharp separation of the mammillary and continuous layers. Wavy growth lines are present in the continuous layer and parallel the

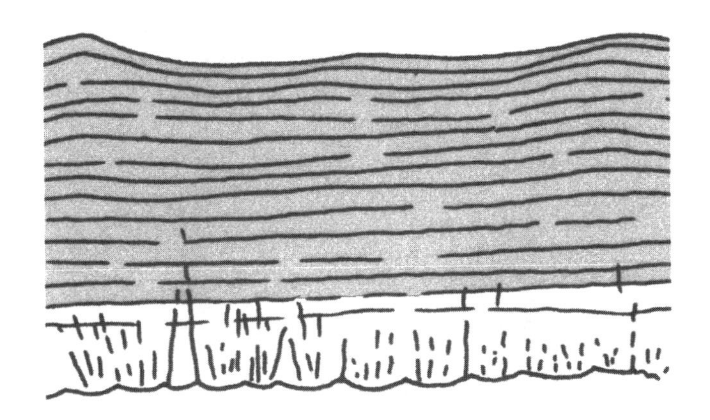

Fig. AII.19. Schematic drawing of Elongatoolithus andrewsi *in thin section.*

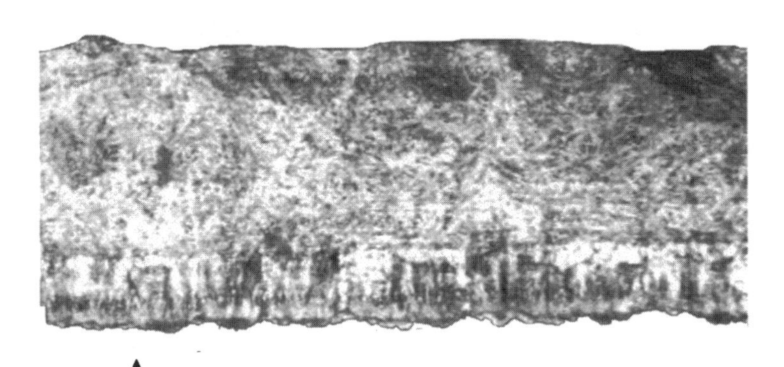

Fig. AII.20. Elongatoolithus *sp.: A, thin section in normal light; B, in polarized light; C, whole egg. (Courtesy of K. Sabath.)*

Fig. AII.21. Elongatoolithus frustrabilis: *A, clutch; B, egg; C, thin section—sl = continuous layer, ml = mammillary layer. Scale in C = 1 mm. (Courtesy of K. Mikhailov.)*

surface of the shell. The mammillary layer is about 1/6–1/2 the shell thickness. Eggs are usually paired in a clutch, with their wider end out. Known from the Lower and Upper Cretaceous of China, Mongolia, and Kyrgyzstan. Originally called *Oolithes elongatus* by Young, 1954.

Elongatoolithus andrewsi Zhao Zi-Kui, 1975. Eggs 13.8–15.1 cm (5 1/2–6") long, 5.5–7.7 cm (2–3") in diameter. Length of eggs is 2–2.5 times diameter. The shell is 1.1–1.5 mm thick, and the mammillary layer is about 1/4 of the shell thickness (Fig. AII.19). Clutch is about a dozen eggs. From the Upper Cretaceous Yuanpu and Pingling Formations, Guangdong Province; and the Upper and Middle parts of the Wangshih Group of Shandong Province, China.

Elongatoolithus elongatus (Young, 1954). Eggs are 14–15 cm (5 1/2–

6") long, and 6.1–6.7 cm (2½") in diameter. The length is 2.25 times diameter. The shell is about 0.7–1.1 mm thick, with the mammillary layer about ⅙ the shell thickness. The clutch contains at least twelve eggs in two circular layers. These eggs were originally named *Oolithes elongatus*. From the Upper Cretaceous Yuanpu and Pingling Formations, Guangdong Province, and Upper Wangshih Group of Shandong Province, China.

Elongatoolithus excellens Mikhailov, 1994a. Smallest species of *Elongatoolithus,* the eggs are 9–11 cm (3½–4") long and 4 cm (1½") in diameter; length about 2.25–2.75 times diameter. The shell is 0.3–0.9 mm thick, of which the equatorial region is 0.4–0.7 mm. The surface has anastomotuberculate ornamentation. At least seven eggs in a clutch. From the Upper Cretaceous Nemegt Formation of Mongolia.

Elongatoolithus frustrabilis Mikhailov, 1994a. Eggs measuring 15–17 cm (6–6¾") in length and 6–7 cm (2¼–2¾") in diameter; length is 2.4–2.5 times diameter (Fig. AII.20). The shell measures 0.4–0.7 mm thick, except on the blunt end where it is 1–1.2 mm thick (Fig. AII.21). From the Upper Cretaceous Djadokhta Formation of Mongolia.

Elongatoolithus magnus Zeng and Zhang, 1979. The eggs range in size from 16.2–17.2 cm (6¼–6¾") long and 6.3–8.2 cm (2½–3¼") in width, and have a length/diameter ratio of 2–2.5. Shell thickness is 0.6–0.9 mm, with about ⅙ of this the mammillary layer. The ornamentation consists of linearituberculate ridges around the midsection and ramotuberculate ornamentation at the poles. From the Upper Cretaceous of Hunan Province, China.

Elongatoolithus sigillarius Mikhailov, 1994a. Eggs are 15–17 cm (6–6¾") long and 6–7 cm (2¼–2¾") in diameter; length is 2.5 times diameter. The ornamentation consists of linearituberculate ornamentation that includes some linearituberculate ridges arranged in a cross-hatched pattern. From the Upper Cretaceous Nemegt Formation of Mongolia.

Elongatoolithus subtitectorius Mikhailov, 1994a. Only known from shell fragments that are 0.5–0.9 mm thick and covered with anastomotuberculate ornamentation. An identical shell type was reported from France by Beetschen and others (1977). These fragments measured 0.7–0.8 mm thick. From the Djadokhta Formation of Mongolia.

Macroelongatoolithus *Li, Yin, and Liu, 1995*

Very large, elongated eggs 39.3–53.6 cm (15½–21") long, 13–17.9 cm (5–7") in diameter, with a length/diameter ratio of 3. The shell surface is covered with very coarse linearituberculate ornamentation, except at the poles where it is discretispherulitic. Some of the nodes are expanded at the top so that they resemble a mushroom.

Macroelongatoolithus xixia Li, Yin, and Liu, 1995. From the Lower Cretaceous Sigou Formation and Zeumagang Formation of Henan Province, China. Shell thickness 2–3.2 mm and mammillary layer ⅓–¼ the shell thickness. The clutch consists of at least thirteen pairs of eggs laid in a large ring.

Macroelongatoolithus (=*Oolithes,* =*Boteluoolithus*) *carlylei* (Jensen, 1970). From the Lower Cretaceous, Cedar Mountain Formation, and Middle Cretaceous Dakota Formation, Utah. Shell thickness 1.38–3.04 mm. Known only from shell fragments that cannot be separated from those of *Macroelongatoolithus xixia.*

Macroolithus *Zhao, 1975*

These eggs are much larger than those of *Elongatoolithus.* The col-

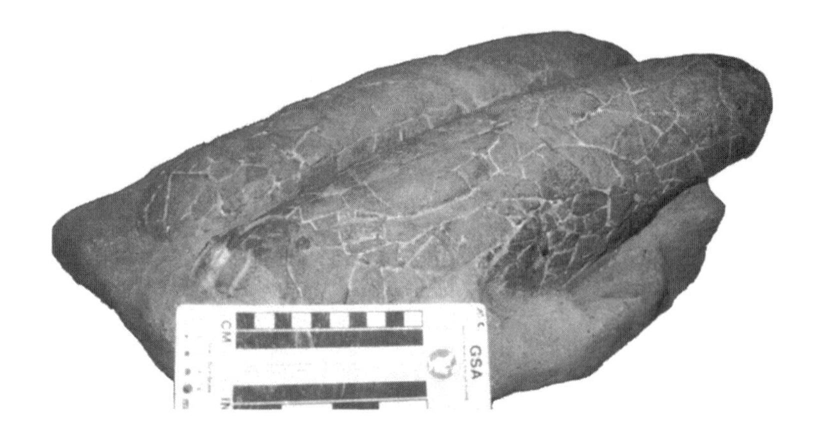

Fig. AII.22. Macroelongatoolithus xixia *paired eggs. Scale: length = 10 cm.*

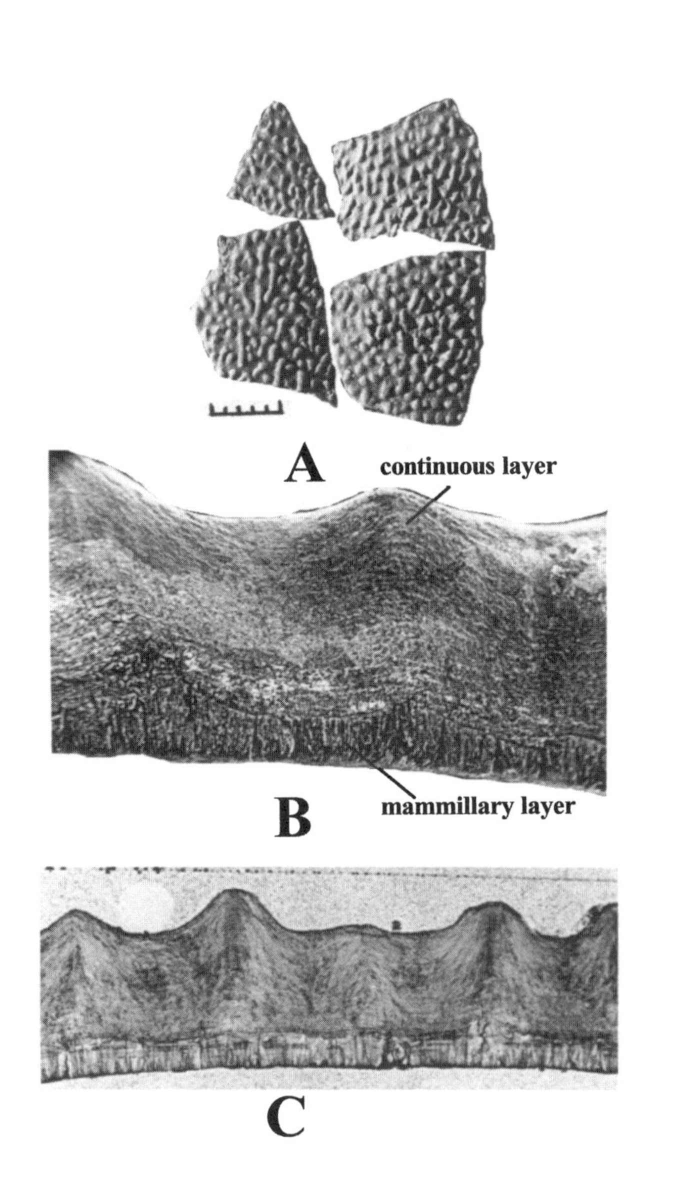

Fig. AII.23. Macroolithus rugustus: *A, shell pieces showing surface texture (scale = mm); B, SEM of etched shell thin section; C, thin section in normal light. (Courtesy of K. Mikhailov.)*

Fig. AII.24. Macroolithus rugustus *showing pore canal. (Courtesy of K. Mikhailov.)*

umns of the shell unit are interlocked, with some fusion along their edges. The mammillary layer is 1/3–1/2 of the shell thickness. The surface ornamentation is similar to that of *Elongatoolithus,* but is much coarser. There are up to twenty eggs in a clutch arranged in two or three concentric layers.

Macroolithus mutabilis Mikhailov, 1994a. Known only from eggshell that is 1.3–2 mm thick. The pore openings are funnel-shaped and the pore canals rimocanaliculate. From the Upper Cretaceous Barun Goyot Formation in southern Mongolia.

Macroolithus rugustus (Young, 1965). Eggs are 16.5–18 cm (6 1/2–7") long and 7.5–8.5 cm (3–3 1/4") in diameter, giving a length/diameter ratio of 1.75–2.2. The shell thickness is 0.8–1.7 mm and in thin section, shows horizontal growth lines (Fig. AII.23, AII.24). The surface ornamentation is coarsely ramotubercular near the middle, becoming pitted or sagenotuberculate toward the ends. A clutch consists of at least twenty eggs arranged in two, perhaps three circular layers. Originally named *Oolithes rugustus* by Young (1965) and renamed by Zhao (1975) for eggs from the Upper Cretaceous Yuanpu and Pingling Formations, Guangdong Province, China. Also known from the Upper Cretaceous Nemegt and Barun Goyot Formations of southern Mongolia, and Zaysan Basin, Kazakhstan.

Macroolithus yaotunensis was named by Zhao Zi-Kui (1975). Eggs measuring 17.5–21 cm (7–8 1/4") long and 6.7–9.4 cm (2 1/2–3 3/4") in diameter, with a length/diameter ratio of 2.2–2.5. There are not many pores on the shell surface considering the size of the egg. In thin section, the shell is between 1.4–1.9 mm thick. The columnar layer shows wavy growth lines. Known from the Upper Cretaceous Yuanpu and Pingling Formations, Guangdong Province, China.

Nanhsiungoolithus *Zhao, 1975*

The eggs are about 13.1–14.5 cm (5 1/2") long, 6.8 cm (2 2/3") in diameter, with a length/diameter ratio of 2, and a shell thickness of 0.8 mm. The surface ornamentation is faint so that the eggshell looks smooth. The distribution of pores is uneven, with most (about nineteen per square millimeter) on the larger end. There are about three per square millimeter on the rest of the shell. Clutches are not known.

Nanhsiungoolithus chuetienensis was named by Zhao Zi-Kui in 1975. The characters that define the species are the same as for the genus. Known

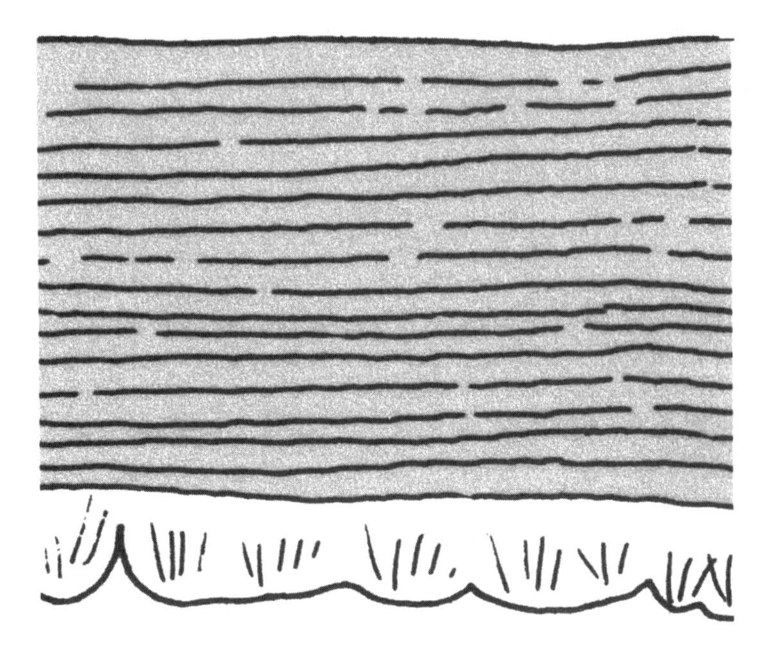

Fig. AII.25. Schematic of Nanhsiungoolithus chuetienensis *thin section.*

from the Upper Cretaceous Yuanpu and Pingling Formations, Guangdong Province, China. Originally called *Oolithes nanhsiungensis* by Young (1965).

Trachoolithus *Mikhailov, 1994a*

Eggshell is thin, about 0.8–1.2 mm thick, with fine linearituberculate ornamentation. The mammillary layer is about 1/3 of the shell thickness. No complete eggs are known.

Trachoolithus faticanus Mikhailov, 1994a. Known from the Upper Cretaceous Dushihin Formation of Mongolia. The characters that define the species are the same as for the genus.

OBLONGOOLITHIDAE Mikhailov, 1996

About the size of *Elongatoolithus* eggs, but differing in having a smooth shell. The shell has a ratite morphotype, and mammillary layer that is 1/3–2/3 shell thickness.

Oblongoolithus *Mikhailov, 1996*

The characters that define the genus are the same as for the family.

Oblongoolithus glaber. Named by Mikhailov (1996) for material from the Barun Goyot(?) Formation, southern Mongolia. The characters that define the species are the same as for the genus.

PHACELOOLITHIDAE Zeng and Zhang, 1979

Oval, slightly flattened eggs (but this could be due to crushing) measuring 16.7–16.8 cm (6½") long, and 14–15 cm (5½–6") in diameter, with length being 1.12–1.2 times width. Surface ornamentation is sagenotuberculate, with numerous pores interspersed. The pore canals are of the prolatocanaliculate type and very irregular in diameter. The shell is thin, about 0.5–0.7 mm thick, with the mammillary layer making up about half of this. The ultrastructure is ratite, with growth lines extending into the ornamentation. The mammillae form clusters of twos or threes.

Phaceloolithus *Zeng and Zhang, 1979*

Fig. AII.26. Schematic of Phaceloolithus hunanensis *thin section.*

The characters that define the genus are the same as for the family.

Phaceloolithus hunanensis Zeng and Zhang, 1979. The characters that define the species are the same as for the genus. No complete clutch is known, but it must have exceeded six eggs; they were apparently not arranged in the nest. Known from the Upper Cretaceous of Hunan Province, China.

As Mikhailov (1997) has noted, Zhao has not mentioned this egg genus in any of his papers since it was named. I suspect the reason is that Zhao has decided that the genus is not valid, possibly because the eggs are some known ornithoid genus that are weathered.

LAEVISOOLITHIDAE Mikhailov, 1991

Small elongated eggs with a ratite morphotype and angusticanaliculate pore canals. The smooth shell is very thin, 0.3–0.6 mm thick. The mammillary layer is $^1/_2$–$^2/_3$ the shell thickness. The family Subtiliolithidae, named by Mikhailov (1991), is no longer considered valid (Mikhailov, 1997). Laevisoolithid eggs were laid by enantiornithine birds and/or small theropods (Mikhailov, 1997).

Laevisoolithus *Mikhailov, 1991*

Eggs about 7 cm (3") long and 3.5–4 cm (1$^1/_4$–1$^1/_2$") in diameter, for a length/diameter ratio of 1.75–2. Shell smooth, less than 1 mm thick, and with mammillary layer $^3/_4$ the shell thickness.

Laevisoolithus sochavai named by Mikhailov in 1991. The characters that define the species are the same as for the genus. The eggs have been found in the Djadokhta and Nemegt formations of southern Mongolia. The ornithoid structure of the shell may indicate that the egg-layer was a small theropod or bird.

Subtiliolithus *Mikhailov, 1991*

Eggshell very thin, 0.3–0.45 mm, and with a smooth surface or very fine tubercles or "bumps."

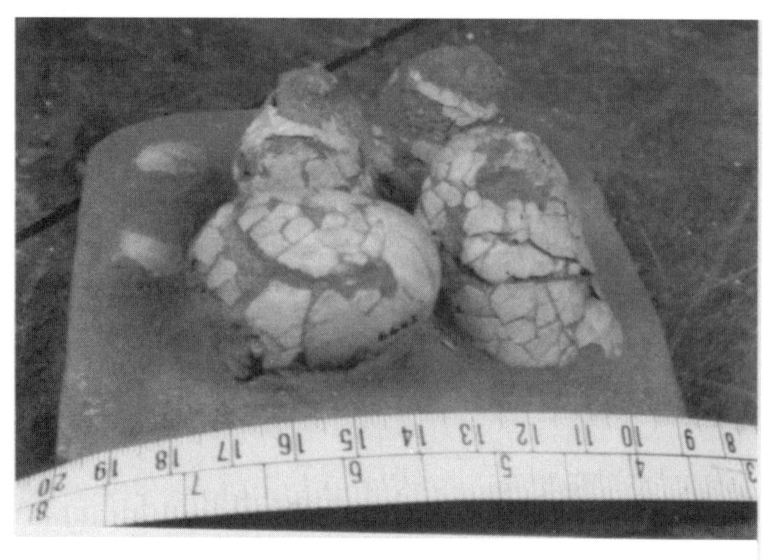

Fig. AII.27. Laevisoolithus sochavai: *A, clutch collected by the Central Asiatic Expedition; B, SEM of thin section—sl = continuous layer, ml = mammillary layer. Scale = 0.1 mm. (B, courtesy of K. Mikhailov.)*

Fig. AII.28. Subtiliolithus kachchhensis *thin section. Scale = 0.5 mm.*

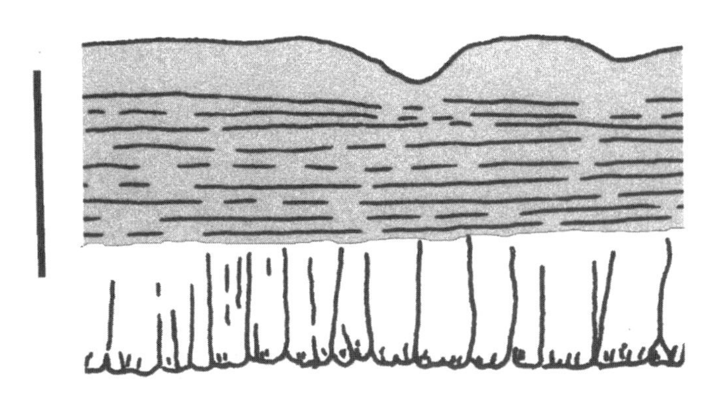

Subtiliolithus kachchhensis Khosla and Sahni, 1995. Known only from very thin eggshell, 0.35–0.45 mm thick. The mammillary layer is $^1/_3$–$^1/_2$ the shell thickness. From the Maastrichtian Lameta Formation of India.

Subtiliolithus microtuberculatus Mikhailov, 1991. Shell very thin, 0.3–0.4 mm. Mammillary layer is thick, $^2/_3$–$^3/_4$ the shell thickness. Lower Maastrichtian, Nemegt Formation of southern Mongolia.

OOFAMILY INCERTAE SEDIS

Ageroolithus *Vianey-Liaud and Lopez-Martinez, 1997*

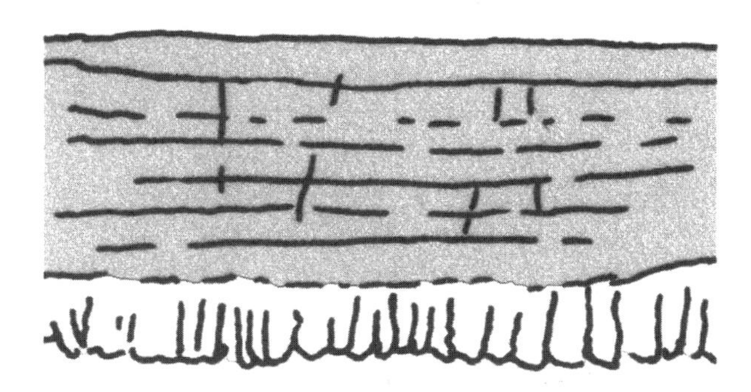

Fig. AII.29. Ageroolithus fontllongensis, *schematic thin section.*

Known only from shell fragments that are thin (0.25–0.36 mm), and with angusticanaliculate pores; shell surface is smooth.

Ageroolithus fontllongensis Vianey-Liaud and Lopez-Martinez, 1997. Known only from the Maastrichtian Aren Sandstone, Tremp Basin, Spain.

Continuoolithus *Zelenitsky, Hills, and Currie, 1996*

Eggs 9.5–12.3 cm ($3^3/_4$–$4^3/_4$") long, 6–7.7 cm ($2^1/_3$–3") in diameter, and a length/diameter ratio of 1.6. The surface is covered with disper-

Fig. AII.30. Continuoolithus canadensis, *whole egg. Scale = 9 cm.*

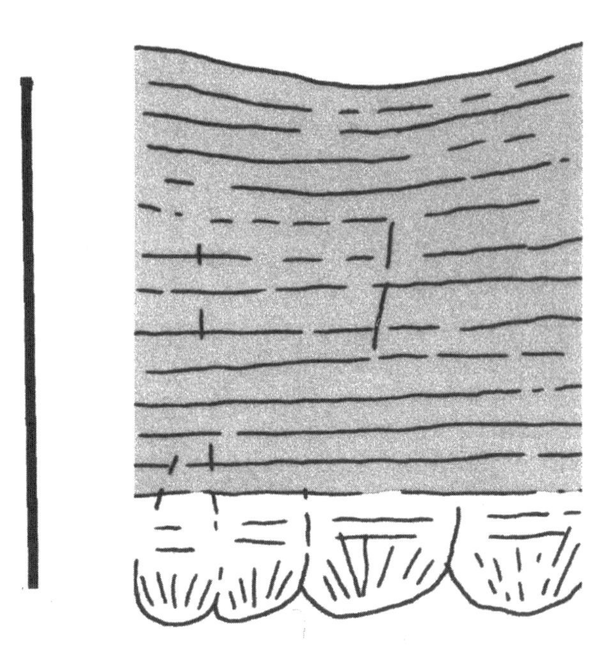

Fig. AII.31. Continuoolithus canadensis, *schematic thin section. Scale = 0.5 mm.*

situberculate ornamentation. Shell thickness 0.94–1.24 mm (excluding the nodes). and with the mammillary layer $^1/_5$–$^1/_6$ the shell thickness (excluding the nodes). These eggs are so different from any other ornithoid egg that they belong in their own family. At one time these eggs were thought to belong to *Troodon,* but we now know this was wrong.

Continuoolithus canadensis Zelenitsky, Hills, and Currie, 1996. Known from the Campanian Oldman Formation of Alberta and the Two Medicine Formation of Montana.

Dispersituberoolithus *Zelenitsky, Hills, and Currie, 1996*

Eggshell 0.26–0.28 mm, not counting the nodes. The mammillary layer is $^2/_5$ the shell thickness, although the boundary between the mammillary and continuous layer is not very sharp. There is a thin external zone on the surface composed of short, vertical calcite prisms; this feature is seen on some modern bird eggshells (see chapter 6). The ornamentation is dispersituberculate and the nodes are tall, about equal to $^1/_3$ to $^1/_2$ the shell thickness.

Dispersituberoolithus exilis Zelenitsky, Hills, and Currie, 1996. Named for eggshell from the Campanian Oldman Formation of Alberta.

Porituberoolithus *Zelenitsky, Hills, and Currie, 1996*

Shell thickness 0.5–0.65 mm, of which $^1/_3$ is the mammillary layer. The boundary between the mammillary and continuous layers is sharp. The surface is covered with dispersituberculate ornamentation. A pore is located atop each node, but offset to one corner. The pore canal is angusticanaliculate.

Porituberoolithus warnerensis named by Zelenitsky, Hills, and Currie (1996) for eggshells from the Oldman Formation, Alberta. The characters that define the species are the same as for the genus.

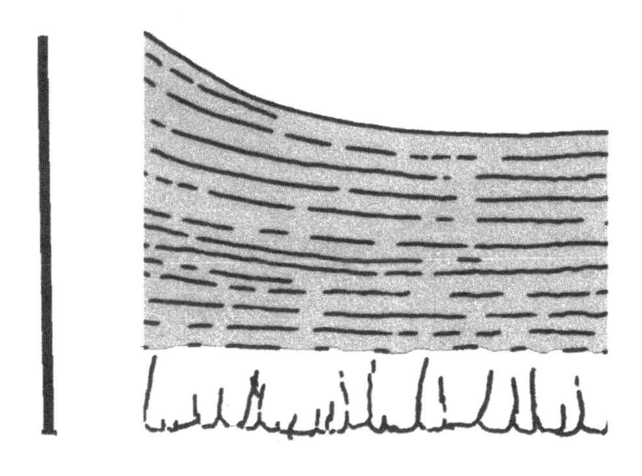

Fig. AII.32. Schematic drawing of Porituberoolithus warnerensis *thin section; note edge of node on left. Scale = 0.25 mm.*

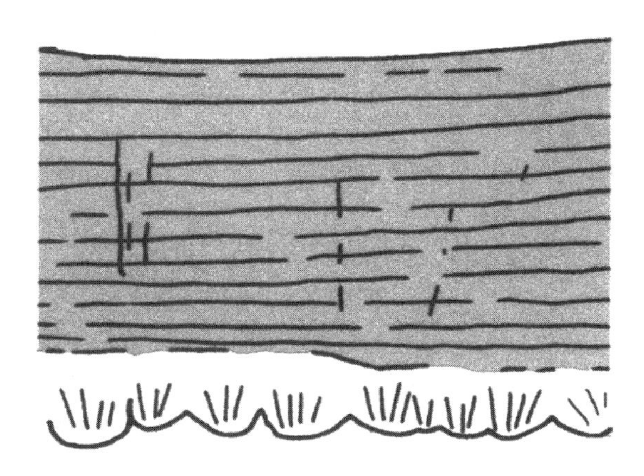

Fig. AII.33. Tristraguloolithus craciodes, *schematic thin section.*

Tristraguloolithus *Zelenitsky, Hills, and Currie, 1996*

Eggshell 0.32–0.36 mm thick, with the mammillary layer $^{2}/_{5}$ the shell thickness. The shell has a thin external zone (see chapter 6). The ornamentation is dispersituberculate.

Tristraguloolithus craciodes Zelenitsky, Hills, and Currie, 1996. Known only from eggshell from the Campanian Oldman Formation of Alberta.

NON-VALID EGG NAME

Parvoolithus *Mikhailov, 1996*

Named for a single small, elongated egg, measuring 4 cm ($1^{1}/_{2}$") long, 2.5 cm in diameter, and having a length/diameter ratio of 1.6. The shell is less than 0.1 mm thick. The eggshell has little structure, except for a small portion, which Mikhailov (1996) claims shows spherulitic morphology. I suspect, however, that the "shell" is a calcite rim of an insect pupal chamber named *Fictovichnus gebiensis* by Johnson and others (1996). The diameter of Mikhailov's "egg" is the same as these chambers.

Parvoolithus tortuosus Mikhailov, 1996. From the Barun Goyot? Formation of Mongolia.

References

Ackerman, R. A. 1980. Physiology and ecological aspects of gas exchange by sea turtle eggs. *American Zoologist* v. 20, pp. 575–583.

Alf, K. 1998. Results of field and laboratory techniques applied to an eggshell site in the Morrison Formation of Garden Park. *Modern Geologist* v. 23, pp. 241–248.

Andrews, R. C. 1932. *The New Conquest of Asia: A Narrative of the Explorations of the Central Asiatic Expeditions in Mongolia and China, 1921–30.* New York: American Museum of Natural History, 678pp.

Andrews, R. C. 1943. *Under a Lucky Star.* New York: Viking Press, 300pp.

Andrews, R. M. 1985. Patterns of growth in reptiles. *Biology of the Reptiles* v. 14, pp. 273–319.

Anonymous. 1995. Dinosaur eggs in Japan for the first time. *Yoiuri* (Newspaper), March 6.

Anonymous. 1997. "Farm fresh" Dorset dinosaur eggs! *Down to Earth* (Geo Supplies Ltd.) no. 21, p. 1.

Ar, A., C. V. Paganelli, R. B. Reeves, D. B. Green, and H. Rahn. 1974. The avian egg: Water vapor conductance, shell thickness, and functional pore area. *Condor* v. 76, pp. 153–158.

Ar, A., H. Rahn, and C. V. Paganelli. 1979. The avian egg: Mass and strength. *Condor* v. 81, pp. 331–337.

Auffenberg, W. 1981. *The Behavioral Ecology of the Komodo Monitor.* Gainesville: University of Florida Press, 406pp.

Barreto, C., R. M. Albrecht, D. E. Bjorling, J. R. Horner, and N. J. Wilsman. 1993. Evidence of the growth plate and growth of long bones in juvenile dinosaurs. *Science* v. 262, pp. 2020–2023.

Barrick, R. E., and W. J. Showers. 1995. Oxygen isotope variability in juvenile dinosaurs (*Hypacrosaurus*): Evidence for thermoregulation. *Paleobiology* v. 21, pp. 552–560.

Barsbold, R., and A. Perle. 1983. On the taphonomy of the joint burial of juvenile dinosaurs and some aspects of their ecology. *Transactions of the Soviet-Mongolian Paleontological Expedition* v. 24, pp. 121–125.

Bazhanov, V. 1961. First discovery of dinosaur eggshells in the USSR. *Trudy Zoologicheskogo Instituta, Akademiya Nauk Kazakstan SSR* v. 15, pp. 177–181.

Beetschen, J.-C., R. Dughi, and F. Sirugue. 1977. Sur la présence de coquilles d'Oiseaux dans le Crétacé supérieur des Corbières occidentales.

Comptes Rendus de l'Académie des Sciences Paris v. 284, pp. 2491–2493.

Bellairs, R. 1991. Overview of early stages of avian and reptilian development. In Deeming, D., and Ferguson, M. (eds.), *Egg Incubation: Its Effects on Embryonic Development in Birds and Reptiles,* pp. 371–383. New York: Cambridge University Press.

Benton, M. 1991. Origin and interrelationships of dinosaurs. In Weishampel, D. B., Dodson, P., and Osmolska, H. (eds.), *The Dinosauria,* pp. 11–30. Berkeley: University of California Press.

Berman, D. S., and S. S. Sumida. 1990. *Limnoscelis dynatis,* a new species (Amphibian, Diadectomorpha) from the Late Pennsylvanian Sangre de Cristo Formation of New Mexico. *Annals of the Carnegie Museum* v. 59, pp. 303–341.

Bertram, B. 1992. *The Ostrich Communal Nesting System.* Princeton: Princeton University Press, 196pp.

Bibler, C. J., and J. G. Schmitt. 1986. Barrier-island coastline deposition and paleogeographic implications of the Upper Cretaceous Horsethief Formation, northern Disturbed Belt, Montana. *The Mountain Geologist* v. 23, pp. 113–127.

Bird, R. T. 1944. A dinosaur walks into the museum. *Natural History* v. 47, pp. 74–81.

Board, R. G. 1982. Properties of avian eggshells and their adaptive value. *Biological Reviews* v. 57, pp. 1–28.

Board, R. G., and V. D. Scott. 1980. Porosity of the avian eggshell. *American Zoologist* v. 20, pp. 339–349.

Board, R. G., and N. H. Sparks. 1991. Shell structure and formation in avian eggs. In Deeming, D. C., and Ferguson, M. J. (eds.), *Egg Incubation: Its Effects on Embryonic Development in Birds and Reptiles,* pp. 71–86. New York: Cambridge University Press.

Boersma, P. D. 1982. Why some birds take so long to hatch. *American Naturalist* v. 120, pp. 733–750.

Bohlin, B. 1953. Fossil reptiles from Mongolia and Kansu. *Vertebrate Palaeontology, The Sino-Swedish Expedition Publication* v. 37, no. 6, pp. 1–67.

Bonaparte, J. F., and M. Vince. 1979. El Hallazgo del primer nido de dinosaurios Triasicos (Saurischia, Prosauropoda), Triasico Superior de Patagonia, Argentina. *Ameghiniana* v. 16, pp. 173–182.

Bond, G. M., R. G. Board, and V. D. Scott. 1988. A comparative study of changes in the fine structure of avian eggshells during incubation. *Zoological Journal of the Linnean Society* v. 92, pp. 105–113.

Bray, E., and K. Hirsch. 1998. Eggshell from the Upper Jurassic Morrison Formation. *Modern Geology* v. 23, pp. 219–240.

Bray, E., and S. Lucas. 1997. Theropod dinosaur eggshell from the Upper Jurassic of New Mexico. In Lucas, S., Williamson, T., and Morgan, G. *New Mexico Museum of Natural History Bulletin* v. 11, pp. 41–43.

Breton, G., R. Fourniet, and J.-P. Watté. 1986. Le lieu de ponte de dinosaures de Rennes-le-Château (aude): Premiers résultats de la camoagne de fouilles 1984. *Annales du Muséum du Havre* v. 32, pp. 1–13.

Brinkman, D. 1986. Microvertebrate sites: Progress and prospects. In *Dinosaur Systematics Symposium, Field Trip Guidebook to Dinosaur Provincial Park,* pp. 24–37. Drumheller, Canada: Tyrrell Museum of Palaeontology.

Britt, B. B., and B. G. Naylor. 1994. An embryonic *Camarasaurus* (Dinosauria, Sauropoda) from the Upper Jurassic Morrison Formation

(Dry Mesa Quarry, Colorado). In Carpenter, K., Hirsch, K. F., and Horner, J. R. (eds.), *Dinosaur Eggs and Babies,* pp. 256–264. New York: Cambridge University Press.
Brown, B., and E. M. Schlaikjer. 1940. The origin of ceratopsian horn-cores. *American Museum Novitates* v. 1065, pp. 1–7.
Brown, R. W. 1950. Cretaceous fish egg capsule from Kansas. *Journal of Paleontology* v. 24, pp. 594–600.
Buckland, W. 1824. Notice on the *Megalosaurus,* or great fossil lizard of Stonesfield. *Transactions of the Geological Society of London* v. 1, pp. 390–396.
Buckman, J., 1859. On some fossil reptilian eggs from the Great Oolite of Chirencester. *Quarterly Journal of the Geological Society, London* v. 16, pp. 107–110.
Buffetaut, E., and J. Le Loeuff. 1991. Late Cretaceous dinosaur faunas of Europe: Some correlation problems. *Cretaceous Research* v. 12, pp. 159–176.
Buffetaut, E., and J. Le Loeuff. 1994. The discovery of dinosaur eggshells in nineteenth-century France. In Carpenter, K., Hirsch, K. F., and Horner, J. (eds.), *Dinosaur Eggs and Babies,* pp. 31–34. New York: Cambridge University Press.
Burghardt, G. M. 1977. Of iguanas and dinosaurs: Social behavior and communication in neonate reptiles. *American Zoologist* v. 17, pp. 177–190.
Burghardt, G. M., H. W. Greene, and A. S. Rand. 1977. Social behavior in hatchling green iguanas: Life at the rookery. *Science* v. 195, pp. 689–693.
Bürgin, T., O. Rieppel, P. M. Sander, and K. Tschanz. 1989. The fossils of Monte San Giorgio. *Scientific American* v. 260, pp. 50–57.
Burley, R., and D. Vadehra. 1989. *The Avian Egg: Chemistry and Biology.* New York: Wiley, 472pp.
Carpenter, K. 1979. Vertebrate fauna of the Laramie Formation (Maastrichtian), Weld County, Colorado. *University of Wyoming Contributions to Geology* v. 17, pp. 37–49.
Carpenter, K. 1982. Baby dinosaurs from the Late Cretaceous Lance and Hell Creek formations and a description of a new species of theropod. *University of Wyoming Contributions to Geology* v. 20, pp. 123–134.
Carpenter, K. 1987. Paleoecological significance of droughts during the Late Cretaceous of the Western Interior. In Currie, P. J., and Koster, E. H. (eds.), *Fourth Symposium on Mesozoic Terrestrial Ecosystems, Short Papers,* pp. 42–47. Drumheller, Canada: Tyrrell Museum of Palaeontology.
Carpenter, K. 1990a. Variation in *Tyrannosaurus rex*. In Carpenter, K., and Currie, P. J. (eds.), *Dinosaur Systematics: Approaches and Perspectives.* New York: Cambridge University Press, pp. 141–145.
Carpenter, K. 1990b. Ankylosaur systematics: Example using *Panoplosaurus* and *Edmontonia* (Ankylosauria, Nodosauridae). In Carpenter, K., and Currie, P. J. (eds.), *Dinosaur Systematics: Approaches and Perspectives,* pp. 281–298. New York: Cambridge University Press.
Carpenter, K. 1992. Behavior of hadrosaurs as interpreted from footprints in the "Mesaverde" Group (Campanian) of Colorado, Utah, and Wyoming. *University of Wyoming Contributions to Geology* v. 29, pp. 81–96.
Carpenter, K. 1994. Baby *Dryosaurus* from the Upper Jurassic Morrison Formation of Dinosaur National Monument. In Carpenter, K., Hirsch,

K. F., and Horner, J. R. (eds.), *Dinosaur Eggs and Babies,* pp. 288–297. New York: Cambridge University Press.

Carpenter, K., K. F. Hirsch, and J. R. Horner. 1994. Introduction. In Carpenter, K., Hirsch, K. F., and Horner, J. R. (eds.), *Dinosaur Eggs and Babies,* pp. 1–11. New York: Cambridge University Press.

Carpenter, K., and J. S. McIntosh. 1994. Upper Jurassic sauropod babies from the Morrison Formation. In Carpenter, K., Hirsch, K. F., and Horner, J. R. (eds.), *Dinosaur Eggs and Babies,* pp. 265–278. New York: Cambridge University Press.

Carruthers, W. 1871. On some supposed vegetable fossils. *Quarterly Journal of the Geological Society, London* v. 27, pp. 443–449.

Case, T. 1978. Speculations on the growth rate and reproduction of some dinosaurs. *Paleobiology* v. 4, pp. 320–328.

Chao T.-K., and Y.-K. Chiang. 1974. Microscopic studies on the dinosaurian egg-shells from Laiyang, Shantung Province. *Scientia Sinica* v. 17, pp. 71–83.

Chen P.-J., Dong Z., and Zhen S. 1998. An exceptionally well-preserved theropod dinosaur from the Yixian Formation of China. *Nature* v. 391, pp. 147–152.

Chiappe, L. M., R. A. Corio, L. Dingus, F. Jackson, A. Chinsamy, and M. Fox. 1998. Sauropod embryos from the Late Cretaceous of Patagonia. *Nature* v. 396, pp. 258–261.

Chinsamy, A. 1990. Physiological implications of the bone histology of *Syntarsus rhodesiensis* (Saurischia, Theropoda). *Palaeontologia Africana* v. 27, pp. 77–82.

Chinsamy, A. 1993. Bone histology and growth trajectory of the prosauropod dinosaur *Massospondylus carinatus* Owen. *Modern Geology* v. 18, pp. 319–329.

Chinsamy, A. 1995. Ontogenetic changes in the bone histology of the Late Jurassic ornithopod *Dryosaurus. Journal of Vertebrate Paleontology* v. 15, pp. 96–104.

Chinsamy, A., and P. Dodson. 1995. Inside a dinosaur bone. *American Scientist* v. 83, pp. 174–180.

Chinsamy, A., T. Rich, and P. Vickers-Rich. 1998. Polar dinosaur bone histology. *Journal of Vertebrate Paleontology* v. 18, pp. 385–390.

Chow, M. M. 1951. Notes on the Late Cretaceous dinosaurian remains and the fossil eggs from Laiyang, Shantung. *Bulletin of the Geological Society of China* v. 31, pp. 89–96.

Chow, M. M. 1954. Additional notes on the microstructure of the supposed dinosaurian eggshells from Laiyang, Shantung. *Acta Palaeontologica Sinica* v. 2, pp. 389–394.

Chure, D., C. Turner, and F. Peterson. 1994. An embryo of *Camptosaurus* from the Morrison Formation (Jurassic: Middle Tithonian) in Dinosaur National Monument, Utah. In Carpenter, K., Hirsch, K. F., and Horner, J. R. (eds.), *Dinosaur Eggs and Babies,* pp. 298–311. New York: Cambridge University Press.

Cody, M. 1966. A general theory of clutch size. *Evolution* v. 20, pp. 174–184.

Cohen, S., A. Cruickshank, K. Joysey, T. Mannin, and P. Upchurch. 1995. The Dinosaur Egg and Embryo Project Exhibition Guide. Leicester, England: Rock Art Publishers.

Colbert, E. H. 1961. *Dinosaurs: Their Discovery and Their World.* New York: Dutton.

Colbert, E. H. 1968. *Men and Dinosaurs: The Search in Field and Laboratory.* New York: Dutton.
Colbert, E. H. 1989. The Triassic dinosaur *Coelophysis. Museum of Northern Arizona Bulletin* v. 57, pp. 1–160.
Colbert, E. H. 1990. Variation in *Coelophysis bauri. In* Carpenter, K., and Currie, P. J. (eds.), *Dinosaur Systematics: Approaches and Perspectives,* pp. 81–90. New York: Cambridge University Press.
Collias, N. E. 1991. Nests. In Brooke, M., and Birkhead, T. (eds.), *The Cambridge Encyclopedia of Ornithology,* pp. 224–226. Cambridge: Cambridge University Press.
Cooke, A. S. 1973. Shell thinning in avian eggs by environmental pollutants. *Environmental Pollution* v. 4, pp. 85–157.
Cooke, A. S. 1975. Pesticides and eggshell formation. *Zoological Society of London Symposium* v. 35, pp. 339–361.
Coombs, W. P. 1982. Juvenile specimens of the ornithischian dinosaur *Psittacosaurus. Palaeontology* v. 25, pp. 89–107.
Coombs, W. P. 1989. Modern analogs for dinosaur nesting and parental behavior. In Farlow, J. O. (ed.), Paleobiology of the dinosaurs. *Geological Society of America Special Paper* v. 238, pp. 21–53.
Cott, H. B. 1961. Scientific results of an inquiry into the ecology and economic status of the Nile Crocodile (*Crocodilus niloticus*) in Uganda and northern Rhodesia. *Transactions of the Zoological Society of London* v. 29, pp. 211–337.
Cousin, R., G. B. Breton, and N. Gomez. 1987. La campagne de "fouilles préliminaires" sur les lieux de ponte de dinosaures de Rousset-sur-Arc (Bouches-du-Rhône). *Bulletin trimestriel de la Société Géologique Normandie et des Amis Muséum du Havre* v. 74, pp. 5–15.
Cousin, R., G. B. Breton, R. Fournier, and J.-P. Watté. 1994. Dinosaur egglaying and nesting in France. In Carpenter, K., Hirsch, K. F., and Horner, J. R. (eds.), *Dinosaur Eggs and Babies,* pp. 56–74. New York: Cambridge University Press.
Currie, P. J. 1983. Hadrosaur trackways from the Lower Cretaceous of Canada. *Acta Palaeontologica Polonica* v. 28, pp. 63–73.
Currie, P. J., and R. L. Carroll. 1984. Ontogenetic changes in the eosuchian reptile *Thadeosaurus. Journal of Vertebrate Paleontology* v. 4, pp. 68–84.
Currie, P. J., and P. Dodson. 1984. Mass death of a herd of ceratopsian dinosaur. In Reif, W. E., and Westphal, F. (eds.), *Third Symposium on Mesozoic Terrestrial Ecosystems, Short Papers,* pp. 61–66. Tübingen, Germany: ATTEMPTO Verlag.
Dal Sasso, C., and M. Signore. 1998. Exceptional soft-tissue preservation in a theropod dinosaur from Italy. *Nature* v. 392, pp. 383–387.
Dantas, P. M., J. J. Moratalla, K. F. Hirsch, and V. F. Santos. 1992. Mesozoic reptile eggs from Portugal. New data. *Dinosaurs and Other Fossil Reptiles of Europe, Second Georges Cuvier Symposium Abstracts.* Montbeliard, France.
Dauphin, Y. 1990. Microstructures et Composition chimique des coquilles d'oeufs d'oiseaux et de Reptiles. I. Oiseaux actuels. *Palaeontographica, Abt. A.* v. 214, pp. 1–12.
Dauphin, Y. 1992. Microstructures et Composition chimique des coquilles d'oeufs d'oiseaux et de Reptiles. II. Dinosauriens du Sud de la France. *Palaeontographica* v. 223, pp. 1–17.
Dauphin, Y., and J.-J. Jaeger. 1990. Géochimie des oeufs de dinosaures du

Bassin d'Aix en Provence (Crétacé): Influence de la méthode d'analyse. *Neues Jahrbuch für Geologie und Paläontologie, Monatshefte,* v. 1990, pp. 479–492.

Deeming, D. 1991. Reason for the dichotomy in egg turning in birds and reptiles. In Deeming, D., and Ferguson, M. (eds.), *Egg Incubation: Its Effects on Embryonic Development in Birds and Reptiles,* pp. 307–323. New York: Cambridge University Press.

Deeming D. C., and M. W. Ferguson. 1990. Morphometric analysis of embryonic development in *Alligator mississippiensis, Crocodylus johnstoni,* and *Crocodylus porosus. Journal of Zoology, London* v. 221, pp. 419–439.

Deeming D. C., and M. W. Ferguson. 1991. Physiological effects of incubation temperature on embryonic development in reptiles and birds. In Deeming, D., and Ferguson, M. (eds.), *Egg Incubation: Its Effects on Embryonic Development in Birds and Reptiles,* pp. 147–171. New York: Cambridge University Press.

Deeming, D., L. B. Halstead, M. Manabe, and D. Unwin. 1993. An ichthyosaur embryo from the Lower Lias (Jurassic: Hettangian) of Somerset, England, with comments on the reproductive biology of ichthyosaurs. *Modern Geology* v. 18, pp. 423–442.

Deitz, D. C., and T. C. Hines. 1980. Alligator nesting in north-central Florida. *Copeia* v. 1980, pp. 249–258.

Dobie, J. L. 1978. A fossil amniote egg from an Upper Cretaceous deposit (Mooreville Chalk of the Selma Group) in Alabama. *Copeia* v. 1978, pp. 460–464.

Dodson, P. 1975a. Functional and ecological significance of relative growth in *Alligator. Journal of Zoology* v. 175, pp. 315–355.

Dodson, P. 1975b. Relative growth in two sympatric species of *Sceloporus. American Naturalist* v. 94, pp. 421–450.

Dodson, P. 1975c. Taxonomic implications of relative growth in lambeosaurine hadrosaurs. *Systematic Zoology* v. 24, pp. 37–54.

Dodson, P. 1976. Quantitative aspects of relative growth and sexual dimorphism in *Protoceratops. Journal of Paleontology* v. 50, pp. 929–940.

Dong Z. 1980. The dinosaurian faunas of China and their stratigraphic distribution. *Journal of Stratigraphy* v. 4, pp. 256–263.

Dong Z., and P. J. Currie. 1996. On the discovery of an oviraptorid skeleton on a nest of eggs at Bayan Mandahu, Inner Mongolia, People's Republic of China. *Canadian Journal of Earth Sciences* v. 33, pp. 631–636.

Dong Z., P. J. Currie, and D. A. Russell. 1989. The 1988 field program of the dinosaur project. *Vertebrata PalAsiatica* v. 27, pp. 233–236.

Dorr, J. 1985. Newfound Early Cretaceous dinosaurs and other fossils in southeastern Idaho and westernmost Wyoming. *University of Michigan, Contributions from the Museum of Paleontology* v. 27, pp. 73–85.

Dowling, H. G., and W. E. Duellman. 1978. Systematic herpetology: A synopsis of families and higher categories. New York: HISS Publications.

Drent, R. 1975. Incubation. *Avian Biology* v. 5, pp. 333–420.

Dufaure, J. P., and J. Hubert. 1961. Table de dévelopement du lézard vivipare *Lacerta (Zootoca) vivipara* Jacquin. *Archives D'anatomie Microscopique et De Morphologie Expérimentale* v. 50, pp. 309–328.

Duvall, D., L. Guillette, and R. Jones. 1982. Environmental control of

reptilian reproductive cycles. *Biology of the Reptiles* v. 13, pp. 201–231.
Elinson, R. P. 1989. Egg evolution. In Wake, D. B., and Roth, G. (eds.), *Complex Organismal Functions: Integration and Evolution in Vertebrates,* pp. 251–262. New York: Wiley.
Elzanowski, A. 1981. Embryonic bird skeletons from the Late Cretaceous of Mongolia. *Palaeontologia Polonica* v. 42, pp. 147–179.
Elzanowski, A., and P. Wellnhofer. 1992. A new link between theropods and birds from the Cretaceous of Mongolia. *Nature* v. 359, pp. 821–823.
Enlow, D. H., and S. O. Brown. 1956. A comparative histological study of fossil and recent bone tissues. Part I. *Texas Journal of Science* v. 8, pp. 403–443.
Enlow, D. H., and S. O. Brown. 1957. A comparative histological study of fossil and recent bone tissues. Part II. *Texas Journal of Science* v. 9, pp. 185–214.
Enlow, D. H., and S. O. Brown. 1958. A comparative histological study of fossil and recent bone tissues. Part III. *Texas Journal of Science* v. 10, pp. 187–230.
Erben, H. K., J. Hoefs, and K. H. Wedepohl. 1979. Paleobiological and isotopic studies of eggshells from a declining dinosaur species. *Paleobiology* v. 5, pp. 380–414.
Estes, R., and E. E. Williams. 1984. Ontogenetic variation in the molariform teeth in lizards. *Journal of Vertebrate Paleontology* v. 4, pp. 96–107.
Ewert, M. 1985. Embryology of turtles. *Biology of the Reptiles* v. 14, pp. 75–267.
Ewert, M. 1991. Cold torpor, diapause, delayed hatching and aestivation in reptiles and birds. In Deeming, D., and Ferguson, M. (eds.), *Egg Incubation: Its Effects on Embryonic Development in Birds and Reptiles,* pp. 173–191. New York: Cambridge University Press.
Faccio, G. 1994. Dinosaurian eggs from the Upper Cretaceous of Uruguay. In Carpenter, K., Hirsch, K. F., and Horner, J. R. (eds.), *Dinosaur Eggs and Babies,* pp. 47–55. New York: Cambridge University Press.
Farlow, J. O. 1975. The behavioral significance of frill and horn morphology in ceratopsian dinosaurs. *Evolution* v. 29, pp. 353–361.
Farlow, J. O., and M. K. Brett-Surman. 1997. *The Complete Dinosaur.* Bloomington: Indiana University Press, 752pp.
Farlow, J. O., P. Dodson, and A. Chinsamy. 1995. Dinosaur biology. *Annual Review of Ecology and Systematics* v. 26, pp. 445–471.
Farlow, J. O., J. G. Pittman, and J. M. Hawthorne. 1989. *Brontopus birdi,* Lower Cretaceous sauropod footprints from the U.S. Gulf Coastal Plain. In Gillette, D. D., and Lockley, M. G. (eds.), *Dinosaur Tracks and Traces,* pp. 371–394. New York: Cambridge University Press.
Farlow, J. O., C. V. Thompson, and D. E. Rosner. 1976. Plates of the dinosaur *Stegosaurus:* Forced convection heat loss? *Science* v. 192, pp. 1123–1125.
Fastovsky, D. E., and D. B. Weishampel. 1996. *The Evolution and Extinction of the Dinosaurs.* New York: Cambridge University Press, 461pp.
Ferguson, M. 1985. Reproductive biology and embryology of crocodilians. In Billet, F., and Gans, C. (eds.), *Biology of the Reptiles.* v. 14, pp. 330–491.
Ferguson, M. 1987. Post-laying stages of embryonic development for crocodilians. In Webb, G., Manolis, S., and Whitehead, P. (eds.),

Wildlife Management: Crocodiles and Alligators, pp. 427–444. Canberra: Surrey.

Ferguson, M. W., and T. Joanen. 1983. Temperature-dependent sex determination in *Alligator mississippiensis. Journal of Zoology* v. 200, pp. 143–177.

Fiorillo, A. R. 1987. Significance of juvenile dinosaurs from Careless Creek Quarry (Judith River Formation), Wheatland County, Montana. In Currie, P. J., and Koster, E. H. (eds.), *Fourth Symposium on Mesozoic Terrestrial Ecosystems, Short Papers,* pp. 88–95. Drumheller, Canada: Tyrrell Museum of Palaeontology.

Fitch, H. S. 1970. Reproductive cycles in lizards and snakes. *University of Kansas Museum of Natural History Miscellaneous Publication* v. 52, pp. 1–247.

Fitch, H. S. 1981. Sexual size differences in reptiles. *University of Kansas Museum of Natural History Miscellaneous Publication* v. 70, pp. 1–72.

Fitch, H. S. 1985. Variation in clutch and litter size in New World reptiles. *University of Kansas Museum of Natural History Miscellaneous Publication* v. 76, pp. 1–60.

Folinsbee, R. E., P. Fritz, H. R. Krouse, and A. R. Robble. 1970. Carbon-13 and Oxygen-18 in dinosaur, crocodile, and bird eggshells indicate environmental conditions. *Science* v. 168, pp. 1353–1356.

Forster, C. A. 1990. Evidence for juvenile groups in the ornithopod dinosaur *Tenontosaurus tilletti* Ostrom. *Journal of Paleontology* v. 64, pp. 164–165.

Freeman, B. M., and M. A. Vince. 1974. *Development of the Avian Embryo.* London: Chapman & Hall, 362pp.

Frith, H. J. 1957. Experiments on the control of temperature in the nesting mounds of Mallee Fowl *Leipoa ocellata. Commonwealth Scientific and Industrial Research Organisation Wildlife Research* v. 2, pp. 101–110.

Frith, H. J. 1980. Incubator bird. In *Birds: Readings from Scientific American,* pp. 142–148. San Francisco: Freeman.

Fritz, S. 1988. *Tyrannosaurus* sex: A love tail. *Omni* pp. 64–68, 78.

Galton, P. M. 1982. Juveniles of the stegosaurian dinosaur *Stegosaurus* from the Upper Jurassic of North America. *Journal of Vertebrate Paleontology* v. 2, pp. 47–62.

Geist, N. R., and T. D. Jones. 1996. Juvenile skeletal structure and the reproductive habits of dinosaurs. *Science* v. 272, pp. 712–714.

Ghevariya, Z. G., and C. Srikarni. 1990. Anjar Formation, its fossils and their bearing on the extinction of dinosaurs. In Sahni, A. (ed.), *Cretaceous Event Stratigraphy and the Correlation of the Indian Nonmarine Strata, Contributions from the Seminar cum Workshop I.G.C.P. 216 and 245,* pp. 106–109. Chandigarh, India: Punjab University.

Gilmore, C. W. 1909. Osteology of the Jurassic reptile *Camptosaurus,* with a revision of the species of the genus, and a description of two new species. *Proceeding of the U.S. National Museum* v. 36, pp. 197–332.

Gilmore, C. W. 1917. *Brachyceratops,* a ceratopsian dinosaur from the Two Medicine Formation of Montana. *U.S. Geological Survey Professional Paper* v. 103, pp. 1–45.

Gore, R. 1993. Dinosaur. *National Geographic Magazine* v. 183, pp. 2–53.

Gould, S. J. 1980. A biological homage to Mickey Mouse. In *The Panda's Thumb,* pp. 95–107. New York: Norton.

Gould, S. J. 1984. Sex, drugs, disasters, and the extinction of the dinosaurs. *Discover* v. 5(3), pp. 67–72.

Granger, W. 1936. The story of the dinosaur eggs. *Natural History* v. 38, pp. 21–25.
Grigorescu, D. 1993. The Latest Cretaceous dinosaur eggs and embryos from the Hateg Basin—Romania. *Revue de Paléobiologie,* Volume Spécial 7, pp. 95–99.
Grigorescu, D., D. Weishampel, D. Norman, M. Seclamen, M. Rusu, A. Baltres, and V. Teodorescu. 1994. Late Maastrichtian dinosaur eggs from the Ha Basin (Romania). In Carpenter, K., Hirsch, K. F., and Horner, J. R. (eds.), *Dinosaur Eggs and Babies,* pp. 75–87. New York: Cambridge University Press.
Grine, F. E., and J. W. Kitching. 1987. Scanning electron microscopy of early dinosaur egg shell structure: A comparison with other rigid sauropsid eggs. *Scanning Microscopy* v. 1, pp. 615–630.
Hamburger, V., and H. L. Hamilton. 1951. A series of normal stages in the development of the chick embryo. *Journal of Morphology* v. 88, pp. 49–92.
Hao Y., and S. Guan. 1984. The Lower–Upper Cretaceous and Cretaceous–Tertiary boundaries in China. *Bulletin of the Geological Society of Denmark* v. 33, pp. 129–138.
Hayward, J. L., K. F. Hirsch, and T. C. Robertson. 1991. Rapid dissolution of avian eggshells buried by Mount St. Helens ash. *Palaios* v. 6, pp. 174–178.
Heinrich, R. E., C. B. Ruff, and D. Weishampel. 1993. Femoral ontogeny and locomotor biomechanics of *Dryosaurus lettowvorbecki* (Dinosauria, Iguanodontia). *Zoological Journal of the Linnean Society* v. 108, pp. 179–196.
Hirsch, K. F. 1979. The oldest vertebrate egg? *Journal of Paleontology* v. 53, pp. 1068–1084.
Hirsch, K. F. 1983. Contemporary and fossil chelonian eggshells. *Copeia* v. 1983, pp. 382–397.
Hirsch, K. F. 1986. Not every "egg" is an egg. *Journal of Vertebrate Paleontology* v. 6, pp. 200–201.
Hirsch, K. F. 1988. Eggshell material. In Mourier, T., Bengtson, P., Bonhomme, M., Buge, E., Cappetta, H., Crochet, J.-Y., Feist, M., Hirsch, K. F., Jaillard, E., Laubacher, G., Lefranc, J. P., Moullade, M., Noblet, C., Pons, D., Rey, J., Sigé, B., Tambareau, Y., and Taquet, P. The Upper Cretaceous–Lower Tertiary marine to continental transition in the Bagua Basin, northern Peru. *Newsletter in Stratigraphy* v. 19, pp. 143–177.
Hirsch, K. F. 1994. Upper Jurassic eggshells from the Western interior of North America. In Carpenter, K., Hirsch, K. F., and Horner, J. R. (eds.), *Dinosaur Eggs and Babies,* pp. 137–150. New York: Cambridge University Press.
Hirsch, K. F. 1996. Parataxonomic classification of fossil chelonian and gecko eggs. *Journal of Vertebrate Paleontology* v. 16, pp. 752–762.
Hirsch, K. F., and L. F. Lopez-Jurado. 1987. Pliocene chelonian fossil eggs from Gran Canaria, Canary Islands. *Journal of Vertebrate Paleontology* v. 7, pp. 96–99.
Hirsch, K. F., and B. Quinn. 1990. Eggs and eggshell fragments from the Upper Cretaceous Two Medicine Formation of Montana. *Journal of Vertebrate Paleontology* v. 10, pp. 491–511.
Hirsch, K. F., A. J. Kihm, and D. K. Zelenitsky. 1997. New eggshell of ratite morphology with predation marks from the Eocene of Colorado. *Journal of Vertebrate Paleontology* v. 17, pp. 360–369.

Hirsch, K. F., K. L. Stadtman, W. E Miller, and J. H. Madsen. 1989. Upper Jurassic dinosaur egg from Utah. *Science* v. 243, pp. 1711–1713.
Holman, J. A. 1995. Pleistocene amphibians and reptiles in North America. *Oxford Monograph on Geology and Geophysics* v. 32, pp. 1–243.
Hopson, J. A. 1975. The evolution of cranial display structures in hadrosaurian dinosaurs. *Paleobiology* v. 1, pp. 21–43.
Horner, J. R. 1982. Evidence of colony nesting and "site fidelity" among ornithischian dinosaurs. *Nature* v. 297, pp. 675–676.
Horner, J. R. 1984. The nesting behavior of dinosaurs. *Scientific American* v. 250, pp. 130–137.
Horner, J. R. 1987. Ecological and behavioral implications derived from a dinosaur nesting site. In Czerkas, S. J., and Olsen, E. C. (eds.), *Dinosaurs Past and Present,* V. II, pp. 51–63. Los Angeles: Natural History Museum of Los Angeles County.
Horner, J. R. 1992. Dinosaur behavior and growth. Fifth North American Paleontological Convention Abstracts and Program. *Paleontological Society Special Publication* v. 6, p. 135.
Horner, J. R., and P. J. Currie. 1994. Embryonic and neonatal morphology and ontogeny of a new species of *Hypacrosaurus* (Ornithischia, Lambeosauridae) from Montana and Alberta. In Carpenter, K., Hirsch, K. F., and Horner, J. R. (eds.), *Dinosaur Eggs and Babies,* pp. 312–336. New York: Cambridge University Press.
Horner, J. R., and E. Dobb. 1997. *Dinosaur Lives.* New York: HarperCollins, 244pp.
Horner, J. R., and J. Gorman. 1988. *Digging Dinosaurs.* New York: Workman, 210pp.
Horner, J. R., and R. Makela. 1979. Nest of juveniles provides evidence of family structure among dinosaurs. *Nature* v. 82, pp. 296–298.
Horner, J. R., and D. B. Weishampel. 1988. A comparative embryological study of two ornithischian dinosaurs. *Nature* v. 332, pp. 256–257.
Horner, J. R., and D. B. Weishampel. 1996. A comparative embryological study of two ornithischian dinosaurs—a correction. *Nature* v. 383, p. 103.
Hotton, N. 1980. An alternative to dinosaur endothermy: The happy wanderers. In Thomas, D. K., and Olson, E. C. (eds.), *A Cold Look at the Warm-Blooded Dinosaurs,* pp. 311–350. American Association for the Advancement of Science Symposium v. 28.
Hutton, J. 1989. Movements, home range, dispersal and the separation of size classes in the Nile crocodile. *American Zoologist* v. 29, pp. 1033–1049.
Immelmann, K. 1971. Ecological aspects of periodic reproduction. *Avian Biology* v. 1, pp. 341–389.
Iverson, J. B., and M. A. Ewert. 1991. Physical characteristics of reptilian eggs and a comparison with avian eggs. In Deeming, D. C., and Ferguson, M. eds.), *Egg Incubation: Its Effects on Embryonic Development in Birds and Reptiles,* pp. 87–100. New York: Cambridge University Press.
Jacobs, L. L., D. A. Winkler, P. A. Murry, and J. M. Maurice. 1994. A nodosaurid scuteling from the Texas shore of the Western Interior Seaway. In Carpenter, K., Hirsch, K. F., and Horner, J. R. (eds.), *Dinosaur Eggs and Babies,* pp. 337–346. New York: Cambridge University Press.
Jain, S. L. 1989. Recent dinosaur discoveries in India, including eggshells, nests, and coprolites. In Gillette, D. D., and Lockley, M. G. (eds.),

Dinosaur Tracks and Traces, pp. 99–108. Cambridge: Cambridge University Press.
Jensen, J. A. 1966. Dinosaur eggs from the Upper Cretaceous North Horn Formation of central Utah. *Brigham Young University Geology Studies* v. 13, pp. 55–67.
Jensen, J. A. 1970. Fossil eggs in the Lower Cretaceous of Utah. *Brigham Young University Geology Studies* v. 17, pp. 51–65.
Jepsen, G. L. 1931. Dinosaur egg shell fragments from Montana. *Science* v. 73, pp. 12–13.
Jepsen, G. L. 1964. Riddle of the terrible lizards. *American Scientist* v. 52, pp. 227–246.
Jerzykiewicz, T., and D. Russell. 1991. Late Mesozoic stratigraphy and vertebrates of the Gobi Basin. *Cretaceous Research* v. 12, pp. 345–377.
Jerzykiewicz, T., P. J. Currie, D. A. Eberth, P. A. Johnston, E. H. Koster, and Z. Zheng. 1993. Djadokhta Formation correlative strata in Chinese Inner Mongolia: An overview of the stratigraphy, sedimentary geology, and paleontology and comparison with the type locality in the pre-Altai Gobi. *Canadian Journal of Earth Science* v. 30, pp. 2180–2195.
Joanen, T. 1969. Nesting ecology of alligators in Louisiana. *Annual Conference, Southeast Association of Game and Fish Commissioners Proceedings,* pp. 141–151.
Joanen, T., and L. L. McNease. 1989. Ecology and physiology of nesting and early development of the American alligator. *American Zoologist* v. 29, pp. 987–998.
Johnston, P. A. 1979. Growth rings in dinosaur teeth. *Nature* v. 278, pp. 635–636.
Johnston, P. A., D. A. Eberth, and P. K. Anderson. 1996. Alleged vertebrate eggs from the Upper Cretaceous redbeds, Gobi Desert, are fossil insect (Coleoptera) pupal chambers: *Fictovichnus* new ichnogenus. *Canadian Journal of Earth Sciences* v. 33, pp. 511–525.
Jollie, M. 1973. *Chordate Morphology.* Huntington, N.Y.: Kreiger, 478pp.
Kérourio, P. 1981. La distribution des "coquilles d'oeufs de dinosauriens multistratifiées" dans le Maestrichtien continental du Sud de la France. *Geobios* v. 14, pp. 533–536.
Kérourio, P., and B. Sigé. 1984. L'apport des coquilles d'oeufs de dinosaures de Laguna Umayo a l'âge de la Formation Vilquechico (Pérou) et à la comprehension de *Perutherium altiplanense. Newsletter in Stratigraphy* v. 13, pp. 133–142.
Khosla, A., and A. Sahni. 1995. Parataxonomic classification of Late Cretaceous dinosaur eggshells from India. *Journal of the Palaeontological Society of India* v. 40, pp. 87–102.
Kiff, L. F., D. B. Peakall, and S. R. Wilbur. 1979. Recent changes in California condor eggshells. *Condor* v. 81, pp. 166–172.
Kirkland, J. I. 1994. Predation of dinosaur nests by terrestrial crocodilians. In Carpenter, K., Hirsch, K. F., and Horner, J. R. (eds.), *Dinosaur Eggs and Babies,* pp. 124–133. New York: Cambridge University Press.
Kitching, J. W. 1979. Preliminary report on a clutch of six dinosaurian eggs from the Upper Triassic Elliot Formation, Orange Free State. *Palaeontographica Africana* v. 22, pp. 41–45.
Kohring, R. 1990a. Fossile Reptil-Eischalen (Chelonia, Crocodilia, Dinosauria) aus dern unteren Barremiurn von Galve (Provinz Teruel, SE-Spanien). *Paleontologische Zeitschrift* v. 64, pp. 329–344.

Kohring, R. 1990b. Upper Jurassic chelonian eggshell fragments from the Guimarota Mine (Central Portugal). *Journal of Vertebrate Paleontology* v. 10, pp. 128–130.

Kohring, R. 1995. Reflections on the origin of the amniote egg in light of reproductive strategies and shell structure. *Historical Biology* v. 10, pp. 259–275.

Kohring, R., K. Bandel, D. Kortum, and S. Parthasararthy. 1996. Shell structure of a dinosaur egg from the Maastrichtian of Ariyalus (southern India). *Neues Jahrbuch für Geologie und Paläontologie, Monatshefte* v. 1996, pp. 48–64.

Kohring, R., and K. F. Hirsch. 1996. Crocodilian and avian eggshells from the Middle Eocene of the Geiseltal, Eastern Germany. *Journal of Vertebrate Paleontology* v. 16, pp. 67–80.

Kolesnikov, C. H., and A. V. Sochava. 1972. A paleobiological study of Cretaceous dinosaur eggshells from the Gobi. *Paleontological Journal* v. 6, pp. 235–245.

Lance, V. 1987. Hormonal control of reproduction in crocodilians. In Webb, G., Manolis, S., and Whitehead, P. (eds.), *Wildlife Management: Crocodiles and Alligators,* pp. 409–415. Canberra: Surrey.

Lapparent, A.-F. de. 1967. Les dinosaures de France. *Sciences* v. 51, pp. 5–18.

Lapparent, A. F., and G. Zybszewski. 1957. Les Dinosauriens du Portugal. *Mémoires du Service géologique de Portugal* v. 2, pp. 1–63.

Lehman, T. M. 1990. The ceratopsian subfamily Chasmosaurinae: Sexual dimorphism and systematics. In Carpenter, K., and Currie, P. J. (eds.), *Dinosaur Systematics: Approaches and Perspectives,* pp. 211–229. New York: Cambridge University Press.

Leuthold, W. 1977. *African Ungulates.* Berlin: Springer Verlag, 307pp.

Li Y., An C., Zhu Y., Zhang Y., Liu Y., Qu L., You L., Liang X., Li X., Qu L., Zhou Z., and Chen Z. 1995. DNA isolation and sequence analysis of dinosaur DNA from Cretaceous dinosaur egg in Xixia, Henan, China. *Acta Scientiarum Naturalium Universitatis Pekinensis* v. 31, pp. 148–152.

Li Y., Yin Z., and Liu Y. 1995. The discovery of a new genus of dinosaur egg from Xixia, Henan, China. *Journal of Wuhan Institute of Chemical Technology* v. 17, pp. 38–41.

Lockley, M. G., B. H. Young, and K. Carpenter. 1983. Hadrosaur locomotion and herding behavior: Evidence from footprints in the Mesaverde Formation, Grand Mesa Coal Field, Colorado. *Mountain Geologist* v. 20, pp. 5–14.

Long, J. A., and K. I. McNamara. 1997. Heterochrony: The key to dinosaur evolution. In D. L. Wolberg, Stump, E., and Rosenberg, G. R. (eds.), *Dinofest International Proceedings,* pp. 113–123. Philadelphia: Academy of Natural Sciences.

Madsen, J. 1976. *Allosaurus fragilis:* A revised osteology. *Utah Geological and Mineralogical Survey Bulletin* v. 1091, pp. 1–163.

Magnusson, W. E. 1982. Mortality of eggs of the crocodile *Crocodylus porosus* in northern Australia. *Journal of Herpetology* v. 16, pp. 121–130.

Mamay, S. 1994. Fossil eggs of probable piscine origin preserved on Pennsylvanian *Sphenopteridium* foliage from the Kinney Quarry, central New Mexico. *Journal of Vertebrate Paleontology* v. 14, pp. 320–326.

Marsh, O. C. 1883. Principal characters of American Jurassic dinosaurs.

Part VI. Restoration of *Brontosaurus*. *American Journal of Science* v. 26, pp. 81–85.
Marsh, O. C. 1888. Notice of a new genus of Sauropoda and other new dinosaurs from the Potomac Formation. *American Journal of Science* v. 35, pp. 89–94.
Martin, V. 1994. Baby sauropods from the Sao Khua Formation (Lower Cretaceous) in northeastern Thailand. *Gaia* v. 10, pp. 147–153.
Maryanska, T., and H. Osmolska. 1975. Protoceratopsidae (Dinosauria) of Asia. *Palaeontologica Polonica* v. 33, pp. 133–181.
Massare, J. A., and J. M. Callaway. 1988. Live birth in ichthyosaurs: Evidence and implications. *Journal of Vertebrate Paleontology, Abstracts with Program* v. 8, p. 21A.
Mateer, N. J. 1989. Upper Cretaceous reptilian eggs from the Zhejiang Province, China. In Gillette, D., and Lockley, M. (eds.), *Dinosaur Tracks and Traces,* pp. 116–118. New York: Cambridge University Press.
Mateus, I., H. Mateus, M. Antunes, O. Mateus, P. Taquet, V. Ribeiro, and G. Manuppella. 1997a. Covée, oeufs, et embryons d'un Dinosaure Theropode du Jurassique supérieur de Lourinhã (Portugal). *C. R. Academie Sciences, Paris, Sciences de la terre et des planètes* v. 325, pp. 71–78.
Mateus, I., H. Mateus, M. Antunes, O. Mateus, P. Taquet, V. Ribeiro, and G. Manuppella. 1997b. Upper Jurassic theropod dinosaur embryos from Lourinhã (Portugal). *Memórias da Acaddemia das Ciênas de Lisboa,* pp. 101–109.
Maxwell, W. D., and J. R. Horner. 1994. Neonate dinosaurian remains and dinosaurian eggshell from the Cloverly Formation, Montana. *Journal of Vertebrate Paleontology* v. 14, pp. 143–146.
McFarland, L. Z., R. L. Garrett, and J. A. Nowell. 1971. Normal eggshells and thin eggshells caused by organochlorine insecticides viewed by the scanning electron microscope. *Scanning Electron Microscopy* v. 1, pp. 377–384.
Melmore, S. 1930. A reptilian egg from the Lias of Whitby. *Proceedings of the Yorkshire Philosophical Society for 1930.*
Mikhailov, K. E. 1987a. Pore complexes of noncarinate avian eggshells and the mechanism of pore formation. *Paleontological Journal* v. 3, pp. 77–86.
Mikhailov, K. E. 1987b. The principal structure of the avian egg-shell: Data of SEM studies. *Acta Zoologica Cracoviensis* v. 36, pp. 193–238.
Mikhailov, K. E. 1988. Some aspects of the structure of the shell of the egg. *Paleontological Journal* v. 1987, pp. 54–61.
Mikhailov, K. E. 1991. Classification of fossil eggshells of amniotic vertebrates. *Acta Palaeontologica Polonica* v. 36, pp. 193–238.
Mikhailov, K. E. 1992. The microstructure of avian and dinosaurian eggshell: Phylogenetic implications. *Natural History Museum of Los Angeles County, Science Series* v. 36, pp. 361–373.
Mikhailov, K. 1994a. Theropod and protoceratopsian dinosaur eggs from the Cretaceous of Mongolia and Kazakstan. *Paleontological Journal* v. 28, pp. 101–120.
Mikhailov, K. 1994b. Eggs of sauropod and ornithopod dinosaurs from the Cretaceous deposits of Mongolia. *Paleontological Journal* v. 28, pp. 141–159.
Mikhailov, K. 1996. New genera of fossil eggs from the Upper Cretaceous of Mongolia. *Paleontologicheskiy Zhurnal* v. 1996, pp. 122–124.

Mikhailov, K. 1997. Fossil and recent eggshell in amniotic vertebrates: Fine structure, comparative morphology, and classification. *Special Papers in Palaeontology* v. 56, pp. 1–80.
Mikhailov, K. E., E. S. Bray, and K. F. Hirsch. 1996. Parataxonomy of fossil egg remains (Veterovata): Principles and applications. *Journal of Vertebrate Paleontology* v. 16, pp. 763–769.
Mikhailov, K., K. Sabath, and S. Kurzanov. 1994. Eggs and nests from the Cretaceous of Mongolia. In Carpenter, K., Hirsch, K. F., and Horner, J. R. (eds.), *Dinosaur Eggs and Babies,* pp. 88–115. New York: Cambridge University Press.
Mohabey, D. M. 1987. Juvenile sauropod dinosaur from Upper Cretaceous Lameta Formation of Panchmahals District, Gujarat, India. *Journal of the Geological Society of India* v. 30, pp. 210–216.
Mohabey, D. M. 1990. Dinosaur eggs from Lameta Formation of western and central India: Their occurrence and nesting behaviour. In Sahni A. (ed.), *Cretaceous Event Stratigraphy and the Correlation of the Indian Nonmarine Strata, Contributions from the Seminar cum Workshop I.G.C.P. 216 and 245,* pp. 86–89. Chandigarh, India: Punjab University.
Mohabey, D. M. 1996. A new oospecies, *Megaloolithus matleyi,* from the Lameta Formation (Upper Cretaceous) of Chandrapur district, Maharashtra, India, and general remarks on the paleoenvironment and nesting behavior of dinosaurs. *Cretaceous Research* v. 17, pp. 183–196.
Mohabey, D. M. 1998. Systematics of Indian Upper Cretaceous dinosaur and chelonian eggshells. *Journal of Vertebrate Paleontology* v. 18, pp. 348–362.
Mohabey, D. M., and U. B. Mathur. 1989. Upper Cretaceous dinosaur eggs from new localities of Gujarat, India. *Journal of the Geological Society of India* v. 33, pp. 32–37.
Mohabey, D. M., and S. G. Udhoji. 1990. Fossil occurrences and sedimentation of Lameta Formation of Nand Area, Maharashtra: Palaeoenvironmental, palaeoecological, and taphonomical implications. In Sahni, A. (ed.), *Cretaceous Event Stratigraphy and the Correlation of the Indian Nonmarine Strata, Contributions from the Seminar cum Workshop I.G.C.P. 216 and 245,* pp. 75–77. Chandigarh, India: Punjab University.
Molnar, R., and M. O'Reagan. 1989. Dinosaur extinctions. *Australian Natural History* v. 22, pp. 562–570.
Mones, A. 1980. Nuevos elementos de la paleoherpetofauna del Uruguay (Crocodilia y Dinosauria). *Actas del Congreso Argentino de Paleontología y Bioestratigrafía* v. 1, pp. 265–277.
Monteith, B. G. 1996. Possible partial egg from the Upper Triassic Chinle Formation of Coconino County, Arizona. *Journal of Vertebrate Paleontology, Abstracts with Program* v. 16, p. 54A.
Moratalla, J. 1992. Dinosaurian and crocodilian eggshells. In Pol, C., Buscalioni, A. D., Carballeira, J., Frances, V., Martinez, N. L., Marandat, B., Moratalla, J. J., Sanz, J. L., Sigé, B., and Villatte, J. Reptiles and mammals from the Late Cretaceous new locality Quintanilia del Coco (Burgos Province, Spain). *Neues Jahrbuch für Geologie und Paläontologie, Abhandlungen* v. 184, pp. 279–314.
Morell, V. 1987. Announcing the birth of a heresy. *Discover* v. 8, pp. 26–50.
Morris, W. J. 1973. A review of Pacific coast hadrosaurs. *Journal of Paleontology* v. 47, pp. 551–561.

Mourier, T., P. Bengtson, M. Bonhomme, E. Buge, H. Cappetta, J.-Y. Crochet, M. Feist, K. F. Hirsch, M. Moullade, C. Noblet, D. Pons, J. Rey, B. Sigé, Y. Tambareau, and P. Taquet. 1988. The Upper Cretaceous–Lower Tertiary marine to continental transition in the Bagua Basin, northern Peru. *Newsletters in Stratigraphy* v. 19, pp. 143–177.

Myhrvold, N. P., and P. J. Currie. 1997. Supersonic sauropods? Tail dynamics in the diplodocids. *Paleobiology* v. 23, pp. 393–409.

Nessov, L. A. 1991. Mesozoic and Palaeogene birds of the USSR and their paleoenvironments. In Campbell, K. E., *Papers in Avian Paoleontology Honoring Pierce Brodkorb*. Los Angeles: Natural History Museum of Los Angeles County, Science Series, pp. 465–478.

Nessov, L. A., and M. N. Kaznyshkin. 1986. Discovery of a site in the USSR with remains of eggs of Early and Late Cretaceous dinosaurs. *Biological Sciences, Zoology* v. 9, pp. 35–49.

Nice, M. M. 1962. Development of behavior in precocial birds. *Transactions of the Linnean Society* v. 8, pp. 1–211.

Nopcsa, F. 1929. Sexual differences in ornithopodous dinosaurs. *Palaeobiologica* v. 2, pp. 187–201.

Norell, M., J. Clark, L. Chiappe, and D. Dashzeveg. 1995. A nesting dinosaur. *Nature* v. 378, pp. 774–776.

Norell, M., J. Clark, D. Dashzeveg, R. Barsbold, L. Chiappe, A. Davidson, M. McKenna, A. Perle, and M. Novacek. 1994. A theropod dinosaur embryo and the affinities of the Flaming Cliffs dinosaur eggs. *Science* v. 266, pp. 779–782.

Norman, D. 1980. On the ornithischian dinosaur *Iguanodon bernissartensis* from the Lower Cretaceous of Bernissart (Belgium). *Institut Royal des Sciences Naturelles de Belgique Mémoire* v. 178, pp. 1–103.

Norman, D. 1986. On the anatomy of *Iguanodon antherfieldensis* (Ornithischia, Ornithopoda). *Bulletin de l'Institut Royal des Sciences Naturelles de Belgique. Sciences de la Terre* v. 56, pp. 281–372.

O'Connor, R. J. 1991a. Eggs. In Brooke, M., and Birkhead, T. (eds.), *The Cambridge Encyclopedia of Ornithology,* pp. 230–236. Cambridge: Cambridge University Press.

O'Connor, R. J. 1991b. Incubation. In Brooke, M., and Birkhead, T. (eds.), *The Cambridge Encyclopedia of Ornithology,* pp. 239–243. Cambridge: Cambridge University Press.

Oelofsen, B. W. 1978. Atmospheric carbon dioxide/oxygen imbalance in the Late Cretaceous, hatching of eggs and the extinction of biota. *Palaeontologia Africana* v. 21, pp. 45–51.

Osborn, H. F. 1912. Integument of the iguanodont dinosaur *Trachodon*. *Memoires of the American Museum of Natural History* v. 1, pp. 33–54.

Osborn, H. 1924. Three new Theropoda, *Protoceratops* zone, central Mongolia. *American Museum of Natural History Novitates* v. 144, pp. 1–12.

Packard, G. C., and M. J. Packard. 1980. Evolution of the cleidoic egg among reptilian antecedents of birds. *American Zoologist* v. 20, pp. 351–362.

Packard, M. J., and K. F. Hirsch. 1989. Structure of shells from eggs of the geckos *Gekko gecko* and *Phelsuma madagascarensis*. *Canadian Journal of Zoology* v. 67, pp. 746–758.

Packard, M. J., and R. S. Seymour. 1997. Evolution of the amniote egg. In Sumida, S., and Martin, K. L. (eds.), *Amniote Origins,* pp. 265–290. San Diego: Academic Press.

Paladino, F. V., P. Dodson, J. L. Hammond, and J. R. Spotila. 1989.

Temperature-dependent sex determination in dinosaurs? Implications for population dynamics and extinction. In Farlow, J. (ed.), Paleobiology of the dinosaurs. *Geological Society of America Special Paper* v. 238, pp. 63–70.

Palmer, B. D., and L. J. Guillette. 1992. Alligators provide evidence of the evolution of an archosaurian mode of oviparity. *Biology of Reproduction* v. 46, pp. 39–47.

Parrish, J. T., A. M. Ziegler, and C. R. Scotese. 1982. Rainfall patterns and the distribution of coals and evaporites in the Mesozoic and Cenozoic. *Palaeogeography, Palaeoclimatology, Palaeoecology* v. 40, pp. 67–101.

Patten, B. M. 1951. *Early Embryology of the Chick*. New York: Blakiston, 244pp.

Paul, G. 1994a. Dinosaur reproduction in the fast lane: Implications for size, success, and extinction. In Carpenter, K., Hirsch, K. F., and Horner, J. R. (eds.), *Dinosaur Eggs and Babies,* pp. 244–251. New York: Cambridge University Press.

Paul, G. 1994b. Thermal environments of dinosaur nestlings: Implications for endothermy and insulation. In Carpenter, K., Hirsch, K. F., and Horner, J. R. (eds.), *Dinosaur Eggs and Babies,* pp. 279–287. New York: Cambridge University Press.

Penner, M. M. 1983. Contribution à l'étude de la microstructure des coquilles d'oeufs de Dinosaures du Crétacé supérieur dans le bassin d'Aix-en-Provence (France): Application Biostratigraphique. *Mémoires des Sciences de la Terre* v. 83, pp. 1–234.

Penner, M. M. 1985. The problem of dinosaur extinction: Contribution of the study of terminal Cretaceous eggshells from southeast France. *Geobios* v. 18, pp. 665–669.

Peterson, O. A., and C. W. Gilmore. 1902. *Elosaurus parvus:* A new genus and species of the Sauropoda. *Annals of the Carnegie Museum* v. 1, pp. 490–499.

Porter, K. R. 1972. *Herpetology.* Philadelphia: Saunders, 524pp.

Powell, J. E. 1987. The Late Cretaceous fauna of Los Alamitos, Patagonia, Argentina. Part VI. The Titanosaurids. *Revista del Museo Argentino de Ciencias Naturales, Paleontología* v. 3, pp. 147–153.

Powell, J. In press. Hallazgo de huevos asignables a Dinosaurios titanosduridos (Saurischia, Sauropoda) de la provincia de Rio Negro, Argentina.

Price, L. 1951. Un ovo de dinosaurio na formacao Bauru do Cretácio de Estado de Minas Gerais. *Departamento Nacional da Producao Mineral, Divisao de Geologia e Mineria, Notas Preliminares e Estudos* v. 53, pp. 1–9.

Qiang, J., P. J. Currie, M. A. Novell, and J. Shu-An. 1998. Two feathered dinosaurs from northeastern China. *Nature* v. 393, pp. 753–761.

Quinn, B. 1994. Fossilized eggshell preparation. In Leiggi, P., and May, P. (eds.), *Vertebrate Paleontological Techniques,* pp. 146–153. New York: Cambridge University Press.

Raath, M. A. 1977. The anatomy of the Triassic theropod *Syntarsus rhodesiensis* (Saurischia, Podokeosaurus) and a consideration of its biology. Unpublished Ph.D. thesis. Grahamstown, S. Africa: Rhodes University.

Raath, M. A. 1990. Morphological variation in small theropods and its meaning in systematics: Evidence from *Syntarsus rhodesiensis*. In Carpenter, K., Hirsch, K. F., and Horner, J. R. (eds.), *Dinosaur Eggs and Babies,* pp. 91–105. New York: Cambridge University Press.

Rahn, H., and A. Ar. 1974. The avian egg and water loss. *Condor* v. 76, pp. 147–152.
Rahn, H., A. Ar, and C. V. Paganelli. 1980. How bird eggs breathe. In *Birds: Readings from Scientific American*, pp. 208–218. San Francisco: Freeman.
Rahn, H., V. Paganelli, and A. Ar. 1975. Relation of the avian egg weight to body weight. *The Auk* v. 92, pp. 750–765.
Rahn, H., V. Paganelli, and A. Ar. 1987. Pores and gas exchange of avian eggs: A review. *Experimental Zoology* Supplement 1, pp. 165–172.
Reid, R. E. 1990. Zonal "growth rings" in dinosaurs. *Modern Geology* v. 15, pp. 19–48.
Reid, R. E. 1993. Apparent zonation and slowed late growth in a small Cretaceous theropod. *Modern Geology* v. 18, pp. 391–406.
Reid, R. E. 1997. How dinosaurs grew. In Farlow, J., and Brett-Surman, M. (eds.), *The Complete Dinosaur,* pp. 403–413. Bloomington: Indiana University Press.
Retallack, G. J. 1988. Field recognition of paleosols. In Reinhart, J., and Sigleo, W. R. (eds.), Paleosols and weathering through geologic time: Principles and applications. *Geological Society of America Special Paper* v. 216, pp. 1–20.
Rich, T., and P. Rich. 1989. Polar dinosaurs and biotas of the Early Cretaceous of southeastern Australia. *National Geographic Research* v. 5, pp. 15–53.
Richardson, K. C. 1985. The secretory phenomena in the oviduct of the fowl, including the process of shell formation examined by the micro-incineration technique. *Philosophical Transaction of the Royal Society of London,* Ser. B, v. 225, pp. 149–194.
Ricklefs, R. 1983. Avian postnatal development. *Avian Biology* v. 7, pp. 1–83.
Ricqlès, A. 1980. Tissue structure of dinosaur bone: Functional significance and possible relation to dinosaur physiology. In Thomas, D. K., and Olson, E. C. (eds.), *A Cold Look at the Warm-Blooded Dinosaurs,* pp. 103–139. American Association for the Advancement of Science Selected Symposium 28.
Rieppel, O. 1993. Studies on skeleton formation in reptiles. V. Patterns of ossification in the skeleton of *Alligator mississippiensis* DAUDIN (Reptilia, Crocodylia). *Zoological Journal of the Linnean Society* v. 109, pp. 301–325.
Romanoff, A. L., and A. J. Romanoff. 1949. *The Avian Egg.* New York: Wiley, 918pp.
Romer, A. S. 1927. The development of the thigh musculature in the chick. *Journal of Morphology* v. 43, pp. 347–385.
Romer, A. S. 1957. Origin of the amniote egg. *Scientific Monthly* v. 85, pp. 57–63.
Romer, A. S., and L. I. Price. 1939. The oldest vertebrate egg. *American Journal of Science* v. 237, pp. 826–829.
Rosa, R. d. l., and R. Armando. 1995. Microestructura de las cáscaras de huevos de "Dinosaurios" de la Formación "El Gallo" (Campaniano) de Baja California Norte, México. *Congreso Nacional de Zoología,* p. 13.
Rozhdestvensky, A. K. 1973. *Animal Kingdom in Ancient Asia.* Tokyo.
Russell, D., and P. Dodson. 1997. The extinction of the dinosaurs: A dialogue between a catastrophist and a gradualist. In Farlow, J., and Brett-Surman, M. K. (eds.), *The Complete Dinosaur,* pp. 662–674. Bloomington: Indiana University Press.

Sabath, K. 1991. Upper Cretaceous amniotic eggs from the Gobi Desert. *Acta Palaeontologica Polonica* v. 36, pp. 151–192.
Sadov, I. A. 1970. Eggshell structure in fossil reptiles and birds. *Paleontology Journal* v. 1970, pp. 535–538.
Sahni, A. 1972. The vertebrate fauna of the Judith River Formation, Montana. *American Museum of Natural History Bulletin* v. 147, pp. 321–412.
Sahni, A., S. Tandon, A. Jolly, S. Bajpai, A. Sood, and S. Srinivasan. 1994. Upper Cretaceous dinosaur eggs and nesting sites from the Deccan volcano-sedimentary province of peninsular India. In Carpenter, K., Hirsch, K. F., and Horner, J. R. (eds.), *Dinosaur Eggs and Babies,* pp. 204–226. New York: Cambridge University Press.
Sahni, M. 1957. A fossil reptilian egg from the Uttatlur (Cenomanian) of south India, being the first record of a vertebrate fossil egg in India. *Record of the Geological Survey of India* v. 87, pp. 671–674.
Sander, P. M. 1988. A fossil reptile embryo from the Middle Triassic of the Alps. *Science* v. 239, pp. 780–783.
Sanz, J. L., J. J. Moratalla, M. Diaz-Molina, N. Lopez-Martinez, O. Källn, and M. Vianey-Liaud. 1995. Dinosaur nests at the sea shore. *Nature* v. 376, pp. 731–732.
Sanz, J., L. Chiappe, B. Pérez-Moreno, J. Moratalla, F. Hernández-Carrasquilla, A. Buscaloni, F. Ortega, F. Poyato-Ariza, D. Rasskin-Gutman, and X. Martinez-Delclòs. 1997. A nestling bird from the Lower Cretaceous of Spain: Implications for avian skull and neck evolution. *Science* v. 276, pp. 1543–1546.
Sarkar, A., S. K. Bhattacharya, and D. M. Mohabey. 1991. Stable-isotope analyses of dinosaur eggshells: Paleoenvironmental implications. *Geology* v. 19, pp. 1068–1071.
Sasso, C., and M. Signore. 1998. Exceptional soft-tissue preservation in a theropod dinosaur from Italy. *Nature* v. 392, pp. 383–387.
Sauer, E. G. 1968. Calculations of struthious egg sizes from measurements of shell fragments and their correlation with phylogenetic aspects. *Cimbebasia,* Ser. A, v. 1, pp. 26–55.
Scheetz, R. 1991. Progress report of juvenile and embryonic *Dryosaurus* remains from the Upper Jurassic Morrison Formation of Colorado. In Averett, W. (ed.), *Guidebook for Dinosaur Quarries and Tracksite Tour,* pp. 27–29. Grand Junction, CO: Grand Junction Geological Society.
Schmidt, W. J. 1943. Uber den Aufbau der Kalkschale bei den Schilkröteneiern. *Zeitschrift für Morphologie und Oekologie der Tiere* v. 40, pp. 1–16.
Schmidt-Nielsen, K. 1972. *How Animals Work.* Cambridge: Cambridge University Press, 114pp.
Seymour, R. 1979. Dinosaur eggs: Gas conductance through the shell, water loss during incubation, and clutch size. *Paleobiology* v. 5, pp. 1–11.
Seymour, R., and R. A. Ackerman. 1980. Adaptations to underground nesting in birds and reptiles. *American Zoologist* v. 20, pp. 437–447.
Shine, R. 1991. Influences of incubation requirements on the evolution of viviparity. In Deeming, D. C., and Ferguson, M. J. (eds.), *Egg Incubation: Its Effects on Embryonic Development in Birds and Reptiles,* pp. 361–369. New York: Cambridge University Press.
Smart, I. H. 1991. Egg-shape in birds. In Deeming, D. C., and Ferguson, M. J. (eds.), *Egg Incubation: Its Effects on Embryonic Development in*

Birds and Reptiles, pp. 101–116. New York: Cambridge University Press.
Smith, A. G., D. G. Smith, and B. M. Funnell. 1994. *Atlas of Mesozoic and Cenozoic Coastlines.* Cambridge: Cambridge University Press, 99pp.
Sochava, A. 1969. Dinosaur eggs from the Upper Cretaceous of the Gobi Desert. *Paleontological Journal* v. 4, pp. 517–527.
Sochava, A. 1971. Two types of eggshell in Senonian dinosaurs. *Paleontological Journal* v. 3, pp. 353–361.
Srivastava, S., D. M. Mohabey, A. Sahni, and S. C. Pant. 1986. Upper Cretaceous dinosaur egg clutches from Kheda District (Gujarat, India). *Paleontographica Abt. A* v. 193, pp. 219–233.
Sternberg, C. M. 1955. A juvenile hadrosaur from the Oldman Formation of Alberta. *Annual Report, National Museum of Canada, Bulletin* v. 136, pp. 120–122.
Straelen, V. van. 1925. The microstructure of the dinosaurian eggshells from the Cretaceous beds of Mongolia. *American Museum Novitates* v. 173, pp. 1–4.
Straelen, V. van. 1928. Les oeufs de reptiles fossiles. *Palaeontologica* v. 1, pp. 295–317.
Sullivan, G. E. 1962. Anatomy and embryology of the wing musculature of the domestic fowl (*Gallus*). *Australian Journal of Zoology* v. 10, pp. 458–518.
Swinton, W. E. 1950. Fossil eggs from Tanganyika. *The Illustrated London News* v. 217, pp. 1082–1083.
Tandon, S. K., A. Sood, J. E. Andrews, and P. F. Dennis. 1995. Palaeoenvironments of the dinosaur-bearing Lameta Beds (Maastrichtian), Narmada Valley, Central India. *Palaeogeography, Palaeoclimatology, Palaeoecology* v. 117, pp. 153–184.
Thulborn, R. A. 1990. *Dinosaur Tracks.* London: Chapman & Hall, 410pp.
Thulborn, R. A., and M. Wade. 1984. Dinosaur trackways in the Winton Formation (Mid-Cretaceous) of Queensland. *Memoirs of the Queensland Museum* v. 21, pp. 413–517.
Tokaryk, T. 1990. A baby *Triceratops* or a very small adult dinosaur. *Saskatchewan Archaeological Society Newsletter* v. 11, pp. 127–128.
Tokaryk, T., and J. E. Storer. 1991. Dinosaur eggshell fragments from Saskatchewan, and evaluation of potential distance of eggshell transport. *Journal of Vertebrate Paleontology, Abstracts with Program* v. 11(3), p. 58A.
Tyler, C., and K. Simkiss. 1959. A study of eggshell of ratite birds. *Proceedings of the Zoological Society of London* v. 133, pp. 201–243.
Varricchio, D. J. 1993. Bone microstructure of the Upper Cretaceous theropod dinosaur *Troodon formosus. Journal of Vertebrate Paleontology* v. 13, pp. 99–104.
Varricchio, D. J., F. Jackson, J. J. Borkpwski, and J. Horner. 1997. Nest and egg clutches of the dinosaur *Troodon formosus* and the reproductive evolution of avian reproductive traits. *Nature* v. 385, pp. 247–250.
Varricchio, D. J., F. Jackson, and C. Trueman. 1999. A nesting trace with eggs for the Cretaceous theropod dinosaur *Troodon formosus. Journal of Vertebrate Paleontology* v. 19, pp. 91–100.
Vianey-Liaud, M., and J.-Y. Crochet. 1993. Dinosaur eggshells from the Late Cretaceous of Languedoc (southern France). *Revue de Paléobiologie* Special Volume 7, pp. 237–249.
Vianey-Liaud, M., S. L. Jain, and A. Sahni. 1987. Dinosaur eggshells

(Saurischia) from the Late Cretaceous Intertrappean and Lameta Formations (Deccan, India). *Journal of Vertebrate Paleontology* v. 7, pp. 408–424.

Vianey-Liaud, M., and N. Lopez-Martinez. 1997. Late Cretaceous dinosaur eggshells from the Tremp Basin, Southern Pyrenees, Lleida, Spain. *Journal of Paleontology* v. 71, pp. 1157–1171.

Vianey-Liaud, M., P. Mallan, O. Buscail, and C. Montgelard. 1994. Review of French dinosaur eggshells: Morphology, structure, mineral, and organic composition. In Carpenter, K., Hirsch, K. F., and Horner, J. R. (eds.), *Dinosaur Eggs and Babies,* pp. 151–183. New York: Cambridge University Press.

Vleck, C. M., and D. F. Hoyt. 1991. Metabolism and energetics of reptilian and avian embryos. In Deeming, D. C., and Ferguson, M. J. (eds.), *Egg Incubation: Its Effects on Embryonic Development in Birds and Reptiles,* pp. 285–306. New York: Cambridge University Press.

Webb, G., S. Manolis, K. Dempsey, and P. Whitehead, 1987. Crocodilian eggs: A functional overview. In Webb, G., Manolis, S., and Whitehead, P. (eds.), *Wildlife Management: Crocodiles and Alligators,* pp. 417–422. Canberra: Surrey.

Weishampel, D. B., and J. R. Horner. 1994. Life history syndromes, heterochrony, and the evolution of the Dinosauria. In Carpenter, K., Hirsch, K. F., and Horner, J. R. (eds.), *Dinosaur Eggs and Babies,* pp. 229–243. New York: Cambridge University Press.

Wellnhofer, P. 1991. *The Illustrated Encyclopedia of Pterosaurs.* New York: Crescent Books, p. 192.

Whitehead, P. 1987. Respiration in *Crocodylus johnstoni* embryos. In Webb, G., Manolis, S., and Whitehead, P. (eds.), *Wildlife Management: Crocodiles and Alligators,* pp. 473–497. Canberra: Surrey.

Wieland, G. R. 1925. Dinosaur extinction. *American Naturalist* v. 69, pp. 557–565.

Williams, D. L. 1981. *Genyornis* eggshell (Dromornithidae, Aves) from the Late Pleistocene of South Australia. *Alcheringa* v. 5, pp. 133–140.

Williams, D.L.G., R. S. Seymour, and P. Kérourio. 1984. Structure of fossil dinosaur eggshell from the Aix Basin, France. *Palaeogeography, Palaeoclimatology, Palaeoecology* v. 45, pp. 23–37.

Winkler, D. A., and P. A. Murry. 1989. Paleoecology and hypsilophodontid behavior at the Proctor Lake dinosaur locality (Early Cretaceous), Texas. In Farlow, J. O. (ed.), Paleobiology of the Dinosaurs. *Geological Society of America, Special Paper* v. 238, pp. 55–61.

Wolberg, D. L., and D. Bellis. 1989. The discovery of dinosaur nests, eggs, and tracks, Fruitland Formation, Late Cretaceous, San Juan Basin, New Mexico. *Geological Society of America Annual Meeting Abstracts,* p. A73.

Wolfe, J. A., and G. R. Upchurch. 1987. North American nonmarine climates and vegetation during the Late Cretaceous. *Palaeogeography, Palaeoclimatology, Palaeoecology* v. 61, pp. 33–77.

Woodward, A., T. Hines, C. Abercrombie, and C. Hope. 1984. Spacing patterns in alligator nests. *Journal of Herpetology* v. 18, pp. 8–12.

Yang H. 1995. Authentication of ancient DNA sequence. *Acta Palaeontologica Sinica* v. 34, pp. 657–673.

Young, C. C. 1954. Fossil reptilian eggs from Laiyang, Shantung, China. *Acta Palaeontologica Sinica* v. 2, pp. 371–388.

Young, C. C. 1959. On a new fossil egg from Laiyang, Shantung. *Vertebrata PalAsiatica* v. 3, pp. 34–35.

Young, C. C. 1965. Fossil eggs from Nanhsiung, Kwangtung, and Kanchou, Kiangsi. *Vertebrata PalAsiatica* v. 9, pp. 141–170.
Young, C. C. 1979. Note on an egg from Ninghsia. *Vertebrata PalAsiatica* v. 17, pp. 35–36.
Zelenitsky, D. K., and L. V. Hills. 1996. An egg clutch of *Prismatoolithus levis* oosp. nov. from the Oldman Formation (Upper Cretaceous), Devil's Coulee, southern Alberta. *Canadian Journal of Earth Sciences* v. 33, pp. 1127–1131.
Zelenitsky, D. K., and L. V. Hills. 1997. Normal and pathological eggshells of *Spheroolithus albertensis,* oosp. nov., from the Oldman Formation (Judith River Group, Late Campanian), southern Alberta. *Journal of Vertebrate Paleontology* v. 17, pp. 167–171.
Zelenitsky, D. K., L. V. Hills, and P. Currie. 1996. Parataxonomic classification of ornithoid eggshell fragments from the Oldman Formation (Judith River Group, Upper Cretaceous), southern Alberta. *Canadian Journal of Earth Sciences* v. 33, pp. 1655–1667.
Zeng D., and Zhang J. 1979. On the dinosaurian eggs from the western Dongting Basin, Hunan. *Vertebrata PalAsiatica* v. 17, pp. 131–136.
Zhao Z. 1975. Microstructure of the dinosaurian eggshells of Nanxiong, Guangdong, and the problems in dinosaur egg classification. *Vertebrata PalAsiatica* v. 13, pp. 105–117.
Zhao Z. 1979a. Progress in the research of dinosaur eggs. In *Mesozoic and Cenozoic Red Beds of South China,* pp. 330–340. Beijing: Science Press.
Zhao Z. 1979b. Discovery of the dinosaurian eggs and footprint from Neixiang County, Henan Province. *Vertebrata PalAsiatica* v. 17, pp. 304–309.
Zhao Z. 1994. Dinosaur eggs in China: On the structure and evolution of eggshells. In Carpenter, K., Hirsch, K. F., and Horner, J. R. (eds.), *Dinosaur Eggs and Babies,* pp. 184–203. New York: Cambridge University Press.
Zhao Z., and Ding S. 1976. Discovery of the dinosaurian eggshells from Alxa, Ningxia and its stratigraphic significance. *Vertebrata PalAsiatica* v. 14, pp. 42–44.
Zhao Z., and Jiang Y. K. 1974. Microscopic studies on the dinosaurian eggshells from Laiyang, Shangdong Province. *Scientia Sinica* v. 17, pp. 73–83.
Zhao Z., and Li R. 1988. A new structural type of the dinosaur eggs from Anlu County, Hubei Province. *Vertebrata PalAsiatica* v. 26, pp. 107–115.
Zhao Z., and Li R. 1993. First record of Late Cretaceous hypsilophontid eggs from Bayan Manduhu, Inner Mongolia. *Vertebrata PalAsiatica* v. 31, pp. 77–84.
Zhao Z., and Ma H. 1997. Biomechanical properties of dinosaur eggshells (VI)—the stability of dinosaur eggshell under pressure. *Vertebrata PalAsiatica* v. 35, pp. 88–101.
Zhao Z., Ye J., Li H., Zhao Z., and Yan Z. 1991. Extinction of the dinosaurs across the Cretaceous–Tertiary boundary in Nanxiong Basin, Guangdong Province. *Vertebrata PalAsiatica* v. 29, pp. 1–20.
Zhao Z., Wang J., Chen S., and Zhong Y. 1993. Amino acid composition of dinosaur eggshells nearby the K/T boundary in Nanxiong Basin, Guangdong Province, China. *Palaeogeography, Palaeoclimatology, Palaeoecology* v. 104, pp. 213–218.

Index

Page numbers in italics refer to a photograph, table or illustration.

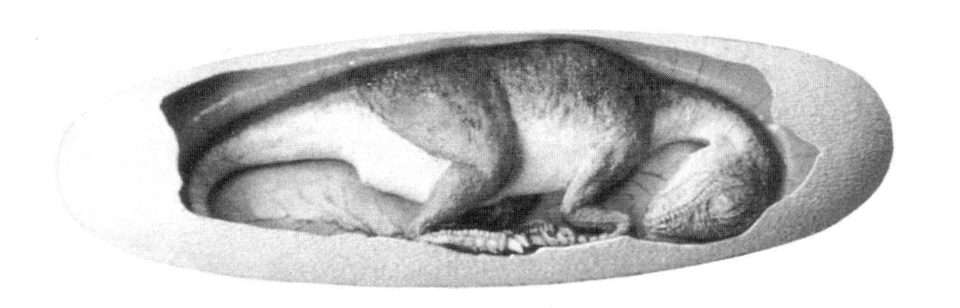

Dinosaurs have long held an interest for Kenneth Carpenter, ever since his mother took him to see *Godzilla* when he was five. He has retained his childhood fascination for these terrible beasts and is now the dinosaur paleontologist at the Denver Museum of Natural History. He is the author of more than a hundred scientific papers and the co-editor of three books, including *Dinosaur Systematics* (1990), *Dinosaur Eggs and Babies* (1994), and *The Upper Jurassic Morrison Formation: An Interdisciplinary Study* (1998).

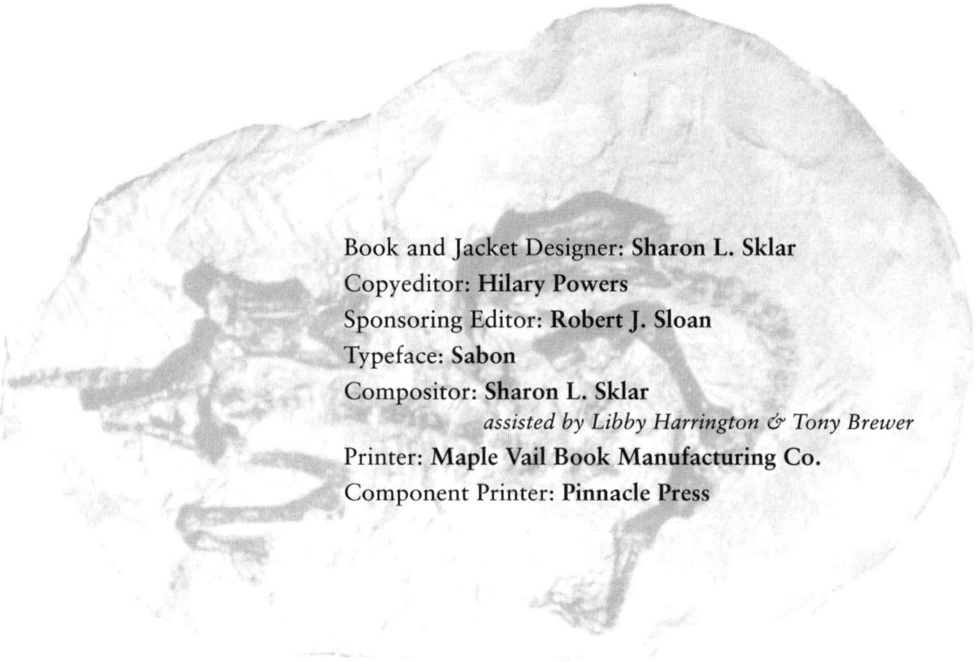

Book and Jacket Designer: **Sharon L. Sklar**
Copyeditor: **Hilary Powers**
Sponsoring Editor: **Robert J. Sloan**
Typeface: **Sabon**
Compositor: **Sharon L. Sklar**
assisted by Libby Harrington & Tony Brewer
Printer: **Maple Vail Book Manufacturing Co.**
Component Printer: **Pinnacle Press**